Monographs in Mathematics
Vol. 88

Roelof W. Bruggeman

Families of Automorphic Forms

1994

Birkhäuser Verlag
Basel · Boston · Berlin

Author:
Roelof W. Bruggeman
Mathematisch Instituut
Universiteit Utrecht
Postbus 80.010
3508 TA Utrecht
The Netherlands

A CIP catalogue record for this book is available from the Library of Congress, Washington D.C., USA

Deutsche Bibliothek Cataloging-in-Publication Data
Bruggeman, Roelof W.:
Families of automorphic forms / Roelof W. Bruggeman. – Basel ; Boston ; Berlin : Birkhäuser, 1994
(Monographs in mathematics ; Vol. 88)
ISBN 3-7643-5046-6 (Basel...)
ISBN 0-8176-5046-6 (Boston)
NE: GT

Camera-ready copy prepared by the author
Printed on acid-free paper produced of chlorine-free pulp
Printed in Germany
ISBN 3-7643-5046-6
ISBN 0-8176-5046-6

9 8 7 6 5 4 3 2 1

Contents

II Examples

Preface

Automorphic forms on the upper half plane have been studied for a long time. Most attention has gone to the holomorphic automorphic forms, with numerous applications to number theory. Maass, [34], started a systematic study of real analytic automorphic forms. He extended Hecke's relation between automorphic forms and Dirichlet series to real analytic automorphic forms. The names Selberg and Roelcke are connected to the spectral theory of real analytic automorphic forms, see, e.g., [50], [51]. This culminates in the trace formula of Selberg, see, e.g., Hejhal, [21].

Automorphic forms are functions on the upper half plane with a special transformation behavior under a discontinuous group of non-euclidean motions in the upper half plane. One may ask how automorphic forms change if one perturbs this group of motions. This question is discussed by, e.g., Hejhal, [22], and Phillips and Sarnak, [46]. Hejhal also discusses the effect of variation of the multiplier system (a function on the discontinuous group that occurs in the description of the transformation behavior of automorphic forms). In [5]–[7] I considered variation of automorphic forms for the full modular group under perturbation of the multiplier system. A method based on ideas of Colin de Verdière, [11], [12], gave the meromorphic continuation of Eisenstein and Poincaré series as functions of the eigenvalue and the multiplier system jointly. The present study arose from a plan to extend these results to much more general groups (discrete cofinite subgroups of $\mathrm{SL}_2(\mathbb{R})$).

To carry this out I look at more general families of automorphic forms than one usually considers. In particular, I admit singularities inside the upper half plane, and relax the usual condition of polynomial growth at the cusps. This led me to reconsider a fairly large part of the theory of real analytic automorphic forms on the upper half plane.

This is done in Part I for arbitrary cofinite discrete groups. Chapters 2–6 discuss real analytic automorphic forms of a rather general type. Most results are known, or are easy extensions of known results. Chapters 7–12 consider families of these automorphic forms, with the eigenvalue and the multiplier system as the parameters. The ideas of Colin de Verdière are worked out in Chapters 8–10. The central result is Theorem 10.2.1; it gives the meromorphic continuation of Eisenstein and Poincaré series. The meromorphic continuation in the eigenvalue is well known; the meromorphy in all parameters jointly is new. Chapters 11 and 12 study singularities of the resulting families of automorphic forms. In Chapter 11 the eigenvalue is the sole variable. I summarize known results, and prepare for the study in Chapter 12 of the singularities in more than one variable. Table 1.1 on p. 18 gives a more detailed description of Part I.

The treatment in Part I is complicated. This is due to the fact that I consider general cofinite discrete groups, without restriction on the dimension of the group of multiplier systems. Chapter 1 is meant as an introduction. It explains the main ideas and results in the context of the full modular group. In the three chapters of Part II I consider three examples of cofinite discrete groups. Chapter 13 extends the discussion for the full modular group in Chapter 1. The other groups considered are the theta group and the commutator subgroup of the modular group. Although the first objective of Part II is to give the reader examples of the concepts, I have included some discussions that did not fit into the general context of Part I; see 1.7.7–1.7.10.

The reader I have had in mind has seen automorphic forms before, holomorphic as well as real analytic ones. For the latter I would suggest to have a look at Chapters IV and V of Maass's lecture notes [35], or at §3.5–7 of Terras's book [57]. The reader should also be prepared to look up facts concerning analytic functions in more than one complex variable, and be not afraid of a modest use of sheaf language when dealing with this subject.

The ideas of Colin de Verdière employed in Part I concern unbounded operators in Hilbert spaces. Kato's book, [25], is consulted for many results from functional analysis. I restrict the discussion of the spectral theory of automorphic forms to those results I need. In particular I do not mention the continuous part of the spectral decomposition.

R. Matthes visited Utrecht in 1989. The discussions we had gave me the stimulus to start this work. I am very grateful for this contribution, and also for the many comments he gave on early versions of this book. At a later stage the interest of D. Zagier has been a great encouragement. I also thank F. Beukers, J. Elstrodt, and B. van Geemen for corrections, comments, and suggestions.

Chapter 1
Modular introduction

To introduce the main ideas of this book, we discuss in this chapter modular forms, i.e., automorphic forms for the modular group $\mathrm{SL}_2(\mathbb{Z})$.

First, we discuss the modular group and its action on the upper half plane. After that, we define various types of modular forms. In Definition 1.5.6 we arrive at real analytic modular forms of arbitrary complex weight. The central result in this chapter is the continuation of the Eisenstein series as a family depending meromorphically on two parameters, see 1.5.8. Section 1.6 sketches a proof. This proof gives in a nutshell the main points of the central Chapters 8–10.

Modular forms may be seen as functions on the upper half plane, as functions on $\mathrm{SL}_2(\mathbb{R})$, or on its universal covering group. The last point of view is taken in the later chapters of this book; in this chapter we consider modular forms as functions on the upper half plane $\{ z \in \mathbb{C} : \mathrm{Im}(z) > 0 \}$.

1.1 The modular group

The action of $\mathrm{SL}_2(\mathbb{Z})$ in the upper half plane forms the geometric base of the study of real analytic modular forms. This section gives a short discussion. Much more information may be found in, e.g., Chapter I of [35], or §3.1 of [57].

1.1.1 *Definition.* The *modular group* $\Gamma_{\mathrm{mod}} = \mathrm{SL}_2(\mathbb{Z})$ consists of the matrices $\begin{pmatrix} a & b \\ c & d \end{pmatrix}$ with $a, \ldots, d \in \mathbb{Z}$, and determinant $ad - bc = 1$. It is a subgroup of

$$G = \mathrm{SL}_2(\mathbb{R}) = \left\{ \begin{pmatrix} a & b \\ c & d \end{pmatrix} : a, b, c, d \in \mathbb{R},\ ad - bc = 1 \right\}.$$

1.1.2 *Action on the upper half plane.* The group G acts on the *upper half plane* $\mathfrak{H} = \{ z \in \mathbb{C} : \mathrm{Im}(z) > 0 \}$ by $\begin{pmatrix} a & b \\ c & d \end{pmatrix} \cdot z = \frac{az+b}{cz+d}$.

As $-\mathrm{Id} = \begin{pmatrix} -1 & 0 \\ 0 & -1 \end{pmatrix}$ acts trivially, this action factors through $\mathrm{PSL}_2(\mathbb{R}) = \mathrm{SL}_2(\mathbb{R})/\{\pm\mathrm{Id}\}$. We put $\bar{\Gamma}_{\mathrm{mod}} = \Gamma_{\mathrm{mod}}/\{\pm\mathrm{Id}\}$.

1.1.3 *Fundamental domain.* $F_{\mathrm{mod}} = \{ z \in \mathfrak{H} : |z| \geq 1, |\mathrm{Re}\, z| \leq 1/2 \}$ is the well known *standard fundamental domain* for the modular group, see, e.g., [52], Ch. I, §5.1, proof of Theorem 13. It is a fundamental domain as it is a reasonable subset of $\mathfrak{H}$ satisfying

i) $\Gamma_{\mathrm{mod}} \cdot F_{\mathrm{mod}} = \mathfrak{H}$,

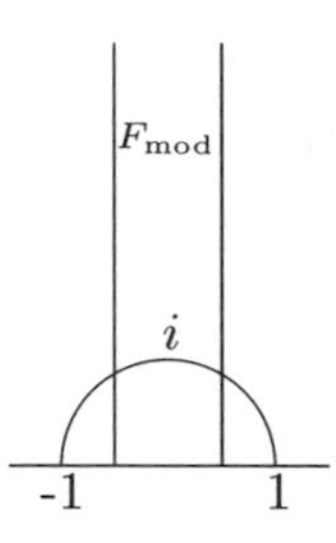

Figure 1.1 The standard fundamental domain of Γ_{mod}.

ii) If $z \in F_{\text{mod}}$, $\gamma \in \Gamma_{\text{mod}}$, $\gamma \cdot z \neq z$ and also $\gamma \cdot z \in F_{\text{mod}}$, then z and $\gamma \cdot z$ are boundary points of F_{mod}.

If we would take

$$F_0 = \left\{ z \in \mathfrak{H} : |z| \geq 1,\ -\frac{1}{2} \leq x < \frac{1}{2},\ \text{ if } |z| = 1 \text{ then } x \leq 0 \right\} \subset F_{\text{mod}},$$

we would have a fundamental domain meeting each Γ_{mod}-orbit exactly once.

1.1.4 *Generators.* Define $U, W \in \Gamma_{\text{mod}}$ by $U = \begin{pmatrix} 1 & 1 \\ 1 & 0 \end{pmatrix}$ and $W = \begin{pmatrix} 0 & 1 \\ -1 & 0 \end{pmatrix}$. These two elements generate Γ_{mod}. To see this take $\begin{pmatrix} a & b \\ c & d \end{pmatrix} \in \Gamma_{\text{mod}}$ with $c \neq 0$. We may arrange $|a| < |c|$ by left multiplication by a power of U. Reverse $\pm a$ and $\pm c$ by left multiplication by W. This leads eventually to $\begin{pmatrix} \pm 1 & m \\ 0 & \pm 1 \end{pmatrix}$. Now use $W^2 = \begin{pmatrix} -1 & 0 \\ 0 & -1 \end{pmatrix}$.

A bit more work shows that the relations are generated by $(UW)^3 = \text{Id}$, $W^4 = \text{Id}$ and $W^2 U = U W^2$. See Theorem 8 on p. 54 of [35].

1.1.5 *Quotient.* The generators U and W of Γ_{mod} give bijective maps between parts of the boundary of F_{mod}:

$$\begin{aligned} U &: \{ -\tfrac{1}{2} + y : y \geq \tfrac{1}{2}\sqrt{3} \} \longrightarrow \{ \tfrac{1}{2} + y : y \geq \tfrac{1}{2}\sqrt{3} \} \\ W &: \{ e^{i\varphi} : \tfrac{\pi}{2} \leq \varphi \leq \tfrac{2\pi}{3} \} \longrightarrow \{ e^{i\varphi} : \tfrac{\pi}{3} \leq \varphi \leq \tfrac{\pi}{2} \}. \end{aligned}$$

Gluing the boundaries of the standard fundamental domain as indicated by these maps gives a model of the topological space $Y_{\text{mod}} = \Gamma_{\text{mod}} \backslash \mathfrak{H}$. This space is homeomorphic to the plane. The *modular invariant* gives an explicit homeomorphism; see [53], Theorem 2.9 for its definition. This function can be used to give Y_{mod} a complex structure.

Let $\overline{\text{pr}}$ be the projection map $\mathfrak{H} \to Y_{\text{mod}}$. For most $z_0 \in \mathfrak{H}$ its restriction gives a homeomorphism between a neighborhood of z_0 in $\mathfrak{H}$ to a neighborhood of $\overline{\text{pr}} z_0$ in Y_{mod}. For such a point $\overline{\text{pr}}$ also carries over the complex structure. This is not the case at the *elliptic points.* These are the elements of the orbits $\Gamma_{\text{mod}} \cdot i$ and

$\Gamma_{\text{mod}} \cdot \frac{1}{2}(1+i\sqrt{3})$. (An elliptic point $z_0 \in \mathfrak{H}$ is characterized by the property that the group $\{\, \gamma \in \Gamma_{\text{mod}} : \gamma \cdot z_0 = z_0 \,\}$ contains more elements than Id and $-$Id. In fact, i is fixed by all powers of W, and $\rho = \frac{1}{2}(1+i\sqrt{3})$ by the powers of UW.) A local coordinate at $\overline{\text{pr}}\, i$ is induced by $z \mapsto \left(\frac{z-i}{z+i}\right)^2$, and at $\overline{\text{pr}}\, \rho$ by $z \mapsto \left(\frac{z-\rho}{z-\bar{\rho}}\right)^3$.

1.1.6 *Cuspidal orbit.* For the modular group we define $\mathfrak{H}^* = \mathfrak{H} \cup \mathbb{Q} \cup \{i\infty\}$. The action $\begin{pmatrix} a & b \\ c & d \end{pmatrix} : z \mapsto \frac{az+b}{cz+d}$ extends to an action of Γ_{mod} in $\mathfrak{H}^*$. The set $\mathbb{Q} \cup \{i\infty\}$ consists of one orbit, the *cusp* $\Gamma_{\text{mod}} \cdot i\infty$.

Put $X_{\text{mod}} = \Gamma_{\text{mod}} \backslash \mathfrak{H}^* = Y_{\text{mod}} \cup \Gamma_{\text{mod}} \cdot i\infty$. Extend $z \mapsto e^{2\pi i z}$ by defining it equal to 0 in $i\infty$. We use this extended function as a local coordinate at the cusp to extend the complex structure to X_{mod}. In this way X_{mod} becomes a complex Riemann surface, isomorphic to the projective line (use the obvious extension of the modular invariant J).

1.1.7 *Metric.* The upper half plane $\mathfrak{H}$ carries the structure of a riemannian space with metric given by $(ds)^2 = y^{-2}(dx)^2 + y^{-2}(dy)^2$. (We have used the convention $x = \text{Re}(z)$, $y = \text{Im}(z)$; this we shall do throughout the book.) The geodesics are the vertical half lines in $\mathfrak{H}$, and the half circles with center on the real axis. The *Laplacian* associated to the metric is $y^2\partial_x^2 + y^2\partial_y^2$. The metric is invariant under the action of G, and hence carries over to Y_{mod}.

1.1.8 *Volume.* The volume form associated to the metric structure is $d\mu(z) = \frac{dx \wedge dy}{y^2}$. It induces a measure on Y_{mod} for which the total volume is $\frac{\pi}{3}$. This can be computed as

$$\int_{\Gamma_{\text{mod}} \backslash \mathfrak{H}} d\mu = \int_{F_{\text{mod}}} d\mu(z) = \int_{x=-1/2}^{1/2} \int_{y=\sqrt{1-x^2}}^{\infty} \frac{dy}{y^2}\, dx = \frac{\pi}{3}.$$

1.2 Maass forms

We define real analytic modular forms, and mention their relation to the spectral decomposition of the Laplacian. See [57], §3.5, for more information.

1.2.1 Γ_{mod}*-invariant functions.* Once the space Y_{mod} has been constructed, it is sensible to study the functions on it. Such functions are functions on $\mathfrak{H}$ that are invariant under the action of Γ_{mod}. Especially interesting are the eigenfunctions of the Laplacian; these are the analogues of the exponential functions on $\mathbb{R}$. There are a lot of such eigenfunctions. A growth condition at the cusp admits only those that are related to the spectral decomposition of the Laplacian.

1.2.2 *Definition.* A *real analytic modular form*, or *Maass form*, is a twice differentiable function $f : \mathfrak{H} \to \mathbb{C}$ such that

i) (Γ_{mod}*-invariance*) $f(\gamma \cdot z) = f(z)$ for all $g \in \Gamma_{\text{mod}}$.

ii) (*Eigenfunction of the Laplacian*) $L_0 f = \lambda f$ for some $\lambda \in \mathbb{C}$, with $L_0 = -y^2\partial_x^2 - y^2\partial_y^2$.

iii) (*Polynomial growth*) There is a real number a such that $f(z) = \mathcal{O}(y^a)$ $(y \to \infty)$, uniformly for $x \in \mathbb{R}$.

Later on we shall call such a function a real analytic modular form *of weight* 0. The name *Maass form*, or *Maass wave form*, is often attached to a wider class of functions. This name honors H. Maass, who first studied these functions systematically, see [34].

All functions satisfying condition i) are determined by their values on a fundamental domain, for instance on F_{mod}. The number λ in condition ii) is called the *eigenvalue* of f.

1.2.3 *Examples.* Trivial example: constant functions, with eigenvalue 0.

To get less trivial examples of real analytic modular forms, consider for $s \in \mathbb{C}$ the function $h_s : z \mapsto y^{s+1/2}$. It satisfies condition ii) (with eigenvalue $\frac{1}{4} - s^2$) and condition iii). But invariance as in condition i) holds only for γ in the subgroup $\Gamma_{\mathrm{mod}}^{\infty} = \{\pm \left(\begin{smallmatrix} 1 & * \\ 0 & 1 \end{smallmatrix}\right)\}$ of Γ_{mod}. So one might hope that

$$e(s; z) = \sum_{\gamma \in \Gamma_{\mathrm{mod}}^{\infty} \backslash \Gamma_{\mathrm{mod}}} h_s(\gamma \cdot z) \tag{1.1}$$

defines a modular form with eigenvalue $\frac{1}{4} - s^2$. Indeed, it is known that for $\operatorname{Re} s > \frac{1}{2}$ the sum converges absolutely, and that Equation (1.1) defines a real analytic modular form $e(s)$; it is called an *Eisenstein series.* This is a special case of Proposition 5.1.6.

1.2.4 *Notation.* I depart from the usual choice of the parameter: $s_{\mathrm{usual}} = s_{\mathrm{here}} + \frac{1}{2}$. Two advantages of this choice are:

- The transformation $s \mapsto -s$ in the functional equation of the Eisenstein family, see 1.4.4, is slightly simpler than $s \mapsto 1 - s$.
- In the Fourier expansion of modular forms one needs Whittaker functions. My normalization leads to simpler expressions for the parameters.

1.2.5 *Analyticity.* The differential operator L_0 is elliptic, and has real analytic coefficients. This implies that all its eigendistributions on $\mathfrak{H}$ are real analytic functions (around each point z_0 they are given by a converging power series in $z - z_0$ and $\overline{z - z_0}$). In particular all modular forms just defined are real analytic functions on $\mathfrak{H}$. (See, e.g., [29], App. 4, §5 and [3], p. 207–210, for a proof of this fundamental analyticity result.)

1.2.6 *Selfadjoint extension of the Laplacian.* Put $H_0 = L^2(F_{\mathrm{mod}}, d\mu)$. One may view this as the Hilbert space of square integrable function on Y_{mod}. If f is a smooth Γ_{mod}-invariant function on $\mathfrak{H}$ that has compact support in Y_{mod}, then both f and $L_0 f$ represent elements of H_0. The map $f \mapsto L_0 f$ has a selfadjoint extension A_0 in H_0 (this will be proved in Section 6.5). One may show that A_0 has a continuous spectrum with multiplicity one, with support $[\frac{1}{4}, \infty)$, and a discrete spectrum consisting of a countable discrete subset of $[0, \infty)$. (Consult, e.g., [57], §3.7. Theorem 1 on p. 254 gives the spectral decomposition; in Theorem 5 on p. 290 one sees that the discrete spectrum is indeed infinite.)

The constant functions form the one-dimensional eigenspace for the eigenvalue 0. All other eigenvalues are strictly positive, in fact, at least $\frac{3}{2}\pi^2$; see, e.g., [57], §3.5, Theorem 3 on p. 226, and §3.7, Theorem 1 on p. 254. The corresponding eigenfunctions are square integrable real analytic modular forms (Proposition 6.6.2 of this book).

1.3 Holomorphic modular forms

Holomorphic modular forms are better known than the real analytic ones discussed in the previous section. When we consider modular forms as functions on $G = \mathrm{SL}_2(\mathbb{R})$ instead of on $\mathfrak{H}$, it will become clear that both types are realizations of the same idea.

1.3.1 *Definition.* A holomorphic modular form of *weight* $k \in 2\mathbb{Z}$ is a function $f : \mathfrak{H} \to \mathbb{C}$ for which

i) $f(\gamma \cdot z) = (cz + d)^k f(z)$ for all $\gamma = \begin{pmatrix} a & b \\ c & d \end{pmatrix} \in \Gamma_{\mathrm{mod}}$,

ii) f is a holomorphic function on $\mathfrak{H}$,

iii) $f(z) = \mathcal{O}(1)$ $(y \to \infty)$, uniformly in x.

1.3.2 *Examples.* We have the holomorphic Eisenstein series

$$G_k(z) = \sum_{n,m \in \mathbb{Z}}{}' (cz + d)^{-k}$$

for even $k \geq 4$. The prime indicates that $(n, m) = (0, 0)$ is to be omitted from the sum. The sum converges absolutely, and defines a holomorphic modular form G_k of weight k, see, e.g., [53], §2.2 (our G_k is Shimura's E_k^*). We may rewrite the sum defining $G_k(z)$ as

$$\begin{aligned} G_k(z) &= 2\zeta(k) \left(1 + \sum_{c=1}^{\infty} \sum_{d \in \mathbb{Z},\, (d,c)=1} (cz + d)^{-k} \right) \\ &= 2\zeta(k) \sum_{\gamma \in \Gamma_{\mathrm{mod}}^{\infty} \backslash \Gamma_{\mathrm{mod}}} \left(\frac{d\gamma z}{dz} \right)^{k/2}, \end{aligned}$$

with $\zeta(\cdot)$ the zeta function of Riemann. The latter sum is similar to that in Equation (1.1) on p. 4 defining the real analytic Eisenstein series.

The set of all holomorphic modular forms is closed under multiplication. This gives a method of constructing more holomorphic modular forms. For instance

$$\Delta = (2\pi)^{-12}\Big((60G_4)^3 - 27(140G_6)^2\Big)$$

defines a holomorphic modular form Δ of weight 12 that is not a multiple of G_{12}. (Shimura's Δ in [53], Theorem 2.9, is $(2\pi)^{12}$ times this Δ.)

1.3.3 *Comparison.* The three conditions above are similar to those in 1.2.2. The factor $(cz+d)^k$ makes condition i) more complicated. The condition of being an eigenfunction of L_0 is replaced by holomorphy. If f satisfies conditions i) and ii) here, then both conditions iii) are equivalent.

A property that both types of modular forms have in common is the possibility to associate interesting Dirichlet series to them. For holomorphic modular forms, one may consult [30], §1.5 and §2.2, or the more general discussion in §4.3 of [38]. Maass's motivation, see [34], to consider real analytic modular forms is the wish to get more functions to which one can associate Dirichlet series.

1.3.4 *Functions on* $\mathrm{SL}_2(\mathbb{R})$. Another way to see that both definitions are natural, and narrowly related, is to look at functions on $G = \mathrm{SL}_2(\mathbb{R})$.

Let $K = \mathrm{SO}_2(\mathbb{R})$, the group of orthogonal matrices in G. The map $g \mapsto g \cdot i$ gives an identification $G/K \cong \mathfrak{H}$. This means that real analytic modular forms may be viewed as functions on G by $f_0(g) = f(g \cdot i)$. In this way one obtains functions that are invariant under $\Gamma_{\mathrm{mod}} = \mathrm{SL}_2(\mathbb{Z})$ on the left, and under K on the right.

Similarly, we may lift functions f that satisfy condition i) in 1.3.1 to G by defining

$$f_k\left(\begin{pmatrix} a & b \\ c & d \end{pmatrix}\right) = (ci+d)^{-k} f\left(\frac{ai+b}{ci+d}\right).$$

Again f_k is Γ_{mod}-invariant on the left, but on the right it transforms according to the character $\begin{pmatrix} \cos\theta & \sin\theta \\ -\sin\theta & \cos\theta \end{pmatrix} \mapsto e^{ik\theta}$ of K.

1.3.5 *Casimir operator.* The space of linear differential operators on G that are homogeneous of order two and that commute with all left and right translations has dimension 1. It is spanned by the *Casimir operator* ω. In the coordinatization $(x, y, \theta) \mapsto \begin{pmatrix} \sqrt{y} & x/\sqrt{y} \\ 0 & 1/\sqrt{y} \end{pmatrix} \begin{pmatrix} \cos\theta & \sin\theta \\ -\sin\theta & \cos\theta \end{pmatrix}$ of G it is given by

$$\omega = -y^2\partial_x^2 - y^2\partial_y^2 + y\partial_x\partial_\theta.$$

For functions that are K-invariant on the right it amounts to the operator L_0

on $\mathfrak{H}$. For a holomorphic modular form f of weight k we get

$$\begin{aligned} &\omega f_k\left(\begin{pmatrix} \sqrt{y} & x/\sqrt{y} \\ 0 & 1/\sqrt{y} \end{pmatrix}\begin{pmatrix} \cos\theta & \sin\theta \\ -\sin\theta & \cos\theta \end{pmatrix}\right) \\ &= \left(-y^2\partial_x^2 - y^2\partial_y^2 + y\partial_x\partial_\theta\right) e^{ik\theta} y^{k/2} f(x+iy) \\ &= \frac{k}{2}\left(1-\frac{k}{2}\right) f_k\left(\begin{pmatrix} \sqrt{y} & x/\sqrt{y} \\ 0 & 1/\sqrt{y} \end{pmatrix}\begin{pmatrix} \cos\theta & \sin\theta \\ -\sin\theta & \cos\theta \end{pmatrix}\right). \end{aligned}$$

Thus we see that all modular forms considered up to now correspond to eigenfunctions of ω on G that are Γ_{mod}-invariant on the left, and transform on the right according to a character of K.

1.3.6 *Representations.* If the function f_0 or f_k is square integrable on $\Gamma_{\mathrm{mod}}\backslash G$, then it generates an irreducible subspace of $L^2(\Gamma_{\mathrm{mod}}\backslash G)$ for the action of G by right translation. If f is a holomorphic modular form, then this irreducible representation belongs to the *discrete series* of representations of G. If f is a square integrable real analytic modular form with positive eigenvalue, the function f_0 is a weight zero vector in an irreducible representation of the *principal series.* For more information on the representational point of view one may consult §2 of [15].

Hecke operators are not discussed in thus book. But they reveal very interesting properties of modular forms. See, e.g., Chapter II of [29] for the holomorphic case, and Chapter V of [35] for Maass forms. The representational point of view incorporates the Hecke operators by working with functions on the adele group of GL_2, see [15].

1.4 Fourier expansion of modular forms

Up till now we have motivated the study of modular forms from harmonic analysis: spectral decomposition of the Laplace operator, and irreducible subspaces for the right representation of G in $L^2(\Gamma_{\mathrm{mod}}\backslash G)$. Number theoretically interesting formulas arise as soon as one writes down the Fourier expansion of modular forms.

1.4.1 *Fourier expansion.* For both types of modular forms discussed thus far, the transformation behavior implies periodicity in $x = \mathrm{Re}(z)$: take $\gamma = \left(\begin{smallmatrix} 1 & 1 \\ 0 & 1 \end{smallmatrix}\right)$ in condition i) to conclude that $f(z+1) = f(z)$. Hence there is a Fourier expansion

$$f(z) = \sum_{n=-\infty}^{\infty} a_n(y) e^{2\pi i n x}.$$

Condition ii) in the definitions above implies that the a_n satisfy ordinary differential equations. All Fourier terms $a_n(y)e^{2\pi i n x}$ inherit the growth condition iii).

We see in the holomorphic case that $a_n(y)$ is a multiple of $e^{-2\pi n y}$, and has to vanish for $n < 0$. Thus we get

$$f(z) = \sum_{n\geq 0} c_n(f) e^{2\pi i n z}.$$

The $c_n(f)$ are called the *Fourier coefficients* of f.

In the real analytic case the differential equation is

$$-y^2 a_n''(y) = (\lambda - 4\pi^2 n^2 y^2) a_n(y).$$

We write $\lambda = \frac{1}{4} - s^2$, $s \in \mathbb{C}$. For $n = 0$ there is a two dimensional space of solutions, with basis $y^{s+1/2}$, $y^{-s+1/2}$ (if $s \neq 0$). For $n \neq 0$ we get a variant of the Whittaker differential equation. The growth condition restricts the possibilities to a one dimensional space, spanned by $W_{0,s}(4\pi|n|y)$; see, e.g., [56], 1.7. (The Whittaker function $W_{0,s}$ decreases exponentially: $W_{0,s}(t) \sim e^{-t/2}$ $(t \to \infty)$.) This leads to

$$f(z) = b_0(f) y^{s+1/2} + c_0(f) y^{-s+1/2} + \sum_{n\neq 0} c_n(f) W_{0,s}(4\pi|n|y) e^{2\pi i n x}.$$

To distinguish between the Fourier coefficients b_0 and c_0, a choice of s such that $\lambda = \frac{1}{4} - s^2$ is necessary.

Often one uses a modified Bessel function in the terms with $n \neq 0$; the function $y \mapsto \sqrt{y} K_s(2\pi|n|y)$ spans the space of possible $a_n(y)$. I prefer Whittaker functions, as they can be used in weights other than 0 as well. When comparing the Fourier expansions below with those at other places, one should keep in mind that $s_{\text{here}} = s_{\text{usual}} - \frac{1}{2}$, and that $W_{0,s}(y) = \sqrt{y/\pi} K_s(y/2)$.

1.4.2 *Holomorphic modular forms.* [53], (2.2.1), on p. 32, gives the Fourier expansion of holomorphic Eisenstein series ($k \geq 4$ even):

$$G_k(z) = 2\zeta(k) + 2\frac{(2\pi i)^k}{(k-1)!} \sum_{n\geq 1} \sigma_{k-1}(n) e^{2\pi i n z},$$

with the *divisor function* $\sigma_u(n) = \sum_{d|n} d^u$.

All coefficients in the expansion of $\frac{1}{2\zeta(k)} G_k(z)$ are rational numbers, even integers if $k = 4$ or 6. This implies (after some computations) that $\Delta(z) = \sum_{n\geq 1} \tau(n) e^{2\pi i n z}$, with all $\tau(n) \in \mathbb{Z}$, $\tau(1) = 1$. In particular, $c_0(\Delta) = 0$.

1.4.3 *Definition.* A *cusp form* is a modular form for which the Fourier term of order zero vanishes. This means that $c_0 = 0$ in the holomorphic case, and $b_0 = c_0 = 0$ in the real analytic case. The holomorphic modular form Δ is a cusp form.

1.4.4 *Real analytic Eisenstein series.* If the integers c and d are relatively prime, then there are a and b such that $\begin{pmatrix} a & b \\ c & d \end{pmatrix} \in \Gamma_{\text{mod}}$, and the coset $\Gamma_{\text{mod}}^{\infty} \begin{pmatrix} a & b \\ c & d \end{pmatrix}$ depends only on $\pm(c,d)$. In this way we get for $\operatorname{Re} s > \frac{1}{2}$:

$$\begin{aligned} e(s;z) &= \frac{1}{2} \sum_{c,d\in\mathbb{Z},\,(c,d)=1} y^{s+1/2} |cz+d|^{-2s-1} \\ &= \frac{y^{s+1/2}}{2\zeta(2s+1)} \sum_{n,m\in\mathbb{Z}}{}' |mz+n|^{-2s-1}. \end{aligned}$$

Hence $e(s;z) = \frac{1}{2} y^{s+1/2} \zeta(2s+1)^{-1} G(z,\bar{z};s+1/2,s+1/2)$, with Maass's Eisenstein series $G(\cdot,\cdot;\alpha,\beta)$ as on p. 207 of [35]. The Fourier expansion is given on p. 210 of *loc. cit.*, and leads to

$$\begin{aligned} e(s;z) &= y^{s+1/2} + \sqrt{\pi}\frac{\Gamma(s)\zeta(2s)}{\Gamma(s+1/2)\zeta(2s+1)} y^{-s+1/2} \\ &\quad + \frac{\pi^{s+1/2}}{\Gamma(s+1/2)\zeta(2s+1)} \sum_{n\neq 0} \frac{\sigma_{2s}(|n|)}{|n|^{s+1/2}} W_{0,s}(4\pi|n|y) e^{2\pi i n x}. \end{aligned}$$

Again we see that the divisor function appears in the Fourier coefficients of Eisenstein series.

This Fourier expansion defines the function $e(s)$ on $\mathfrak{H}$ for all $s \in \mathbb{C}$ that satisfy $s \neq 0, 1$ and $\Gamma(s+1/2)\zeta(2s+1) \neq 0$. The Γ_{mod}-invariance is preserved, as are the other conditions in Definition 1.2.2. In this way we get e as a meromorphic family on $\mathbb{C}$ of real analytic modular forms. The singularities have order one, and the residues are again modular forms. Moreover, the functional equation of the zeta function of Riemann implies the *functional equation* of e:

$$e(-s;z) = c_0(e(-s)) \cdot e(s;z).$$

1.4.5 *Cuspidal Maass forms.* In 1.2.6 we mentioned that there is a countable set $\psi_0, \psi_1, \ldots$ of square integrable real analytic modular forms that constitute an orthonormal basis of the part of $L^2(\Gamma_{\text{mod}}\backslash\mathfrak{H})$ in which the selfadjoint extension A_0 of the Laplacian has a discrete spectrum. We may arrange the ψ_j such that their eigenvalues λ_j increase. Take $\psi_0 = \sqrt{3/\pi}$. For $j \geq 1$ one knows that $\lambda_j > \frac{1}{4}$, hence $\lambda_j = \frac{1}{4} - s_j^2$ with $s_j \in i\mathbb{R}$. The square integrability is inherited by the Fourier coefficients: $\int_1^\infty |a_n(y)|^2 y^{-2}\, dy < \infty$. Hence $b_0(\psi_j) = c_0(\psi_j) = 0$ for $j \geq 1$. So $\psi_1, \psi_2, \ldots$ are *cusp forms.*

We can choose all ψ_j to be real-valued. (Use that $f \mapsto \bar{f}$ preserves the space of real analytic cusp forms for a real eigenvalue.) We may even arrange that each ψ_j is an eigenfunction of all Hecke operators. Much more information can be found in §3.5 of [57]. We only mention the *Ramanujan-Petersson conjecture for real analytic modular forms* (not proved up till now):

$$c_n(\psi_j) = \mathcal{O}\left(|n|^{-1/2+\varepsilon}\right) \quad (|n| \to \infty) \qquad \text{for each } j \geq 1, \text{ for each } \varepsilon > 0.$$

1.5 More modular forms

There are more general types of modular forms. First we consider real analytic modular forms of even weight. Next we introduce a multiplier system to be able to define modular forms of arbitrary complex weight. This opens the possibility to consider families of modular forms for which the weight varies continuously.

1.5.1 *Real analytic modular forms of even weight.* We have seen in 1.3.5 that the modular forms considered up till now correspond to functions on $\Gamma_{\text{mod}}\backslash G$ that are eigenfunctions of the Casimir operator. These forms transform according to a character of K. Usually, one calls all such functions modular forms, provided a growth condition at the cusp is satisfied.

All characters of K are of the form $\left(\begin{smallmatrix}\cos\theta & \sin\theta \\ -\sin\theta & \cos\theta\end{smallmatrix}\right) \mapsto e^{ik\theta}$ with $k \in \mathbb{Z}$. But as $-\text{Id} \in \Gamma_{\text{mod}} \cap K$ is central in G, only characters with even k admit non-zero functions with the prescribed transformation properties.

The correspondence in 1.3.4 between holomorphic modular forms and functions on G is not the most convenient one if one wants to study real analytic modular forms. This is caused by the fact that the factor $(cz+d)^k$ in the transformation behavior of holomorphic modular forms does not have absolute value 1. We follow the convention to relate functions f on $\mathfrak{H}$ and functions F on G by

$$F\left(\begin{pmatrix} a & b \\ c & d \end{pmatrix}\right) = e^{-ik\arg(ci+d)} f\left(\frac{ai+b}{ci+d}\right).$$

As $k \in 2\mathbb{Z}$ the choice of the argument does not matter. This leads to the following definition.

1.5.2 *Definition.* A *real analytic modular form* of *even weight* $k \in 2\mathbb{Z}$ with *eigenvalue* $\lambda \in \mathbb{C}$ is a function $f : \mathfrak{H} \to \mathbb{C}$ that satisfies the conditions

i) $f(\gamma \cdot z) = e^{ik\arg(cz+d)} f(z)$ for all $\gamma = \left(\begin{smallmatrix} a & b \\ c & d \end{smallmatrix}\right) \in \Gamma_{\text{mod}}$.

ii) $L_k f = \lambda f$, with $L_k = -y^2\partial_x^2 - y^2\partial_y^2 + iky\partial_x$.

iii) There is a real number a such that $f(z) = \mathcal{O}(y^a)$ $(y \to \infty)$, uniformly in $x \in \mathbb{R}$.

1.5.3 *Examples* The real analytic modular forms defined in 1.2.2 have weight 0. If h is a holomorphic modular form of weight $k \in \mathbb{Z}$, then $z \mapsto y^{k/2}h(z)$ is a real analytic modular form in the sense just defined, of weight k, with eigenvalue $\frac{k}{2}(1-\frac{k}{2})$.

For each $k \in 2\mathbb{Z}$ there are Eisenstein series of weight k with eigenvalue $\frac{1}{4} - s^2$. They have a meromorphic extension, and satisfy a functional equation. If $k \geq 4$, then the value at $s = \frac{1}{2}(k-1)$ corresponds to a multiple of G_k.

There are countably many cuspidal real analytic modular forms of weight k with eigenvalues $\lambda_1, \lambda_2, \ldots$, obtained from the ψ_j by differential operators (see Proposition 4.5.3). Those differential operators are multiples of the operators described on p. 177 of [35], often called *Maass operators.*

1.5.4 *The eta function of Dedekind.* If $k \notin 2\mathbb{Z}$ the definition above admits only the zero function as a modular form. But there are modular forms with other weight, even holomorphic ones. The best known example is the *eta function* of Dedekind:

$$\eta(z) = e^{\pi i z/12} \prod_{n\geq 1} (1 - e^{2\pi i n z}).$$

See, e.g., [30], Chapter IX, §1. It satisfies $\eta(\gamma \cdot z) = v_{1/2}(\gamma)(cz+d)^{1/2}\eta(z)$ for all $\gamma = \begin{pmatrix} a & b \\ c & d \end{pmatrix} \in \Gamma_{\mathrm{mod}}$. Here $v_{1/2}$ is a multiplier system; that is a map from Γ_{mod} into $\mathbb{C}^*$. It depends on the choice of the argument in $(cz+d)^{1/2}$. In this book the argument is taken in $(-\pi, \pi]$ (this is the standard choice). The function η is called a holomorphic modular form of weight $\frac{1}{2}$ for the multiplier system $v_{1/2}$. It is known that $\eta^{24} = \Delta$.

The multiplier system $v_{1/2}$ is almost a character of Γ_{mod}, but not completely. It has to compensate for the fact that $(c_{\gamma\delta} z + d_{\gamma\delta})^{1/2}$ is not always equal to $(c_\gamma \delta \cdot z + d_\gamma)^{1/2}(c_\delta z + d_\delta)^{1/2}$.

1.5.5 *Powers of the eta function.* As $\log \eta(z)$ is well defined for $z \in \mathfrak{H}$, we have a holomorphic modular form $z \mapsto \eta(z)^{2r}$ for each $r \in \mathbb{C}$, with multiplier system $v_r = v_{1/2}^{2r}$. See [30], Chapter IX, §1, for an explicit description of $\log v_{1/2}$ in terms of Dedekind sums. The transformation behavior under elements of the modular group is easily written down. Note that the growth at the cusp is not polynomial in y if $\mathrm{Re}\, r < 0$.

Let us consider $\eta_r : z \mapsto y^{r/2}\eta(z)^{2r}$, in the spirit of the relation between holomorphic and real analytic modular forms in 1.5.3. This function satisfies

$$\eta_r(\gamma \cdot z) = v_r(\gamma) e^{ir \arg(c_\gamma z + d_\gamma)} \eta_r(z) \quad \text{for all } \gamma \in \Gamma_{\mathrm{mod}},$$
$$L_r \eta_r = \frac{r}{2}\left(1 - \frac{r}{2}\right) \quad \text{with } L_r = -y^2\partial_x^2 - y^2\partial_y^2 + iry\partial_x.$$

As $\eta_r(z+1) = e^{\pi i r/6}\eta_r(z)$, there is a kind of Fourier series expansion

$$\eta_r(z) = \sum_{\nu \geq 0} p_\nu(r) y^{r/2} e^{2\pi i(\nu + r/12)z},$$

with polynomials $p_\nu(r)$ defined by $\prod_{m=1}^\infty (1-q^m)^{2r} = \sum_{\nu=0}^\infty p_\nu(r) q^\nu$. Hence $p_0 = 1$, and p_ν has degree ν.

We can omit a finite number of terms from this expansion in such a way that the remainder has polynomial growth in y. For $-12 < \mathrm{Re}\, r < 0$ it suffices to omit only the term with the factor $e^{\pi i r z/6}$. This gives an example of the following type of modular form:

1.5.6 *Definition.* Let $l, \lambda \in \mathbb{C}$, let $r \equiv l \bmod 2$ with $|\mathrm{Re}\, r| < 12$. A *real analytic modular form* of weight l, for the multiplier system v_r, with eigenvalue λ, is a function $f : \mathfrak{H} \to \mathbb{C}$ satisfying the following conditions:

i) $f(\gamma \cdot z) = v_r(\gamma) e^{il \arg(c_\gamma z + d_\gamma)} f(z)$ for all $\gamma \in \Gamma_{\mathrm{mod}}$,

ii) $L_l f = \lambda f$, with $L_l = -y^2\partial_x^2 - y^2\partial_y^2 + ily\partial_x$,

iii) $e^{-\pi irx/6} f(iy+x) - \int_0^1 e^{-\pi irx'/6} f(iy+x+x')\, dx' = \mathcal{O}\left(y^a\right)$ $(y \to \infty)$ uniformly in x, for some $a \in \mathbb{R}$.

The modular forms defined in 1.5.2 all come under this definition, with $l = k$ and $r = 0$. For these examples it is not necessary to subtract the 'constant term' in the Fourier expansion before stating the growth condition. The same holds for the modular forms obtained by applying Roelcke's definition, see [50], Definition 1.1, to the modular case.

We have already seen the example η_r, with $-12 < \mathrm{Re}\, r < 12$; take $l = r$. For $-12 < \mathrm{Re}\, r < 0$ we need to subtract the 'constant Fourier term' before applying the growth condition.

This definition can be made more general by omitting more terms from the Fourier expansion before imposing the growth condition. The present definition suffices in this introductory chapter. We restrict ourselves to the case $l = r$ in the sequel.

1.5.7 *Selfadjoint extension of L_r.* We take $-12 < r < 12$. Let H_r be the Hilbert space of (classes of) functions that satisfy the transformation behavior in condition i), with $l = r$, and that are square integrable on F_{mod}. Note that $\eta_r \in H_r$ for $0 \leq r < 12$. The differential operator L_r has a selfadjoint extension A_r in H_r, see, e.g., [50], Satz 3.2.

If $r \in (-12, 12) \setminus \{0\}$ there is no continuous spectrum, see, e.g., [51], Satz 8.2. This is due to the fact that $v_r\left(\begin{smallmatrix} 1 & 1 \\ 0 & 1 \end{smallmatrix}\right) \neq 1$ if $r \neq 0$. For these r there is an orthonormal Hilbert basis of H_r of eigenfunctions of A_r. These eigenfunctions are modular forms in the sense of Definition 1.5.6, and moreover, are square integrable. The lowest eigenvalue $\frac{|r|}{2}\left(1 - \frac{|r|}{2}\right)$ has multiplicity one.

Let r run through $(0, 12)$. The lowest eigenvalue $\frac{r}{2}\left(1 - \frac{r}{2}\right)$ depends on r in a real analytic way, and the corresponding eigenfunctions are given by the *family* $r \mapsto \eta_r$ *of modular forms.* Propositions 2.14 and 2.15 in [7] show that the other eigenvalues λ are also real analytic in $r \in (0, 12)$, satisfy $\lambda(r) > \frac{1}{4}$, and that each eigenvalue $\lambda : (0, 12) \to \mathbb{R}$ has an eigenfamily of square integrable modular forms that is unique up to multiples. In this way the spectral theory of A_r leads to families of modular forms.

1.5.8 *Extension of the Eisenstein series.* The Eisenstein series discussed in 1.2.3 gives another example of a family of modular forms. The parameter of this family is not the weight, but the spectral parameter s.

Proposition 2.19 in [6] states that there is a meromorphic family E of functions on $\mathfrak{H}$ with the following properties:

i) E is defined on $U \times \mathbb{C} \subset \mathbb{C}^2$, with U an open neighborhood of $(-12, 12)$ in $\mathbb{C}$.

ii) For each $z \in \mathfrak{H}$ the function $(r,s) \mapsto E(r,s)$ is meromorphic on $U \times \mathbb{C}$, with singularities contained in a fixed analytic set of codimension 1.

iii) At each (r,s) outside this singular set, the function $z \mapsto E(r,s;z)$ is a real analytic modular form of weight r, for the multiplier system v_r, with eigenvalue $\frac{1}{4} - s^2$, as defined in 1.5.6.

iv) There is a Fourier expansion

$$E(r,s;z) = \mu_r^0(r,s;z) + C(r,s)\mu_r^0(r,-s;z) + \sum_{\nu \neq 0} C^\nu(r,s)\omega_r^\nu(r,s;z),$$

with

$$\begin{aligned} \mu_r^\nu(r,s;z) &= e^{2\pi i(\nu+r/12)x} y^{s+1/2} e^{-2\pi(\nu+r/12)y} \\ &\quad \cdot {}_1F_1\left[\begin{array}{c} 1/2+s-r/2 \\ 1+2s \end{array} \middle| 4\pi(\nu+r/12)y\right] \\ \omega_r^\nu(r,s;z) &= e^{2\pi i(\nu+r/12)x} W_{\varepsilon r/2,s}(4\pi\varepsilon(\nu+r/12)y) \end{aligned}$$

$(\varepsilon = \operatorname{sign}\nu)$.

The family E is the unique family of modular forms with a Fourier expansion of this type.

v) The uniqueness implies the *functional equations*

$$\begin{aligned} E(r,-s;z) &= C(r,-s)E(r,s;z), \\ E(r,s;-\bar{z}) &= E(-r,s;z). \end{aligned}$$

vi) The restriction $s \mapsto E(0,s)$ exists, and coincides with the Eisenstein family $s \mapsto e(s)$ discussed in 1.4.4. The existence of this restriction is not trivial: E might have had a singularity along the complex line $\{0\} \times \mathbb{C}$.

These properties suggest to call E the *Eisenstein family* for Γ_{mod}.

The multiples of ω_r^ν are the only possibilities for quickly decreasing Fourier terms; ω_r^ν is not defined for $\nu + \frac{r}{12} = 0$. There we need another description. The $\mu_r^0(r,s)$ and $\mu_r^0(r,-s)$ form a basis of the space of functions that may occur in the Fourier term with $e^{\pi i r x/6}$. In general these functions are exponentially increasing in y. The restriction of $(r,s) \mapsto \mu_r^0(r,s)$ along the line $\{0\} \times \mathbb{C}$ is given by the function $s \mapsto y^{s+1/2}$, which we have seen in the Fourier expansion of $e(s)$.

We have seen in 1.4.4 that $s \mapsto E(0,s)$ and $s \mapsto C(0,s)$ are holomorphic at all points of $i\mathbb{R}$. This does not exclude the possibility that E and C may have singularities at points $(0,it)$, $t \in \mathbb{R}$. If C is singular at such a point, then its zero set and its polar set intersect at such a point, otherwise the restriction $s \mapsto C(0,s)$ would be singular at it. Proposition 2.19 in [7] states that C is singular at $(0,it)$,

$t \in \mathbb{R}$, if and only if the space of Maass forms of weight 0 with eigenvalue $\frac{1}{4} + t^2$ is not spanned by the values at $r = 0$ of families of square integrable modular forms depending analytically on the weight $r \in (-12, 12)$.

1.5.9 *Poincaré series.* Fix $r \in (-12, 12)$, $r \neq 0$. The restriction $s \mapsto E(r, s)$ is a family of modular forms of weight r. It can be related to a Poincaré series. That are modular forms constructed in the same way as Eisenstein series. For instance, the series

$$\sum_{\gamma \in \Gamma_{\mathrm{mod}}^\infty \backslash \Gamma_{\mathrm{mod}}} v_r(\gamma)^{-1} e^{-ir \arg(c_\gamma z + d_\gamma)} \mu_r^0(r, s; \gamma \cdot z)$$

can be shown to converge absolutely for $\mathrm{Re}\, s > \frac{1}{2}$, and to define a real analytic modular form $P(r, s)$ for v_r of weight r with eigenvalue $\frac{1}{4} - s^2$. There is a meromorphic function ψ on a neighborhood of $(0, 12) \times \mathbb{C}$ in $\mathbb{C}^2$ such that $\psi(r, s)E(r, s)$ is equal to $P(r, s)$ for all (r, s) with $r \in (0, 12)$ and $\mathrm{Re}\, s > \frac{1}{2}$ at which $\psi \cdot E$ is holomorphic. The function ψ cannot be extended meromorphically to $U \times \mathbb{C}$, it behaves badly along the line $\{0\} \times \mathbb{C}$; see 13.5.7 for more explicit formulas.

The Fourier coefficients of $P(r, s)$ can be expressed in terms of Dirichlet series with coefficients containing Dedekind sums. This gives expressions for the Fourier coefficients $C^\nu(r, s)$ of the Eisenstein family, valid for $0 < r < 12$ and $\mathrm{Re}\, s$ large. This suffices to get information on the singularities of the derivatives $\partial_r^m C^\nu(r, s)|_{r=0}$. In [8] we expressed these derivatives in terms of the Dirichlet series $\sum_{c=1}^\infty c^{-s} \sum_{d \bmod c}^* S(d, c)^k e^{2\pi i n d/c}$, for $k \in \mathbb{N}$, $n \in \mathbb{Z}$, with Dedekind sums $S(d, c)$. This led to two distribution results for $\left(\frac{d}{c}, \frac{S(d,c)}{c}\right)$. The same method can be used to get distribution results for other quantities, see Section 13.6 of this book.

1.6 Truncation and perturbation

We indicate how to prove the results on the Eisenstein family E, discussed in the previous section, by applying ideas of Colin de Verdière, [12]. The main purpose of Part I of this book is to give a generalization of this proof. The present section sketches the main ideas in the context of the modular group.

Here we look at a neighborhood of $r = 0$ only; we shall be content to get the family E on a set $U \times \mathbb{C}$, where U is a small neighborhood of 0 in $\mathbb{C}$.

1.6.1 *Transformation.* We shall apply analytic perturbation theory for linear operators, as discussed in Chapter VII of Kato's book [25]; see especially §4. This theory studies families of operators in a fixed Hilbert space. Here we have not one Hilbert space, but infinitely many: the H_r for r running from -12 to 12. The first step is to transform everything to $H = H_0$. Of course all H_r are isomorphic to $L^2(F_{\mathrm{mod}}, d\mu)$, but we need a more explicit isomorphism, under which differential operators correspond to differential operators, and that behaves very nicely near the cusp.

In Lemma 3.2 of [5] we used the logarithm of the eta function of Dedekind to construct a real valued function $t \in C^\infty(\mathfrak{H})$ with the properties

i) $t(z) = \frac{1}{6}\pi x$ for all $y \geq 5$.

ii) $t(\gamma \cdot z) = \alpha(\gamma) + \arg(c_\gamma z + d_\gamma) + t(z)$ for each $\gamma \in \Gamma_{\text{mod}}$.

The function $\alpha : \Gamma \to \frac{\pi}{6}\mathbb{Z}$ satisfies $v_r(\gamma) = e^{ir\alpha(\gamma)}$. It can be expressed in terms of Dedekind sums. Lemma 8.2.1 will give a construction of such a function t that is not based on $\log \eta$.

Condition ii) implies that e^{irt} has the transformation behavior of elements of H_r. The transformation $f \mapsto e^{-irt} f$ gives a unitary isomorphism $H_r \to H$. Condition i) implies that near the cusp the Fourier series expansion is transformed term by term, and that the term with a special status in condition iii) in Definition 1.5.6 is mapped to the constant term of $e^{-irt} f$.

This transformation makes sense for $r \in \mathbb{C} \smallsetminus \mathbb{R}$ as well. For all $r \in \mathbb{C}$ it maps functions transforming under Γ_{mod} as in condition i) of Definition 1.5.6 onto Γ_{mod}-invariant functions.

Under this transformation, the differential operator L_r corresponds to the differential operator $L(r) = e^{-irt} \circ L_r \circ e^{irt}$ on the Γ_{mod}-invariant functions. Of course, one can express $L(r)$ in terms of t and its derivatives. A reader who carries out the computation will find differential operators $L^{(1)}$ and $L^{(2)}$ such that $L(r) = L_0 + rL^{(1)} + r^2 L^{(2)}$.

As $L(r)$ corresponds to L_r, it has a selfadjoint extension in H. But the family of selfadjoint operators thus obtained has properties that are not as nice as one would want. The idea in [12] is to work in a subspace of H.

1.6.2 *Truncation.* Let $c > 0$. Each $f \in H_0$ has a Fourier coefficient $F_0 f \in L^2\left((c,\infty), y^{-2}dy\right)$ given by $F_0 f(y) = \int_0^1 f(x+iy)\, dx$. (Section 6.3 gives a more careful definition of $F_0 f$.)

We fix a real number $a > 5$. Define ${}^aH$ as the Hilbert subspace of H characterized by the condition that $F_0 f$ vanishes on (a,∞). The differential operators $L(r)$ act in the space $C_c^\infty(\Gamma_{\text{mod}}\backslash\mathfrak{H}) \cap {}^aH$ of smooth compactly supported functions on $\Gamma_{\text{mod}}\backslash\mathfrak{H}$ for which the zero order Fourier term vanishes above a.

The great advantage of this 'truncated' setup is that for r near 0 the differential operators $L(r)$ have selfadjoint extensions ${}^aA(r)$ in ${}^aH$ with *compact resolvent*, and, moreover, that there is a neighborhood U of 0 in $\mathbb{C}$ on which $r \mapsto {}^aA(r)$ is a *selfadjoint holomorphic family* of operators *of type (B)* in the sense of Kato, [25], Chapter VII, §4.2. Proposition 9.2.2 gives the compactness of the resolvent. We shall give in Section 9.1 all ingredients needed to see that ${}^aA$ is a family of type (B) in Kato's sense. (In the general treatment in Chapter 9 we may have more than one parameter; hence the family ${}^aA$ discussed there does not come completely under Kato's type (B).)

We may draw strong conclusions concerning the holomorphy of the eigenvalues of the family ${}^aA$; see [25], Chapter VII, Theorem 3.9 and Remark 4.22. The eigenvalues and eigenfunctions are not considered in this book; I do not know how to extend those results to the situation of more than one complex parameter.

1.6.3 *Relation to modular forms.* The eigenfunctions of ${}^aA(r)$ are related to modular forms in the sense of Definition 1.5.6. We define the truncation ${}^{(a)}F$ of a smooth Γ_{mod}-invariant function F on $\mathfrak{H}$ to be the Γ_{mod}-invariant function on $\mathfrak{H}$ that satisfies

$$ {}^{(a)}F(z) = \begin{cases} F(z) & \text{if } z \in F_{\mathrm{mod}} \text{ and } y \le a \\ F(z) - \int_0^1 F(z+x')\,dx' & \text{if } z \in F_{\mathrm{mod}} \text{ and } y > a. \end{cases} $$

Proposition 9.2.6 states that $\ker({}^aA(r) - \frac{1}{4} + s^2)$ consists of the functions ${}^{(a)}(e^{-irt}f)$, where f runs trough the space of all modular forms for v_r of weight r with eigenvalue $\frac{1}{4} - s^2$ that satisfy $\int_0^1 e^{-\pi i r x/6} f(ia + x)\,dx = 0$.

In particular, the cuspidal Maass forms of weight zero are eigenfunctions of ${}^aA(0)$. Moreover, ${}^{(a)}e(s)$ is an eigenfunction of ${}^aA(0)$ for those values of s that satisfy $a^{2s} + \Lambda(2s)/\Lambda(2s+1) = 0$, with $\Lambda(u) = \pi^{-u/2}\Gamma(u/2)\zeta(u)$.

1.6.4 *The resolvent of* ${}^aA$. Proving the existence of the Eisenstein family E amounts to solving $(L_r - \frac{1}{4} + s^2)f = 0$ with a family f that depends meromorphically on (r,s), and has a Fourier expansion of a prescribed form. This is difficult. Let us look at the transformed differential equation

$$ \left(L(r) - \frac{1}{4} + s^2 \right) {}^a\tilde{E}(r,s) = 0. $$

We try to solve this equation with a meromorphic family ${}^a\tilde{E}$ of Γ_{mod}-invariant functions. We further impose the condition that $F_0{}^a\tilde{E}(r,s;a) = 1$.

Consider a Γ_{mod}-invariant function h that satisfies for $z \in F_{\mathrm{mod}}$:

$$ \begin{aligned} h(z) &= 0 && \text{for } y \le 5 \\ h(z) &= h_0(y) && \text{for } y \ge 5, \end{aligned} $$

for $h_0 \in C^\infty(5,\infty)$ of the following form: on the interval $((5+a)/2, \infty)$ we take $h_0(y)$ equal to a linear combination of $\mu_r^0(r,s;iy)$ and $\mu_r^0(r,-s;iy)$, such that $h_0(a) = 1$. This can be arranged with coefficients meromorphic in $(r,s) \in U$. (Holomorphic coefficients would be more difficult, as the μ_r^0 do not behave nicely at the lines $\mathbb{C} \times \{s\}$ with $s \in \frac{1}{2}\mathbb{Z}$.) On $(5, (5+a)/2)$ we let h_0 go to zero smoothly. We define $k = (L(r) - \frac{1}{4} + s^2)h$. Hence $k \in C_c^\infty(\Gamma_{\mathrm{mod}}\backslash\mathfrak{H})$ with support in F_{mod} below a. Moreover, k is meromorphic in (r,s). Its polar set does not contain vertical lines $\{r\} \times \mathbb{C}$. We may view k as a meromorphic family of elements of ${}^aH$.

In this way we have one solution h of $(L(r) - \frac{1}{4} + s^2)h = k$. If we find another solution of this equation, then their difference solves the homogeneous equation. We use the resolvent $\mathrm{R}(r,s) = ({}^aA(r) - \frac{1}{4} + s^2)^{-1}$ to get another solution. This is a meromorphic family of bounded operators in ${}^aH$, see [25], Chapter IV, Theorem 3.12. So $G(r,s) = \mathrm{R}(r,s)k(r,s)$ defines a meromorphic family of elements of ${}^aH$, without singularities along vertical lines, and $({}^aA(r)-\frac{1}{4}+s^2)G(r,s) = k(r,s)$. The fact that $G(r,s) \in \mathrm{dom}({}^aA(r))$ implies that $F_0G(r,s;a)$ has a meaning, and moreover, that $F_0G(r,s;a) = 0$ as identity between meromorphic functions. This in turn implies that $(L(r)-\frac{1}{4}+s^2)G(r,s) = k(r,s)$. See Propositions 8.4.5 and 9.2.5 for a proof.

Thus we have obtained ${}^a\tilde{E}(r,s) = h(r,s) - G(r,s)$, with the desired properties. Moreover, these properties determine ${}^a\tilde{E}$ uniquely. The difference of two such families would be an eigenfunction of ${}^aA(r)$ for all (r,s) at which it is holomorphic (look at the Fourier expansion to see the square integrability). The compactness of the resolvent implies that ${}^aA(r)$ has a discrete spectrum. We conclude that the difference vanishes.

1.6.5 *Construction of E.* Now $\tilde{E}(r,s) = e^{irt}\,{}^a\tilde{E}(r,s)$ is a meromorphic family of modular forms of the right type, but condition iv) in 1.5.8 is replaced by the condition that the Fourier coefficient with $e^{\pi irx/6}$ is equal to 1 at $y = a$. The square integrability of $G(r,s)$ implies that all other Fourier coefficients of $\tilde{E}$ are square integrable, hence multiples of $\omega_r^\nu(r,s)$.

The Fourier term with $e^{\pi irx/6}$ is of the form

$$p(r,s)\mu_r^0(r,s) + q(r,s)\mu_r^0(r,-s),$$

with p and q meromorphic on U without singularities along vertical lines. Suppose that the restriction $s \mapsto p(0,s)$ is the zero meromorphic function. Then $\tilde{E}(0,s)$ would be a square integrable modular form for many non-real s with $\mathrm{Re}\, s > 0$. This contradicts the fact that the extension A_0 of L_0 is selfadjoint. Hence $s \mapsto p(0,s)$ is non-zero, and $E(r,s) = \frac{1}{p(r,s)}\tilde{E}(r,s)$ is a meromorphic family with the properties discussed in 1.5.8. The difference $s \mapsto E(0,s) - e(s)$ has to vanish, again by the selfadjointness of A_0.

1.6.6 *Remark.* This completes the sketch of a proof of the meromorphic continuation of the Eisenstein series jointly in weight and spectral parameter. I want to emphasize that the method comes from [12]. Colin de Verdière proves the continuation with s as the only parameter, but the method is the same.

1.6.7 *Overview of Part I.* At this point the reader may appreciate a look at Table 1.1. Except for the use of the term *automorphic form* instead of modular form, I have tried to stay close to the terminology of this section in the description of the contents of the chapters, and have ignored many generalizations. In particular, Part I considers a general type of Poincaré series, of which the Eisenstein series in the table are only a very special case. Chapters 11 and 12 have a much wider scope than indicated here. The partition in *known* and *new* is rough. The concept of automorphic form defined in Chapter 4 is wider than is usual in, e.g., [50, 51] and [21]. The chapters called *new* contain many ideas that are present in the literature. The systematic study of meromorphic families of automorphic forms of this general type is new, as far as I know.

Chapter		indication of the contents
2	known	Automorphic forms on the universal covering group of $\mathrm{SL}_2(\mathbb{R})$
3	known	Discrete subgroups of $\mathrm{SL}_2(\mathbb{R})$
4	known	Automorphic forms and Fourier expansions
5	known	Eisenstein series in the domain of absolute convergence
6	known	Spectral theory; automorphic forms as eigenfunctions of a selfadjoint operator
7	new	Families of automorphic forms; various types of holomorphy and meromorphy
8	new	Transformation and truncation
9	new	Families of automorphic forms and eigenfunctions of a selfadjoint family of operators in the Hilbert space of truncated functions
10	new	Eisenstein family, meromorphic dependence on multiplier system and spectral parameter jointly
11	known	Singularities of the restriction $s \mapsto E(0,s)$
12	new	Singularities of $(r,s) \mapsto E(r,s)$ at points $(0,s)$

Table 1.1 Overview of Part I of this book.

1.7 Further remarks

1.7.1 *Modular functions.* The modular invariant J is a Γ_{mod}-invariant function on $\mathfrak{H}$ with an expansion of the form

$$e^{-2\pi i z} + \sum_{n=0}^{\infty} c_n e^{2\pi i n z}.$$

It satisfies all properties in 1.5.6, except condition iii). If we ignore the Fourier term with $e^{-2\pi i x}$ in the formulation of the growth condition, we can call it a holomorphic modular form of weight 0 for the trivial multiplier system. That is what we shall do in Definition 4.3.4. Usually, J is called a *modular function*, as are all rational functions in J. These need not be defined on the whole of $\mathfrak{H}$, for instance, $\frac{1}{J}$ has a pole at each point of the orbit $\Gamma_{\mathrm{mod}} \cdot \frac{1+i\sqrt{3}}{2}$. We shall allow in Definition 4.3.4 automorphic forms to have singularities in a discrete subset of $\mathfrak{H}$. At each of these exceptional points we impose some growth condition that amounts to meromorphy in the complex analytic case. This generalized growth condition admits all modular forms we shall need, but saves us from studying Γ_{mod}-invariant functions such as $e^{1/J}$.

1.7.2 *Poincaré series* have already been discussed in 1.5.9. We may as well use μ^ν instead of μ^0. See, e.g., Niebur, [41], or Hejhal, [21], Theorem 4.1 on p. 254. The resulting Poincaré series $P^\nu(r,s)$ have a meromorphic continuation. The modular invariant J can be expressed in the value at $s=\frac{1}{2}$ of the family $s\mapsto P^{-1}(0,s)$.

1.7.3 *Resolvent kernel.* The selfadjoint extension A_0 of the Laplacian, discussed in 1.2.6, has a resolvent $\mathrm{R}(s)=\left(A_0-\frac{1}{4}+s^2\right)^{-1}$. The resolvent $\mathrm{R}(0,s)$ of the pseudolaplacian ${}^aA(0)$, as used in 1.6.4, is a compact operator. The resolvent $\mathrm{R}(s)$ of A_0 is not. But it has a kernel $k(s;z,w)$, given by a series converging absolutely for $\mathrm{Re}\, s>\frac{1}{2}$; see, e.g., [29], Chapter XIV, §5. If we keep fixed $w\in\mathfrak{H}$, this kernel satisfies $L_0k(s;\cdot,w)=(\frac{1}{4}-s^2)k(s;\cdot,w)$ on $\mathfrak{H}\smallsetminus\Gamma_{\text{mod}}\cdot w$. We view it as a Poincaré series; it has a logarithmic singularity at the points of $\Gamma_{\text{mod}}\cdot w$. It is known that this kernel has a meromorphic continuation in s. Definition 4.3.4 of automorphic forms includes the resolvent kernel. We shall see that this kernel has a meromorphic continuation in (r,s).

1.7.4 *Other weights.* The Eisenstein series in 1.5.8 has weight equal to the parameter r determining the multiplier system. We could have used any weight $r\mapsto r+\nu$ with $\nu\in 2\mathbb{Z}$. The family of multiplier systems $r\mapsto v_r$ is the only one for the modular group. The results on Eisenstein families with more general weights are almost the same as those discussed above.

The map $r\mapsto v_r$ has period 12, but the Eisenstein families are definitely not periodic in r.

1.7.5 *Other discrete subgroups of* $\mathrm{SL}_2(\mathbb{R})$. The modular group is only one example of the groups with which we shall work: discrete subgroups of $\mathrm{SL}_2(\mathbb{R})$ containing $\left(\begin{smallmatrix}-1&0\\0&-1\end{smallmatrix}\right)$, with finite covolume. These groups are discussed in Chapter 3. They may have more than one cusp. That does not matter much. But, in general, the multiplier systems form a complex variety of dimension larger than 1. This makes application of perturbation theory more complicated than in the one-dimensional case.

1.7.6 *Eisenstein series.* In general, there is an Eisenstein series at each cusp for which the multiplier system is 'singular', i.e., the multiplier system equals 1 on the matrices fixing that cusp. The meromorphic continuation in s is known; see, e.g., [21], Theorem 11.6 on p. 128, and Theorem 11.8 on p. 130. In general, this continuation cannot be obtained by looking at the Fourier expansion. One needs the resolvent of the selfadjoint extension of the Laplacian, or the Casimir operator.

The unique cusp of the modular group is singular for v_r if and only if $r\equiv 0 \bmod 12$. Hence for $r\in(-12,12)\smallsetminus\{0\}$ the restriction $s\mapsto E(r,s)$ is not described by Eisenstein series on the region $\mathrm{Re}\, s>\frac{1}{2}$.

1.7.7 *Part II* discusses three example groups. Part I refers for most examples to either this chapter, or to Part II.

Chapter 13	Modular group
13.1–13.4	examples
13.5	singularities of E
13.6	distribution results
Chapter 14	Theta group, examples
Chapter 15	Commutator subgroup
15.1–15.5	examples
15.6	harmonic automorphic forms
15.7	perturbation of cusp forms

Table 1.2 Overview of Part II.

We also use Part II to explain some ideas that go further than Part I.

1.7.8 *Distribution results.* We have mentioned in 1.5.9 that the meromorphic continuation of the Eisenstein family, jointly in weight and spectral parameter, can be used to get distribution results for Dedekind sums. In Section 13.6 we use the same method to get distribution results for another quantity. The method is based on the fact that the Eisenstein family in two variables has rather unexpected singularities at the points $(0, \frac{1}{2}l)$, with $l = 2, 3, 4, \ldots$. We study these singularities in Section 13.5.

1.7.9 *Harmonic automorphic forms.* The multiplier systems suitable for weight 0 form a group, i.e., the character group of the discrete subgroup Γ. For the commutator subgroup of the modular group this character group has dimension 2. The theory in Part I gives a meromorphic Eisenstein family depending on the character and the spectral parameter. On the region $\operatorname{Re} s > \frac{1}{2}$, and for all unitary characters, this Eisenstein family is given by (absolutely converging) Eisenstein series. On the two-dimensional complex variety given by $s = \frac{1}{2}$, the family is not given by Eisenstein series. But the restriction of the family to this complex plane is a meromorphic family of automorphic forms that can be related to Jacobi theta functions. With the help of D. Zagier I could get a rather explicit description. See Section 15.6.

1.7.10 *Perturbation of square integrable automorphic forms.* In the modular case, there are results on the *global* behavior of eigenvalues of the selfadjoint operators A_r and the corresponding families of square integrable modular forms; see Proposition 2.8 in [5], and Theorem 2.21 in [7]. By global I mean here 'holomorphic on a neighborhood of $(-12, 12)$ or $(0, 12)$'. This is equivalent to real analyticity on the real interval. I have refrained from trying to generalize those results in this book. Of course, once one has the results in Chapter 9, one may work on a one-dimensional subspace of the variety of multiplier systems, and try to generalize [5]–[7].

The *local* behavior of square integrable automorphic forms under variation of the discrete group or the multiplier system has been studied, see Hejhal, [22], or Venkov, [59], Ch. 7. (Venkov's nice survey [60] has the same title, but is not identical to [59].) Analytic variation is hinted at in §10, (iii) of [22].

Analytic variation of the Riemannian metric is considered by Colin de Verdiére, [12], and Phillips and Sarnak, [46]. In [46] the non-vanishing of certain L-series at a point determined by the eigenvalue is shown to imply the 'annihilation' of the corresponding cusp form under variation of the group. We do not go deeply into this type of question for perturbation of the multiplier system. In Section 15.7 we mention one result, for the commutator subgroup of the modular group. This shows that 'annihilation' of cusp forms is related to the presence of singularities of the Eisenstein family.

Part I

General theory

Chapter 2
Automorphic forms on the universal covering group

One may view automorphic forms with general complex weight as functions on the upper half plane, or as functions on the universal covering group of $\mathrm{SL}_2(\mathbb{R})$. In this book the latter point of view is taken.

This chapter discusses the universal covering group, and defines automorphic forms on it. Sections 2.1 and 2.3 discuss automorphic forms. The universal covering group is introduced in Section 2.2.

2.1 Automorphic forms on the upper half plane

This section recalls some definitions already given in Chapter 1 for the modular case. It discusses the transformation behavior and differential equation for automorphic forms considered as functions on the upper half plane. We ignore growth conditions for the moment.

2.1.1 *Notations.* We denote the *upper half plane* $\{ z \in \mathbb{C} : \mathrm{Im}\, z > 0 \}$ by $\mathfrak{H}$. We use $x = \mathrm{Re}\, z$, $y = \mathrm{Im}\, z$, for the coordinate z on $\mathfrak{H}$.

The group $G = \mathrm{SL}_2(\mathbb{R}) = \{ \left(\begin{smallmatrix} a & b \\ c & d \end{smallmatrix}\right) : a, b, c, d \in \mathbb{R},\ ad - bc = 1 \}$ acts in $\mathfrak{H}$ by fractional linear transformations $\left(\begin{smallmatrix} a & b \\ c & d \end{smallmatrix}\right) : z \mapsto \dfrac{az+b}{cz+d}$. This action factors through the center
$vz$$\pm\mathrm{Id}$, hence it is an action of $\bar{G} = \mathrm{PSL}_2(\mathbb{R}) = \mathrm{SL}_2(\mathbb{R})/\{\pm\mathrm{Id}\}$.

2.1.2 *Discontinuous group.* We consider a subgroup Γ of G that acts discontinuously in $\mathfrak{H}$, i.e., each $z \in \mathfrak{H}$ has a neighborhood U such that $\gamma \cdot z \in U$ for only finitely many $\gamma \in \Gamma$. The modular group $\Gamma_{\mathrm{mod}} = \mathrm{SL}_2(\mathbb{Z})$ is an example. Other examples are the theta group Γ_θ, see Section 14.1, and the commutator subgroup Γ_{com} of Γ_{mod}, see Section 15.1. Chapter 3 gives a discussion of the class of subgroups to be used in this book.

We denote the quotient $\Gamma/\{\pm\mathrm{Id}\}$ by $\bar{\Gamma}$. The central element $-\mathrm{Id} = \left(\begin{smallmatrix} -1 & 0 \\ 0 & -1 \end{smallmatrix}\right)$ acts trivially in $\mathfrak{H}$. The group $\bar{\Gamma}$ is isomorphic to the group of transformations in $\mathfrak{H}$ determined by Γ.

2.1.3 *Definition.* We define an *automorphic form* on $\mathfrak{H}$ for the group Γ with *weight* $l \in \mathbb{C}$, *multiplier system* $v : \Gamma \mapsto \mathbb{C}^*$, and *eigenvalue* $\lambda \in \mathbb{C}$, as a function $f \in C^\infty(\mathfrak{H})$ satisfying

i) $f(\gamma \cdot z) = v(\gamma) e^{il \arg(cz+d)} f(z)$ for all $\gamma = \begin{pmatrix} a & b \\ c & d \end{pmatrix} \in \Gamma$;

ii) $L_l f = \lambda f$, with $L_l = (-y^2\partial_y^2 - y^2\partial_x^2 + ily\partial_x)$.

Throughout we use the *argument convention* $-\pi < \arg(u) \leq \pi$ for all $u \in \mathbb{C}^*$. The multiplier system v is a function $v : \Gamma \to \mathbb{C}^*$. It has to satisfy a relation expressing $v(\gamma\delta)$ in terms of $v(\gamma)$ and $v(\delta)$ for all $\gamma, \delta \in \Gamma$; otherwise i) would allow the zero solution only. The relation is best understood after reformulation in terms of functions on the universal covering group, see 2.3.6.

This definition of automorphic form is very wide. Usually one imposes growth conditions at the cusps. We introduce our concept of growth condition in 4.3.4. That definition includes the usual automorphic forms. It will also allow automorphic forms to have singularities at points of a discrete subset of $\mathfrak{H}$. The more strict the growth condition, the less automorphic forms satisfying it exist. The rarest are the *cusp forms,* which decrease exponentially at all cusps (see Definition 4.3.9).

Examples of *modular forms*, i.e., automorphic forms for Γ_{mod}, occur in 1.2.3 and 1.5.5. Even unusual functions like e^J, with J the modular invariant, come under this definition. In fact, e^J is an automorphic form for Γ_{mod}, with weight 0, trivial multiplier system, and eigenvalue 0.

2.1.4 *Analyticity.* Condition ii) imposes an elliptic differential equation with analytic coefficients. Thus all automorphic forms are real analytic functions on $\mathfrak{H}$, and in the definition we could have replaced $f \in C^\infty(\mathfrak{H})$ by the condition that f is a distribution on $\mathfrak{H}$; see, e.g., [29], App. 4, §5 and [3], p. 207–210.

2.1.5 *Holomorphic automorphic forms.* The well known *holomorphic automorphic forms* on $\mathfrak{H}$ of weight l, with multiplier system v satisfy

i)′ $F(\gamma \cdot z) = v(\gamma)(cz+d)^l F(z)$ for all $\gamma = \begin{pmatrix} a & b \\ c & d \end{pmatrix} \in \Gamma$;

ii)′ $\partial_{\bar{z}} F = 0$.

The differential operator $\partial_{\bar{z}} = \frac{1}{2}(\partial_x + i\partial_y)$ is simpler than the the differential operator L_l above. Condition ii)′ states that F is holomorphic on $\mathfrak{H}$.

If F satisfies these conditions, then $z \mapsto y^{l/2} F(z)$ is an automorphic form of weight l, with multiplier system v and eigenvalue $\frac{1}{2}l(1-\frac{1}{2}l)$. On the other hand, condition ii) with $\lambda = \frac{1}{2}l(1-\frac{1}{2}l)$ need not imply condition ii)′.

2.1.6 *Examples.* In 1.3.2 we have seen the holomorphic Eisenstein series G_k. As we do not impose a growth condition here, the modular invariant J, and also e^J, fall under this definition. Usually a function like J is called an automorphic function, and e^J is not mentioned at all.

See 14.3.3 for an example of a non-holomorphic form of weight 0 with eigenvalue $0 = \frac{1}{2} \cdot 0(1 - \frac{1}{2} \cdot 0)$.

2.1.7 *Maass's definition.* If we take an automorphic form F of type $\{\Gamma, \alpha, \beta, v\}$ in the sense of Maass, [35], p. 185 (so $\alpha, \beta \in \mathbb{C}$, $\alpha - \beta \in \mathbb{R}$, v a multiplier system for

Γ), then $z \mapsto y^{(\alpha+\beta)/2} F(z)$ is an automorphic form for Γ as defined in 2.1.3. The weight is $\alpha-\beta$, the multiplier system is v, and the eigenvalue equals $\frac{\alpha+\beta}{2}(1-\frac{\alpha+\beta}{2})$.

2.1.8 *Line bundle.* In particular in the holomorphic case it may be useful to consider automorphic forms as sections of a suitable line bundle on the quotient space $\Gamma\backslash\mathfrak{H}$. This has the advantage of working on a smaller space. The disadvantage is that we no longer study functions.

In this book we make the opposite choice. We prefer to view automorphic forms as functions on the universal covering group of $\mathrm{SL}_2(\mathbb{R})$. In that way, the description of multiplier systems becomes simpler; the same holds for certain coordinate systems. We pay for it by working on a larger space.

2.2 The universal covering group

In 1.3.4 we have considered modular forms of even weight as functions on $G = \mathrm{SL}_2(\mathbb{R})$, left-invariant under Γ_{mod}, transforming on the right according to a character of the maximal compact subgroup $K = \mathrm{SO}_2(\mathbb{R}) = \left\{ \begin{pmatrix} \cos\theta & \sin\theta \\ -\sin\theta & \cos\theta \end{pmatrix} : \theta \in \mathbb{R} \right\}$ of orthogonal matrices in G. This has the advantage that condition i) in Definition 2.1.3 becomes the simpler condition of left-invariance. This can be done for modular forms of odd weight as well. Then the multiplier system has to satisfy $v(-\mathrm{Id}) = -1$. But for other weights there is no corresponding character of K. This is due to the fact that $\begin{pmatrix} \cos\theta & \sin\theta \\ -\sin\theta & \cos\theta \end{pmatrix}$ has period 2π in θ. We avoid this problem by using the universal covering group.

In this section, we construct the universal covering group of $\mathrm{SL}_2(\mathbb{R})$. It is a Lie group, and as such it has a Lie algebra. We describe the corresponding differential operators. From 2.2.4 on, one finds some technicalities to be used later on. The reader may want to skip these at first reading.

2.2.1 *Universal covering group.* The universal covering group of $G = \mathrm{SL}_2(\mathbb{R})$ is a simply connected Lie group $\tilde{G}$ together with a surjective continuous homomorphism $\tilde{G} \rightarrow \mathrm{SL}_2(\mathbb{R}) : g \mapsto \hat{g}$ that is locally a homeomorphism; the kernel of the homomorphism is discrete and central in $\tilde{G}$. This universal covering group is unique up to isomorphism. We shall give an explicit construction.

We use the *Iwasawa decomposition* of G to construct the universal covering group. Each element of $\mathrm{SL}_2(\mathbb{R})$ has a unique decomposition

$$\begin{pmatrix} \sqrt{y} & x/\sqrt{y} \\ 0 & 1/\sqrt{y} \end{pmatrix} \begin{pmatrix} \cos\theta & \sin\theta \\ -\sin\theta & \cos\theta \end{pmatrix},$$

with $x + iy \in \mathfrak{H}$ and $\theta \in \mathbb{R} \bmod 2\pi\mathbb{Z}$. This Iwasawa decomposition shows that $G \cong \mathfrak{H} \times (\mathbb{R} \bmod 2\pi)$ as analytic varieties. To make $\tilde{G}$ simply connected, we define the underlying analytic variety of $\tilde{G}$ to be $\mathfrak{H}\times\mathbb{R}$. The projection $\tilde{G} \rightarrow G : g \mapsto \hat{g}$ is given by $(z,\theta) \mapsto (z, \theta \bmod 2\pi)$. We define $p(z)k(\theta) \in \tilde{G}$ as the element corresponding to $(z,\theta) \in \mathfrak{H} \times \mathbb{R}$.

We write, for $z = x + iy \in \mathfrak{H}$, $p(z) = n(x)a(y)$. So every element of $\tilde{G}$ can be written in exactly one way as $n(x)a(y)k(\theta)$, with $x \in \mathbb{R}$, $y > 0$, and $\theta \in \mathbb{R}$. The projection $\tilde{G} \to G$ is given by

$$\widehat{n(x)} = \begin{pmatrix} 1 & x \\ 0 & 1 \end{pmatrix}, \quad \widehat{a(y)} = \begin{pmatrix} \sqrt{y} & 0 \\ 0 & 1/\sqrt{y} \end{pmatrix}, \quad \widehat{k(\theta)} = \begin{pmatrix} \cos\theta & \sin\theta \\ -\sin\theta & \cos\theta \end{pmatrix}.$$

We define the group structure on $\tilde{G}$ by the following multiplication rules:

$$\begin{aligned} n(x_1)n(x_2) &= n(x_1 + x_2) \\ a(y_1)a(y_2) &= a(y_1 y_2) \\ k(\theta_1)k(\theta_2) &= k(\theta_1 + \theta_2) \\ a(y)n(x) &= n(yx)a(y) \\ k(\theta)p(z) &= p\left(\frac{z\cos\theta + \sin\theta}{-z\sin\theta + \cos\theta}\right) k\left(\theta - \arg\left(e^{i\theta}(-z\sin\theta + \cos\theta)\right)\right). \end{aligned}$$

As mentioned before, the argument is taken in $(-\pi, \pi]$.

This amounts to the following: The compact subgroup K of G is covered by the subgroup $\tilde{K}$ of $\tilde{G}$, isomorphic to the additive group $\mathbb{R}$. The kernel of $\tilde{K} \to K$ is a subgroup $\tilde{Z}$ of $\tilde{K}$, corresponding to $\pi\mathbb{Z}$. The subgroup $P = \{ \left(\begin{smallmatrix} a & b \\ 0 & 1/a \end{smallmatrix}\right) : a > 0 \}$ of G has an isomorphic copy $\tilde{P} = \{ p(z) : z \in \mathfrak{H} \}$ in $\tilde{G}$. This group $\tilde{P}$ is a semidirect product of the subgroups $\tilde{A} = \{ a(y) : y > 0 \} \cong \mathbb{R}^*_{>0}$ and $\tilde{N} =$ $vzmn(x)x \in \mathbb{R} \cong \mathbb{R}$. This leaves only the definition of the multiplication of $k(\theta)$ and $p(z)$. The definition above lifts the product of $\widehat{p(z)}$ and $\widehat{k(\theta)}$ to $\tilde{G}$. I invite the reader to check that we now have a Lie group $\tilde{G}$, and that $g \mapsto \hat{g}$ gives a surjective homomorphism $\tilde{G} \to G$, with kernel central in $\tilde{G}$.

The group $\tilde{Z}$ is the *center* of $\tilde{G}$. It is generated by $\zeta = k(\pi)$. We denote the quotient $\tilde{G}/\tilde{Z} \cong G/\{\pm\mathrm{Id}\}$ by $\bar{G}$. If we use $\mathfrak{H}$ as a model of hyperbolic plane geometry, then $\bar{G}$ is the group of motions. The group $\tilde{G}$ acts in $\mathfrak{H}$ via $\bar{G}$.

We refer to [4], 2.2, for more information. The notation there is a bit different. $\tilde{G}, \tilde{K}, \tilde{N}, \ldots$ here correspond to $G_0, K_0, N, \ldots$ in *loc. cit.*

2.2.2 *Standard section.* In (2.2.12) of [4] a section $\left(\begin{smallmatrix} a & b \\ c & d \end{smallmatrix}\right) \mapsto \widetilde{\left(\begin{smallmatrix} a & b \\ c & d \end{smallmatrix}\right)}$ of the projection $\tilde{G} \to G$ is given. It is determined by

$$\widetilde{\begin{pmatrix} a & b \\ c & d \end{pmatrix}} p(z) = p\left(\frac{az+b}{cz+d}\right) k(-\arg(cz+d)). \tag{2.1}$$

This section, restricted to P, gives an isomorphism $P \to \tilde{P}$. If $c \neq 0$ or $d > 0$, then $x = \widetilde{\left(\begin{smallmatrix} a & b \\ c & d \end{smallmatrix}\right)}$ satisfies $\widetilde{x^{-1}} = \tilde{x}^{-1}$. Note that $k(\theta) = \widetilde{\widehat{k(\theta)}}$ for $-\pi \le \theta < \pi$.

2.2.3 *Invariant differential operators.* The *Lie algebra* $\mathfrak{g}_r$ of $\tilde{G}$ is the same as the Lie algebra of G. It has the $\mathbb{R}$-basis

$$\mathbf{X} = \begin{pmatrix} 0 & 1 \\ 0 & 0 \end{pmatrix}, \quad \mathbf{W} = \begin{pmatrix} 0 & 1 \\ -1 & 0 \end{pmatrix}, \quad \mathbf{H} = \begin{pmatrix} 1 & 0 \\ 0 & -1 \end{pmatrix},$$

with $n(x) = \exp(x\mathbf{X})$, $k(\theta) = \exp(\theta\mathbf{W})$ and $a(y) = \exp(\frac{1}{2}\log(y)\mathbf{H})$. See, e.g., [29], Ch. VI, §1.

For $f \in C^\infty(\tilde{G})$ and $\mathbf{Z} \in \mathfrak{g}_r$ we have the right differentiation

$$\mathbf{Z}f(g) = \frac{d}{dt} f(g \exp(t\mathbf{Z}))\Big|_{t=0}.$$

This can be extended $\mathbb{C}$-linearly to give an action of the complexified Lie algebra $\mathfrak{g} = \mathbb{C} \otimes_{\mathbb{R}} \mathfrak{g}_r$ in $C^\infty(\tilde{G})$ by left invariant differential operators of first order. All left invariant differential operators are obtained from these first order operators. So one obtains an action of the enveloping algebra $U(\mathfrak{g})$ of $\mathfrak{g}$, see, e.g., [29], Ch. X. The center of $U(\mathfrak{g})$ is the polynomial ring $\mathbb{C}[\omega]$ generated by the *Casimir operator*

$$\omega = -\frac{1}{4}\mathbf{H}^2 - \mathbf{X}^2 + \frac{1}{2}(\mathbf{XW} + \mathbf{WX}) = -\frac{1}{4}\mathbf{E}^+\mathbf{E}^- + \frac{1}{4}\mathbf{W}^2 - \frac{i}{2}\mathbf{W}$$

with $\mathbf{E}^- = \mathbf{H} - 2i\mathbf{X} + i\mathbf{W}$, and $\mathbf{E}^+ = \mathbf{H} + 2i\mathbf{X} - i\mathbf{W}$. As $[\mathbf{E}^-, \mathbf{E}^+] = \mathbf{E}^-\mathbf{E}^+ - \mathbf{E}^+\mathbf{E}^- = 4i\mathbf{W}$, the enveloping algebra $U(\mathfrak{g})$ is generated by $\mathbf{E}^+$ and $\mathbf{E}^-$.

2.2.4 *Iwasawa coordinates.* The isomorphism of analytic varieties $\tilde{G} \cong \mathfrak{H} \times \mathbb{R}$ corresponds to the *Iwasawa decomposition* $\tilde{G} = \tilde{N}\tilde{A}\tilde{K}$. It leads to the *Iwasawa coordinates* on $\tilde{G}$:

$$p(z)k(\theta) \mapsto (x, y, \theta).$$

In these coordinates

$$\begin{aligned} \mathbf{W} &= \partial_\theta \\ \mathbf{E}^\pm &= e^{\pm 2i\theta}\left(\pm 2iy\partial_x + 2y\partial_y \mp i\partial_\theta\right) \\ \omega &= -y^2\partial_x^2 - y^2\partial_y^2 + y\partial_x\partial_\theta. \end{aligned}$$

2.2.5 *Polar coordinates.* The *polar decomposition* $\tilde{G} = \tilde{K}\tilde{A}\tilde{K}$ leads to the *polar coordinates* on $\tilde{G} \smallsetminus \tilde{K}$:

$$k(\eta)a(t_u)k(\psi) \mapsto (u, \eta, \psi)$$

with η, $\psi \in \mathbb{R}$, $\eta + \psi \bmod \pi\mathbb{Z}$, $u \in (0, \infty)$ and

$$t_u = 1 + 2u + 2\sqrt{u^2 + u} > 1, \quad u = \frac{(t_u - 1)^2}{4t_u} > 0.$$

We have $\mathbf{W} = \partial_\psi$. To express $\mathbf{H}$ and $\mathbf{X}$ in these coordinates, we use

$$\frac{d}{dx} f\left(k(\eta)a(t)\exp(x\mathbf{H})k(\psi)\right)\Big|_{x=0} = 2t\partial_t f(u,\eta,\psi),$$
$$k(\psi)^{-1}\mathbf{H}\cdot k(\psi) = \cos 2\psi \cdot \mathbf{H} + \sin 2\psi(2\mathbf{X} - \mathbf{W}).$$

Hence $\cos 2\psi \cdot \mathbf{H} + \sin 2\psi(2\mathbf{X} - \mathbf{W}) = 2t\partial_t$, which may be expressed in ∂_u. Proceeding in a similar way for ∂_η from $\exp(x\mathbf{W})a(t_u)k(\psi)$, and inverting the resulting relations, one finds the following description of the action of $U(\mathfrak{g})$ in polar coordinates:

$$\begin{aligned}
\mathbf{W} &= \partial_\psi \\
\mathbf{E}^\pm &= e^{\pm 2i\psi}\left(2\sqrt{u^2+u}\,\partial_u \mp \frac{i}{2}\frac{1}{\sqrt{u^2+u}}\partial_\eta \pm \frac{i}{2}\frac{1+2u}{\sqrt{u^2+u}}\partial_\psi\right) \\
\omega &= -(u^2+u)\partial_u^2 - (2u+1)\partial_u - \frac{1}{16}\frac{1}{u^2+u}\partial_\eta^2 \\
&\quad + \frac{2u+1}{8(u^2+u)}\partial_\eta\partial_\psi - \frac{1}{16}\frac{1}{u^2+u}\partial_\psi^2.
\end{aligned}$$

2.2.6 *Coordinate transformations.* The relations between polar and Iwasawa coordinates are

$$\begin{cases}
z = x + iy = \dfrac{it_u\cos\eta + \sin\eta}{-it_u\sin\eta + \cos\eta} = \dfrac{i - 2\sqrt{u^2+u}\sin 2\eta}{1 + 2u - 2\sqrt{u^2+u}\cos 2\eta} \\
\theta = \psi + \eta - \arg\left(1 - e^{2i\eta}\sqrt{\dfrac{u}{u+1}}\right)
\end{cases}$$

$$\begin{cases}
u = \dfrac{|z-i|^2}{4y} \\
\eta = \dfrac{1}{2}\arg\dfrac{z-i}{z+i} \\
\psi = \theta - \frac{1}{2}\arg(-1-z^2).
\end{cases}$$

In the latter transformation, the argument gives rise to a discontinuity. But $\eta + \psi = \theta - \arg(1 - iz)$ is continuous in (z, θ).

2.2.7 *Left invariance.* Let g_0 be a fixed element of $\tilde{G}$. We can apply the left translation $\tilde{G} \to \tilde{G} : g \mapsto g_0 g$, before taking the coordinates. As the differential operators in $U(\mathfrak{g})$ are all left invariant, the operators $\mathbf{W}$, $\mathbf{E}^\pm$ and ω are still described by the formulas given above.

2.2.8 *Haar measure.* The group $\tilde{G}$, being unimodular, carries a measure invariant under left and right translations. This *Haar measure* is unique up to a constant factor. We fix it by

$$dg = dx \wedge \frac{dy}{y^2} \wedge \frac{d\theta}{\pi} \quad \text{in Iwasawa coordinates,}$$

$$= \quad 4\pi\, du \wedge \frac{d\eta}{\pi} \wedge \frac{d\psi}{\pi} \quad \text{in polar coordinates.}$$

We decompose the Haar measure on $\tilde{G} \cong \mathfrak{H} \times \tilde{K}$ as the product of $d\mu(z) = \frac{dx \wedge dy}{y^2}$ on $\mathfrak{H}$, and $\frac{1}{\pi} d\theta$ on $\tilde{K}$. The latter measure gives volume 1 to $\tilde{K}/\tilde{Z}$.

2.3 Automorphic forms on $\tilde{G}$

We transform Definition 2.1.3 into a definition of automorphic form on the universal covering group $\tilde{G}$ in exactly the same way as in 1.3.4. In some sense this is the opposite of the procedure followed by Maass; see [35], Ch. IV, §1.

2.3.1 *Discrete group.* We consider a discrete subgroup $\tilde{\Gamma}$ of $\tilde{G}$ such that $\tilde{Z} \subset \tilde{\Gamma}$ and such that $\Gamma = \{\hat{\gamma} : \gamma \in \tilde{\Gamma}\}$ acts discontinuously on $\mathfrak{H}$. This also gives an action of $\tilde{\Gamma}$ on $\mathfrak{H}$. The actual group of transformations in $\mathfrak{H}$ corresponding to $\tilde{\Gamma}$ is $\bar{\Gamma} = \tilde{\Gamma}/\tilde{Z} \cong \Gamma/\{\pm\mathrm{Id}\}$.

To discrete subgroups Γ of $G = \mathrm{SL}_2(\mathbb{R})$ acting discontinuously in $\mathfrak{H}$, we may associate the full original $\tilde{\Gamma}$ under $g \mapsto \tilde{g}$ in $\tilde{G}$. There is a bijective correspondence between the groups $\tilde{\Gamma}$ indicated above, and those groups Γ considered in 2.1.2 that contain $-\mathrm{Id}$.

The full original $\tilde{\Gamma}_{\mathrm{mod}} \subset \tilde{G}$ of the modular group $\Gamma_{\mathrm{mod}} = \mathrm{SL}_2(\mathbb{Z})$ is discussed in Section 13.1. Chapters 14 and 15 give other examples. In the next chapter we shall be more precise concerning the conditions on $\tilde{\Gamma}$.

2.3.2 *Definition.* A function f on an open subset U of $\tilde{G}$ has *weight* $l \in \mathbb{C}$ if $f(gk(\theta)) = f(g)e^{il\theta}$ for all $g \in U$ and all $\theta \in \mathbb{R}$. The domain U of f should satisfy $U\tilde{K} = U$.

A function f on $\tilde{G}$ with weight l is fully determined by the function $f_l : \mathfrak{H} \to \mathbb{C} : z \mapsto f(p(z))$; similarly if the domain of f is a subset of $\tilde{G}$.

2.3.3 *Notations.* The *character group* $\mathcal{X} = \mathcal{X}(\tilde{\Gamma})$ of $\tilde{\Gamma}$ is the group of *characters* of $\tilde{\Gamma}$. Its elements are the group homomorphisms $\tilde{\Gamma} \to \mathbb{C}^*$. We denote by $\mathcal{X}_u = \mathcal{X}_u(\tilde{\Gamma})$ the subgroup of *unitary characters*, i.e., characters with values in $\{ t \in \mathbb{C}^* : |t| = 1 \}$.

We say that a character $\chi \in \mathcal{X}$ *belongs to the weight* l if $\chi(\zeta) = e^{\pi i l}$. The set of weights to which a given character belongs has the form $l_0 + 2\mathbb{Z}$.

We discuss the character group of $\tilde{\Gamma}_{\mathrm{mod}}$ in 13.1.3.

2.3.4 *Definition.* An *automorphic form* for the group $\tilde{\Gamma}$ is a function $f \in C^\infty(\tilde{G})$ satisfying

i) $f(\gamma g) = \chi(\gamma) f(g)$ for all $g \in \tilde{G}$, $\gamma \in \tilde{\Gamma}$, for some $\chi \in \mathcal{X}$.

ii) f has weight l, for some $l \in \mathbb{C}$.

iii) $\omega f = \lambda f$ for some $\lambda \in \mathbb{C}$.

We call f an automorphic form of weight l, for the character χ, with *eigenvalue* λ.

We also apply this definition to $f \in C^\infty(U)$ if U is a dense subset of $\tilde{G}$ satisfying $\tilde{\Gamma} U \tilde{K} = U$.

Even when the character, weight, and eigenvalue are fixed, the space of automorphic forms is huge. In Chapter 4 we shall impose a growth condition to get finite dimensional spaces.

In 13.1.7 we discuss the modular forms (i.e., automorphic forms for the modular group) on $\tilde{G}$ corresponding to powers of the eta function of Dedekind. For another example see 14.2.3.

2.3.5 *Action of the center.* We have assumed $\tilde{Z} \subset \tilde{\Gamma}$. Hence 0 is the only automorphic form, unless the character belongs to the weight. We shall always assume that this condition is fulfilled.

If we consider a discrete subgroup Δ of G that does not contain $-\mathrm{Id}$, then its full original $\tilde{\Delta}$ in $\tilde{G}$ contains only $\{\, k(2m\pi) : m \in \mathbb{Z} \,\} \subset \tilde{Z}$. Let $\tilde{\Gamma} = \tilde{\Delta} \cup (-\mathrm{Id})\tilde{\Delta}$. The automorphic forms for $\tilde{\Delta}$ with weight l for the character χ of $\tilde{\Delta}$ correspond to the automorphic forms for $\tilde{\Gamma}$ of the same weight and eigenvalue, for the character $\chi_1 : k(m\pi)\delta \mapsto e^{\pi i m l}\chi(\delta)$ of $\tilde{\Gamma}$. So the condition $-\mathrm{Id} \in \Gamma$ is not really a restriction.

2.3.6 *Relation to functions on $\mathfrak{H}$.* Let l be the weight of the function f on $\tilde{G}$. Then condition i) above is equivalent to condition i) in 2.1.3 on f_l for the multiplier system $\begin{pmatrix} a & b \\ c & d \end{pmatrix} \mapsto \widetilde{\begin{pmatrix} a & b \\ c & d \end{pmatrix}}$. As a multiplier system is a map $\Gamma \to \mathbb{C}^*$ for which i) in 2.1.3 has non-zero solutions, all multiplier systems for Γ come from characters of $\tilde{\Gamma}$ in this way.

If the character belongs to an integral weight, then $\{\, k(2m\pi) : m \in \mathbb{Z} \,\} \subset \tilde{Z}$ is contained in its kernel. In that case multiplier systems are characters of $\Gamma \cong \tilde{\Gamma}/\{\, k(2\pi m) : m \in \mathbb{Z} \,\}$. In the case of an even weight, multiplier systems are characters of $\bar{\Gamma} = \Gamma/\{\pm\mathrm{Id}\}$.

If condition ii) holds, condition iii) is equivalent to condition ii) in 2.1.3 for f_l with the same λ.

In this way we can go easily from automorphic forms on $\tilde{G}$ to automorphic forms on $\mathfrak{H}$.

2.3.7 *Holomorphic automorphic forms.* A function f of weight l corresponds to a holomorphic automorphic form (of weight l) if condition i) above holds, and $\mathbf{E}^- f = 0$. The corresponding holomorphic automorphic form is given by $z \mapsto y^{-l/2} f(p(z))$.

Chapter 3
Discrete subgroups

This chapter describes the class of discrete subgroups of the universal covering group that we shall consider: the cofinite discrete subgroups. The covering $\tilde{\Gamma}_{\mathrm{mod}}$ of the modular group and all its subgroups of finite index belong to this class. This includes the groups $\tilde{\Gamma}_\theta$ and $\tilde{\Gamma}_{\mathrm{com}}$ considered in the Chapters 14 and 15. There are many more cofinite discrete groups; most of them have nothing to do with the modular group.

There is a well-developed spectral theory of automorphic forms for cofinite discrete groups, see, e.g., Roelcke, [50], [51], or Hejhal, [21].

We discuss the group of characters of a cofinite discrete group $\tilde{\Gamma}$ in Section 3.4. As we prefer simply connected spaces, we describe the character group with help of its Lie algebra, the vector space of group homomorphisms $\tilde{\Gamma} \to \mathbb{C}$. We use canonical generators, discussed in Section 3.3, to give that description in a unified way. Petersson, [42], has done this in the language of multiplier systems. We include the computations to give a feeling of how things look when one works on the universal covering group $\tilde{G}$.

The geometrical object underlying automorphic forms is the quotient $\tilde{\Gamma}\backslash\mathfrak{H} \cong \tilde{\Gamma}\backslash\tilde{G}/\tilde{K}$. We discuss this quotient in Sections 3.2 and 3.5. Section 3.5 serves mainly to fix some notations to be used in the sequel.

I thank J. Elstrodt for several remarks on an earlier version of this chapter. He also pointed out pitfalls in the terminology concerning Fuchsian group.

3.1 Cofinite groups

3.1.1 *Conditions.* In Part I of this book $\tilde{\Gamma}$ stands for a subgroup of $\tilde{G}$ such that

i) $\tilde{\Gamma}$ is discrete in $\tilde{G}$,

ii) $\tilde{Z} \subset \tilde{\Gamma}$,

iii) $\int_{\tilde{\Gamma}\backslash\tilde{G}} dg < \infty$.

Conditions i) and iii) determine the *cofinite discrete subgroups* of $\tilde{G}$. Condition ii) is not essential. It serves to avoid some complications.

An example is the covering $\tilde{\Gamma}_{\mathrm{mod}}$ of the modular group, see Section 13.1. In Chapters 14 and 15 we consider two other examples.

The Haar measure dg on $\tilde{G}$ gives a quotient measure on $\tilde{\Gamma}\backslash\tilde{G}$, also denoted by dg.

3.1.2 *Fundamental domain.* The action of $\mathrm{SL}_2(\mathbb{R})$ in $\mathfrak{H}$ factors through the quotient group $\bar{G} = \mathrm{PSL}_2(\mathbb{R}) = \mathrm{SL}_2(\mathbb{R})/\{\pm\mathrm{Id}\} \cong \tilde{G}/\tilde{Z}$. So $\bar{G}$ acts in $\mathfrak{H}$, and $\bar{\Gamma} = \tilde{\Gamma}/\tilde{Z}$ is a group of transformations in $\mathfrak{H}$.

A *fundamental domain* F for $\tilde{\Gamma}$ is a reasonable set in $\mathfrak{H}$ such that

i) $\tilde{\Gamma} \cdot F = \mathfrak{H}$,

ii) If $z \in F$, $\gamma \in \tilde{\Gamma}$ satisfy $z \neq \gamma z \in F$, then z and γz are boundary points of F.

All fundamental domains we shall use are very reasonable: they will be connected and the boundary will be the union of finitely many smooth curves. But one could allow more complicated sets. At least F should be the closure of an open set.

In Figure 1.1 on p. 2 we have seen a fundamental domain for the modular group. For other examples see 14.1.4 and 15.1.1.

3.1.3 *Finite volume.* Let $d\mu(z) = y^{-2} dx \wedge dy$. This gives a measure on $\mathfrak{H}$ that is invariant under the action of $\tilde{G}$.

Consider a fundamental domain F for $\tilde{\Gamma}$. The quantity $\int_F d\mu(z)$ does not change if we replace F by another fundamental domain for $\tilde{\Gamma}$. Moreover, $\tilde{G} = \tilde{\Gamma} \cdot (p(F)k([0,\pi)))$ and except for a set of boundary points, this decomposition is unique. We have seen in 2.2.8 that $dg = d\mu(z) \wedge \frac{d\theta}{\pi}$ in Iwasawa coordinates, and we get

$$\int_{\tilde{\Gamma}\backslash\tilde{G}} dg = \int_F d\mu(z) \int_0^\pi \frac{d\theta}{\pi} = \int_F d\mu(z).$$

Condition iii) is equivalent to $\int_F d\mu(z) < \infty$ for some fundamental domain F of $\tilde{\Gamma}$, if this can be found.

3.1.4 *Existence of fundamental domains.* For groups $\tilde{\Gamma} \subset \tilde{G}$ containing $\tilde{Z}$, discreteness of $\tilde{\Gamma}$ in $\tilde{G}$ is equivalent to discreteness of Γ in $G = \mathrm{SL}_2(\mathbb{R})$, and hence to $\bar{\Gamma} = \tilde{\Gamma}/\tilde{Z}$ acting discontinuously in $\mathfrak{H}$; see Maass, [35], Ch. I, Theorem 1 on p. 8. The construction in *loc. cit.*, Ch. I, §3, shows that $\bar{\Gamma}$ has a fundamental domain with a boundary consisting of, possibly countably many, hyperbolic lines or line segments. The finiteness of the volume of such a fundamental domain implies that its boundary consists of finitely many hyperbolic line segments. This follows from the proof of Theorem 5 in Siegel, [54]. So our $\bar{\Gamma}$ is finitely generated. (See also Lehner, [32], Ch. IV, 5.D, p. 135.) As $\zeta = k(\pi)$ generates $\tilde{Z}$, the group $\tilde{\Gamma}$ is *finitely generated* as well.

3.1.5 *Fuchsian groups.* Now we know that $\tilde{\Gamma}/\tilde{Z}$ satisfies the condition on p. 25 in Maass, [35]. The reasoning on p. 29–30 of *loc. cit.* shows that it is a *principal circle group*, or Fuchsian group of the *first kind.*

The group $\bar{\Gamma} = \tilde{\Gamma}/\tilde{Z}$ is a Fuchsian group if it is a discontinuous group of fractional linear transformations leaving $\mathfrak{H}$ invariant. $\bar{\Gamma}$ is of the first kind if it is not discontinuous at any point of the boundary $\mathbb{R} \cup \{\infty\}$ of $\mathfrak{H}$ in $\mathbb{C} \cup \{\infty\}$.

Condition iii) in 3.1.1 gives the additional property of having finite covolume. So our $\Gamma = \tilde{\Gamma}/\{\, k(2\pi m) : m \in \mathbb{Z}\,\} \subset \mathrm{SL}_2(\mathbb{R})$ is a *finitely generated* Fuchsian group of the first kind. Miyake, [38], §1.7, p. 28, calls this a Fuchsian group of the first kind; Shimura does the same, see [53], §1.5, p. 19.

3.1.6 *Elliptic points* for $\tilde{\Gamma}$ are points in $\mathfrak{H}$ which are fixed by an element of $\tilde{\Gamma}$ that is not in $\tilde{Z}$.

Each element $\gamma \in \tilde{\Gamma}$, $\gamma \notin \tilde{Z}$, that fixes $z \in \mathfrak{H}$ is of the form $p(z)k(\varphi)p(z)^{-1}$ with $\varphi \in \pi\mathbb{Q}$, $\varphi \notin \pi\mathbb{Z}$. Such elements are called *elliptic*. The subgroup of $\tilde{\Gamma}$ fixing an elliptic point z is of the form $\{ p(z)k(l\pi/v)p(z)^{-1} : l \in \mathbb{Z} \}$ for some $v \in \mathbb{N}$, $v \geq 2$. We call v the *order* of the elliptic point z. The order is the same for all points in the $\tilde{\Gamma}$-orbit of an elliptic point.

Elliptic elements $\gamma \in \tilde{\Gamma}$ are characterized by the condition $|\text{trace}(\hat{\gamma})| < 2$ on $\hat{\gamma} \in \mathrm{SL}_2(\mathbb{R})$; see, e.g., Shimura, [53], §1.2. For examples see 13.1.10.

3.1.7 *Cuspidal points* of $\tilde{\Gamma}$ are elements of the boundary $\mathbb{R} \cup \{\infty\}$ of $\mathfrak{H}$ which are fixed by parabolic transformations in $\tilde{\Gamma}$. An element $\gamma \in \tilde{\Gamma}$ is *parabolic* if $|\text{trace}(\hat{\gamma})| = 2$ and $\gamma \notin \tilde{Z}$. Parabolic elements are of the form $\pi = gn(x)g^{-1}k(l\pi)$ with $x \neq 0$, $l \in \mathbb{Z}$, and $g \in \tilde{G}$ such that $g \cdot \infty$ is the cuspidal point that π leaves fixed.

The subgroup of $\tilde{\Gamma}$ fixing a cuspidal point $g \cdot \infty$ may be written in the form $\{ gn(r)g^{-1}k(l\pi) : r, l \in \mathbb{Z} \}$. It may be necessary to replace $g \in \tilde{G}$ by $ga(t)$ with $t > 0$ suitable to get $\mathbb{Z}$ as the set of values of r in that representation.

The presence of parabolic elements in $\tilde{\Gamma}$ is visible in each fundamental domain F for $\tilde{\Gamma}$. If $g \cdot \infty$ is a cuspidal point for $\tilde{\Gamma}$, then there is a $\gamma \in \tilde{\Gamma}$ such that $\gamma g \cdot \infty$ is an end point of two boundary segments of F. See, e.g., Maass, [35], Ch. I, Theorems 3 and 4, on p. 15 and p. 24.

3.2 The quotient

We mention some facts concerning $\tilde{\Gamma}\backslash\mathfrak{H}$ and the compact Riemann surface obtained from it. One finds a more complete discussion in, e.g., §1.7–9 in Miyake, [38], or the first chapter of Shimura, [53].

3.2.1 *Quotient and fundamental domain.* As $\bar{\Gamma}$ acts discontinuously in $\mathfrak{H}$, the quotient $\tilde{\Gamma}\backslash\mathfrak{H} = \bar{\Gamma}\backslash\mathfrak{H}$ is a Hausdorff topological space.

Let F be a fundamental domain for $\tilde{\Gamma}$. We may use it as a model for $\tilde{\Gamma}\backslash\mathfrak{H}$, if we pairwise identify the boundary segments of F by transformations in $\bar{\Gamma}$. Connected neighborhoods in $\tilde{\Gamma}\backslash\mathfrak{H}$ of boundary points of F look a bit unconnected in this way.

3.2.2 *Cusps.* One may also consider the action of $\tilde{\Gamma}$ in

$$\mathfrak{H}^* = \mathfrak{H} \cup \left\{ \text{cuspidal points for } \tilde{\Gamma} \right\}.$$

Each $\tilde{\Gamma}$-orbit of cuspidal points corresponds to at least two boundary segments of a given fundamental domain, hence the number of $\tilde{\Gamma}$-orbits in the cuspidal points is finite. We call such a $\tilde{\Gamma}$-orbit a *cusp*.

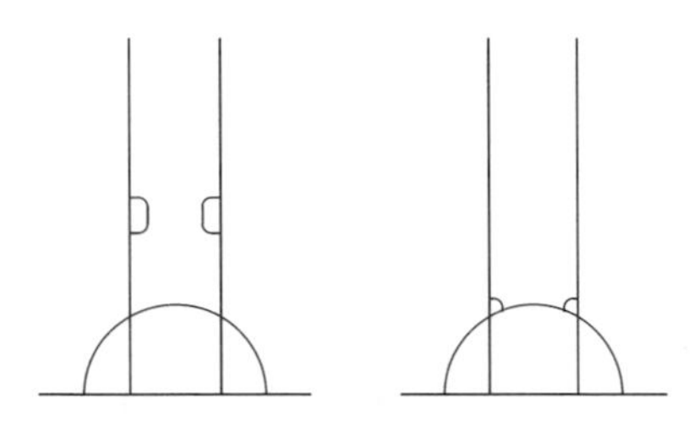

Figure 3.1 Connected neighborhoods of $\tilde{\Gamma}_{\rm mod}\cdot(\frac{1}{2}+2i)$ and $\tilde{\Gamma}_{\rm mod}\cdot(\frac{1}{2}+\frac{i}{2}\sqrt{3})$ in $\tilde{\Gamma}_{\rm mod}\backslash\mathfrak{H}$, as represented in the standard fundamental domain.

Miyake, [38], §1.7, describes how to put a topology on $\mathfrak{H}^*$ such that $\tilde{\Gamma}\backslash\mathfrak{H}^*$ is a Hausdorff topological space. $\tilde{\Gamma}\backslash\mathfrak{H}^*$ is compact, as one sees in Theorem 1.9.1 in [38]. (Miyake calls a cofinite discrete group a Fuchsian group of the first kind.)

3.2.3 *Notation.* $Y=\tilde{\Gamma}\backslash\mathfrak{H}\cong\tilde{\Gamma}\backslash\tilde{G}/\tilde{K}$ and $X=\tilde{\Gamma}\backslash\mathfrak{H}^*$.

X is a compact space, Y an open subspace. $X\setminus Y$ is finite; we denote it by X^∞. It is the set of cusps.

$\mathrm{pr}:\tilde{G}\to Y$ and $\overline{\mathrm{pr}}:\mathfrak{H}^*\to X$ denote the projection maps.

3.2.4 ∞ *or* $i\infty$. Up till here we have considered $\bar{\Gamma}$ as a group of fractional linear transformations acting on the projective line over $\mathbb{C}$ leaving $\mathfrak{H}$ invariant. From this point of view it is natural to use ∞. If one works in the upper half plane, one usually writes $i\infty$. This I shall do in the sequel.

3.2.5 *Complex structure.* X may be given a complex structure. Then it becomes a compact Riemann surface. See, e.g., Miyake, [38], §1.8. Most points of X are of the form $\tilde{\Gamma}\cdot z_0$ with $z_0\in\mathfrak{H}$ not an elliptic point for $\tilde{\Gamma}$. Then z_0 has a neighborhood U in $\mathfrak{H}$ on which $\overline{\mathrm{pr}}$ is a homeomorphism. On $\overline{\mathrm{pr}}\,U$ we use $\tilde{\Gamma}\cdot z\mapsto z-z_0$ as a local coordinate. If z_0 is an elliptic point of order v we use $\tilde{\Gamma}\cdot z\mapsto\left(\frac{z-z_0}{z-\bar{z}_0}\right)^v$ as a local coordinate at $\overline{\mathrm{pr}}(z_0)$.

If $g\cdot i\infty$ is a cuspidal point of $\tilde{\Gamma}$ with $\{\,gn(k)g^{-1}k(\pi l):k,l\in\mathbb{Z}\}$ as its fix-group in $\tilde{\Gamma}$, then $\xi=\tilde{\Gamma}g\cdot i\infty\in X^\infty$ has a neighborhood U such that $\dot{U}=U\setminus\{\xi\}$ is covered by $z\mapsto\tilde{\Gamma}g\cdot e^{2\pi iz}$, with $\overline{\mathrm{pr}}(z)=\overline{\mathrm{pr}}(z_1)$ if and only if $z\equiv z_1 \bmod \mathbb{Z}$. Then

$$\begin{cases}\xi & \mapsto & 0\\ \tilde{\Gamma}g\cdot z & \mapsto & e^{2\pi iz}\end{cases}$$

is a local coordinate on U.

In §1.8 of [38] the map $z\mapsto e^{2\pi iz/h}$ defines a local coordinate at a cusp. This $h\in\mathbb{Z}$ gives more freedom in normalizing the fix-group. That is an advantage when one compares automorphic forms for $\tilde{\Gamma}$ with those for a subgroup of finite index. I prefer to give all cusps 'width 1' in the context of this book.

The Riemann surface X is the natural space to work on when one considers holomorphic automorphic forms.

3.2.6 *Metric structure.* $\mathfrak{H}$ is also a riemannian space with metric given by $(ds)^2 = y^{-2}(dx)^2 + y^{-2}(dy)^2$. This metric is invariant under the action of $\tilde{G}$, and hence carries over to Y, except at the elliptic orbits, where it degenerates. So from the point of view of differential geometry, Y is a riemannian surface with singularities at the elliptic points. The cusps are infinitely far away, so there is no reason to consider the compactification X. If there are cusps, the surface Y has 'tentacles' with infinite length but finite area.

3.2.7 *Compact quotient.* There are groups $\tilde{\Gamma}$ without cusps. For example the *triangle groups*; see, e.g., Lehner, [32], Ch. VII, 1.G, p. 227. Other examples are the unit groups of norm 1 in indefinite division quaternion algebras over $\mathbb{Q}$; see, e.g., Miyake, [38], Chapter 5, especially Theorem 5.2.13.

For such groups $Y = X$, and Y itself is compact.

3.3 Canonical generators

When we shall study the group of characters, it will be convenient to have a description of a general discrete subgroup $\tilde{\Gamma}$ in terms of generators and relations. Such a description is available; we discuss it in this section.

3.3.1 *Canonical fundamental domain.* In Lehner, [32], Ch. VII.4, p. 241, or Petersson, [42], §3, one finds a canonical fundamental domain F for $\tilde{\Gamma}$ with the following properties:

i) The boundary ∂F of F consists of finitely many oriented smooth arcs, near their end points given by non-euclidean geodesics. We enumerate these boundary arcs in counterclockwise succession ($g, p, q \geq 0$):

$$\begin{gathered} l_1', l_1, l_2', l_2, \ldots, l_p', l_p, \\ h_1', h_1, h_2', h_2, \ldots, h_q', h_q, \\ a_1', b_1, a_1, b_1', a_2', b_2, a_2, b_2', \ldots, a_g', b_g, a_g, b_g'. \end{gathered}$$

ii) For $1 \leq j \leq p$ there exists $\pi_j = g_j n(1) g_j^{-1} \in \tilde{\Gamma}$ such that l_j' is $\pi_j l_j$ with opposite orientation; $g_j \cdot i\infty$ is the initial point of l_j, and $g_j \in \tilde{G}$.

iii) For $1 \leq j \leq q$ there exists $\varepsilon_j = p(z_j) k(\pi/v_j) p(z_j)^{-1} \in \tilde{\Gamma}$, with $v_j > 1$ integral, such that h_j' is $\varepsilon_j h_j$ with opposite orientation.

iv) For $1 \leq j \leq g$ there are $\gamma_j, \eta_j \in \tilde{\Gamma}$ with traces satisfying $|\mathrm{trace}(\hat{\gamma}_j)| > 2$, $|\mathrm{trace}(\hat{\eta}_j)| > 2$, such that a_j' is $\gamma_j a_j$ and b_j' is $\eta_j b_j$, both with opposite orientation. The $\hat{\gamma}_j$ and $\hat{\eta}_j$ are hyperbolic matrices.

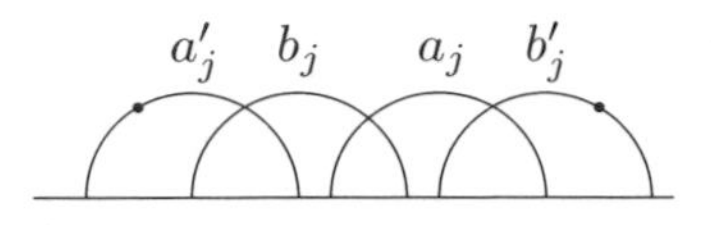

Figure 3.2 The part of the boundary corresponding to γ_j and η_j.

$g_1 \cdot i\infty, \ldots, g_p \cdot i\infty$ are *cuspidal* fixed points of $\tilde{\Gamma}$, and $z_1, \ldots, z_q$ are *elliptic* fixed points. These points represent the cuspidal and the elliptic orbits of $\tilde{\Gamma}$. The number g is the *genus* of $\tilde{\Gamma}$.

3.3.2 *Examples.* 13.1.8 gives a canonical fundamental domain for $\tilde{\Gamma}_{\mathrm{mod}}$. The standard fundamental domain of the modular group in Figure 1.1 on p. 2 is not canonical. See 14.2.4 and 15.1.8 for other examples.

3.3.3 *Generators and relations.* The advantage of canonical fundamental domains is the resulting uniform description of all cofinite discrete groups.

We see in Lehner, [32], Ch. VII.2, p. 230, or Petersson, [42], §3, that $\bar{\Gamma}$ is generated by $\overline{\pi_j}\,(1 \le j \le p)$, $\overline{\varepsilon_j}\,(1 \le j \le q)$, and $\overline{\gamma_j}, \overline{\eta_j}\,(1 \le j \le g)$, with explicitly given relations. These relations imply that as an abstract group $\tilde{\Gamma}$ is generated by

$$\{\, \pi_j : 1 \le j \le p \,\} \cup \{\, \varepsilon_j : 1 \le j \le q \,\} \cup \{\, \gamma_j : 1 \le j \le g \,\} \cup \{\, \eta_j : 1 \le j \le g \,\} \cup \{\zeta\}$$

with $\zeta = k(\pi)$. The relations are

$$\begin{aligned} &\zeta \text{ is central },\\ &\varepsilon_j^{v_j} = \zeta \quad \text{for } 1 \le j \le q,\\ &\pi_1 \cdots \pi_p \varepsilon_1 \cdots \varepsilon_q \gamma_1 \eta_1 \gamma_1^{-1} \eta_1^{-1} \gamma_2 \eta_2 \gamma_2^{-1} \eta_2^{-1} \cdots \gamma_g \eta_g \gamma_g^{-1} \eta_g^{-1}\\ &\qquad = \zeta^{2g-2+p+q}. \end{aligned}$$

The relations in $\bar{G}$ immediately imply those in $\tilde{G}$, except for the exponents of ζ. In the second relation the exponent 1 follows from the definition of ε_j. The exponent $2g - 2 + p + q$ in the last relation requires more work. Petersson discusses it in terms of multiplier systems, see [42], p. 64, and [43], p. 192, Satz 8. Here we shall relate it to the formula

$$\mathrm{vol}(\Gamma\backslash\mathfrak{H}) = \int_F \frac{dx \wedge dy}{y^2} = 2\pi \left(2g - 2 + p + q - \sum_{j=1}^{q} \frac{1}{v_j} \right), \tag{3.1}$$

which can be proved by considering divisors of automorphic forms on the compact Riemann surface related to $\tilde{\Gamma}\backslash\mathfrak{H}$. See, e.g., Shimura, [53], Thm. 2.20, p. 42.

We discuss in 13.1.9 a choice of canonical generators for the modular group: ζ, $n(1)$, $k(\pi/2)$, and $k(\pi/2)n(-1)$. But 13.1.2 shows that $n(1)$ and $k(\pi/2)$ already generate $\tilde{\Gamma}_{\text{mod}}$. This is a general phenomenon: $\tilde{\Gamma}$ can be generated with less generators, but with canonical generators one gets a uniform description.

3.3.4 *Proof of the main relation.* In the remainder of this section we check the relation

$$\pi_1 \cdots \pi_p \varepsilon_1 \cdots \varepsilon_q \gamma_1 \eta_1 \gamma_1^{-1} \eta_1^{-1} \gamma_2 \eta_2 \gamma_2^{-1} \eta_2^{-1} \cdots \gamma_g \eta_g \gamma_g^{-1} \eta_g^{-1} = \zeta^{2g-2+p+q}.$$

The idea is to lift a walk around the canonical fundamental domain F. In $\mathfrak{H}$ the completion of such a walk brings us back to the point of departure. This need not be the case if we lift this walk to $\tilde{G}$. Of course, the discrepancy depends on the way in which the boundary curves in $\mathfrak{H}$ are lifted to $\tilde{G}$.

3.3.5 *Choice of reference points.* Actually, we do not walk around the canonical fundamental domain F itself, but along the boundary of

$$F(a) = F \smallsetminus \bigcup_{j=1}^{p} \{ g_j \cdot z \in \mathfrak{H} : y > a \}.$$

So we truncate F at the cusps. If a is large enough, then the truncation involves only the boundary components l'_j and l_j for $1 \le j \le p$.

We define on $\partial F(a)$ the points $u_1, \ldots, u_{p+q+g}$ by

$$u_j \text{ is the initial point of } \begin{cases} l'_j & \text{if } 1 \le j \le p \\ h'_{j-p} & \text{if } p+1 \le j \le p+q \\ a'_{j-p-q} & \text{if } p+q+1 \le j \le p+q+g. \end{cases}$$

Put $u_{p+q+g+1} = u_1$. The conditions in 3.3.1 imply

$$u_{j+1} \text{ is the end point of } \begin{cases} l_j & \text{if } 1 \le j \le p \\ h_{j-p} & \text{if } p+1 \le j \le p+q \\ b'_{j-p-q} & \text{if } p+q+1 \le j \le p+q+g, \end{cases}$$

and $u_j = \rho_j \cdot u_{j+1}$, where $\rho_1, \ldots, \rho_{p+q+g} \in \tilde{\Gamma}$ are defined by

$$\rho_j = \begin{cases} \pi_j & \text{if } 1 \le j \le p \\ \varepsilon_{j-p} & \text{if } p+1 \le j \le p+q \\ \gamma_{j-p-q} \eta_{j-p-q} \gamma_{j-p-q}^{-1} \eta_{j-p-q}^{-1} & \text{if } p+q+1 \le j \le p+q+g. \end{cases}$$

The relation to be proved holds after projection to $\bar{G}$. So $\rho_1 \cdots \rho_{p+q+g} = k(\pi m)$ for some $m \in \mathbb{Z}$. We want to show that $m = 2g-2+p+q$.

We need not lift the whole curve $\partial F(a)$ to $\tilde{G}$, but only the u_j. We define $\tilde{u}_1, \ldots, \tilde{u}_{p+q+g+1} \in \tilde{G}$, by $\tilde{u}_1 = p(u_1)$ and $\tilde{u}_{j+1} = \rho_j^{-1} \tilde{u}_j$ for $1 \le j \le p+q+g$.

Then $\tilde{u}_j$ is a lift of u_j, and in general $\tilde{u}_{p+q+g+1} \neq \tilde{u}_1$. Define $\theta_1, \ldots, \theta_{p+q+g+1}$ by $\tilde{u}_j = p(u_j)k(\theta_j)$. This gives $-\pi m = \theta_{p+q+g+1} - \theta_1 = \theta_{p+q+g+1}$.

Equation (3.1) relates $2g - 2 + p + q$ to the volume of F. We consider

$$\mathrm{vol}(\tilde{\Gamma}\backslash\mathfrak{H}) = \lim_{a\to\infty} \int_{F(a)} \frac{dx \wedge dy}{y^2} = \lim_{a\to\infty} \int_{\partial F(a)} \frac{dx}{y}.$$

Let $B_j(a)$ be the part of $\partial F(a)$ between u_j and u_{j+1}. If $j > p$, then $B_j(a) = B_j$ does not depend on a. It suffices to show that

$$\int_{B_j(a)} \frac{dx}{y} = \begin{cases} 2\theta_j - 2\theta_{j+1} + o(1) \quad (a \to \infty) & \text{if } 1 \le j \le p \\ 2\theta_j - 2\theta_{j+1} - 2\pi/v_{j-p} & \text{if } p+1 \le j \le p+q \\ 2\theta_j - 2\theta_{j+1} & \text{if } p+q+1 \le j \le p+q+g. \end{cases}$$

We shall use the relation

$$\frac{dx}{y} \circ \begin{pmatrix} \alpha & \beta \\ \gamma & \delta \end{pmatrix} = \frac{dx}{y} + 2\, d \arg(\gamma z + \delta) \quad \text{for } \begin{pmatrix} \alpha & \beta \\ \gamma & \delta \end{pmatrix} \in \mathrm{SL}_2(\mathbb{R}).$$

3.3.6 *Cuspidal pieces.* Let $1 \le j \le p$. Put $v = g_j^{-1} \cdot u_j$, $w = g_j^{-1} \cdot u_{j+1}$. Then $w = v - 1$, and $\tilde{B}_j(a) = g_j^{-1} \cdot B_j(a)$ consist of three pieces: from v via $ia + \xi + 1$ and $ia + \xi$ to w. The point $g_j \cdot (ia + \xi)$ is the intersection of l_j with the line that truncates F.

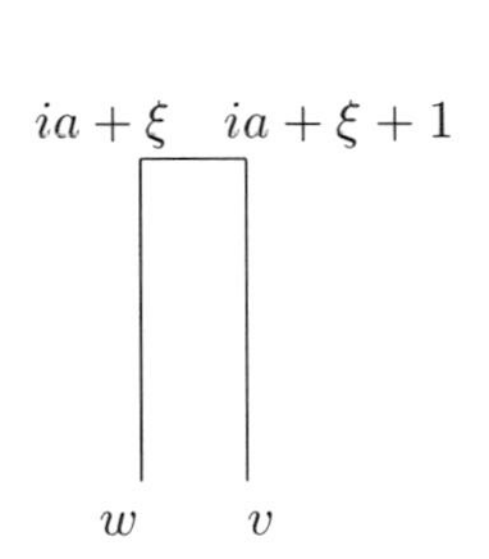

Figure 3.3 The path $\tilde{B}_j(a)$. The vertical lines need not be straight in reality, but they are related by a horizontal translation over 1.

Multiplying g_j by an element of the center does not change the action in $\mathfrak{H}$ or the relation $\pi_j = g_j n(1) g_j^{-1}$. So we assume $g_j = \widetilde{\begin{pmatrix} \alpha & \beta \\ \gamma & \delta \end{pmatrix}}$ and $g_j^{-1} = \widetilde{\begin{pmatrix} \delta & -\beta \\ -\gamma & \alpha \end{pmatrix}}$.

Now

$$\begin{aligned}\int_{B_j(a)} \frac{dx}{y} = \int_{\tilde{B}_j(a)} \frac{dx}{y} \circ \begin{pmatrix} \alpha & \beta \\ \gamma & \delta \end{pmatrix} = \int_{\tilde{B}_j(a)} \frac{dx}{y} + 2\arg(\gamma z + \delta)\Big|_v^w \\ = -\int_w^{ia+\xi} \frac{dx}{y} + o(1) + \int_v^{ia+\xi+1} \frac{dx}{y} + 2\arg(\gamma w + \delta) - 2\arg(\gamma v + \delta) \\ = 2\arg\left(\frac{\gamma w + \delta}{\gamma v + \delta}\right) + o(1) \quad (a \to \infty).\end{aligned}$$

When combining the arguments, we note that $\gamma w + \delta$ and $\gamma v + \delta$ are either both in the upper half plane, or both in the lower half plane. So we do not need a correction term $\pm 2\pi$.

On the other hand $\pi_j p(u_{j+1}) k(\theta_{j+1}) = p(u_j) k(\theta_j)$, hence $n(1) g_j^{-1} p(g_j \cdot w)$ $k(\theta_{j+1}) = g_j^{-1} p(u_j) k(\theta_j)$. We use Equation (2.1) on p. 28 to obtain

$$-\arg\left(-\gamma \frac{\alpha w + \beta}{\gamma w + \delta} + \alpha\right) + \theta_{j+1} = -\arg\left(-\gamma \frac{\alpha v + \beta}{\gamma v + \delta} + \alpha\right) + \theta_j.$$

Again, combining the arguments poses no problem. We arrive at

$$\theta_j - \theta_{j+1} = \arg\left(\frac{1}{\gamma v + \delta} \Big/ \frac{1}{\gamma w + \delta}\right) = \arg\left(\frac{\gamma w + \delta}{\gamma v + \delta}\right).$$

So $\int_{B_j(a)} \frac{dx}{y} = 2\theta_j - 2\theta_{j+1} + o(1)$, as we set out to show.

3.3.7 *Elliptic pieces.* Let $1 \le j \le q$, and put $l = p + j$. Equation (2.1) implies

$$\begin{aligned}\varepsilon_j &= p(z_j) k(\pi / v_j) p(z_j)^{-1} \\ &= p(\varepsilon_j \cdot i) k\left(-\arg\left(\frac{x_j - i}{y_j} \sin\frac{\pi}{v_j} + \cos\frac{\pi}{v_j}\right)\right).\end{aligned}$$

Hence ε_j is of the form $\widetilde{\begin{pmatrix} \alpha & \beta \\ \gamma & \delta \end{pmatrix}}$, and $\gamma z_j + \delta = e^{-\pi i / v_j}$.

$$\begin{aligned}\int_{B_j} \frac{dx}{y} &= \int_{h_j} \left(\frac{dx}{y} - \frac{dx}{y} \circ \varepsilon_j\right) = -2\arg(\gamma z + \delta)\Big|_{z_j}^{u_{l+1}} \\ &= -2\arg(\gamma u_{l+1} + \delta) - 2\pi \frac{1}{v_j}.\end{aligned}$$

We finish this case by noting that $\varepsilon_j \tilde{u}_{l+1} = \tilde{u}_l$ implies $\theta_{l+1} - \theta_l = \arg(\gamma u_{l+1} + \delta)$.

3.3.8 *Hyperbolic pieces.* Let $1 \le j \le g$, and put $l = p + q + j$. As $\hat{\gamma}_j$ and $\hat{\eta}_j$ are hyperbolic elements of G, we call the remaining pieces of $\partial F(a)$ hyperbolic.

Now we have the situation of Figure 3.2 on p. 38. Let v be the end point of b_j, then we have the following scheme:

boundary part	initial point	end point
a_j'	$\gamma_j \eta_j v = u_l$	$\gamma_j v$
b_j	$\gamma_j v$	v
a_j	v	$\eta_j v$
b_j'	$\eta_j v$	$\eta_j \gamma_j v = u_{l+1}$

For the relation it does not matter whether we multiply γ_j or η_j by an element of $\tilde{Z}$. We assume

$$\gamma_j^{-1} = \widetilde{\begin{pmatrix} \alpha & \beta \\ \gamma & \delta \end{pmatrix}}, \qquad \eta_j^{-1} = \widetilde{\begin{pmatrix} \varepsilon & \theta \\ \kappa & \lambda \end{pmatrix}}.$$

The equality $\gamma_j^{-1}\eta_j^{-1}p(\eta_j\gamma_j v)k(\theta_{l+1}) = \eta_j^{-1}\gamma_j^{-1}p(\gamma_j\eta_j v)k(\theta_l)$ implies

$$\begin{aligned} & -\arg\left(\gamma(\gamma_j \cdot v) + \delta\right) - \arg\left(\kappa(\eta_j\gamma_j \cdot v) + \lambda\right) + \theta_{l+1} \\ = \quad & -\arg\left(\kappa(\eta_j \cdot v) + \lambda\right) - \arg\left(\gamma(\gamma_j\eta_j \cdot v) + \delta\right) + \theta_l. \end{aligned}$$

So the contribution to the integral has the right value:

$$\begin{aligned} \int_{B_j} \frac{dx}{y} &= \int_{a_j'} \left(\frac{dx}{y} - \frac{dx}{y} \circ \gamma_j^{-1} \right) + \int_{b_j'} \left(\frac{dx}{y} - \frac{dx}{y} \circ \eta_j^{-1} \right) \\ &= -2\arg(\gamma z + \delta)\big|_{a_j'} - 2\arg(\kappa z + \lambda)\big|_{b_j'} = 2\left(\theta_l - \theta_{l+1}\right). \end{aligned}$$

3.4 Characters

We want to study families of automorphic forms depending on the spectral parameter and the character. (Characters of $\tilde{\Gamma}$ are in bijective correspondence to multiplier systems for Γ; see 2.3.6.) So we need to understand the group of characters of $\tilde{\Gamma}$. This is no problem, as we have available the canonical generators, and the relations between them. We shall describe the characters with help of the vector space of group homomorphisms $\tilde{\Gamma} \to \mathbb{C}$.

3.4.1 *Notation.* $\mathcal{X} = \hom(\tilde{\Gamma}, \mathbb{C}^*)$ is the group of characters of Γ. So a character is a group homomorphism from $\tilde{\Gamma}$ to the multiplicative group of $\mathbb{C}$.

We denote the subgroup of unitary characters (i.e., $|\chi(\gamma)| = 1$ for all $\gamma \in \tilde{\Gamma}$) by $\mathcal{X}_u$. This is a (real) compact subgroup of the (complex) abelian Lie group $\mathcal{X}$.

3.4.2 *Modular group.* In 13.1.3 we give an explicit description for the modular case: $\mathcal{X}(\tilde{\Gamma}_{\text{mod}}) \cong \mathbb{C} \bmod 12\mathbb{Z}$. It is convenient to use $r \mapsto \chi_r = e^{ir\alpha}$ to parametrize $\mathcal{X}(\tilde{\Gamma}_{\text{mod}})$, see 13.1.7. The α in the exponent is a group homomorphism $\tilde{\Gamma}_{\text{mod}} \to \mathbb{R}$. This indicates that it might be useful to consider the homomorphisms into the additive group of $\mathbb{C}$:

3.4.3 *Notation.* $\mathcal{V} = \hom(\tilde{\Gamma}, \mathbb{C})$, $\mathcal{V}_r = \hom(\tilde{\Gamma}, \mathbb{R})$.

$\mathcal{V}$ is a vector space over $\mathbb{C}$, and $\mathcal{V}_r$ a vector space over $\mathbb{R}$. They are related by $\mathcal{V} = \mathbb{C} \otimes_{\mathbb{R}} \mathcal{V}_r$. The space $\mathcal{V}$ is the Lie algebra of the Lie group $\mathcal{X}$. As $\mathcal{X}$ is an abelian group, all Lie brackets in $\mathcal{V}$ are zero.

In the *modular case,* the element α discussed in 13.1.7 is an element of $\mathcal{V}_r(\tilde{\Gamma}_{\mathrm{mod}})$. From the relations in 13.1.2 it is easily seen that α spans $\mathcal{V}(\tilde{\Gamma}_{\mathrm{mod}})_r$ over $\mathbb{R}$, and $\mathcal{V}(\tilde{\Gamma}_{\mathrm{mod}})$ over $\mathbb{C}$. These facts also follow from the next result:

3.4.4 Lemma. $\dim_{\mathbb{C}} \mathcal{V} = \dim_{\mathbb{R}} \mathcal{V}_r = 2g + p$.

$\varphi \in \mathcal{V}$ is determined by its values on γ_j, η_j and π_l, with $1 \le j \le g$, $1 \le l \le p$. The values on the other generators are

$$\varphi(\zeta) = \frac{2\pi}{\mathrm{vol}(\tilde{\Gamma}\backslash\mathfrak{H})} \sum_{j=1}^{p} \varphi(\pi_j), \qquad \varphi(\varepsilon_j) = \frac{1}{v_j}\varphi(\zeta) \quad \textit{for } 1 \le j \le q.$$

Proof. See the relations in 3.3.3 and Equation 3.1.

3.4.5 *Notation.* $\mathbf{exp} : \mathcal{V} \to \mathcal{X} : \varphi \mapsto e^{i\varphi}$ is a group homomorphism. It sends $\mathcal{V}_r$ to $\mathcal{X}_u$; its kernel is $\hom(\tilde{\Gamma}, 2\pi\mathbb{Z}) \subset \mathcal{V}_r$.

3.4.6 *Remarks.* In the modular case this amounts to $\mathbf{exp} : r\alpha \mapsto e^{ir\alpha}$. In 14.2.5 and 15.1.9 we discuss the space $\mathcal{V}$ for two other example groups.

It may happen that $\mathcal{V} = \{0\}$. This is, for example, the case for those triangle groups that are generated by elliptic elements. See, e.g., Lehner, [32], Ch. VII, 1.G, p. 228. For these groups, one even has $\mathcal{X} = \{1\}$. This means that these groups are not interesting in the study of families of automorphic forms that depend on the character in a holomorphic way.

For $\tilde{\Gamma}_{\mathrm{mod}}$ we have seen that $\mathcal{X} = \mathbf{exp}(\mathcal{V})$. For the two other example groups (discussed in the final chapters of this book) one has the same equality. We give an example of groups for which this is not the case:

3.4.7 *Hecke groups.* In §5 of [20] Hecke discusses a family of cofinite groups, indexed by $q \in \mathbb{N}$, $q \ge 3$. Define Γ_q as the group generated by $\begin{pmatrix} 1 & 2\cos\pi/q \\ 0 & 1 \end{pmatrix}$ and $\begin{pmatrix} 0 & 1 \\ -1 & 0 \end{pmatrix}$. For $q = 3$ one obtains Γ_{mod}.

This leads to $\tilde{\Gamma}_q$, the subgroup of $\tilde{G}$ generated by $\pi_1 = n(2\cos(\pi/q))$ and $w = k(\pi/2)$. One may check that $g = 0$, $p = 1$ and that there are two elliptic orbits. Canonical generators are π_1, $\varepsilon_1 = w$ and $\varepsilon_2 = p(z_2)k(\pi/q)p(z_2)^{-1}$ with $z_2 = e^{\pi i/q}$. The character χ_0, determined by $\chi_0(\pi_1) = 1$ and $\chi_0(w) = e^{2\pi i/(q-2)}$ is not in $\mathbf{exp}(\mathcal{V})$ if q is even.

3.4.8 *Notation.* $\mathcal{X}_0$ is the subgroup of $\mathcal{X}_u$ of characters with value 1 on

$$\gamma_1, \ldots, \gamma_g, \eta_1, \ldots, \eta_g, \pi_1, \ldots, \pi_p$$

and with $\chi(\zeta)$ a root of unity.

For $\chi \in \mathcal{X}_0$ all $\chi(\varepsilon_j)$ are roots of unity as well.

The character χ_0 of $\tilde{\Gamma}_q$ is an element of $\mathcal{X}_0$. This shows that $\mathcal{X}_0$ need not be trivial, and that it need not be contained in $\mathbf{exp}\,\mathcal{V}$.

3.4.9 Lemma. *$\mathcal{X}_0$ is a finite group; $\mathcal{X} = \mathcal{X}_0 \cdot \mathbf{exp}\,\mathcal{V}$.*

Remark. This shows that to study the holomorphic dependence of a family of automorphic forms on the character it suffices to work on components $\chi_0 \cdot \mathbf{exp}\,(\mathcal{V})$ of $\mathcal{X}$, with $\chi_0 \in \mathcal{X}_0$.

Proof. Let $\chi \in \mathcal{X}$ be given. Choose $p_j \in \mathbb{C}$ such that $e^{ip_j} = \chi(\pi_j)$, $r \in \mathbb{C}$ such that $e^{ir} = \chi(\zeta)$ and $b_j \in 2\mathbb{Z}$ such that $e^{i(r+\pi b_j)/v_j} = \chi(\varepsilon_j)$. Then

$$\frac{r}{\pi}\sum_{j=1}^{q}\frac{1}{v_j} + \sum_{j=1}^{q}\frac{b_j}{v_j} + \frac{1}{\pi}\sum_{j=1}^{p} p_j \equiv (2g-2+p+q)\frac{r}{\pi} \bmod 2.$$

We may arrange equality if we choose p_1 or b_1 suitably. If $p = q = 0$, then $g > 1$ by Equation (3.1) on p. 38, and hence $\varphi(\zeta) = 0$. So

$$\frac{1}{\pi}\sum_{j=1}^{p} p_j = \left(2g-2+p+q-\sum_{j=1}^{q}\frac{1}{v_j}\right)\frac{r}{\pi} + \xi$$

with $\xi \in \mathbb{Q}$.

Define $\varphi \in \mathcal{V}$ such that $e^{i\varphi(\gamma_j)} = \chi(\gamma_j)$, $e^{i\varphi(\eta_j)} = \chi(\eta_j)$ for $1 \le j \le g$ and $\varphi(\pi_j) = p_j$ for $1 \le j \le p$. Then $\mathbf{exp}\,\varphi$ and χ coincide on the generators γ_j, η_j and π_j. Further

$$\left(2g-2+p+q-\sum_{j=1}^{q}\frac{1}{v_j}\right)\varphi(\zeta) = \left(2g-2+p+q-\sum_{j=1}^{q}\frac{1}{v_j}\right) r + \xi\pi.$$

So $\varphi(\zeta) \in r + \pi\mathbb{Q}$. So $\chi(\zeta).(\mathbf{exp}\,\varphi)^{-1}(\zeta)$ is a root of unity, and $\chi \cdot (\mathbf{exp}\,\varphi)^{-1} \in \mathcal{X}_0$.

Let $\chi \in \mathcal{X}_0$. In the computations above we have $p_j = 0$ for $1 \le j \le p$. We can impose the restrictions $0 \le b_j < 2v_j$ for $1 \le j \le q$, and $0 \le r < 2\pi$. The computation shows that

$$\sum_{j=1}^{q}\frac{b_j}{v_j} \equiv \left(2g-2+p+q-\sum_{j=1}^{q}\frac{1}{v_j}\right)\frac{r}{\pi} \bmod 2.$$

This has only finitely many solutions that satisfy the restrictions.

3.5 Notations

We consider automorphic forms as functions on $\tilde{G}$. But we have to keep in mind that their transformation behavior implies that they are determined by their values on $p(z)$, where z runs through a fundamental domain. So in a sense we work above the quotient $Y = \tilde{\Gamma}\backslash\mathfrak{H}$. In this section we fix some conventions that will be useful in the sequel. The choices are based on a fixed canonical fundamental domain.

3.5.1 *Exceptional points.* We choose a finite set $\mathcal{P} \subset X$ containing at least the cusps. We call it the set of *exceptional points*, above which we shall allow automorphic forms to have singularities; see also 1.7.1. In the context of meromorphic automorphic forms, the set $\mathcal{P}$ indicates where we allow poles. We leave $\mathcal{P}$ fixed in Part I.

We put $\mathcal{P}_Y = \mathcal{P} \cap Y$, and call it the set of *interior exceptional points.* Thus $\mathcal{P}$ is the disjoint union of two finite — possibly empty — subsets of X: the set X^∞ of cusps, and the set $\mathcal{P}_Y$.

3.5.2 *Notation.* We fix a canonical fundamental domain F, and corresponding canonical generators.

For each cusp $P \in X^\infty$ there is a unique representative $g_j \cdot i\infty$ in the closure of F in $\mathfrak{H}^*$. We put $g_P = g_j$ and $\pi_P = \pi_j = g_P n(1) g_P^{-1} \in \tilde{\Gamma}$. We arrange g_P to have the form $g_P = \widetilde{\begin{pmatrix} a & b \\ c & d \end{pmatrix}}$, $g_P^{-1} = \widetilde{\begin{pmatrix} d & -b \\ -c & a \end{pmatrix}}$, for some $\begin{pmatrix} a & b \\ c & d \end{pmatrix} \in G$. This is convenient when we want to view automorphic forms as functions on $\mathfrak{H}$.

Next consider $P \in \mathcal{P}_Y$. We choose $z_P \in F$ such that $P = \tilde{\Gamma} \cdot z_P$. (If z_P is a boundary point of F, but not an elliptic point, we have some freedom of choice.) We put $g_P = p(z_P)$. This implies that again $g_P = \widetilde{\begin{pmatrix} a & b \\ c & d \end{pmatrix}}$, $g_P^{-1} = \widetilde{\begin{pmatrix} d & -b \\ -c & d \end{pmatrix}}$, with $\begin{pmatrix} a & b \\ c & d \end{pmatrix} = \begin{pmatrix} \sqrt{y_P} & x_P/\sqrt{y_P} \\ 0 & 1/\sqrt{y_P} \end{pmatrix} \in G$. If P is an elliptic orbit $\tilde{\Gamma} \cdot z_j$, we take $v_P = v_j$ and $\varepsilon_P = g_P k(\pi/v_P) g_P^{-1}$. Otherwise we put $\varepsilon_P = \zeta = k(\pi)$ and $v_P = 1$. In both cases $\varepsilon_P = g_P k(\pi/v_P) g_P^{-1}$.

Thus we have selected for each $P \in \mathcal{P}$ an element $g_P \in \tilde{G}$ such that either $g_P \cdot i\infty$ or $g_P \cdot i$ corresponds to P under $\overline{\mathrm{pr}}$, and we have chosen π_P, respectively ε_P, in the subgroup of $\tilde{\Gamma}$ that leaves $g_P \cdot i\infty$, respectively $g_P \cdot i$, invariant.

3.5.3 *Standard neighborhoods* of exceptional points are obtained from simple neighborhoods of $i\infty$ and i in $\mathfrak{H} \cup \mathbb{R} \cup \{i\infty\}$ by projection. For $\sigma > 0$ we put

$$U_P(\sigma) = \begin{cases} \overline{\mathrm{pr}}\,\{\, g_P z \in \mathfrak{H} : y > \sigma \,\} \cup \{P\} & \text{if } P \in X^\infty \\ \overline{\mathrm{pr}} \left\{ g_P z \in \mathfrak{H} : \left| \dfrac{z-i}{z+i} \right| < \sqrt{\dfrac{\sigma}{\sigma+1}} \right\} & \text{if } P \in \mathcal{P}_Y \end{cases}$$

with $\sigma \in (0, \infty)$. If we take $u = \frac{|z-i|^2}{4y}$ (see 2.2.6), then the condition $\left|\frac{z-i}{z+i}\right| < \sqrt{\frac{\sigma}{\sigma+1}}$ is equivalent to $u < \sigma$.

We put $\dot{U}_P(\sigma) = U_P(\sigma) \smallsetminus \{P\}$.

3.5.4 *Coordinates.* On

$$\mathrm{pr}^{-1}(\dot{U}_P(\sigma)) \cap p(F)\tilde{K} = \mathrm{pr}^{-1}(\dot{U}_P(\sigma)) \cap \{\, p(z)k(\theta) : z \in F,\, \theta \in \mathbb{R} \,\}$$

we may use the following coordinates:

$$\begin{array}{llcll} P \in X^\infty & g_P p(z) k(\theta) & \mapsto & (x,y,\theta) & \text{(Iwasawa coordinates)} \\ P \in \mathcal{P}_Y & g_P k(\eta) a(t_u) k(\psi) & \mapsto & (u,\eta,\psi) & \text{(polar coordinates),} \end{array}$$

provided σ is suitably chosen. Suitably means large enough for $P \in X^\infty$, and small enough for $P \in \mathcal{P}_Y$.

If $P \in \mathcal{P}_Y$ and σ is small enough, the set $\mathrm{pr}^{-1}\dot{U}_P(\sigma)$ consists of infinitely many connected components. Let us consider the component C of $g_P a(t_{\sigma/2})$. The polar coordinates may be used to map C bijectively onto $\{\,(u,\eta,\psi) : 0 < u < \sigma,\ \eta \in \mathbb{R}, 0 \leq \psi < \pi\,\}$. Left translation by ε_P maps C into itself, and amounts to $(u,\eta,\psi) \mapsto (u, \eta + \pi/v_P, \psi)$. As a set of points, $\mathrm{pr}^{-1}\dot{U}_P(\sigma)$ is isomorphic to C modulo the action of ε_P, provided σ is small enough.

For $P \in X^\infty$ and σ large enough, the Iwasawa coordinates map the connected component C of $g_P a(2\sigma)$ onto $\{\,(x,y,\theta) : x \in \mathbb{R}, y > \sigma, \theta \in \mathbb{R}\,\}$. Left translation by π_P and by ζ both leave C invariant; π_P corresponds to $(x,y,\theta) \mapsto (x+1,y,\theta)$, and ζ to $(x,y,\theta) \mapsto (x,y,\theta+\pi)$. If σ is large enough, then $\mathrm{pr}^{-1}\dot{U}_P(\sigma)$ is isomorphic to C modulo the group generated by the actions of π_P and ζ.

3.5.5 *Choice of the A_P.* We choose $\tilde{A}_P \in (0,\infty)$ suitable in the sense just mentioned, such that in addition all $U_P(\tilde{A}_P)$ for $P \in \mathcal{P}$ are pairwise disjoint. To see that this is possible, note that the set $\mathcal{P}$ is finite, and that the $U_P(\sigma)$ are a basis for the neighborhoods of P in the Hausdorff space X.

The set $Y \setminus \bigcup_{P \in \mathcal{P}} U_P(\tilde{A}_P)$ is compact.

To have some freedom in the construction of a transformation function in Section 8.2, we take $A_P = \tilde{A}_P + 1$ if $P \in X^\infty$ and $A_P = \frac{1}{2}\tilde{A}_P$ if $P \in \mathcal{P}_Y$. The $U_P(A_P)$ are strictly smaller than the $U_P(\tilde{A}_P)$. We shall glue functions on the sets $U_P(\tilde{A}_P) \setminus U_P(A_P)$.

If we consider $\mathcal{P} = \{P\}$, $P = \tilde{\Gamma}_{\mathrm{mod}} \cdot i\infty$, in the *modular case*, then any positive $A_P \leq 2$ is right. If we take $\mathcal{P}$ larger, other choices of the A_* should be made. For instance, if $\mathcal{P} = \{P, Q_i\}$, with P as before, and $Q_i = \tilde{\Gamma}_{\mathrm{mod}} \cdot i$, we need to be more careful; see 13.1.12. For other examples see 14.2.7, and 15.1.11.

3.5.6 *Notations.* $Y_{\mathcal{P}} = Y \setminus \mathcal{P}$, $\tilde{G}_{\mathcal{P}} = \mathrm{pr}^{-1}(Y_{\mathcal{P}})$. We also define

$$\begin{array}{lcll} \tilde{I}_P = (\tilde{A}_P, \infty) & \supset & I_P = (A_P, \infty) & \text{if } P \in X^\infty \\ \tilde{I}_P = (0, \tilde{A}_P) & \supset & I_P = (0, A_P) & \text{if } P \in \mathcal{P}_Y. \end{array}$$

Chapter 4
Automorphic forms

Definition 2.3.4 allows too huge a space of automorphic forms to be convenient. In this chapter we formulate our concept of 'automorphic form with growth condition'. In 1.7.1 we have already indicated that we want to allow singularities above a finite set in $Y = \tilde{\Gamma}\backslash\mathfrak{H}$. In 3.5.1 we have fixed this set as $\mathcal{P}_Y$, the set of 'interior exceptional points'. Moreover, we want to impose growth conditions at all points of $\mathcal{P} = \mathcal{P}_Y \cup \{\text{cusps}\}$.

In the *cuspidal case*, $P \in X^\infty = \{\text{cusps}\}$, condition iii) in 1.5.2 imposes the usual condition of polynomial growth near the cusp. But it is better to allow exponential growth to have for instance the Poincaré series in 1.5.9 and 1.7.2 come under our definition. The idea is present in 1.5.6: disregard a finite number of terms in the Fourier expansion, and impose a growth condition on the remaining sum. The condition of square integrability turns out to be more convenient in the context of this book than that of polynomial growth. That is the *regularity* defined in 4.1.8.

For *interior points* $P \in \mathcal{P}_Y$ we proceed in an analogous way. The Fourier expansion is formed with respect to η in the polar coordinates (u, η, ψ) at P. Here we define regularity as smoothness.

We use the term 'growth condition' to denote a rule that gives for each $P \in \mathcal{P}$ a finite number of Fourier terms to be left free.

To carry this out, we consider in Section 4.1 the Fourier expansion at each $P \in \mathcal{P}$, and define regularity. Section 4.3 discusses the definition of 'automorphic form with growth condition'. The Maass-Selberg relation in Section 4.6 gives a condition that is automatically satisfied by the Fourier terms of automorphic forms.

In 1.4.4 and 1.5.8 we have seen that Whittaker functions and confluent hypergeometric functions turn up in the explicit description of Fourier terms at a cusp. In Section 4.2 we discuss special functions needed to describe the Fourier terms.

Maass, see [35], Chapter IV, (12) and (13) on p. 177, introduces differential operators that map automorphic forms to automorphic forms, and shift the weight over 2. This can be seen as the action of elements of the Lie algebra $\mathfrak{g}$ on automorphic forms, see Section 4.5. In Section 4.4 we consider the action of these differential operators on the Fourier terms.

The ideas we follow in this chapter may be found in Chapter IV of [35], except that Maass allows no singularities in $\mathfrak{H}$, and less growth at the cusps.

4.1 Fourier expansion

In Section 3.5 we have fixed a finite set $\mathcal{P} \subset X = \tilde{\Gamma}\backslash\mathfrak{H}^*$ of exceptional points, containing at least the cusps. We have called the elements of $\mathcal{P}$ exceptional points. We consider in this section the Fourier expansion of an automorphic form at an exceptional point $P \in \mathcal{P}$, and we define the condition 'regular at P'. Actually, we do this for a wider class of functions, although we shall mainly use it for automorphic forms.

We fix a character χ of the cofinite discrete group $\tilde{\Gamma}$ and a weight l satisfying $e^{\pi il} = \chi(\zeta)$.

4.1.1 *Definition.* Let $U \subset \tilde{G}$ satisfy $\tilde{\Gamma}U\tilde{K} = U$. For example, $U = \tilde{G}_{\mathcal{P}}$, or U is the full original $U = \mathrm{pr}^{-1}\dot{U}_P(\sigma)$ in $\tilde{G}$ of the open set $U_P(\sigma) \subset Y$ for some $P \in \mathcal{P}$ and some $\sigma > 0$. A function $f : U \to \mathbb{C}$ is called *χ-l-equivariant* if it satisfies

$$f(\gamma g k(\theta)) = \chi(\gamma) f(g) e^{il\theta} \quad \text{for all } \gamma \in \tilde{\Gamma},\ g \in U,\ \theta \in \mathbb{R}.$$

The assumption $e^{\pi il} = \chi(\zeta)$ is necessary if we want to have non-zero equivariant functions. The property of χ-l-equivariance combines conditions i) an ii) in the definition of automorphic form in 2.3.4.

In 4.1.2 and 4.1.3 we consider a χ-l-equivariant function $f \in C^\infty(U)$ with $\tilde{\Gamma}U\tilde{K} = U$.

4.1.2 *Fourier expansion at a cusp.* Let $P \in X^\infty$, and assume that $\dot{U}_P(\sigma)$ is contained in $\mathrm{pr}\, U$ for some $\sigma \geq \tilde{A}_P$. The element π_P of $\tilde{\Gamma}$ maps $\mathrm{pr}^{-1}\dot{U}_P(\sigma)$ into itself, see 3.5.4. As $f(\pi_P g) = \chi(\pi_P) f(g)$, we have the following Fourier series expansion in x, valid on $\mathrm{pr}^{-1}\dot{U}_P(\sigma)$:

$$f(g_P p(z) k(\theta)) = \sum_n e^{2\pi inx} F_{P,n} f(y) e^{il\theta},$$

where n runs over the set $\mathcal{C}_P(\chi) = \{\, n \in \mathbb{C} : e^{2\pi in} = \chi(\pi_P) \,\}$, and where $F_{P,n} f \in C^\infty(\sigma, \infty)$ is given by

$$F_{P,n} f(y) = \int_0^1 f(g_P p(x + iy)) e^{-2\pi inx}\, dx.$$

The $n \in \mathcal{C}_P(\chi)$ need not be integers.

$F_{P,n} f$ is a function of one variable, with domain (σ, ∞). It determines a term in the Fourier expansion. We call

$$\tilde{F}_{P,n} f : \mathrm{pr}^{-1}\dot{U}_P(\sigma) \to \mathbb{C} : g_p p(z) k(\theta) \mapsto e^{2\pi inx} F_{P,n} f(y) e^{il\theta}$$

the *Fourier term* of *order* (P, n) of f.

Automorphic forms are eigenfunctions of the Casimir operator ω. Let us consider the action of ω on Fourier terms. In the integral

$$\tilde{F}_{P,n}f(g) = \int_0^1 e^{-2\pi i n x} f(g_P n(x) g_P^{-1} g)\, dx,$$

we differentiate under the integral sign; this gives $\omega \tilde{F}_{P,n} f = \tilde{F}_{P,n}\omega f$. (Here we use the invariance of ω under left translations.) The description of ω in Iwasawa coordinates, obtained in 2.2.4, gives $F_{P,n}\omega f = l_{P,n}(l) F_{P,n} f$ with

$$l_{P,n}(l) = -y^2\partial_y^2 + 4\pi^2 n^2 y^2 - 2\pi n l y.$$

4.1.3 *Fourier expansion at interior exceptional points.* Let $P \in \mathcal{P}_Y$, and assume that $\dot{U}_P(\sigma) \subset U$ for some $\sigma \in (0, \tilde{A}_P]$. As $f(\varepsilon_P g) = \chi(\varepsilon_P) f(g)$, we have the following Fourier series expansion in η, valid on $\mathrm{pr}^{-1}\dot{U}_P(\sigma)$:

$$f(g_P k(\eta) a(t_u) k(\psi)) = \sum_n e^{in\eta} F_{P,n} f(u) e^{il\psi},$$

where n runs over $\mathcal{C}_P(\chi) = \{ n \in \mathbb{C} : e^{in\pi/v_P} = \chi(\varepsilon_P) \}$, and where $F_{P,n}f \in C^\infty(0, \tilde{A}_P)$ is given by

$$F_{P,n}f(u) = \int_0^\pi f(g_P k(\eta) a(t_u)) e^{-in\eta} \frac{d\eta}{\pi}.$$

The Fourier term of order (P, n) is $\tilde{F}_{P,n} f(g_P k(\eta) a(t_u) k(\psi)) = e^{in\eta} F_{P,n} f(u) e^{il\psi}$, for $0 < u < \sigma$, $\eta, \psi \in \mathbb{R}$. Again $\omega \tilde{F}_{P,n} f = \tilde{F}_{P,n}\omega f$, and $F_{P,n}\omega f = l_{P,n}(l) F_{P,n} f$, with

$$l_{P,n}(l) = -(u^2+u)\partial_u^2 - (2u+1)\partial_u + \frac{(n-l)^2}{16(u^2+u)} - \frac{nl}{4(u+1)}.$$

4.1.4 *Differential equation.* The domain of $F_{P,n}f$ is a subset of $(0, \infty)$, depending on the domain of f. It always contains an interval (σ, ∞) if $P \in X^\infty$, and an interval $(0, \sigma)$ if $P \in \mathcal{P}_Y$.

If f satisfies $\omega f = \lambda f$ on U, all $F_{P,n}f$ are solutions of the second order linear differential equation $l_{P,n}(l) F = \lambda F$. This differential equation is regular on $(0, \infty)$ with analytic coefficients. We shall discuss its 2-dimensional space of solutions in Section 4.2.

4.1.5 *Examples.* In Section 1.4, and also in 1.5.5, we have seen Fourier expansions at the cusp of several modular forms. (Note that $g_P = 1$ in these examples.) One finds explicit Fourier expansions of automorphic forms for other groups in Sections 14.3 and 15.2.

Fourier expansions at interior points generalize the concept of power series expansion; see 13.2.4 for an example.

4.1.6 *Dependence on the group.* Our choice of the g_P has the disadvantage that if a modular form is considered as an automorphic form for a subgroup of finite index, the Fourier expansions are written down in a different way; see 14.3.2.

4.1.7 *Domain of convergence.* the Fourier series expansion at each $P \in \mathcal{P}$ of a χ-l-equivariant $f \in C^\infty(\tilde G)$ converges on $\tilde G$. But if f has singularities on $\mathrm{pr}^{-1}\mathcal{P}_Y$, the series at P converges on a smaller set, containing $\mathrm{pr}^{-1}\dot U_P(\tilde A_P)$, whereas the individual Fourier terms may be extended to $\tilde G$, respectively $\tilde G \smallsetminus (p(z_P)\tilde K)$. But whatever the domain of a given automorphic form is, we may extend its Fourier terms at P as real analytic functions on $\tilde G$ if $P \in X^\infty$, and on $\tilde G \smallsetminus \overline{\mathrm{pr}}^{-1}\{P\}$ if $P \in \mathcal{P}_Y$.

4.1.8 *Regularity.* Let $P \in \mathcal{P}_Y$. If an automorphic form f is in $C^\infty(\tilde G)$, then each Fourier term $\tilde F_{P,n} f$ is a C^∞-function on $\tilde G$ (as is seen from the integral representation). But if f is singular at the points of $\mathrm{pr}^{-1}\{P\}$, then its Fourier terms may have a singularity at the points of $\mathrm{pr}^{-1}\{P\}$ as well. See for instance 13.2.4. It seems natural to single out those Fourier terms that correspond to C^∞-functions at points of $\mathrm{pr}^{-1}\{P\}$. We shall call them *regular*.

For $P \in X^\infty$ it is less clear when a Fourier term at P should be considered regular. In this book square integrability will turn out to be a useful criterion.

We give the definition of regularity for more general functions than just Fourier terms. In particular, we define it for automorphic forms as well.

4.1.9 *Definition.* Let $P \in \mathcal{P}$, and let f be defined on a set $\mathrm{pr}^{-1}\dot U_P(\sigma)$, and suppose that it satisfies on this set

$$\begin{array}{rcll} f(gk(\theta)) & = & f(g)e^{il\theta} & \text{for all } \theta \in \mathbb{R} \\ f(\pi_P g) & = & \chi(\pi_P) f(g) & \text{if } P \in X^\infty \\ f(\varepsilon_P g) & = & \chi(\varepsilon_P) f(g) & \text{if } P \in \mathcal{P}_Y. \end{array}$$

If $P \in \mathcal{P}_Y$ we call f *regular at* P, if it is the restriction of a function in $C^\infty(\mathrm{pr}^{-1}U_P(\sigma))$.

If $P \in X^\infty$ choose $\nu \in \mathcal{C}_P(\chi)$. We call f *regular at* P, if it is an element of $C^\infty(\mathrm{pr}^{-1}\dot U_P(\sigma))$ and

$$\int_{y=\sigma}^{\infty} \int_{x=0}^{1} \left| e^{-2\pi i\nu x} f(g_P p(z)) \right|^2 \, dx \, \frac{dy}{y^2} < \infty.$$

If the conditions are met on some set $\mathrm{pr}^{-1}\dot U_P(\sigma)$, they are met on smaller sets of this form as well. For $P \in X^\infty$ the choice of ν in its class in $\mathbb{C} \bmod \mathbb{Z}$ does not influence the condition. If $|\chi(\pi_P)| = 1$, which is the case if χ is unitary, then we do not need the factor $e^{-2\pi i\nu x}$ in the integral; but otherwise this factor is really necessary to obtain a function $z \mapsto \left| e^{-2\pi i\nu x} f(g_P p(z)) \right|$ that is invariant under $z \mapsto z+1$ for $y > \sigma$.

The condition of polynomial growth at the cusp in Definitions 1.2.2 and 1.5.2 does not imply regularity at that cusp. Lemma 4.3.7 will imply that modular cusp forms like Δ (see 1.3.2 and 1.4.2), and the cuspidal Maass forms (see 1.4.5) are regular at $\tilde{\Gamma}_{\text{mod}} \cdot i\infty$.

4.1.10 Lemma. *If f is regular at $P \in \mathcal{P}$, then all its Fourier terms at P are regular as well.*

Proof. Let $n \in \mathcal{C}_P(\chi)$. For $P \in \mathcal{P}_Y$ we conclude that $\tilde{F}_{P,n}f$ is regular by differentiation under the integral sign. For $P \in X^\infty$ we take $\nu = n$.

$$\begin{aligned}
&\int_{y=\sigma}^{\infty} \int_{x=0}^{1} \left| e^{-2\pi i n x} \tilde{F}_{P,n} f(g_P p(z)) \right|^2 dx \, \frac{dy}{y^2} \\
&\quad = \int_{y=\sigma}^{\infty} \left| \tilde{F}_{P,n} f(g_P a(y)) \right|^2 \frac{dy}{y^2} \leq \int_{y=\sigma}^{\infty} \int_0^1 \left| e^{-2\pi i n x} f(g_P p(z)) \right|^2 dx \, \frac{dy}{y^2} < \infty.
\end{aligned}$$

Remark. The lemma states that taking a Fourier term does not destroy regularity. On the other hand, it is in general not clear that the sum of a converging series of regular Fourier terms is regular. In Lemma 4.3.7 we shall see that this is almost always true for eigenfunctions of ω.

4.2 Spaces of Fourier terms

In this section we discuss the spaces of functions that can occur as Fourier terms of automorphic forms. These spaces have dimension 2. We shall give a basis for most values of the parameters.

We have collected the explicit formulas used in the sequel. That makes this section rather technical. The main points are the definitions of the spaces $W_l(P, n, s)$ and the Wronskian below, the definitions of the standard elements $\mu_l(P, n, s)$ and $\omega_l(P, n, s)$ in 4.2.5, 4.2.6, 4.2.8 and 4.2.9, and Proposition 4.2.11. The reader may prefer to skip the rest now, and refer to it when needed.

4.2.1 *Definition.* Let $l, n, s \in \mathbb{C}$, and $P \in \mathcal{P}$. If $P \in \mathcal{P}_Y$, then we suppose $n \equiv l \bmod 2$ throughout this section. We define $W_l(P, n, s)$ as the space of functions F on $\tilde{G}$ if $P \in X^\infty$, respectively on $\tilde{G} \setminus \overline{\text{pr}}^{-1}\{P\}$ if $P \in \mathcal{P}_Y$, that satisfy $\omega F = \left(\frac{1}{4} - s^2\right) F$, and

$$\begin{aligned}
F(g_P n(x) g_P^{-1} g k(\theta)) &= e^{2\pi i n x} F(g) e^{il\theta} && \text{if } P \in X^\infty \\
F(g_P k(\eta) g_P^{-1} g k(\psi)) &= e^{in\eta} F(g) e^{il\psi} && \text{if } P \in \mathcal{P}_Y.
\end{aligned}$$

If f is an automorphic form of weight l, with character χ and eigenvalue $\frac{1}{4} - s^2$, and $n \in \mathcal{C}_P$, then $\tilde{F}_{P,n}f$ is a function on $\text{pr}^{-1}\dot{U}_P(A_P)$ that is the restriction of an element of $W_l(P, n, s)$.

In this definition of W_l, we do not use the eigenvalue λ as a parameter, but a complex number s such that $\lambda = \frac{1}{4} - s^2$. We have seen already in 1.2.3 that this parameter is more convenient than λ. We often call s the *spectral parameter*. Note that $W_l(P,n,s) = W_l(P,n,-s)$.

In the literature one often meets the parametrization $\lambda = s(1-s)$. In comparing results one then needs to make the translation $s \mapsto s+\frac{1}{2}$. Our parametrization has the advantage that principal series representations of $\mathfrak{g}$ and $\tilde{G}$ correspond to purely imaginary numbers; see also the discussion in 1.2.3. The choice $\lambda = s(1-s)$ has the advantage of placing the critical line at $\operatorname{Re} s = \frac{1}{2}$, which reminds us of the Riemann zeta function.

Maass, [35], uses another parametrization, in which the spectral parameter and the weight occur in a mixed way; see 2.1.7.

4.2.2 *Differential equation.* A function F as in 4.2.1 is determined by a function f of one variable:

$$\begin{aligned} F(g_P p(z) k(\theta)) &= e^{2\pi i n x} f(y) e^{il\theta} && \text{if } P \in X^\infty \\ F(g_P k(\eta) a(t_u) k(\psi)) &= e^{in\eta} f(u) e^{il\psi} && \text{if } P \in \mathcal{P}_Y. \end{aligned}$$

The condition $\omega F = \left(\frac{1}{4} - s^2\right) F$ amounts to $l_{P,n} f = \left(\frac{1}{4} - s^2\right) f$, with $l_{P,n}$ as in 4.1.2–4.1.3. This second order linear differential operator $l_{P,n}$ has real analytic coefficients, so any function f corresponding to an element of $W_l(P,n,s)$ is a real analytic function on $(0,\infty)$. Moreover, $\dim W_l(P,n,s) = 2$.

4.2.3 *Definition.* The *Wronskian* of $F_1, F_2 \in W_l(P,n,s)$ is given by

$$\operatorname{Wr}(F_1,F_2) = \begin{cases} f_1'(y) f_2(y) - f_1(y) f_2'(y) & \text{if } P \in X^\infty \\ -(u^2+u)\left(f_1'(u) f_2(u) - f_1(u) f_2'(u)\right) & \text{if } P \in \mathcal{P}_Y, \end{cases}$$

where $f_j \in C^\infty(0,\infty)$ corresponds to F_j as indicated above.

One easily checks that $\operatorname{Wr}(f,g)$ does not depend on y, respectively u. So

$$\operatorname{Wr} : W_l(P,n,s) \times W_l(P,n,s) \to \mathbb{C}$$

is a skew symmetric bilinear form. $\operatorname{Wr}(F_1,F_2) = 0$ if and only if F_1 and F_2 are linearly dependent. So the form Wr is non-degenerate.

4.2.4 *Choice of a basis.* We shall give various pairs of elements of $W_l(P,n,s)$ that form a basis for general values of the parameters. The Wronskian of such a pair will be a non-zero meromorphic function of l and s. By 'general case' we mean those values of the parameters at which both elements are holomorphic, and moreover the Wronskian is non-zero. In the Lemmas 7.6.13–7.6.15, we shall consider how to obtain a suitable basis in the special cases.

Elements of $W_l(P,n,s)$ may be singled out either by their asymptotic behavior near P, or by their behavior far away from P. The concept of regularity, defined in 4.1.9, is based on the behavior near P. Here we select first elements of $W_l(P,n,s)$ by their behavior far away from P.

4.2.5 *Cuspidal case.* Let $P \in X^\infty$. If $n = 0$ the functions f corresponding to $F \in W_l(P, 0, s)$ satisfy $-y^2 f''(y) = (\frac{1}{4} - s^2) f(y)$. One solution of this differential equation is given by $y \mapsto y^{s+1/2}$. We put

$$\mu_l(P, 0, s; g_P p(z) k(\theta)) = y^{s+1/2} e^{il\theta}.$$

$\mu_l(P, 0, s)$ and $\mu_l(P, 0, -s)$ form a basis of $W_l(P, 0, s)$ if $s \neq 0$. The asymptotic behavior of $\mu_l(P, 0, s)$ is simple far away from P (i.e., as $y \downarrow 0$), and also near P (i.e., as $y \to \infty$).

For general n, there is an element of $W_l(P, n, s)$ with the same asymptotic behavior far away from P. We write f corresponding to $F \in W_l(P, n, s)$ in the form $f(y) = y^{s+1/2} e^{-2\pi n y} h(4\pi n y)$. Then h satisfies

$$t h''(t) + (1 + 2s - t) h'(t) - \left(\frac{1}{2} + s - \frac{1}{2} l \right) h(t) = 0.$$

This is the confluent hypergeometric differential equation. In [33], 7.4, we find a solution for $s \notin -\frac{1}{2}\mathbb{N}$; this leads to

$$\mu_l(P, n, s; g_P p(z) k(\theta)) = y^{s+1/2} e^{\mp 2\pi n y + 2\pi i n x} {}_1F_1 \left[\begin{matrix} \frac{1}{2} + s \mp \frac{1}{2} l \\ 1 + 2s \end{matrix} \middle| \pm 4\pi n y \right] e^{il\theta}.$$

This extends the definition of $\mu_l(P, 0, s)$ given above, and preserves the behavior as $y \downarrow 0$. In [33], 7.4 (8), one sees that the choice of $\pm$ does not matter.

$\mu_l(P, n, s)$ and $\mu_l(P, n, -s)$ are elements of $W_l(P, n, s)$. One easily checks that their Wronskian is given by

$$\mathrm{Wr}(\mu_l(P, n, s), \mu_l(P, n, -s)) = 2s \qquad \text{for } \pm s \notin \frac{1}{2}\mathbb{N}.$$

Hence they form a basis if $s \notin \frac{1}{2}\mathbb{Z}$.

4.2.6 *Interior case.* Let $P \in \mathcal{P}_Y$. Write f corresponding to $F \in W_l(P, n, s)$ in the form

$$f(u) = u^{(n-l)/4} (u+1)^{-(n-l)/4 - s - 1/2} h\left(\frac{1}{u+1} \right).$$

The differential equation for h is

$$\begin{aligned} & t(1-t) h''(t) + \left(2s + 1 - \left(\frac{n-l}{2} + 2s + 2 \right) \right) h'(t) \\ & \quad - \left(\frac{1}{2} + s + \frac{1}{2} n \right) \left(\frac{1}{2} + s - \frac{1}{2} l \right) h(t) = 0. \end{aligned}$$

This is the hypergeometric differential equation. For $s \notin -\frac{1}{2}\mathbb{N}$ it leads to the following solution:

$$\begin{aligned} & \mu_l(P, n, s; g_P a(t_u)) \\ & = \left(\frac{u}{u+1} \right)^{(n-l)/4} (u+1)^{-s-1/2} {}_2F_1 \left[\begin{matrix} \frac{1}{2} + s + \frac{n}{2}, \frac{1}{2} + s - \frac{l}{2} \\ 1 + 2s \end{matrix} \middle| \frac{1}{u+1} \right], \end{aligned}$$

see [33], 6.3 (1) and (5). The asymptotic behavior far from P is

$$\mu_l(P,n,s;g_P a(t_u)) \sim (u+1)^{-s-1/2} \quad (u \to \infty).$$

$\mathrm{Wr}(\mu_l(P,n,s),\mu_l(P,n,-s)) = 2s$ for $\pm s \notin \frac{1}{2}\mathbb{N}$, so we have a basis $\mu_l(P,n,s)$, $\mu_l(P,n,-s)$ of $W_l(P,n,s)$ for $s \notin \frac{1}{2}\mathbb{Z}$.

4.2.7 *Another basis.* Next we look for elements of $W_l(P,n,s)$ with a simple behavior near P. With the exception of the case $P \in X^\infty$, $\mathrm{Re}\, n = 0$, we shall find a solution $\omega_l(P,n,s)$ that is 'small' near P, and may be expressed in the form $\omega_l(P,n,s) = \sum_\pm v_l(P,n,\pm s)\mu_l(P,n,\pm s)$, with $v_l(P,n,s)$ an explicitly given function. We shall also define a solution $\hat\omega_l(P,n,s) = \sum_\pm \hat v_l(P,n,\pm s)\mu_l(P,n,\pm s)$ that is not a multiple of $\omega_l(P,n,s)$, and has a larger growth than $\omega_l(P,n,s)$ near P. For $\hat\omega_l$ there is no canonical choice. In the cuspidal case one choice will do; in the interior case we choose two possibilities.

In all cases, we choose elements that are invariant under $s \mapsto -s$.

4.2.8 *Cuspidal case.* Let $P \in X^\infty$, and suppose that $\mathrm{Re}\, n \neq 0$. Put $\varepsilon = \mathrm{sign}(\mathrm{Re}\, n)$. If we write f corresponding to $F \in W_l(P,n,s)$ in the form $f(y) = h(4\pi\varepsilon n y)$, we obtain for h the Whittaker differential equation

$$h''(t) + \left(-\frac{1}{4} + \frac{1}{2}\varepsilon l\frac{1}{t} + (\frac{1}{4} - s^2)\frac{1}{t^2}\right) h(t) = 0.$$

It has solutions $t \mapsto W_{\varepsilon l/2,s}(t)$ and $t \mapsto W_{-\varepsilon l/2,s}(-t)$, see [56], 1.7. We take the branch of the Whittaker function $W_{\kappa,s}(z)$ with $-\frac{\pi}{2} < \arg(z) < \frac{3\pi}{2}$. We define ω_l and $\hat\omega_l$ by

$$\omega_l(P,n,s;g_P a(y)) = W_{\varepsilon l/2,s}(4\pi\varepsilon n y), \quad \hat\omega_l(P,n,s;g_P a(y)) = W_{-\varepsilon l/2,s}(-4\pi\varepsilon n y).$$

The choice of $\hat\omega$ is not very canonical, we could have taken the argument in another way as well.

The relation with $\mu_l(P,n,\pm s)$ is given by

$$\begin{aligned} v_l(P,n,s &= (4\pi\varepsilon n)^{s+1/2}\Gamma(-2s)\Gamma(\frac{1}{2} - s - \frac{1}{2}\varepsilon l)^{-1} \\ \hat v_l(P,n,s) &= i e^{\pi i s}(4\pi\varepsilon n)^{s+1/2}\Gamma(-2s)\Gamma(\frac{1}{2} - s + \frac{1}{2}\varepsilon l)^{-1} \\ \omega_l(P,n,s) &= \sum_\pm v_l(P,n,\pm s)\mu_l(P,n,\pm s) \\ \hat\omega_l(P,n,s) &= \sum_\pm \hat v_l(P,n,\pm s)\mu_l(P,n,\pm s). \end{aligned}$$

In [56], 4.1.3, we find the asymptotic behavior near P:

$$\begin{aligned} \omega_l(P,n,s;g_P a(y)) &\sim (4\pi\varepsilon n y)^{\varepsilon l/2} e^{-2\pi\varepsilon n y} && y \to \infty \\ \hat\omega_l(P,n,s;g_P a(y)) &\sim e^{-\pi i \varepsilon l/2}(4\pi\varepsilon n y)^{-\varepsilon l/2} e^{2\pi\varepsilon n y} && y \to \infty. \end{aligned}$$

The Wronskians are

$$\begin{aligned}
&\mathrm{Wr}(\omega_l(P,n,s), \hat{\omega}_l(P,n,s)) \\
&\quad = \sum_{\pm} \pm 2s v_l(P,n,s)\hat{v}_l(P,n,-s) = -4\pi\varepsilon n\, e^{-\pi i \varepsilon l/2} \\
&\mathrm{Wr}(\mu_l(P,n,s), \omega_l(P,n,s)) \\
&\quad = 2s v_l(P,n,-s) \;=\; (4\pi\varepsilon n)^{-s+1/2} \frac{\Gamma(2s+1)}{\Gamma(\frac{1}{2}+s-\frac{1}{2}\varepsilon l)}.
\end{aligned}$$

4.2.9 *Interior case.* Let $P \in \mathcal{P}_Y$. We put $p = \frac{|n-l|}{4}$, and take $\varepsilon = \pm 1$ such that $4\varepsilon p = n - l$. Inspection of the differential equation shows that there is a solution $f(u) \sim u^p$ as $p \downarrow 0$, and also solutions $f(u) \sim u^{-p}$ if $p \neq 0$, and $f(u) \sim \log u$ if $p = 0$. The first case leads to

$$\begin{aligned}
&\omega_l(P,n,s; g_P a(t_u)) \\
&\quad = \left(\frac{u}{u+1}\right)^p (u+1)^{-s-1/2} {}_2F_1\left[\begin{matrix} \frac{1}{2}+s+\frac{\varepsilon}{2}n,\ \frac{1}{2}+s-\frac{\varepsilon}{2}l \\ 1+2p \end{matrix} \,\middle|\, \frac{u}{u+1}\right] \\
&\quad = \sum_{\pm} v_l(P,n,\pm s)\mu_l(P,n,\pm s; g_P a(t_u)), \\
&v_l(P,n,s) = \frac{(2p)!\Gamma(-2s)}{\Gamma(\frac{1}{2}-s+\frac{\varepsilon}{2}n)\Gamma(\frac{1}{2}-s-\frac{\varepsilon}{2}l)}.
\end{aligned}$$

We use [33], 6.4 (9), 6.5 (5) if $\varepsilon = 1$, and 6.4 (14), 6.5 (6) if $\varepsilon = -1$. Note that $\omega_l(P,n,s)$ is well defined for all $s \in \mathbb{C}$. Often $\mu_l(P,n,s)$, $\omega_l(P,n,s)$ may be used as a basis:

$$\begin{aligned}
\mathrm{Wr}(\mu_l(P,n,s), \omega_l(P,n,s)) &= 2s v_l(P,n,-s) \\
&= \frac{(2p)!\Gamma(1+2s)}{\Gamma(\frac{1}{2}+s+\frac{\varepsilon}{2}n)\Gamma(\frac{1}{2}+s-\frac{\varepsilon}{2}l)}.
\end{aligned}$$

It is not obvious what is the right choice of $\hat{\omega}_l(P,n,s)$. We shall use two versions, depending on a parameter $\zeta \in \{1,-1\}$. For a given ζ we put $q = \zeta(n+l)/4$, and define

$$\begin{aligned}
&\hat{\omega}_l(P,n,\zeta,s; g_P a(t_u)) \\
&\quad = e^{\pi i s}\left(\frac{u}{u+1}\right)^{-q} u^{-s-1/2} {}_2F_1\left[\begin{matrix} \frac{1}{2}+s+q+p,\ \frac{1}{2}+s+q-p \\ 1+2q \end{matrix} \,\middle|\, 1+\frac{1}{u}\right],
\end{aligned}$$

provided $q \notin -\frac{1}{2}\mathbb{N}$. We choose the branch of ${}_2F_1[.,.;.|z]$ that is holomorphic in z for $\arg(1-z) \in [0,\pi]$. To see that this indeed defines an element of $W_l(P,n,s)$, we express it in the $\mu_l(P,n,\pm s)$. Remark first that $\{p+q,p-q\}=\{\frac{\varepsilon}{2}n,-\frac{\varepsilon}{2}l\}$. In [33], 6.5 (7), 6.4 (11) and (16) we find

$$\begin{aligned}
\hat{\omega}_l(P,n,\zeta,s;u) &= e^{\pi is}u^{-s-1/2-q}(u+1)^q \\
&\cdot\left(\frac{\Gamma(-2s)\Gamma(1+2q)}{\Gamma(\frac12-s+q-p)\Gamma(\frac12-s+q+p)}\left(\frac{u+1}{u}\right)^{-1/2-s-q-p}\right. \\
&\quad\cdot {}_2F_1\left[\begin{matrix}\frac12+s+q+p,\ \frac12+s-q+p\\ 1+2s\end{matrix}\,\middle|\,\frac{1}{u+1}\right] \\
&\quad+\frac{\Gamma(2s)\Gamma(1+2q)u^{1/2+s+q+p}}{\Gamma(\frac12+s+q+p)\Gamma(\frac12+s+q-p)}(u+1)^{-1/2+s-q-p} \\
&\quad\left.\cdot\, e^{-2\pi is}{}_2F_1\left[\begin{matrix}\frac12-s+q-p,\ \frac12-s-q-p\\ 1-2s\end{matrix}\,\middle|\,\frac{1}{u+1}\right]\right) \\
&= e^{\pi is}u^p(u+1)^{-1/2-s-p}\frac{\Gamma(-2s)\Gamma(1+2q)}{\Gamma(\frac12-s+q-p)\Gamma(\frac12-s+q+p)} \\
&\quad\cdot {}_2F_1\left[\begin{matrix}\frac12+s+\frac{\varepsilon}{2}n,\ \frac12+s-\frac{\varepsilon}{2}l\\ 2s+1\end{matrix}\,\middle|\,\frac{1}{u+1}\right] \\
&\quad+e^{-\pi is}u^p(u+1)^{-1/2+s-p}\frac{\Gamma(2s)\Gamma(1+2q)}{\Gamma(\frac12+s+q+p)\Gamma(\frac12+s+q-p)} \\
&\quad\cdot {}_2F_1\left[\begin{matrix}\frac12-s+\frac{\varepsilon}{2}n,\ \frac12-s-\frac{\varepsilon}{2}l\\ 1-2s\end{matrix}\,\middle|\,\frac{1}{u+1}\right].
\end{aligned}$$

Use [33] 6.4 (2) to see that this equals $\sum_{\pm}\hat{v}_l(P,n,\zeta,\pm s)\mu_l(P,n,\pm s)$, with

$$\hat{v}_l(P,n,\zeta,s)=e^{\pi is}\frac{\Gamma(-2s)\Gamma(1+2q)}{\Gamma(\frac12-s+\frac{\zeta}{2}n)\Gamma(\frac12-s+\frac{\zeta}{2}l)}.$$

The Wronskian is

$$\begin{aligned}
&\mathrm{Wr}(\hat{\omega}_l(P,n,\zeta,s),\omega_l(P,n,s)) \\
&\quad=\sum_{\pm}\pm 2s\hat{v}_l(P,n,\zeta,\pm s)v_l(P,n,\mp s) \\
&\quad=-ie^{\pi i(p-q)}\frac{(2p)!\Gamma(1+2q)}{\Gamma(\frac12+s+p+q)\Gamma(\frac12-s+p+q)}.
\end{aligned}$$

4.2.10 *Regular elements.* We define $W_l^0(P,n,s)$ as the subspace of those elements of $W_l(P,n,s)$ that are regular at P.

4.2.11 Proposition. *Let l, n, $s \in \mathbb{C}$.*

i) Suppose $P \in X^\infty$.

a) If $s \in i\mathbb{R}$ and $n = 0$, then $W_l^0(P, n, s) = \{0\}$.

b) If $\pm \mathrm{Re}\, s > 0$, $n = 0$, then $\dim_\mathbb{C} W_l^0(P, 0, s) = 1$. A basis element is $\mu_l(P, 0, \mp s)$.

c) If $\mathrm{Re}\, n \neq 0$, then $\dim_\mathbb{C} W_l^0(P, n, s) = 1$. A basis element is $\omega_l(P, n, s)$.

ii) Suppose $P \in \mathcal{P}_Y$, and $n \equiv l \bmod 2$. Then $\dim_\mathbb{C} W_l^0(P, n, s) = 1$. A basis element is $\omega_l(P, n, s)$.

Remark. From this explicit form of the basis elements, we see that regular Fourier terms of automorphic forms are in general bounded on a neighborhood of the point P at which the Fourier expansion is centered. The exception occurs for $P \in X^\infty$ and $\mathrm{Re}\, n = 0$. Of course, if n or l is not real, then boundedness should be understood for (x, θ), respectively (η, ψ), in compact sets.

Proof. The statements in part i) are clear from the explicitly given elements in $W_l(P, n, s)$; see 4.2.5 and 4.2.8.

We have seen that in the case $P \in \mathcal{P}_Y$ the differential equation in u determining $W_l(P, n, s)$ has solutions of the form

$$\begin{array}{ll} u^p h_1(u) + u^{-p} h_2(u) & \text{if } p = \frac{1}{4}|n-l| \neq 0 \\ h_1(u) + h_2(u) \ln u & \text{if } p = \frac{1}{4}|n-l| = 0, \end{array}$$

with h_1 and h_2 holomorphic on a neighborhood of $u = 0$ and either $h_j = 0$ or $h_j(0) \neq 0$. The regularity condition at points of $\mathrm{pr}^{-1}\{P\}$ is not easy to check in polar coordinates at P. Regularity is equivalent to smoothness at $z = i$ of $z \mapsto F(g_P p(z))$. We obtain for f corresponding to F (use 2.2.6):

$$\begin{aligned} F(g_P p(z)) &= e^{\frac{1}{2} in \arg((z-i)/(z+i))} f\left(\frac{|z-i|^2}{4y}\right) e^{-\frac{1}{2} il \arg(-1-z^2)} \\ &= f\left(\frac{|z-i|^2}{4y}\right) e^{\frac{1}{2} i(n-l) \arg(-1-iz) - \frac{1}{2} i(n+l) \arg(1-iz)}. \end{aligned}$$

The discontinuity of $\arg(-1 - iz)$ does not give any problem, as $n \equiv l \bmod 2$. Hence

$$F(g_P p(z)) = f\left(\frac{|z-i|^2}{4y}\right) \left(\frac{-1-iz}{|-1-iz|}\right)^{(n-l)/2} \left(\frac{1-iz}{|1-iz|}\right)^{-(n+l)/2}.$$

The only smooth possibilities in the case $n = l$ are of the form

$$h_1\left(\frac{|z-i|^2}{4y}\right) \left(\frac{1-iz}{|1-iz|}\right)^{-(n+l)/2}.$$

We obtain for $n \neq l$

$$F(g_P p(z)) = \begin{cases} f\left(\frac{|z-i|^2}{4y}\right) |z-i|^{-2p}(-1-iz)^{2p} \cdot (*) & \text{if } n > l \\ f\left(\frac{|z-i|^2}{4y}\right) |z-i|^{-2p}(-1+i\bar{z})^{2p} \cdot (*) & \text{if } n < l \end{cases}$$

where $(*)$ is a smooth factor without zeros. Only if f is of the form $u^p h_1(u)$, we obtain a smooth function at $z = i$. So the one-dimensionality of $W_l^0(P, n, s)$ is clear, and $\omega_l(P, n, s)$ is a non-zero element in it, see 4.2.9.

4.2.12 *Isomorphism.* The differential equation in y, respectively u, defining the space $W_l(P, n, s)$ is invariant under $(n, l) \mapsto (-n, -l)$. So we have a linear bijection

$$\iota : W_l(P, n, s) \longrightarrow W_{-l}(P, -n, s).$$

This isomorphism is not very canonical. Its definition depends on our choice of coordinates.

In all cases:

$$\begin{aligned} \iota\mu_l(P, n, s) &= \mu_{-l}(P, -n, s) \\ \iota\omega_l(P, n, s) &= \omega_{-l}(P, -n, s). \end{aligned}$$

Note that $\iota W_l^0(P, n, s) = W_{-l}^0(P, -n, s)$.

4.3 Automorphic forms with growth condition

We are ready to define 'automorphic forms satisfying a growth condition'. The idea is to replace the condition of polynomial growth in the definitions in Chapter 1 by regularity, and to apply this condition to the function minus a finite number of Fourier terms. The property of being an eigenfunction of the Casimir operator is strong enough to make this amount to a condition on the growth of the function.

Let $\mathcal{P} \subset X$ as before.

4.3.1 *Definition.* A *growth condition* $\mathbf{c}$ for $\tilde{\Gamma}$, $\mathcal{P}$ and χ, is a map $\mathbf{c}$ assigning a finite subset $\mathbf{c}(P)$ of $\mathcal{C}_P(\chi)$ to each $P \in \mathcal{P}$, with the additional condition that for each $P \in X^\infty$ the set $\mathcal{C}_P(\chi) \cap i\mathbb{R}$ is contained in $\mathbf{c}(P)$.

The set $\mathcal{C}_P(\chi)$ specifies which terms occur in the Fourier expansion at P, see 4.1.2 and 4.1.3. The additional condition ensures that for all $\in \mathcal{C}_P(\chi) \setminus \mathbf{c}(P)$ the spaces $W_l^0(P, n, s)$ of regular Fourier terms are spanned by $\omega_l(P, n, s)$; see Proposition 4.2.11.

The *minimal growth condition* for a given character χ is $\mathbf{c}^{(\chi)}$, determined by $\mathbf{c}^{(\chi)}(P) = \{0\}$ if $P \in X^\infty$ and $\chi(\pi_P) = 1$, and $\mathbf{c}^{(\chi)}(P) = \emptyset$ otherwise. By $\mathbf{c}_0$ we denote the growth condition $\mathbf{c}_0(P) = \{0\}$ if $P \in X^\infty$, and $\mathbf{c}_0(P) = \emptyset$ if $P \in \mathcal{P}_Y$. Hence $\mathbf{c}_0 = \mathbf{c}^{(1)}$, the minimal growth condition for the trivial character.

By $|\mathbf{c}(P)|$ we denote the number of elements of $\mathbf{c}(P)$, and by $|\mathbf{c}|$ the number $\sum_{P\in\mathcal{P}} |\mathbf{c}(P)|$.

4.3.2 *Comment.* We shall employ a growth condition $\mathbf{c}$ to indicate which Fourier terms to ignore in the regularity condition, when defining automorphic forms with growth condition. If we deal with eigenfunctions of ω, this implies a condition on the growth of the function at P. For general functions in $C^\infty(\tilde{G}_{\mathcal{P}})$, the term growth condition is misleading. Nevertheless, as this book deals with automorphic forms, I feel justified to use it.

4.3.3 *Definition.* Consider a growth condition $\mathbf{c}$, and a χ-l-equivariant function $f \in C^\infty(U)$, with U satisfying $\tilde{\Gamma} U \tilde{K} = U$. If $P \in \mathcal{P}$ satisfies $\dot{U}_P(\sigma) \subset \operatorname{pr} U$ for some σ, we define its $\mathbf{c}$-*remainder* $f[\mathbf{c}, P]$ *at* P to be the function in $C^\infty(\operatorname{pr}^{-1}\dot{U}_P(\sigma))$ given by $f[\mathbf{c}, P] = f - \sum_{n\in\mathbf{c}(P)} \tilde{F}_{P,n} f$. We shall apply this with $U = \tilde{G}_{\mathcal{P}}$; then $f[\mathbf{c}, P]$ is well defined for all $P \in \mathcal{P}$.

4.3.4 *Definition.* Let $\tilde{\Gamma}$, χ, s and l be as before. An *automorphic form* of weight l, with character χ and eigenvalue $\frac{1}{4} - s^2$ *for the growth condition* $\mathbf{c}$ is a function $f \in C^\infty(\tilde{G}_{\mathcal{P}})$ such that

i) f is χ-l-equivariant.

ii) $\omega f = (\frac{1}{4} - s^2) f$ on $\tilde{G}_{\mathcal{P}}$.

iii) $f[\mathbf{c}, P]$ is regular at P for each $P \in \mathcal{P}$.

We denote the space of all such automorphic forms by $A_l(\chi, \mathbf{c}, s)$.

4.3.5 *Comments.* As before, we parametrize the eigenvalue by $s \mapsto \frac{1}{4} - s^2$.

All Fourier terms of $f \in A_l(\chi, \mathbf{c}, s)$ with $n \in \mathcal{C}_P(\chi) \setminus \mathbf{c}(P)$ are regular at P, and are hence at least bounded (boundedness understood as in the remark to Proposition 4.2.11). So the growth of f near P is determined by the growth of the Fourier terms at P with $n \in \mathbf{c}(P)$. As $\mathbf{c}(P)$ is finite, the growth is at most exponential in y at $P \in X^\infty$, and at most polynomial in u at $P \in \mathcal{P}_Y$ — hence the name *growth condition.* One may view growth conditions as a generalization of divisors on X, as used in the study of meromorphic automorphic forms, see, e.g., 2.4 in [53].

In the case $P \in X^\infty$, $n \in \mathcal{C}_P(\chi)$ with $\operatorname{Re} n = 0$, the space $W_l^0(P, n, s)$ may be zero. By stipulating that such n are always an element of $\mathbf{c}(P)$, we ensure that the space of possible $F_{P,n} f$ has exactly dimension 1 for all $n \in \mathcal{C}_P(\chi) \setminus \mathbf{c}(P)$.

4.3.6 *Comparison.* The Definitions 1.2.2 and 1.5.2 of modular forms concern the case $\tilde{\Gamma} = \tilde{\Gamma}_{\mathrm{mod}}$, $\chi = 1$, and $l \in 2\mathbb{Z}$, with the minimal growth condition $\mathbf{c}_0$. Those definitions are the usual ones, see for instance Maass, [35], p. 185. It takes a bit of work to see that the present definition defines the same concept:

Consider the function $\tilde{f}$ on $\tilde{G}$ corresponding to a modular form f on $\mathfrak{H}$, as defined in 1.5.2 (this includes Definition 1.2.2). Put $P = \tilde{\Gamma}_{\mathrm{mod}} \cdot \infty$, and take $\lambda = \frac{1}{4} - s^2$. As the $\mathbf{c}_0$-remainder $\tilde{f}[\mathbf{c}_0, P]$ is smooth on $\operatorname{pr}^{-1}\dot{U}_P(\sigma)$ for each $\sigma > 0$, it has an absolutely converging Fourier expansion. All terms in it have polynomial growth, so are multiples of some $\omega_l(P,n,s)$. Lemma 4.3.7 below shows that $\tilde{f}[\mathbf{c}_0,P]$ is quickly decreasing, hence regular.

For the converse, use Lemma 4.1.10 and Proposition 4.2.11 to obtain a Fourier expansion of $\tilde{f}[\mathbf{c}_0,P]$ to which we can apply Lemma 4.3.7. Together with the known structure of $F_{P,0}\tilde{f}$, this leads to polynomial growth of $\tilde{f}$ near P.

Definition 1.5.6 corresponds to the case $\chi=\chi_r$ (the character χ_r is described in 13.1.3), and $\mathbf{c}(\tilde{\Gamma}_{\mathrm{mod}}\cdot i\infty)=\{\frac{r}{6}\}$.

We have indeed extended the definitions in Chapter 1. If we transform the examples of modular forms in that chapter into functions on $\tilde{G}_{\mathcal{P}}$, we obtain examples of automorphic forms as defined above.

4.3.7 Lemma. *Let $P\in X^\infty$, $\nu\in\mathbb{C}$, $\sigma>0$, $N>0$. Suppose the following series converges absolutely for $y>\sigma$:*

$$F(z)=\sum_n c_n\omega_l(P,n,s;g_P p(z))$$

with $n\equiv\nu \bmod 1$ and $|\operatorname{Re} n|\geq N$. Then

$$|F(z)|\ll y^{|\operatorname{Re} l|/2}e^{-2\pi N y}e^{-2\pi\operatorname{Im}(\nu)x} \qquad \textit{for } y\to\infty,$$

with the implied constant depending on ν, s, l, σ, and F.

Proof. Use the asymptotic estimate, see 4.2.8,

$$\omega_l(P,n,s;g_P a(y))\sim(4\pi\varepsilon n y)^{\varepsilon l/2}e^{-2\pi\varepsilon n y} \qquad \text{as } y\to\infty,$$

with $\varepsilon=\operatorname{sign}\operatorname{Re} n$. Take $t>\sigma$; the convergence of the series at $z=it$ gives

$$c_n=\mathcal{O}\left(|n|^{-\varepsilon l/2}e^{2\pi|\operatorname{Re} n|t}\right).$$

So for $y>t+1$:

$$\begin{aligned}|F(z)| &\ll \sum_n y^{|\operatorname{Re} l|/2}e^{-2\pi|\operatorname{Re} n|(y-t)}e^{-2\pi x\operatorname{Im} n}\\ &\ll y^{|\operatorname{Re} l|/2}e^{-2\pi x\operatorname{Im}\nu}e^{-2\pi N(y-t)}.\end{aligned}$$

4.3.8 *Further remarks.* If $\mathbf{c}$ and $\mathbf{c}_1$ are growth conditions and $\mathbf{c}(P)\subset\mathbf{c}_1(P)$ for all $P\in\mathcal{P}$, then $A_l(\chi,\mathbf{c},s)\subset A_l(\chi,\mathbf{c}_1,s)$.

If $f\in A_l(\chi,\mathbf{c},s)$ and $n\in\mathcal{C}_P(\chi)\smallsetminus\mathbf{c}(P)$, then $F_{P,n}f\in W_l^0(P,n,s)$, see Lemma 4.1.10.

The criterion in 1.4.3 when to call a modular form a cusp form, is the vanishing of the term of order zero in the Fourier expansion at the cusp. This is a special case of the following definition.

4.3.9 *Definition.* $f \in A_l(\chi, \mathbf{c}, s)$ is called a *cusp form for the growth condition* $\mathbf{c}$, if $\tilde{F}_{P,n} f = 0$ for all $P \in \mathcal{P}$ and all $n \in \mathbf{c}(P)$. We denote the corresponding subspace of $A_l(\chi, \mathbf{c}, s)$ by $S_l(\chi, \mathbf{c}, s)$.

4.3.10 *Remarks.* Take $\mathbf{c} = \mathbf{c}_0$ to get the usual definition of cusp form.

We have seen an *example* of a cusp form (for the growth condition $\mathbf{c}_0$) in 1.4.2. The cuspidal Maass forms, see 1.4.5, are cusp forms as well. See 15.2.3 for an unusual example.

If $\mathbf{c}(P) \subset \mathbf{c}_1(P)$ for all $P \in \mathcal{P}$, then $S_l(\chi, \mathbf{c}, s) \supset S_l(\chi, \mathbf{c}_1, s)$.

Fix $P_0 \in \mathcal{P}$ and define $\mathbf{d}_N(P_0) = \{ n \in \mathcal{C}_{P_0}(\chi) : |n| \leq N \}$, $\mathbf{d}_N(P)$ fixed for all $P \in \mathcal{P}$, $P \neq P_0$. Then

$$\bigcap_{N \geq 1} S_l(\chi, \mathbf{d}_N, \sigma) = \{0\} .$$

Indeed, if f is an element of this intersection, it vanishes on $\mathrm{pr}^{-1} \dot{U}_P(\tilde{A}_P)$. As f is a real analytic function on the connected set $\tilde{G}_{\mathcal{P}}$, it vanishes everywhere.

4.3.11 *Holomorphic automorphic forms.* To $f \in A_l\left(\chi, \mathbf{c}, \frac{l-1}{2}\right)$ corresponds a function $\hat{f}(z) = y^{-l/2} f(p(z))$ on $\overline{\mathrm{pr}}^{-1} Y_{\mathcal{P}}$ with transformation behavior

$$\hat{f}\left(\frac{az+b}{cz+d}\right) = \chi\left(\widetilde{\begin{pmatrix} a & b \\ c & d \end{pmatrix}}\right) (cz+d)^l \hat{f}(z) \quad \text{for all } \begin{pmatrix} a & b \\ c & d \end{pmatrix} \in \Gamma.$$

It is holomorphic on $\overline{\mathrm{pr}}^{-1} Y_{\mathcal{P}} \subset \mathfrak{H}$ if and only if $\mathbf{E}^- f = 0$; this follows from the description of $\mathbf{E}^-$ in Iwasawa coordinates in 2.2.4. In 4.5.5 we shall discuss holomorphic Fourier terms.

4.4 Differentiation of Fourier terms

We have just remarked that the action of the differential operator corresponding to the element $\mathbf{E}^-$ of the Lie algebra $\mathfrak{g}$ singles out the holomorphic automorphic forms. The action of $\mathfrak{g}$ gives useful information for other automorphic forms as well. Maass introduced differential operators K_α and Λ_β between spaces of automorphic forms, see, e.g., [35], p. 177 and p. 186. These operators shift the weight by 2. They correspond to $\frac{1}{2}\mathbf{E}^+$ and $\frac{1}{2}\mathbf{E}^-$. In this section we discuss the action of $\mathbf{E}^+$ and $\mathbf{E}^-$ on the Fourier terms. In the next section we consider their effect on automorphic forms.

4.4.1 *The action of* $\mathbf{E}^\pm$. Suppose that $f \in C^\infty(\Omega)$ with $\Omega \subset \tilde{G}$ open, $\Omega \tilde{K} = \Omega$, satisfies $\omega f = \left(\frac{1}{4} - s^2\right) f$, and $f(gk(\theta)) = f(g) e^{il\theta}$. Then f is an analytic function. Indeed, the Casimir operator ω corresponds to an elliptic differential operator with analytic coefficients on $\mathfrak{H}$ for functions of a fixed weight, see 2.2.4. So $\mathbf{E}^\pm f$ exists. The relation $[\mathbf{W}, \mathbf{E}^\pm] = \pm 2i \mathbf{E}^\pm$ implies that $\mathbf{E}^\pm f$ has weight $l \pm 2$. Moreover,

$\omega\mathbf{E}^\pm f = \mathbf{E}^\pm \omega f = \left(\frac{1}{4} - s^2\right)\mathbf{E}^\pm f$; this means that $\mathbf{E}^\pm f$ is an eigenfunction of ω as well, with the same eigenvalue.

As $\mathbf{E}^+$ and $\mathbf{E}^-$ generate the enveloping algebra $U(\mathfrak{g})$, the function f generates a vector space in which $U(\mathfrak{g})$ acts.

4.4.2 *Action on Fourier terms.* This section discusses the action of these differential operators on elements of $W_l(P,n,s)$. As $\mathbf{E}^\pm$ commutes with left translations, we get linear maps:

$$\mathbf{E}^\pm : W_l(P,n,s) \longrightarrow W_{l\pm 2}(P,n,s).$$

Table 4.1 gives the action of $\mathbf{E}^\pm$ on the standard basis elements μ_l and ω_l. It turns out that $\mathbf{E}^\pm W_l^0(P,n,s) \subset W_{l\pm 2}^0(P,n,s)$ in all cases considered. (If $P \in X^\infty$, $n \neq 0$, $\mathrm{Re}\, n = 0$, one may check that this does not hold. This fact is behind the additional condition in Definition 4.3.1.)

4.4.3 *Lie algebra modules.* For $r \in \mathbb{C}^*$, one may form the infinite dimensional space $\bigoplus_l W_l(P,n,s)$, where l runs over the coset modulo 2 of the numbers satisfying $e^{\pi i l} = r$. This is a module for $U(\mathfrak{g})$; all weight spaces have dimension 2. If we are not in the case $P \in X^\infty$, $\mathrm{Re}\, n = 0$, $n \neq 0$, there is a submodule $\bigoplus_l W_l^0(P,n,s)$.

4.4.4 *The action of ι.* The bijective maps $\iota : W_l(P,n,s) \to W_l(P,-n,s)$, defined in 4.2.12, induce a bijection

$$\iota : \bigoplus_l W_l(P,n,s) \longrightarrow \bigoplus_l W_{-l}(P,-n,s).$$

This is *not* an isomorphism of $U(\mathfrak{g})$-modules. It satisfies $\iota(\mathbf{X}f) = j(\mathbf{X})\iota f$, with j the involution in $\mathfrak{g}$ given by

$$j\mathbf{W} = -\mathbf{W}, \qquad j\mathbf{E}^\pm = \mathbf{E}^\mp.$$

This involution j comes from the automorphism $p(z)k(\theta) \mapsto p(-\bar{z})k(-\theta)$ of $\tilde{G}$, which covers the exterior automorphism

$$\begin{pmatrix} a & b \\ c & d \end{pmatrix} \mapsto \begin{pmatrix} a & -b \\ -c & d \end{pmatrix} = \begin{pmatrix} -1 & 0 \\ 0 & 1 \end{pmatrix} \begin{pmatrix} a & b \\ c & d \end{pmatrix} \begin{pmatrix} -1 & 0 \\ 0 & 1 \end{pmatrix}$$

of $\mathrm{SL}_2(\mathbb{R})$.

4.4.5 *Holomorphic Fourier terms.* $\mathbf{E}^+\mathbf{E}^-$ acts in $W_l(P,n,s)$ as multiplication by $-(1-4s^2+l^2-2l) = (1+2s-l)(-1+2s+l)$. This implies that $\mathbf{E}^- : W_l(P,n,s) \to W_{l-2}(P,n,s)$ can have a non-zero kernel only if $s = \pm\frac{l-1}{2}$. Application of the formulas in 2.2.4 and 2.2.5 shows that the kernel has dimension 1. It is spanned by functions determined by

$$\begin{cases} y \mapsto y^{l/2} e^{-2\pi n y} & \text{if} \quad P \in X^\infty \\ u \mapsto u^{(n-l)/4}(u+1)^{-(n+l)/4} & \text{if} \quad P \in \mathcal{P}_Y. \end{cases}$$

$$
\begin{aligned}
\text{For } P \in X^\infty, &\quad s \notin -\tfrac{1}{2}\mathbb{N} : \\
\mathbf{E}^\pm \mu_l(P,n,s) &= (1+2s\pm l)\mu_{l\pm 2}(P,n,s).
\end{aligned}
$$

See [33], 7.3.1 (3).

$$
\begin{aligned}
\text{For } P \in X^\infty, &\quad \pm \mathrm{Re}\, n > 0 : \\
\mathbf{E}^\pm \omega_l(P,n,s) &= -2\omega_{l\pm 2}(P,n,s), \\
\mathbf{E}^\pm \hat{\omega}_l(P,n,s) &= 2(\tfrac{1}{4}(l\mp 1)^2 - s^2)\hat{\omega}_{l\pm 2}(P,n,s), \\
&\quad \pm \mathrm{Re}\, n < 0 : \\
\mathbf{E}^\pm \omega_l(P,n,s) &= 2(\tfrac{1}{4}(l\pm 1)^2 - s^2)\omega_{l\pm 2}(P,n,s), \\
\mathbf{E}^\pm \hat{\omega}_l(P,n,s) &= -2\hat{\omega}_{l\pm 2}(P,n,s).
\end{aligned}
$$

See [56], (2.4.21) and (2.4.24).

$$
\begin{aligned}
\text{For } P \in \mathcal{P}_Y, &\quad s \notin -\tfrac{1}{2}\mathbb{N} : \\
\mathbf{E}^\pm \mu_l(P,n,s) &= -(1+2s\pm l)\mu_{l\pm 2}(P,n,s).
\end{aligned}
$$

See [33], 6.2.1 (3) and 6.2.2 (2).

$$
\begin{aligned}
P \in \mathcal{P}_Y, &\quad |n-l\mp 2| = |n-l|-2 : \\
\mathbf{E}^\pm \omega_l(P,n,s) &= |n-l|\omega_{l\pm 2}(P,n,s), \\
&\quad |n-l\mp 2| = |n-l|+2 : \\
\mathbf{E}^\pm \omega_l(P,n,s) &= -\frac{(l\pm 1)^2 - 4s^2}{|n-l|+2}\omega_{l\pm 2}(P,n,s).
\end{aligned}
$$

See [33], 6.2.1 (8) and (9), or express ω_l in the μ_l.

$$
\begin{aligned}
\text{For } P \in \mathcal{P}_Y, &\quad \pm(n+l)+2 \neq 0 : \\
\mathbf{E}^\pm \hat{\omega}_l(P,n,\pm 1,s) &= \tfrac{4s^2-(1\pm l)^2}{2\pm(n+l)}\hat{\omega}_{l\pm 2}(P,n,\pm 1,s), \\
&\quad \text{without additional condition:} \\
\mathbf{E}^\pm \hat{\omega}_l(P,n,\mp 1,s) &= \mp(n+l)\hat{\omega}_{l\pm 2}(P,n,\mp,s).
\end{aligned}
$$

Express $\hat{\omega}_l$ in the μ_l.

Table 4.1 Differentiation results.

4.4.6 *Holomorphic Fourier terms near a cusp.* Consider $P \in X^\infty$. Let F be the Fourier term determined by the function indicated above. Under the correspondence discussed in 1.3.4 it gives the holomorphic function $g_P \cdot z \mapsto e^{2\pi i n z}$ on $\mathfrak{H}$.

The basis element indicated above is regular at P if $\mathrm{Re}\, n > 0$, and also if $\mathrm{Re}\, n = 0$ and $\mathrm{Re}\, l < 1$. For $\mathrm{Re}\, n > 0$ this basis element should be a multiple of $\omega_l(P, n, \frac{l-1}{2})$. From the asymptotic behavior in 4.2.8 follows that it is equal to $(4\pi n \,\mathrm{sign}(\mathrm{Re}\, n))^{-l/2}\omega_l(P, n, \frac{l-1}{2})$.

This shows that $W_{l,(l-1)2}(t) = t^{l/2}e^{t/2}$, and hence the basis element is equal to $e^{\pi i l/2}(-4\pi n)^{-l/2}\hat{\omega}_l(P, n, \frac{l-1}{2})$ in the case $\mathrm{Re}\, n < 0$.

For $n = 0$ we have obtained $\mu_l(P, 0, \frac{l-1}{2})$.

4.4.7 *Holomorphic Fourier term at an interior point.* Take $P \in \mathcal{P}_Y$. The function $F : g_P k(\eta) a(t_u) k(\psi) \mapsto e^{in\eta} u^{(n-l)/4}(u+1)^{-(n+l)/4} e^{il\psi}$ corresponds to the function $\hat{F} : z \mapsto y^{-l/2} F(p(z))$, which should be holomorphic.

We want to write $g_P p(z)$ in the form $k(\eta)a(t_u)k(\psi)$. Note that $g_P^{-1}p(z) = p(z_P)^{-1}p(z) = p\left(\frac{z - x_P}{y_P}\right)$. The relations in 2.2.6 give

$$\begin{aligned} u &= |z - z_P|^2/(4yy_P) \\ u+1 &= |z - \overline{z_P}|^2/(4yy_P) \\ e^{2i\eta} &= \frac{z - z_P}{z - \overline{z_P}} \left| \frac{z - \overline{z_P}}{z - z_P} \right| \\ \eta + \psi &= \frac{\pi}{2} - \arg(z - \overline{z_P}). \end{aligned}$$

Remember that $n \equiv l \bmod 2$. We obtain

$$\begin{aligned} \hat{F}(z) &= y^{-l/2} e^{i(n-l)/2} \eta e^{il(\eta+\psi)} u^{(n-l)/4} (u+1)^{-(n+l)/4} \\ &= y^{-l/2} \left(\frac{z - z_P}{z - \overline{z_P}} \right)^{(n-l)/2} e^{\pi i l/2} e^{-il \arg(z - \overline{z_P})} \\ &\quad \cdot (4yy_P)^{l/2} |z - z_P|^0 |z - \overline{z_P}|^{-l} \\ &= 2^l y_P^{l/2} e^{\pi i l/2} (z - \overline{z_P})^{-l} \left(\frac{z - z_P}{z - \overline{z_P}} \right)^{(n-l)/2} \end{aligned}$$

We conclude that $(z - \overline{z_P})^l \hat{f}(z)$ is a multiple of $w^{(n-l)/2}$, where w denotes the coordinate $(z - z_P)/(z - \overline{z_P})$ on $\mathfrak{H}$ near z_P.

This basis element is regular at P if and only if $n \geq l$. It is equal to $\mu_l(P, n, \frac{l-1}{2})$ for $l \notin -\mathbb{N} \cup \{0\}$.

4.5 Differentiation of automorphic forms

We consider the action of $\mathbf{E}^+$ and $\mathbf{E}^-$ on automorphic forms.

4.5.1 *Discussion.* Let $f \in A_l(\chi, \mathbf{c}, s)$. Is $\mathbf{E}^\pm f$ an element of $A_{l\pm 2}(\chi, \mathbf{c}, s)$?

We apply the conditions in Definition 4.3.4 to $\mathbf{E}^\pm f$ (with l replaced by $l \pm 2$). Conditions i) and ii) are clearly satisfied. As each Fourier term of f is described by an integral over a compact set, we have $\mathbf{E}^\pm F_{P,n} f = F_{P,n} \mathbf{E}^\pm f$. Hence also $\mathbf{E}^\pm (f[\mathbf{c}, P]) = (\mathbf{E}^\pm f)[\mathbf{c}, P]$. So we have to show that for elements of $A_l(\chi, \mathbf{c}, s)$, the action of $\mathbf{E}^\pm$ preserves regularity. This will also imply the corresponding statement for cusp forms.

For unitary χ and real weight this may be proved from a local form of the convolution representation theorem of Harish Chandra, [18], Theorem 1 on p. 18. This implies that an eigenfunction f of ω, with some weight r, may be written as $f = f * \alpha$ with $\alpha \in C_c^\infty(\tilde{G})$, with $\mathrm{supp}(\alpha)$ contained in a small neighborhood of 1 in $\tilde{G}$. In *loc. cit.* this is proved for f defined on $\tilde{G}$, but one may adapt the proof to the situation in which f is defined on a suitable open subset. As $\mathbf{E}^\pm f = f * \mathbf{E}^\pm \alpha$, the growth of $\mathbf{E}^\pm f$ near a cusp is bounded by an average of the growth of f. In the case of $\chi \in \mathcal{X}_u$ and real weight, one may derive regularity of $f * \mathbf{E}^\pm \alpha$ from regularity of f; but for general χ and l this seems to fail. We derive the desired result with help of the next lemma and Lemma 4.3.7.

4.5.2 Lemma. *Let $P \in \mathcal{P}$ and suppose that a function f on $\mathrm{pr}^{-1} U_P(\sigma)$ as in Definition 4.1.9 is regular at P and is an eigenfunction of ω. If $P \in X^\infty$ we also suppose that $\tilde{F}_{P,n} f = 0$ for $n \in \mathcal{C}_P(\chi)$ with $\mathrm{Re}\, n = 0$, $n \neq 0$. Then $\mathbf{E}^+ f$ and $\mathbf{E}^- f$ are regular at P.*

Proof. For $P \in \mathcal{P}_Y$ this is clear from the definition of regularity. So consider $P \in X^\infty$. We have seen in Lemma 4.1.10 that all $\tilde{F}_{P,n} f$ with $n \in \mathcal{C}_P(\chi)$ are regular. For individual Fourier terms which may occur, we have seen that regularity is preserved under $\mathbf{E}^\pm$. As $\mathbf{E}^\pm f \in C^\infty(\mathrm{pr}^{-1} U_P(\sigma))$, it has a converging Fourier expansion on $\mathrm{pr}^{-1} U_P(\sigma)$. As $\tilde{F}_{P,n} \mathbf{E}^\pm f = \mathbf{E}^\pm \tilde{F}_{P,n} f$, this expansion consists of regular Fourier terms. So the proof may be completed by application of Lemma 4.3.7.

4.5.3 Proposition. *Differentiation gives maps*

$$\begin{aligned} \mathbf{E}^\pm &: A_l(\chi, \mathbf{c}, s) \longrightarrow A_{l\pm 2}(\chi, \mathbf{c}, s) \\ \mathbf{E}^\pm &: S_l(\chi, \mathbf{c}, s) \longrightarrow S_{l\pm 2}(\chi, \mathbf{c}, s). \end{aligned}$$

If $1 - 4s^2 \neq -l^2 \mp 2l$, these maps are invertible.

Remark. Holomorphic automorphic forms correspond to elements of the kernel of $\mathbf{E}^-$.

Proof. Lemma 4.5.2 implies that $\mathbf{E}^{\pm}$ preserves the regularity of $f[\mathbf{c}, P]$, hence the maps end up in the spaces indicated. On $A_l(\chi, \mathbf{c}, s)$ the map $\mathbf{E}^{\mp}\mathbf{E}^{\pm}$ acts as multiplication by $-(1 - 4s^2 + l^2 \pm 2l)$.

4.5.4 *Lie algebra modules.* One may form the $U(\mathfrak{g})$-modules $\bigoplus_l A_l(\chi, \mathbf{c}, s)$ and $\bigoplus_l S_l(\chi, \mathbf{c}, s)$, with l satisfying $\chi(k(\pi)) = e^{\pi i l}$. In the sequel we shall rarely take this representational point of view. Here we note that

$$\tilde{F}_{P,n} : \bigoplus_l A_l(\chi, \mathbf{c}, s) \longrightarrow \bigoplus_l W_l(P, n, s) \qquad \text{if} \quad n \in \mathbf{c}(P)$$

$$\tilde{F}_{P,n} : \bigoplus_l A_l(\chi, \mathbf{c}, s) \longrightarrow \bigoplus_l W_l^0(P, n, s) \qquad \text{if} \quad n \notin \mathbf{c}(P)$$

are $U(\mathfrak{g})$-linear. So if we have explicitly computed the Fourier coefficients at P of some automorphic form, we easily obtain the Fourier coefficients of the forms $(\mathbf{E}^{\pm})^n \cdot f$ in other weights.

4.5.5 *Holomorphic automorphic forms.* The meromorphic automorphic forms of weight l correspond to the kernel of $\mathbf{E}^- : A_l(\chi, \mathbf{c}, s) \to A_{l-2}(\chi, \mathbf{c}, s)$. Let f be an element of this kernel, and let $\hat{f}(z) = y^{-l/2} f(p(z))$. For $P \in X^\infty$ we have a Fourier expansion near $g_P \cdot i\infty$ of the form $(cz + d)^{-l} \hat{f}(g_P \cdot z) = \sum_{n \geq k} c_n e^{2\pi i n z}$, with $g_P = \widetilde{\begin{pmatrix} a & b \\ c & d \end{pmatrix}}$. At the end of the previous section we saw that regular terms can be non-zero only for $\mathrm{Re}\, n \geq 0$. The growth condition $\mathbf{c}$ allows a finite number of terms with $\mathrm{Re}\, n < 0$ to occur. So $\hat{f}$ is meromorphic at the cusp P.

For an interior point, $P \in \mathcal{P}_Y$, a similar reasoning shows that $\hat{f}$ is meromorphic at z_P.

4.6 Maass-Selberg relation

An important theorem concerning real analytic modular forms is the Maass-Selberg relation; see Maass, [35], Theorem 20 on p. 200. In Theorem 4.6.5 it is stated for automorphic forms with growth conditions. It gives a bilinear relation that the non-regular parts of the Fourier expansions of two automorphic forms have to satisfy, if their weight, eigenvalue, and character are related in a special way. The Maass-Selberg relation can be used to obtain an upper bound for the dimension of the spaces $A_l(\chi, \mathbf{c}, s)$ mod $S_l(\chi, \mathbf{c}, s)$.

4.6.1 *Functions on Y.* If $\mathbf{c}$ is a growth condition for χ, then $-\mathbf{c} : P \mapsto -\mathbf{c}(P)$ is one for χ^{-1}.

Consider $f \in A_l(\chi, \mathbf{c}, s)$ and $h \in A_{-l}(\chi^{-1}, -\mathbf{c}, s)$. Clearly the product fh is left-$\tilde{\Gamma}$-invariant and right-$\tilde{K}$-invariant; hence it determines an element of $C^\infty(Y_{\mathcal{P}})$.

In a similar (but a bit more complicated) way one may construct 1-forms on $Y_{\mathcal{P}}$.

4.6.2 Lemma. *Let $f, h \in C^\infty(\tilde G_{\mathcal P})$ satisfy*

$$\left.\begin{array}{rcl} f(\gamma g k(\theta)) &=& \chi(\gamma) f(g) e^{il\theta} \\ h(\gamma g k(\theta)) &=& \chi(\gamma)^{-1} f(g) e^{im\theta} \end{array}\right\} \quad \textit{for all } \gamma \in \tilde\Gamma,\ g \in \tilde G_{\mathcal P},\ \theta \in \mathbb{R}.$$

Put

$$\begin{array}{rcll} \{f,h\}^+(z) &=& f(p(z))h(p(z))\frac{dz}{y} & \textit{if } l+m=2, \\ \{f,h\}^-(z) &=& f(p(z))h(p(z))\frac{d\bar z}{y} & \textit{if } l+m=-2, \\ [f,h](z) &=& f(p(z))h(p(z))\frac{dx\wedge dy}{y^2} & \textit{if } l+m=0. \end{array}$$

Then $\{f,h\}^\pm$ are smooth 1-forms on $Y_{\mathcal P}$, and $[f,h]$ is a smooth 2-form.
If $l+m=\pm 2$, then

$$d\{f,h\}^\pm = -\frac{1}{2}[\mathbf{E}^\mp f, h] - \frac{1}{2}[f, \mathbf{E}^\mp h].$$

Proof. The product fh is a function on $\tilde\Gamma\backslash\tilde G_{\mathcal P}$ of weight $l+m$. It is easily checked that the corresponding differential forms on $\mathfrak{H} \setminus \overline{\mathrm{pr}}^{-1}\mathcal{P}$ are $\tilde\Gamma$-invariant. Use that for F of weight r

$$\begin{array}{rcl} (\mathbf{E}^+ F)(p(z)) &=& (4iy\partial_z + r)\, F(p(z)) \\ (\mathbf{E}^- F)(p(z)) &=& -\,(4iy\partial_{\bar z} + r)\, F(p(z)). \end{array}$$

4.6.3 *Definition.* Suppose that $f \in C^\infty(\tilde G_{\mathcal P})$ is χ-l-equivariant, and that $h \in C^\infty(\tilde G_{\mathcal P})$ is χ^{-1}-$(-l)$-equivariant. We put

$$\eta(f,h) = \{\mathbf{E}^+ f, h\}^+ - \{f, \mathbf{E}^- h\}^-.$$

This is a smooth 1-form on $Y_{\mathcal P}$. Lemma 4.6.2 shows that $d\eta(f,h) = 2[\omega f, h] - 2[f, \omega h]$.

4.6.4 *Closed differential form.* If we take $f \in A_l(\chi, \mathbf{c}, s)$, $h \in A_{-l}(\chi^{-1}, -\mathbf{c}, s)$, then $\eta(f,h)$ is a closed 1-form on $Y_{\mathcal P}$, asking to be integrated over cycles representing homology classes of $Y_{\mathcal P}$.

To obtain the Maass-Selberg relation we integrate it over a representative of the trivial homology class, the cycle consisting of curves $C(P)$ for each $P \in \mathcal{P}$, where $C(P)$ is a path inside $\dot U_P(A_P)$ encircling P once in positive direction. This gives

$$\sum_{P\in\mathcal P} \int_{C(P)} \eta(f,h) = 0.$$

The integrals $\int_{C(P)} \eta(f,h)$ may be expressed in terms of the Fourier expansions of f and h. We replace f, $\mathbf{E}^+ f$, h and $\mathbf{E}^- h$ by their Fourier expansions at P. On the compact path $C(P)$ we can take the integral inside the double sum.

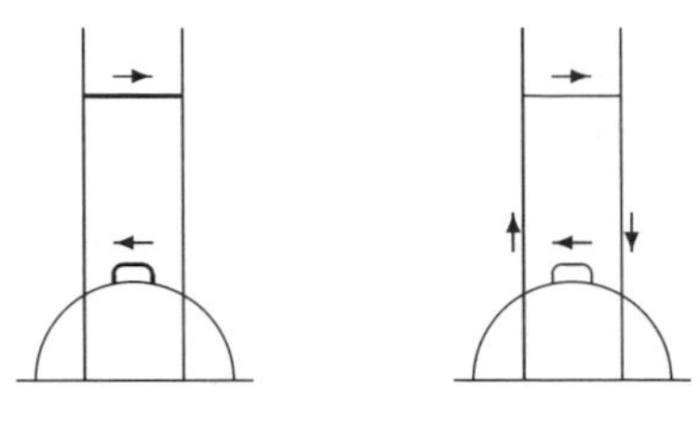

Consider first $P \in X^\infty$, and take $C(P)$ of the form $y = y_0$ in Iwasawa coordinates. As g_P is of the form $\widetilde{\begin{pmatrix} \alpha & \beta \\ \gamma & \delta \end{pmatrix}}$, we obtain

$$\begin{aligned}
&\{\mathbf{E}^+ f, h\}^+(g_P \cdot z) \\
&= \mathbf{E}^+ f(g_P p(z) k(\arg(\gamma z + \delta)) h(g_P p(z) k(\arg(\gamma z + \delta)) \frac{\gamma \bar{z} + \delta}{\gamma z + \delta} \frac{dz}{y} \\
&= \mathbf{E}^+ f(g_P p(z)) h(g_P p(z)) \frac{dz}{y},
\end{aligned}$$

and a similar result for $\{f, \mathbf{E}^- h\}^-(g_P \cdot z)$. This gives

$$\begin{aligned}
&\int_{C(P)} \eta(f, g) \\
&= \int_{iy_0}^{iy_0+1} \left((\mathbf{E}^+ f \cdot h)(g_P p(z)) \frac{dz}{y} - (f \cdot \mathbf{E}^- h)(g_P p(z)) \frac{d\bar{z}}{y} \right) \\
&= \sum_{n \in \mathcal{C}_P(\chi)} \frac{1}{y_0} \Big(F_{P,n} \mathbf{E}^+ f(y_0) F_{P,-n} h(y_0) \\
&\qquad - F_{P,n} f(y_0) F_{P,-n} \mathbf{E}^- h(y_0) \Big) \\
&= \sum_{n \in \mathcal{C}_P(\chi)} \frac{1}{y_0} \Big(\big(2y_0 (F_{P,n} f)'(y_0) \\
&\qquad + (l - 4\pi n y_0) F_{P,n} f(y_0) \big) F_{P,-n} h(y_0) \\
&\quad - F_{P,n} f(y_0) \left(2y_0 (F_{P,-n} h)'(y_0) - (-l - 4\pi(-n) y_0) F_{P,-n} h(y_0) \right) \Big) \\
&= \sum_{n \in \mathcal{C}_P(\chi)} 2 \mathrm{Wr}(F_{P,n} f, \iota F_{P,-n} h),
\end{aligned}$$

see 4.2.3 for the Wronskian Wr.

For $P \in \mathcal{P}_Y$ we take a path $[0, \pi/v_P] \to \mathfrak{H} : \eta \mapsto g_P k(\eta) \cdot it_0$, with $t_0 = t_{u_0}$ and $g_P = p(z_P)$.

$$\begin{aligned}
&\int_{C(P)} \eta(f,h) \\
&= \int_{\eta \bmod \pi/v_P} \Bigg((\mathbf{E}^+ f \cdot h)(g_P p(k(\eta) \cdot it_0)) \frac{d g_P k(\eta) \cdot it_0}{d\eta} \\
&\qquad - (f \cdot \mathbf{E}^- h)(g_P p(k(\eta) \cdot it_0)) \frac{\overline{d g_P k(\eta) \cdot it_0}}{d\eta} \Bigg) \frac{d\eta}{\operatorname{Im}(g_P k(\eta) \cdot it_0)} \\
&= \int_{\eta=0}^{\pi/v_P} \left((\mathbf{E}^+ f \cdot h)(g_P k(\eta) a(t_0)) - (f \cdot \mathbf{E}^- h)(g_P k(\eta) a(t_0)) \right) \\
&\qquad \frac{(1-t_0^2) \operatorname{Im} z_P}{|-it_0 \sin\eta + \cos\eta|^2} \frac{|-it_0 \sin\eta + \cos\eta|^2}{t_0 \operatorname{Im} z_P} d\eta \\
&= \sum_{n \in \mathcal{C}_P(\chi)} \Bigg((F_{P,n} \mathbf{E}^+ f)(u_0) F_{P,-n} h(u_0) \\
&\qquad - F_{P,n} f(u_0) (F_{P,-n} \mathbf{E}^- h)(u_0) \Bigg) \frac{\pi}{v_P} \frac{1-t_0^2}{t_0} \\
&= \frac{\pi}{v_P} \sum_{n \in \mathcal{C}_P(\chi)} \Bigg(2\sqrt{u_0^2+u_0} (F_{P,n} f)'(u_0) F_{P,-n} h(u_0) \\
&\qquad - F_{P,n} f(u_0) 2\sqrt{u_0^2+u_0} (F_{P,-n} h)'(u_0) \Bigg) (-4)\sqrt{u_0^2+u_0} \\
&= \frac{8\pi}{v_P} \sum_{n \in \mathcal{C}_P(\chi)} \operatorname{Wr}(F_{P,n} f, \iota F_{P,-n} h).
\end{aligned}$$

For all $n \in \mathcal{C}_P(\chi) \smallsetminus \mathbf{c}(P)$ the Fourier terms are regular. Hence the Wronskians vanish for these n. Thus we have obtained the following result:

4.6.5 Theorem. Maass-Selberg relation. *Put $u_P = 2$ for $P \in X^\infty$ and $u_P = 8\pi/v_P$ for $P \in \mathcal{P}_Y$. Let $f \in A_l(\chi, \mathbf{c}, s)$ and $h \in A_{-l}(\chi^{-1}, -\mathbf{c}, s)$. Then*

$$\sum_{P \in \mathcal{P}} u_P \sum_{n \in \mathbf{c}(P)} \operatorname{Wr}\left(\tilde{F}_{P,n} f, \iota \tilde{F}_{P,-n} h \right) = 0.$$

4.6.6 *Remarks.* This formula seems to depend on special choices (of g_P in the definition of the Fourier terms, and of the isomorphism ι). But in the proof we have seen that the result is completely natural.

Define $W_l(\mathbf{c}, s) = \bigoplus_P \bigoplus_n W_l(P, n, s)$ with $P \in \mathcal{P}$ and $n \in \mathbf{c}(P)$. The form

$$\operatorname{Wr} : W_l(\mathbf{c}, s) \times W_{-l}(-\mathbf{c}, s) \longrightarrow \mathbb{C},$$

defined by $\mathrm{Wr}(a,b) = \sum_P u_P \sum_n \mathrm{Wr}(a_{P,n}, \iota b_{P,-n})$, gives a non-degenerate bilinear pairing of the $2|\mathbf{c}|$-dimensional spaces $W_l(\mathbf{c},s)$ and $W_{-l}(-\mathbf{c},s)$.

Define $F(\mathbf{c}) : A_l(\chi,\mathbf{c},s) \to W_l(\mathbf{c},s)$ by $(F(\mathbf{c})f)_{P,n} = \tilde{F}_{P,n}f$. Then $\ker F(\mathbf{c}) = S_l(\chi,\mathbf{c},s)$.

The Maass-Selberg relation states that the spaces

$$F(\mathbf{c})A_l(\chi,\mathbf{c},s) \subset W_l(\mathbf{c},s) \text{ and } F(-\mathbf{c})A_{-l}(\chi^{-1},-\mathbf{c},s) \subset W_{-l}(-\mathbf{c},s)$$

are Wr-orthogonal. Hence

$$\dim\left(A_l(\chi,\mathbf{c},s)/S_l(\chi,\mathbf{c},s)\right) + \dim\left(A_{-l}(\chi^{-1},-\mathbf{c},s)/S_{-l}(\chi^{-1},-\mathbf{c},s)\right) \le 2|\mathbf{c}|.$$

4.6.7 *Unitary case.* For $\frac{1}{4} - s^2 \in \mathbb{R}$, a unitary character (and hence a real weight), we can consider

$$(f,h) \mapsto \sum_P u_P \sum_n \mathrm{Wr}(\tilde{F}_{P,n}f, \overline{\iota \tilde{F}_{P,n}h})$$

as a sesquilinear form on $A_l(\chi,\mathbf{c},s)$. This has the advantage of dealing with only one space of automorphic forms.

In the case $\chi^{-1} = \chi$, $l = 0$, we conclude

$$\dim\left(A_l(\chi,\mathbf{c},s)/S_l(\chi,\mathbf{c},s)\right) \le |\mathbf{c}|.$$

We shall prove that in many cases one has equality. In 14.3.3 we show equality in a very special case by explicitly exhibiting two automorphic forms.

Chapter 5
Poincaré series

In the previous chapter we defined automorphic forms of a rather general type. We have seen examples for the modular case in Chapter 1. In this chapter we discuss the sole general method to construct automorphic forms; it works for a unitary character χ and a spectral parameter s with $\operatorname{Re} s > \frac{1}{2}$. It is the construction of Poincaré series.

Eisenstein series, as discussed in 1.2.3, are a special case of Poincaré series. More general Poincaré series are mentioned in 1.5.9 and 1.7.2. The series representing a resolvent kernel, c.f. 1.7.3, also comes under the concept of Poincaré series discussed in this chapter.

The results in this chapter are known, or are easy extensions of known results. In Poincaré's paper [47] one finds the essential ideas; he works on the unit disk and considers holomorphic automorphic forms. Real analytic Poincaré series have been studied by, e.g., Niebur, [41], and Neunhöffer, [40].

Actually, these Poincaré series are not individual automorphic forms; they occur in families, depending continuously on the unitary character χ and the spectral parameter s. It is our aim to prove their meromorphic continuation in (χ, s). We shall attain this aim in Theorem 10.2.1.

We discuss Poincaré series in Section 5.1, and their Fourier terms in Section 5.2.

Throughout this chapter we fix a weight $l \in \mathbb{R}$, and a character $\chi \in \mathcal{X}_u$. (This is essential for the convergence of the Poincaré series.)

5.1 Construction of Poincaré series

The idea behind Poincaré series is simple: to obtain χ-l-equivariant functions, start with an arbitrary function f of weight l, and consider $g \mapsto \sum_{\gamma \in \tilde{\Gamma}/\tilde{Z}} f(\gamma g)$. With luck, this converges, and defines a χ-l-equivariant function. This idea is worked out in 5.1.1–5.1.4. It leads, in Proposition 5.1.6, to Poincaré series that are automorphic forms in the sense of the previous chapter.

5.1.1 *Construction of* $\tilde{\Gamma}$*-equivariant functions.* Suppose Δ is a subgroup of $\tilde{\Gamma}$ containing $\tilde{Z}$. Let $f : U \to \mathbb{C}$ with $U \subset \tilde{G}$, $\tilde{\Gamma} U \tilde{K} = U$ satisfy the condition

$$f(\delta g k(\theta)) = \chi(\delta) f(g) e^{il\theta} \quad \text{for all } \delta \in \Delta,\ g \in U,\ \theta \in \mathbb{R}.$$

This function f is χ-l-equivariant, but on the left for the group Δ only. Then

$$T_{\Delta}^{\tilde{\Gamma}} f(g) = \sum_{\gamma \in \Delta \backslash \tilde{\Gamma}} \chi(\gamma)^{-1} f(\gamma g)$$

defines $T_{\Delta}^{\tilde{\Gamma}} f = Tf : U \to \mathbb{C}$ satisfying

$$Tf(\gamma g k(\theta)) = \chi(\gamma) Tf(g) e^{il\theta} \quad \text{for all } \gamma \in \tilde{\Gamma},\ g \in U,\ \theta \in \mathbb{R},$$

provided the sum converges absolutely for each $g \in U$. We shall use $U = \tilde{G}$, and also $U = \tilde{G} \smallsetminus \left(\tilde{\Gamma} g_0 \tilde{K}\right)$ for some fixed g_0.

If, moreover, the absolute convergence is uniform for g in sets of the form $C\tilde{K}$ with $C \subset U$ compact, then

$$\left(\omega - \frac{1}{4} + s^2\right) f = 0 \implies \left(\omega - \frac{1}{4} + s^2\right) T_{\Delta}^{\tilde{\Gamma}} f = 0.$$

To see this, note first that $(\omega - \frac{1}{4} + s^2) f = 0$ implies that all terms $g \mapsto \chi(\gamma)^{-1} f(\gamma g)$ are eigenfunctions of ω. So we might try to differentiate inside the sum. But it is easier to consider the function $z \mapsto Tf(p(z))$ on the subset $\hat{U} = \overline{\mathrm{pr}}^{-1} U$ of $\mathfrak{H}$. We have to show that it is an eigenfunction of the elliptic differential operator $-y^2 \partial_y^2 - y^2 \partial_x^2 + ily\partial_x$. It suffices to prove this weakly, c.f. the discussion in 2.1.4. Integration against a test function in $C_c^\infty(\hat{U})$ can be taken inside the sum, as the absolute convergence is uniform on compact sets. The individual terms are eigenfunctions of ω, hence the sum represents an eigendistribution with the same eigenvalue, which has to be an analytic eigenfunction.

5.1.2 *Poincaré's series.* Take $\chi = 1$, $l \geq 4$ even, and let h be a holomorphic function on $\mathfrak{H}$. Put $f(p(z)k(\theta)) = y^{l/2} h(z) e^{il\theta}$, and apply the discussion above with $\Delta = \tilde{Z}$. We obtain

$$y^{-l/2} T_{\tilde{Z}}^{\tilde{\Gamma}} f(p(z)) = \sum_{\gamma} \frac{h(\gamma \cdot z)}{(cz+d)^l}$$

where $\gamma = \pm \begin{pmatrix} a & b \\ c & d \end{pmatrix}$ runs through the group of transformations of $\mathfrak{H}$ corresponding to $\tilde{Z} \backslash \tilde{\Gamma}$. Go over to the unit disk by $z = i\frac{1+w}{1-w}$, $w = \frac{z-i}{z+i}$, and take h of the form $h(z) = H(w)(1-w)^l$. Then

$$(1-w)^{-l} \sum_{\gamma} \frac{h(\gamma \cdot z)}{(cz+d)^l} = \sum_{V} H(V \cdot w) \left(\frac{dV \cdot w}{dw}\right)^{l/2}.$$

Here V runs through the corresponding group of transformations of the unit disk. Under some conditions on H, the right hand side is a *série thétafuchsienne*, as introduced by Poincaré; see [47], §1, p. 208.

5.1.3 *Choice of* Δ. We shall apply the construction in 5.1.1 with f given by $\mu_l(P, n, s)$ (see 4.2.5 and 4.2.6 for the definition), $P \in \mathcal{P}$ and

$$\Delta = \Delta_P = \begin{cases} \{\zeta_k \pi_P^m : k, m \in \mathbb{Z}\} & \text{if } P \in X^\infty \\ \tilde{Z} & \text{if } P \in \mathcal{P}_Y. \end{cases}$$

If $P \in \mathcal{P}_Y$ happens to be an elliptic orbit, we could have taken Δ_P a bit larger; the present choice will keep some formulas simpler.

5.1.4 Lemma.

i) Suppose the bounded function $h : \tilde{G} \to \mathbb{C}$ *satisfies*

$$h\left(k(\eta)a(t_u)k(\psi)\right) = \mathcal{O}\left(u^{-\sigma}\right) \quad (u \to \infty)$$

uniformly in η *and* ψ, *for some* $\sigma > 1$. *Then* $F(g) = \sum_{\gamma \in \tilde{Z}\backslash\tilde{\Gamma}} h(\gamma g)$ *converges absolutely, uniformly for* g *in compact sets, and* F *is a bounded function on* $\tilde{G}$.

ii) Let $P \in X^\infty$. *Suppose the left-*Δ_P*-invariant function* $h : \tilde{G} \to \mathbb{C}$ *satisfies*

$$h\left(g_P n(x)a(y)k(\theta)\right) = \mathcal{O}\left(y^\sigma\right) \quad (y \downarrow 0)$$

uniformly in x *and* θ, *for some* $\sigma > 1$. *Then* $F(g) = \sum_{\gamma \in \Delta_P\backslash\tilde{\Gamma}} h(\gamma g)$ *converges absolutely, uniformly for* g *in compact sets. For* $Q \in X^\infty$,

$$F\left(g_Q n(x)a(y)k(\theta)\right) = \delta_{P,Q} h\left(g_P n(x)a(y)k(\theta)\right) + \mathcal{O}(1) \quad (y \to \infty),$$

uniformly in x *and* θ.

Proof. These results are well known. The idea is to estimate the sum by an integral, see Poincaré's proof in §1 of [47], or Satz 1 and 2 of Petersson, [44], p. 38 and 40.

For i) we refer to Hejhal, [21], Ch. VI, Corollary 5.2 on p. 28. (Note that if $z = k(\eta)a(t_u) \cdot i$, then $1 - \left|\frac{z-i}{z+i}\right|^2 = \frac{1}{u+1}$, and apply the corollary to the group Γ.)

We show how to derive ii) from i). We may assume that $h(g_P n(x)a(y)k(\theta))$ does not depend on x and θ, and that $h \geq 0$. First we consider the case $h(g_P a(y)) = 0$ for $y > A_P$. Let

$$h_0(g_P p(z)k(\theta)) = \begin{cases} h(g_P a(y)) & \text{if} -\frac{1}{2} < x \leq \frac{1}{2} \\ 0 & \text{otherwise.} \end{cases}$$

Then $g \mapsto h_0(g_P g)$ satisfies the assumptions of i). This gives the absolute convergence, uniform for g in compact sets, of

$$\sum_{\gamma \in \tilde{Z}\backslash\tilde{\Gamma}} h_0(\gamma g) = \sum_{\gamma \in \Delta_P\backslash\tilde{\Gamma}} \sum_{\delta \in Z\backslash\Delta_P} h_0(\delta\gamma g) = \sum_{\gamma \in \Delta_P\backslash\tilde{\Gamma}} h(\gamma g) \leq C$$

for some constant C.

For general h, write $h = h_1 + h_2$ with $h_2(g_P a(y)) = \tau(y) h(g_P a(y))$ and τ the characteristic function of (A_P, ∞). For g in a fixed compact set, there are only finitely many $\gamma \in \Delta_P \backslash \tilde{\Gamma}$ such that $h_2(\gamma g) \neq 0$. Hence the absolute convergence as stated in the lemma is no problem.

Let $Q \in X^\infty$. Consider $F_2(g) = \sum_{\gamma \in \Delta_P \backslash \tilde{\Gamma}} h_2(\gamma g)$ as $y \to \infty$, where $g = g_Q p(z) k(\theta)$. Let $y > A_Q$, then $\mathrm{pr}(g) \in U_Q(A_Q)$. On the other hand, if $\gamma g \in \mathrm{supp}(h_2)$, then $\mathrm{pr}(g) \in U_P(A_P)$. Hence $F_2(g) = 0$ if $Q \neq P$. If $P = Q$, then $\gamma g \in \mathrm{supp}(h_2)$ only if $\gamma \in \Delta_P$, and hence $F_2 = h_2$. So we obtain uniformly for $g = g_P p(z) k(\theta)$ with $y > A_Q$

$$F(g) = \sum_{\gamma \in \Delta_P \backslash \tilde{\Gamma}} h_1(\gamma g) + \delta_{P,Q} h_2(g) = \mathcal{O}(1) + \delta_{P,Q} h(g).$$

5.1.5 *Notation.* Put $\mathbb{C}_{1/2} = \{ s \in \mathbb{C} : \mathrm{Re}\, s > \frac{1}{2} \}$.

We are ready to state the convergence of Poincaré series. We use $\mu_l(P, n, s)$ (see 4.2.5 and 4.2.6) as the function h in the discussion above. Among the elements of $W_l(P, n, s)$ this function is characterized by its asymptotic behavior far away from P, and that behavior controls the convergence of the Poincaré series.

5.1.6 Proposition. *Let $\chi \in \mathcal{X}_u$, $l \in \mathbb{R}$, $s \in \mathbb{C}$, $\mathrm{Re}\, s > \frac{1}{2}$, $P \in \mathcal{P}$, and $n \in \mathcal{C}_P(\chi)$. Then*

$$P_l(P, n, \chi, s; g) = \sum_{\gamma \in \Delta_P \backslash \tilde{\Gamma}} \chi(\gamma)^{-1} \mu_l(P, n, s; \gamma g)$$

converges absolutely, uniformly for (χ, s, g) in compact subsets of $\mathcal{X}_u \times \mathbb{C}_{1/2} \times \left(\tilde{G} \smallsetminus \mathrm{pr}^{-1}\{P\}\right)$. It defines the automorphic form $P_l(P, n, \chi, s) \in A_l(\chi, \mathbf{c}, s)$ for each growth condition $\mathbf{c}$ for χ that satisfies $n \in \mathbf{c}(P)$.

Remarks. If $P \in X^\infty$, the automorphic form $P_l(P, n, \chi, s)$ is in $C^\infty(\tilde{G})$, whereas it has a singularity along $\mathrm{pr}^{-1}\{P\}$ if $P \in \mathcal{P}_Y$.

If $P \in X^\infty$, and $n = 0 \in \mathcal{C}_P(\chi)$, then $\mu_l(P, 0, s; g_P p(z) k(\theta)) = y^{s+1/2} e^{il\theta}$. In this case $P_l(P, 0, \chi, s)$ is called an *Eisenstein series.* The condition $0 \in \mathcal{C}_P(\chi)$ amounts to $\chi(\pi_P) = 1$. If this holds, then the cusp P is called *singular* for χ; if $\chi_P(\pi_P) \neq 0$, then P is called a *regular cusp.*

If we vary χ in $\mathcal{X}_u$, then l and n have to vary as well. We tacitly assume that the set over which χ varies is small enough to admit n and l as continuous functions of χ satisfying $e^{2\pi i n} = \chi(\pi_P)$ or $e^{i\pi n / v_P} = \chi(\varepsilon_P)$, and $e^{\pi i l} = \chi(k(\pi))$.

From the uniform convergence for (χ, s, g) in compact sets, we conclude that $P_l(P, n, \chi, s; g)$ is continuous in (χ, s, g).

Proof. First we consider $P \in X^\infty$. The function $\mu_l(P, n, s)$ satisfies the estimate in ii) of Lemma 5.1.4, uniformly for s and χ in compact sets. So we have convergence of the series on $\tilde{G}$, uniformly for (χ, s, g) in compact sets. Further $(\omega - \frac{1}{4} + s^2)P_l(P, n, \chi, s) = 0$, as we have discussed in 5.1.1. So $P_l(P, n, \chi, s)$ is analytic on $\tilde{G}$, hence regular at each $Q \in \mathcal{P}_Y$. The boundedness in the lemma gives regularity at $Q \in X^\infty$, $Q \neq P$. At P itself, $P_l(P, n, \chi, s)[\mathbf{c}, P]$ is regular for each growth condition $\mathbf{c}$ satisfying $n \in \mathbf{c}(P)$.

For $P \in \mathcal{P}_Y$ the function $\mu_l(P, n, s)$ may be written as $H_1 + H_2$, with $H_2(g_P k(\eta) a(t_u) k(\psi)) = 0$ for all $u \geq A_P$, and H_1 bounded. To H_1 we apply part i) of the lemma to obtain absolute convergence of $P_1(g) = \sum_{\gamma \in \tilde{Z} \backslash \tilde{\Gamma}} \chi(\gamma)^{-1} H_1(\gamma g)$ uniform for (χ, s, g) in compact sets. The resulting estimate of P_1 is uniform for (χ, s) in compact sets. Consider $H_2(\gamma g)$; it can only be non-zero for γ in a finite number of classes of $\tilde{Z} \backslash \tilde{\Gamma}$ (more than one class only if z_P is an elliptic point). This implies the desired convergence of the whole series. On $\tilde{G} \setminus \mathrm{pr}^{-1}\{P\}$ the sum again satisfies $(\omega - \frac{1}{4} + s^2)P_l(P, n, \chi, s) = 0$. This gives regularity at $Q \in \mathcal{P}_Y$, $Q \neq P$. Moreover, the sum is bounded outside $\mathrm{pr}^{-1} U_P(A_P)$. Hence we obtain regularity at $Q \in X^\infty$. Near P itself

$$\begin{aligned} &P_l(P, n, \chi, s; g) \\ &\quad = \sum_{\gamma \in \tilde{Z} \backslash \tilde{\Gamma},\, \gamma \cdot z_P \neq z_P} \chi(\gamma)^{-1} \mu_l(P, n, s; \gamma g) + \sum_{k=1}^{v_P} \chi(\varepsilon_P)^{-k} \mu_l(P, n, s; \varepsilon_P^k g). \end{aligned}$$

The first sum gives a C^∞-function near $P(z_P)\tilde{K}$; the second sum contributes $v_P \mu_l(P, n, s; g)$. So removal of the Fourier term $\tilde{F}_{P,n} P_l(P, n, \chi, s)$ makes the sum regular at P.

5.1.7 *Comments.* We follow the practice of using the name *Poincaré series* for this general type of series. This type includes Eisenstein series and resolvent kernels.

Neunhöffer [40] and Good [16] have studied these Poincaré series in weight 0. Good defined another type of Poincaré series as well; these are not associated to a point of X, but to a closed geodesic.

Our definition of the Poincaré series $P_l(P, n, \chi, s)$ depends on the choice of g_P. This is not visible in the notation.

5.1.8 *Resolvent kernel.* Let $P \in \mathcal{P}_Y$. We check that the Poincaré series $P_l(P, l, \chi, s)$ is a multiple of the resolvent kernel used in the proof of the trace formula of Selberg, see Hejhal, [21], p. 350. We put $w = z_P = g_P \cdot i$ and compute $G(z) = P_l(P, l, \chi, s; p(z)) = \frac{1}{2} \sum_{\gamma \in \Gamma} \chi(\tilde{\gamma})^{-1} \mu_l(P, l, s; \tilde{\gamma} p(z))$. We shall show that

$$G(z) = \frac{-4\pi \Gamma(2s+1)}{\Gamma(\frac{1}{2} + s + \frac{1}{2}l)\Gamma(\frac{1}{2} + s - \frac{1}{2}l)} G_{s+1/2}(z, w; v),$$

with G_s as in (5.3) of p. 350 of *loc. cit.* and v the multiplier system of Γ given by $v(\gamma) = \chi(\tilde{\gamma})$. We assume that $\Gamma(\frac{1}{2} + s \pm \frac{1}{2}l) \neq 0$, and, as before, that $\mathrm{Re}\, s > \frac{1}{2}$.

Put

$$\tilde{k}_{s+1/2}(z,w) = \left(1 - \left|\frac{z-w}{z-\bar{w}}\right|^g 2\right)^{s+1/2} e^{-il\arg(i\bar{w}-iz)}$$
$$\cdot {}_2F_1\left[\begin{matrix}\frac{1}{2}+s+\frac{1}{2}l,\ \frac{1}{2}+s-\frac{1}{2}l \\ 1+2s\end{matrix}\ \middle|\ 1-\left|\frac{z-w}{z-\bar{w}}\right|^2\right].$$

Up to the factors mentioned above, this is the function $k_{.}(.,.)$ in (5.3) on p. 350 of *loc. cit.* We want to show that

$$\sum_{\gamma\in\Gamma} \chi(\tilde{\gamma})^{-1}\mu_l(P,l,s;\tilde{\gamma}p(z)) = \sum_{\gamma\in\Gamma} v(\gamma)\tilde{k}_{s+1/2}(z,\gamma w)e^{il\arg(cw+d)}.$$

We may replace $\tilde{\gamma} = \widetilde{\begin{pmatrix} a & b \\ c & d\end{pmatrix}}$ by $\tilde{\gamma}^{-1}$ in the sum in the left hand side. As $\tilde{\gamma}^{-1}p(z) = p\left(\frac{dz-b}{-cz+a}\right)k\left(\arg\left(\frac{1}{a-cz}\right)\right)$ we have to show that

$$\mu_l(P,l,s;p(\gamma^{-1}z))e^{il\arg(1/(a-cz))} = \tilde{k}_{s+1/2}(z,\gamma w)e^{il\arg(cw+d))}.$$

The equality $g_P = p(w)$ implies the relation

$$\begin{aligned}
&\mu_l(P,l,s;p(z)) \\
&= \mu_l\left(P,l,s;g_P p\left(\frac{z-\operatorname{Re} w}{4y\operatorname{Im} w}\right)\right) \\
&= \exp\left(\frac{1}{2}il\left(\arg\left(\frac{z-w}{z-\bar{w}}\right) - \arg\left(\frac{\bar{w}-z}{z-w}\right)\right)\right)\mu_l\left(P,l,s;g_P a(t_u)\right),
\end{aligned}$$

with $u = \dfrac{|z-w|^2}{4y\operatorname{Im} w}$. As

$$\left(\frac{|z-w|^2}{4y\operatorname{Im} w}+1\right)^{-1} = \frac{4y\operatorname{Im} w}{|z-w|^2} = 1 - \left|\frac{z-w}{z-\bar{w}}\right|^2,$$

we obtain $\mu_l(P,l,s;p(z)) = \tilde{k}_{s+1/2}(z,w)$. We still have to check that

$$\tilde{k}_{s+1/2}(\gamma^{-1}z,w)e^{il\arg(1/(a-cz))} = \tilde{k}_{s+1/2}(z,\gamma w)e^{il\arg(cw+d)}.$$

A computation shows that $\left|\frac{\gamma^{-1}z-w}{\gamma^{-1}z-\bar{w}}\right| = \left|\frac{z-\gamma w}{z-\overline{\gamma w}}\right|$, and careful consideration of the arguments gives

$$\begin{aligned}
&-\arg(i\bar{w}-i\gamma^{-1}z) + \arg(i\overline{\gamma w}-iz) \\
&= \arg\left(\frac{\overline{\gamma w}-z}{\bar{w}-\gamma^{-1}z}\right) = \arg\left(\frac{a-cz}{c\bar{w}+d}\right) = -\arg\left(\frac{1}{a-cz}\right) + \arg(cw+d).
\end{aligned}$$

This completes the computation.

Of course, both w and z are variables in *loc. cit.*, whereas in our context $w = g_P \cdot i$ is fixed. Nevertheless, Theorem 10.2.1 will give the meromorphic continuation of resolvent kernels in multiplier system and spectral parameter.

The factor $-4\pi\Gamma(2s+1)\Gamma(\frac{1}{2}+s+\frac{1}{2}l)^{-1}\Gamma(\frac{1}{2}+s-\frac{1}{2}l)^{-1}$ arises from a difference in the normalization. A resolvent kernel should have the right logarithmic singularity along the diagonal to represent the resolvent of the differential operator corresponding to ω. We have normalized $\mu_l(P,l,s)$ by imposing a simple behavior as $u \to \infty$. Our choice has the advantage that $\mu_l(P,l,s)$ has no singularities in $\mathrm{Re}\, s > \frac{1}{2}$.

5.1.9 *Examples.* The Eisenstein series in 1.2.3 is an example of a Poincaré series. In 1.5.9 we have seen another Poincaré series for the modular group. Section 15.4 discusses Poincaré series for another cofinite discrete group.

5.1.10 *Dimension.* Let $\mathbf{c}$ be a growth condition for χ. The $P_l(P,n,\chi,s)$ with $P \in \mathcal{P}$, $n \in \mathbf{c}(P)$, correspond to linearly independent elements of $A_l(\chi,\mathbf{c},s)/S_l(\chi,\mathbf{c},s)$ provided $\mu_l(P,n,s) \notin W_l^0(P,n,s)$. The exception may occur if $2sv_l(P,n,-s) = 0$ and $n \neq 0$; see 4.2.8 and 4.2.9.

So the quotient spaces $A_l(\chi,\mathbf{c},s)/S_l(\chi,\mathbf{c},s)$ have at least dimension $|\mathbf{c}|$ for general s with $\mathrm{Re}\, s > \frac{1}{2}$, and $\chi \in \mathcal{X}_u$. From the Maass-Selberg relation, see Theorem 4.6.5, we conclude that the dimension is exactly equal to $|\mathbf{c}|$.

5.1.11 *Differentiation.* The results in Section 4.4 imply the equality

$$\mathbf{E}^{\pm} P_l(P,n,\chi,s) = \theta(1+2s \pm l) P_{l\pm 2}(P,n,\chi,s),$$

with $\theta = 1$ if $P \in X^\infty$ and $\theta = -1$ otherwise, provided we justify differentiation under the sum defining the Poincaré series. This is no problem, as it may be proved weakly; see the discussion in 5.1.1.

5.1.12 *Representational point of view.* Let $H(s,r)$ be the abstract space $\bigoplus_l \mathbb{C} \cdot \varphi_l$, with l running through the solutions of $e^{\pi i l} = r$. The φ_l are not functions on $\tilde{G}$, but just symbols describing an algebraic basis of $H(r,s)$. We make $H(r,s)$ into a module for the Lie algebra $\mathfrak{g}$ by defining

$$\mathbf{W}\varphi_l = il\varphi_l, \quad \mathbf{E}^{\pm}\varphi_l = (1+2s \pm l)\varphi_{l\pm 2}.$$

Let $\mathrm{Re}\, s > \frac{1}{2}$, $P \in \mathcal{P}$, $n \in \mathcal{C}_P$. Take $\alpha = 0$ if $P \in X^\infty$, and $\alpha = 1$ otherwise. The differentiation result in 5.1.11 means that $\varphi_l \mapsto e^{\alpha\pi i l/2} P_l(P,n,\chi,s)$ may be extended linearly to give a Lie algebra morphism

$$P(P,n,\chi,s) : H(s,\chi(k(\pi))) \to \bigoplus_l A_l(\chi,\mathbf{c},s),$$

with $\bigoplus_l A_l(\chi,\mathbf{c},s)$ as discussed in 4.5.4, and $\mathbf{c}$ a growth condition for χ such that $n \in \mathbf{c}(P)$.

The $\mathfrak{g}$-module $H(s,\chi(k(\pi)))$ is contained in larger spaces, on which not only $\mathfrak{g}$ acts, but the group $\tilde{G}$ as well. One may extend $P(P,n,\chi,s)$ to some of these larger spaces. This is discussed by Miatello and Wallach, [39]. They prove the meromorphic continuation in s of the extended $P(P,n,\chi,s)$ for $P \in X^\infty$. They employ another method (spectral decomposition) than we shall use in this book.

We have considered $H(r,s)$ as an abstract space. One may realize it as a space of functions on $\tilde{G}$: identify φ_l with the function $p(z)k(\theta) \mapsto y^{s+1/2}e^{il\theta}$, and extend this linearly. One can obtain this space as an induced representation.

5.2 Fourier coefficients

The continuity in (χ,s) of Poincaré series implies continuity of the coefficients that express the Fourier terms of $P_l(P,n,\chi,s)$ in the basis elements $\mu_l(Q,m,s)$ and $\omega_l(Q,m,s)$ discussed in Section 4.2. This is the content of Proposition 5.2.1.

The major part of this section is devoted to some computations of Fourier coefficients of Poincaré series. We shall refer to these computations in the examples in Part II. In Part I the explicit description of the Fourier coefficients is irrelevant.

5.2.1 Proposition. *Let $P,Q \in \mathcal{P}$, $n \in \mathcal{C}_P(\chi)$, $m \in \mathcal{C}_Q(\chi)$. There exists a continuous function $c_l(Q,m;P,n) : \mathcal{X}_u \times \mathbb{C}_{1/2} \longrightarrow \mathbb{C}$ such that*

$$\begin{aligned} \tilde{F}_{Q,m}P_l(P,n,\chi,s) &= w_P\delta_{P,Q}\delta_{n,m}\mu_l(P,n,s) \\ &+ \quad c_l(Q,m;P,n;\chi,s)\begin{cases} \mu_l(Q,0,-s) & \text{if } Q \in X^\infty \text{ and } m=0 \\ \omega_l(Q,m,s) & \text{otherwise,} \end{cases} \end{aligned}$$

with $w_P = 1$ if $P \in X^\infty$, $w_P = v_P$ if $P \in \mathcal{P}_Y$.

Proof. In the proof of Proposition 5.1.6 we saw that $P_l(P,n,\chi,s)$ is regular at Q if $Q \neq P$. In the case $Q = P$ we have regularity if we omit the term with $\gamma \in \Delta_P$, respectively the terms with $\gamma \cdot z_P = z_P$. So Proposition 4.2.11 implies that $\tilde{F}_{Q,m}P_l(P,n,\chi,s)$ has the form indicated in the proposition. The continuity of $P_l(P,n,\chi,s)$ in (χ,s,g) gives the continuity of $(\chi,s) \mapsto c_l(Q,m;P,n;\chi,s)$, as one sees by taking $g \in \mathrm{pr}^{-1}U_Q(A_Q)$ such that $\mu_l(Q,0,-s)$, respectively $\omega_l(Q,m,s)$ is not zero for (χ,s) in a small neighborhood of some (χ_0,s_0).

5.2.2 *Computations.* One can describe the Fourier coefficients c_l more explicitly. As the series defining the Poincaré series converges absolutely on compact sets, we obtain for $Q \in \mathcal{P}_Y$ and $0 < u < \tilde{A}_Q$

$$\begin{aligned} &c_l(Q,m;P,n;\chi,s)\omega_l\left(Q,m,s;g_Q a(t_u)\right) \\ &\quad= \sum_{\gamma\in\tilde{Z}\backslash\tilde{\Gamma}}{}' \chi(\gamma)^{-1}\int_0^\pi e^{-im\eta}\mu_l(P,n,s;\gamma g_Q k(\eta)a(t_u))\frac{d\eta}{\pi}, \end{aligned}$$

and for $Q \in X^\infty$ and $y > \tilde{A}_Q$

$$c_l(Q,m;P,n;\chi,s)\,(\omega_l\,(Q,m,s;g_Q a(y)) \text{ or } \mu_l\,(Q,0,-s;g_Q a(y))) = \sum_{\gamma\in\Delta_P\backslash\tilde{\Gamma}/\Delta_Q}{}' \chi(\gamma)^{-1}\int_{-\infty}^{\infty}\mu_l(P,n,s;\gamma g_Q n(x)a(y))\,dx.$$

The prime means that we restrict the summation to $\gamma \notin \Delta_P$, respectively $\gamma \cdot z_P \neq z_P$, in the case $Q = P$.

In §7 of [16], Good gives a unified treatment, for $\chi = 1$, and $l = 0$. He gives integral representations of basis elements of $W_0(P,n,s)$ with more or less the same structure for the three cases $P \in X^\infty$, $P \in \mathcal{P}_Y$, and the case coming from closed geodesics (not treated in this book). He derives all properties he needs from those integral representations, whereas we are content to take the functions in $W_l(P,n,s)$, and their properties, from books on special functions.

Good remarks that $c_0(Q,m;P,n;\chi,s)$ has the following structure:

$$(\text{gamma factors and exponentials}) \sum_c (\text{generalized Kloosterman sum for } c) \cdot J(c),$$

where the sum is over a discrete set of positive numbers depending on $\tilde{\Gamma}$, and on P and Q. The function J may be just $c \mapsto c^{-1-2s}$, but is in general more complicated: for instance a Bessel function. See Lemma 9 on p. 61 of [16].

5.2.3 *Four combinations.* One can work this out for four cases: P and Q may each be in X^∞ and in $\mathcal{P}_Y$.

In the mixed case, Hejhal shows in [21], Ch. VI, p. 39–41 that the Fourier coefficients at $Q \in X^\infty$ of a Poincaré series at $P \in \mathcal{P}_Y$ are values at P of Poincaré series for the cusp Q.

For the case P, $Q \in \mathcal{P}_Y$ the Fourier coefficients are again Poincaré series, in other weights, obtained by the Maass operators $\mathbf{E}^+$ and $\mathbf{E}^-$. This is not surprising. The Fourier series expansion at an interior point may be viewed as a generalized power series expansion. Hence the derivatives of the function expanded should turn up in the Fourier coefficients.

5.2.4 *The case $P, Q \in X^\infty$* is considered in the remaining part of this chapter. These computations are used nowhere in Part I. Nevertheless, I have included them, as they are used in applications of this theory.

Let $n \in \mathcal{C}_P(\chi)$, $m \in \mathcal{C}_Q(\chi)$. Take $y > A_Q$, put $\eta(y) = \mu_l(Q,0,-s;g_Q a(y))$ if $m = 0$, and $\eta(y) = \omega_l(Q,m,s;g_Q a(y))$ otherwise. The proof of Proposition 5.1.6 implies that

$$c_l(Q,m;P,n;\chi,s)\eta(y) = \sum_\gamma{}' \chi(\gamma)^{-1}\int_0^1 e^{-2\pi i m x}\mu_l(P,n,s;\gamma g_Q n(x)a(y))\,dx,$$

where $\sum_\gamma'$ denotes the sum over $\gamma \in \Delta_P\backslash\tilde{\Gamma}$, $\gamma \notin \Delta_P$ if $P = Q$. So $g_P^{-1}\gamma g_Q \cdot i\infty \neq i\infty$ for all γ that occur. This means that $\widetilde{g_P^{-1}\gamma g_Q}$ is of the form $\begin{pmatrix} * & * \\ \neq 0 & * \end{pmatrix}$, or in other words, $g_P^{-1}\gamma g_Q$ is an element of the big cell $\tilde{N}\tilde{A}\tilde{Z}w\tilde{N}$ of the Bruhat decomposition $\tilde{G} = \tilde{N}\tilde{A}\tilde{Z} \sqcup \tilde{N}\tilde{A}\tilde{Z}w\tilde{N}$ (disjoint union). One can check that for these γ the cosets $\Delta_P\gamma \cdot g_Q n(k) g_Q^{-1}$, for varying $k \in \mathbb{Z}$, are all different. Hence

$$
\begin{aligned}
&c_l(Q,m;P,n;\chi,s)\eta(y) \\
&= \sum_{\gamma\in\Delta_P\backslash\tilde{\Gamma}/\Delta_Q}' \chi(\gamma)^{-1} \sum_{\delta\in\tilde{Z}\backslash\Delta_Q} \chi(\delta)^{-1} \int_0^1 e^{-2\pi i m x} \mu_l(P,n,s;\gamma\delta g_Q n(x)a(y))\,dx \\
&= \sum_{\gamma\in\Delta_P\backslash\tilde{\Gamma}/\Delta_Q}' \chi(\gamma)^{-1} \int_{-\infty}^{\infty} e^{-2\pi i m x} \mu_l(P,n,s;\gamma g_Q n(x)a(y))\,dx,
\end{aligned}
$$

where in the case $P = Q$ the double coset Δ_P is omitted.

For each term there is some freedom in the choice of γ. If we write $g_P^{-1}\gamma g_Q = \widetilde{\begin{pmatrix} a & b \\ c & d \end{pmatrix}} k(\pi\nu)$ with $\begin{pmatrix} a & b \\ c & d \end{pmatrix} \in \mathrm{SL}_2(\mathbb{R})$, $c > 0$, $\nu \in \mathbb{Z}$, we may arbitrarily choose ν and replace a by $a + pc$ and d by $d + qc$ with $p, q \in \mathbb{Z}$. Let us take $\nu = 0$, and for instance $-a$, $d \in (-\frac{1}{2}c, \frac{1}{2}c]$. This gives

$$
\begin{aligned}
&c_l(Q,m;P,n;\chi,s)\eta(y) \\
&= \sum_{\gamma\in\Delta_P\backslash\tilde{\Gamma}/\Delta_Q}' \chi(\gamma)^{-1} \\
&\qquad \cdot \int_{-\infty}^{\infty} e^{-2\pi i m(x-d/c)} \mu_l\left(P,n,s;g_P \widetilde{\begin{pmatrix} a & b \\ c & d \end{pmatrix}} p\left(x - \frac{d}{c} + iy\right)\right) dx \\
&= \sum_\gamma' \chi(\gamma)^{-1} e^{2\pi i m d/c} \\
&\qquad \cdot \int_{-\infty}^{\infty} e^{-2\pi i m x} \mu_l\left(P,n,s;g_P p\left(\frac{a}{c} - \frac{1}{c(cx+ciy)}\right)\right) e^{-il\arg(x+iy)}\,dx \\
&= \sum_\gamma' e^{2\pi i(md+na)/c} \\
&\qquad \cdot \int_{-\infty}^{\infty} e^{-2\pi i m x} \mu_l(P,n,s;g_P p(-1/c^2(x+iy)) e^{-il\arg(x+iy)}\,dx.
\end{aligned}
$$

5.2.5 *Definition.* This leads to the definition of the *generalized Kloosterman sum*

$$
\mathrm{S}_\chi(Q,m;P,n;c) = \sum_{d \bmod c} \chi(\gamma_{(c,d)})^{-1} e^{2\pi i(na+md)/c}, \tag{5.1}
$$

where $c > 0$, and $d \bmod c$ occur in $\gamma_{(c,d)} = g_P \widetilde{\begin{pmatrix} a & b \\ c & d \end{pmatrix}} g_Q^{-1} \in \tilde{\Gamma}$.

In 13.2.6 we see that $\mathrm{S}_1(\Gamma_{\mathrm{mod}} \cdot i\infty, m; \Gamma_{\mathrm{mod}} \cdot i\infty, n; c)$ is the usual Kloosterman sum.

5.2.6 *General form of Fourier coefficient.* We have obtained

$$c_l(Q, m; P, n; \chi, s)\eta(y) = \sum_c \mathrm{S}_\chi(Q, m; P, n; c)$$
$$\cdot \int_{-\infty}^{\infty} e^{-2\pi i m x} \mu_l(P, n, s; g_P p(-1/c^2(x+iy))) e^{-il \arg(x+iy)}\, dx,$$

where c runs over the positive numbers occurring in $g_P \widetilde{\begin{pmatrix} a & b \\ c & d \end{pmatrix}} g_Q^{-1} \in \tilde{\Gamma}$. We are left with the computation of the integral. We want to write the Fourier coefficient in the form

$$c_l(Q, m; P, n; \chi, s) = G_l(Q, m; P, n; s) \sum_c \mathrm{S}_\chi(Q, m; P, n; c) J(Q, m; P, n; s; c),$$

where J does not depend on l, and where G_l consists of gamma-factors and exponentials and does not depend on c.

Take $\varepsilon \in \{1, -1\}$, and take $x = -\varepsilon y \xi$ in the integral. This gives

$$\begin{aligned} &G_l(Q, m; P, n; s) J(Q, m; P, n; s; c)\eta(y) \\ &= y e^{-\pi i l/2} \int_{-\infty}^{\infty} e^{2\pi i \varepsilon m y \xi} \mu_l(P, n, s; g_P p(-1/c^2 y(-\varepsilon\xi + i)) e^{-i\varepsilon l \arctan \xi}\, d\xi. \end{aligned}$$

5.2.7 *Case* $n = 0$. We put $J(Q, m; P, n; s; c) = c^{-1-2s}$. We obtain, with $|m| = -\varepsilon m$,

$$\begin{aligned} &G_l(Q, m; P, 0; s)\eta(y) \\ &= e^{-\pi i l/2} y^{-s+1/2} \int_{-\infty}^{\infty} e^{-2\pi i |m| y \xi} (1+\xi^2)^{-s-1/2} e^{-il\varepsilon \arctan \xi} d\xi. \end{aligned}$$

For $m = n = 0$ the first and the second formula on p. 9 of [36] imply

$$G_l(Q, 0; P, 0; s) = \frac{2^{1-2s} \pi \Gamma(2s) e^{-\pi i l/2}}{\Gamma(\frac{1}{2} + s + \frac{1}{2}l)\Gamma(\frac{1}{2} + s - \frac{1}{2}l)}.$$

For $m \neq 0$ use Corollary 6.7.5 in [4] to obtain

$$G_l(Q, m; P, 0; s) = \frac{\pi^{s+1/2} e^{-il\pi/2}}{\Gamma(\frac{1}{2} + s + \frac{1}{2}l \operatorname{sign}(m))} |m|^{s-1/2}.$$

5.2.8 *Case $n \neq 0$.* Take $\varepsilon = \text{sign}(n)$. If we apply Lemma 5.5 on p. 357 of [21], we find

$$\frac{1}{\Gamma(1+2s)} \int_{-\infty}^{\infty} e^{2\pi i |n| y^{-1} c^{-2} \xi (1+\xi^2)^{-1}} (4\pi|n|)^{s+1/2}$$
$$\cdot \mu_l \left(P, n, s; g_P a \left(\frac{1}{yc^2(1+\xi^2)} \right) \right) e^{-i\varepsilon l \arg(1+i\xi)} e^{2\pi i \varepsilon m y \xi} \, d\xi$$
$$= \begin{cases} \dfrac{2\pi}{\Gamma(\frac{1}{2}+s+\frac{1}{2}\varepsilon l)} W_{\frac{1}{2}\varepsilon l, s}(4\pi|m|y) y^{-1} c^{-1} \\ \qquad \cdot \left|\frac{n}{m}\right|^{1/2} J_{2s}\left(4\pi \frac{\sqrt{mn}}{c}\right) & \text{if } mn > 0 \\ \dfrac{2\pi^{s+3/2}}{s\Gamma(\frac{1}{2}+s-\frac{1}{2}l)\Gamma(\frac{1}{2}+s+\frac{1}{2}l)} y^{-s-1/2} c^{-1-2s} |n|^{s+1/2} \\ & \text{if } m = 0 \\ \dfrac{2\pi}{\Gamma(\frac{1}{2}+s-\frac{1}{2}\varepsilon l)} W_{-\frac{1}{2}\varepsilon l, s}(4\pi|m|y) y^{-1} c^{-1} \\ \qquad \cdot \left|\frac{n}{m}\right|^{1/2} I_{2s}\left(4\pi \frac{\sqrt{|mn|}}{c}\right) & \text{if } mn < 0 \end{cases}$$

with the Bessel functions

$$J_{2s}(t) = \sum_{v=0}^{\infty} \frac{(-1)^v (t/2)^{2s+2v}}{v!\Gamma(2s+1+v)}$$
$$I_{2s}(t) = \sum_{v=0}^{\infty} \frac{(t/2)^{2s+2v}}{v!\Gamma(2s+1+v)}.$$

If we impose $J(Q,m;P,n;s;c) \sim c^{-1-2s}$ as $c \to \infty$, we obtain the same factors G_l as in the case $n = 0$. We summarize the result of our computations as a proposition:

5.2.9 Proposition. *For $P, Q \in X^\infty$ the Fourier coefficients of Poincaré series satisfy*

$$c_l(Q,m;P,n;\chi,s) = G_l(Q,m;P,n;s) \sum_c \mathrm{S}_\chi(Q,m;P,n;c) J(Q,m;P,n;s;c). \quad (5.2)$$

The variable c runs over the positive numbers occurring in $g_P^{-1}\gamma g_Q \in \widetilde{\left(\begin{smallmatrix} \cdot & \cdot \\ c & \cdot \end{smallmatrix}\right)} Z$ with $\gamma \in \tilde{\Gamma}$. The generalized Kloosterman sum S_χ has been defined in (5.1). The other *factors are*

$$
\begin{aligned}
&J(Q,m;P,n;s;c)\\
&= \begin{cases} c^{-1-2s} & \text{if } mn=0\\ c^{-1}J_{2s}\left(\dfrac{4\pi\sqrt{mn}}{c}\right)\dfrac{\Gamma(1+2s)}{(2\pi)^{2s}(mn)^s} & \text{if } mn>0\\ c^{-1}I_{2s}\left(\dfrac{4\pi\sqrt{|mn|}}{c}\right)\dfrac{\Gamma(1+2s)}{(2\pi)^{2s}|mn|^s} & \text{if } mn<0 \end{cases}
\end{aligned}
$$

$$
\begin{aligned}
&G_l(Q,m;s)\\
&= e^{-il\pi/2}\begin{cases} \dfrac{\pi 2^{1-2s}\Gamma(2s)}{\Gamma(\frac12+s+\frac12 l)\Gamma(\frac12+s-\frac12 l)} & \text{if } m=0\\ \dfrac{\pi^{s+1/2}}{\Gamma(\frac12+s+\frac12 l\,\mathrm{sign}(m))}|m|^{s-1/2} & \text{if } m\neq 0 \end{cases}\\
&= \frac{e^{-il\pi/2}\pi 2^{1-2s}\Gamma(2s)}{\Gamma(\frac12+s+\frac12 l)\Gamma(\frac12+s-\frac12 l)}\begin{cases} 1 & \text{if } m=0\\ v_l(Q,m,-s)^{\;1} & \text{if } m\neq 0. \end{cases}
\end{aligned}
$$

Remarks. $G_l(Q,m;P,n;s)$ does not depend on P and n; we have written it as $G_l(Q,m;s)$.

In general, this description of the Fourier coefficients is as far as one can get. But if Γ is a congruence subgroup of $\mathrm{SL}_2(\mathbb{Z})$, and if the character χ can be described by congruences as well, then more explicit formulas are possible. In 13.2.5 we consider this for the modular case. That gives the expansion in 1.4.4. See also 14.4.5 and 15.4.5.

5.2.10 *Maass-Selberg relation.* A check is provided by the Maass-Selberg relation, see Theorem 4.6.5. If $(P,n)\neq(Q,m)$ we should have

$$
\begin{aligned}
&\mathrm{Wr}\left(\tilde F_{P,n}P_l(P,n,\chi,s),\iota\tilde F_{P,-n}P_{-l}(Q,-m,\chi^{-1},s)\right)\\
&\quad+\ \mathrm{Wr}\left(\tilde F_{Q,m}P_l(P,n,\chi,s),\iota\tilde F_{Q,-m}P_{-l}(Q,-m,\chi^{-1},s)\right)=0.
\end{aligned}
$$

All other (P',n') cannot contribute. This condition gives

$$
\begin{aligned}
&c_{-l}(P,-n;Q,-m;\chi^{-1},s)\mathrm{Wr}(\mu_l(P,n,s),\iota\eta_{-l}(Q,-n))\\
&\quad+\ c_l(Q,m;P,n,\chi,s)\mathrm{Wr}(\eta_l(P,m),\iota\mu_{-l}(Q,-m,s))=0,
\end{aligned}
$$

with the same interpretation of $\eta_\cdot(\cdot,\cdot)$ as above.

If we define, in this computation, $v_l(P,n,-s)=1$ for $n=0$, then we have to check (see 4.2.8 and 4.2.9):

$$
c_{-l}(P,-n;Q,-m;\chi^{-1},s)v_l(P,n,-s)=c_l(Q,m;P,n;\chi,s)v_l(P,m,-s).
$$

As $G_{-l}(P,-n;Q,-m;s)v_l(P,n,-s) = e^{\pi il}G_l(Q,m;P,n;s)v_l(Q,m,-s)$, we may concentrate on the series with J and the Kloosterman sums.

The c occurring in $g_P^{-1}\gamma g_Q \in \widetilde{\begin{pmatrix} a & b \\ c & d \end{pmatrix}}Z$ with $\gamma \in \tilde{\Gamma}$, occur also in $g_Q^{-1}\gamma^{-1}g_P \in \widetilde{\begin{pmatrix} -d & b \\ c & -a \end{pmatrix}}Z$. Indeed, $\widetilde{\begin{pmatrix} a & b \\ c & d \end{pmatrix}}^{-1} = \widetilde{\begin{pmatrix} -d & b \\ c & -a \end{pmatrix}}k(\pi)$ if $c > 0$. Further, if we choose the $\gamma \in \tilde{\Gamma}$ for fixed c with $d \in (-\frac{1}{2}c, \frac{1}{2}c]$ such that $-\frac{1}{2}c \leq a < \frac{1}{2}c$, then $\gamma_{(c,d)}^{-1}k(-\pi) = g_Q\widetilde{\begin{pmatrix} -d & b \\ c & -a \end{pmatrix}}g_P^{-1}$. So

$$\begin{aligned} \mathrm{S}_{\chi^{-1}}(P,-n;Q,-m;c) &= \sum_d \chi(\gamma_{(c,d)}^{-1}k(-\pi))e^{2\pi i(-m(-d)-n(-a))/c} \\ &= \chi(k(\pi))^{-1}\sum_d \chi(\gamma_{(c,d)})^{-1}e^{2\pi i(na+md)/c} = e^{-\pi il}\mathrm{S}_{\chi}(Q,m;P,n;c). \end{aligned}$$

The factor $e^{-\pi il}$ cancels the factor in the relation for G_l. As $J(Q,m;P,n;n;s)$ and $J(P,-n;Q,-m;s)$ are equal, the check is complete.

Chapter 6
Selfadjoint extension of the Casimir operator

In 1.2.6 and 1.5.7 we discussed the selfadjoint extension of the differential operator L_r. This concerned the modular case. The extension was an operator in a Hilbert space $H(r)$, for $r \in \mathbb{R}$. Its eigenfunctions were stated to be modular forms, and $\frac{|r|}{2}(1 - \frac{|r|}{2})$ its smallest eigenvalue. In this chapter we prove these statements, in the more general setting of Part I. We work in a Hilbert space $H(\chi, l)$ depending on a unitary character χ of $\tilde{\Gamma}$, and a (real) weight l suitable for χ. In Section 6.1 we define this Hilbert space as a completion of the space of all smooth χ-l-equivariant functions with compact support in Y.

The differential operator L_r, discussed in 1.5.7, corresponds to the action of the Casimir operator ω in the functions of weight r. The existence of the selfadjoint extension $A(\chi, l)$ of the Casimir operator, as an operator in $H(\chi, l)$, can be proved in several ways. Roelcke, [50], §3, follows a more direct method than the one used here. We shall construct $A(\chi, l)$ as the Friedrichs extension of the operator ω in the compactly supported smooth χ-l-equivariant functions. This idea may be found in Colin de Verdière, [12], and also in Lax & Phillips, [31]. Precisely the same method will be used in Chapter 9 in the truncated situation; for the modular case the result has been discussed in 1.6.2. Although the results will be needed later, I view this chapter primarily as an introduction to the methods used in Chapters 8 and 9.

In Section 6.5 the selfadjoint extension $A(\chi, l)$ of ω will be determined by a sesquilinear form $\mathfrak{s}$ in $H(\chi, l)$. The domain $D(\chi, l)$ of $\mathfrak{s}$ is a dense linear subspace of $H(\chi, l)$. It is a Hilbert space itself for another norm, the energy norm, which takes into account the first order derivatives as well. We discuss this Hilbert space $D(\chi, l)$ in Section 6.2.

If all cusps are regular, i.e., $\chi(\pi_P) \neq 1$ for all $P \in X^\infty$, we shall see in Theorem 6.5.5 that $A(\chi, l)$ has a compact resolvent. This is a strong result. It implies that the spectrum of $A(\chi, l)$ is discrete. To prove compactness, we give in Section 6.3 estimates of the Fourier coefficients of elements of $D(\chi, l)$. We use these estimates in Section 6.4 to prove that the inclusion $D(\chi, l) \to H(\chi, l)$ is compact. This will imply Theorem 6.5.5. The estimates in Section 6.3 will be used also in Chapter 8.

If there are singular cusps (i.e., $\chi(\pi_P) = 1$ for some $P \in X^\infty$), the spectrum of $A(\chi, l)$ has a continuous part. This part can be described with help of the Eisenstein series, after they have been meromorphically continued as a function of the spectral parameter s. So, for the spectral decomposition of $A(\chi, l)$ we need square integrable automorphic forms to get the discrete spectrum, and the Eisenstein series to get the continuous part. This is the subject of the spectral theory of automorphic forms, culminating in the trace formula of Selberg, see, e.g., Hejhal, [21]. In this book we do not discuss the continuous part of the spectrum of $A(\chi, l)$. The possible presence of singular cusps will force us to use another selfadjoint operator. In the modular case this is ${}^{a}A(r)$, see 1.6.2. We shall treat the general case in Section 9.2.

Section 6.6 gives the relation between eigenfunctions of the selfadjoint operator $A(\chi, l)$ and automorphic forms: the kernel of $A(\chi, l) - \frac{1}{4} + s^2$ consists of those automorphic forms in $A_l(\chi, \mathbf{c}^{(\chi)}, s)$ that represent elements of $H(\chi, l)$. So the study of $A(\chi, l)$ is related to the study of square integrable automorphic forms for the minimal growth condition $\mathbf{c}^{(\chi)}$ for χ.

The fact that $A(\chi, l)$ is selfadjoint implies that solely for $s^2 \in \mathbb{R}$ there may be square integrable automorphic forms of eigenvalue $\frac{1}{4} - s^2$. We shall see that there are further restrictions on s. For some of these restrictions we need the representational point of view discussed in 4.5.4. That is the subject of Section 6.7.

In this chapter χ is a unitary character of $\tilde{\Gamma}$, and $l \in \mathbb{R}$ satisfies $e^{\pi i l} = \chi(k(\pi))$. We shall often omit the dependence on χ from the notation.

6.1 The Hilbert space of equivariant functions

We define the Hilbert space of square integrable χ-l-equivariant functions, and consider under what conditions automorphic forms are square integrable.

6.1.1 *Definition.* Let $K(\chi, l) = K(l)$ consist of the χ-l-equivariant functions $h \in C^\infty(\tilde{G})$ that vanish outside a set of the form $\tilde{\Gamma} C \tilde{K}$ with $C \subset \tilde{G}$ compact. Functions with the latter property we call *compactly supported in Y*.

We do not use the abbreviation K for $K(l)$, to avoid confusion with the subgroup $K = \mathrm{SO}_2(\mathbb{R}) \subset G$, and its cover $\tilde{K}$.

6.1.2 *Differentiation.* The differential operators discussed in 2.2.3 give linear maps

$$\mathbf{E}^{\pm} : K(l) \to K(l \pm 2), \qquad \omega : K(l) \to K(l).$$

6.1.3 *Definition.* We put a scalar product on $K(l)$ by defining

$$(k_1, k_2)_l = \int_{\Gamma \backslash \mathfrak{H}} k_1(p(z)) \overline{k_2(p(z))}\, d\mu(z).$$

The integration may be carried out on the fundamental domain F. This scalar product also make sense if both functions are continuous and χ-l-equivariant, and only one of them is compactly supported in Y.

6.1.4 Lemma. *For $k_1, k_3 \in K(l)$, $k_2 \in K(l \pm 2)$:*

$$(\mathbf{E}^{\pm} k_1, k_2)_{l \pm 2} = -(k_1, \mathbf{E}^{\mp} k_2)_l$$

$$(\omega k_1, k_3)_l = (k_1, \omega k_3)_l = \frac{1}{8} \sum_{\pm} (\mathbf{E}^{\pm} k_1, \mathbf{E}^{\pm} k_3)_{l \pm 2} - \frac{1}{4} l^2 (k_1, k_3)_l.$$

The same formulas hold if one of the k_j is not compactly supported in Y, but is smooth and χ-l- respectively χ-$(l \pm 2)$-equivariant.

Proof. Note that

$$(\mathbf{E}^{\pm} k_1, k_2)_{l\pm 2} + (k_1, \mathbf{E}^{\mp} k_2)_l = \int_F d\eta$$

with $\eta = -2k_1(p(z))\overline{k_2(p(z))}\frac{dx \mp i dy}{y}$. We use a truncated version $F(a)$ of the fundamental domain F, like we did in 3.3.5. Take a such that $p(F \setminus F(a))$ is not contained in $\mathrm{supp}(k_1 k_2)$. As η is $\tilde{\Gamma}$-invariant, we obtain $\int_{F(a)} d\eta = \int_{\partial F(a)} \eta = 0$. Use $\omega = -\frac{1}{4}\mathbf{E}^{\pm}\mathbf{E}^{\mp} + \frac{1}{4}\mathbf{W}^2 \mp \frac{i}{2}\mathbf{W}$ for the assertion concerning $(\omega k_1, k_3)_l$.

6.1.5 *Definition.* $H(\chi, l) = H(l)$ is the completion of $K(l)$ for the norm $\|.\|_l$ corresponding to the scalar product $(.,.)_l$.

6.1.6 *Remarks.* One may identify $H(l)$ with $L^2(F, d\mu)$. To see this, note that the $\varphi \in C_c^\infty(\mathfrak{H})$ of which the support is contained in the interior of F form a dense subspace in $L^2(F, d\mu)$. Define $\tilde{\varphi} \in C^\infty(\tilde{G})$ by $\tilde{\varphi}(\gamma p(z) k(\theta)) = \chi(\gamma)\varphi(z)e^{il\theta}$ for $\gamma \in \tilde{\Gamma}$, $z \in F$, $\theta \in \mathbb{R}$. The ambiguity in the choice of γ, z and θ for a given $g = \gamma p(z) k(\theta) \in \tilde{G}$ does not matter, as $\chi(\zeta) = e^{\pi i l}$, and $\varphi(z) = 0$ for z near the boundary of F. Now $\tilde{\varphi} \in K(l)$, and $\|\tilde{\varphi}\|_l^2 = \int_F |\varphi(z)|^2 \, d\mu$. The map $\varphi \mapsto \tilde{\varphi}$ induces an isometric injection $L^2(F, d\mu) \to H(l)$. As the $\tilde{\varphi}$ are dense in $K(l)$, that map is an isomorphism of Hilbert spaces.

I prefer to define $H(l)$ as the completion of $K(l)$, as this does not depend on the choice of a fundamental domain, and avoids the introduction of $H(l)$ as a space of equivalence classes of functions. Moreover, with this definition it is immediately clear that we can view $H(l)$ as the space of continuous antilinear forms on $K(l)$, by associating to $f \in H(l)$ the map $K(l) \to \mathbb{C} : k \mapsto (f, k)_l$.

6.1.7 *Examples of elements of $H(l)$.* An easy example, for $\chi = 1$, $l = 0$, is the constant function $\mathbf{1} : g \mapsto 1$. As $\tilde{\Gamma}\backslash\mathfrak{H}$ has finite volume, $\mathbf{1}$ represents an element of $H(1, 0)$.

We use the identification between $H(l)$ and $L^2(F, d\mu)$ to give another example. For any unitary χ and real l with $\chi(\zeta) = e^{\pi i l}$, the function $F \to \mathbb{R}$ that is 1 on the interior of F and 0 on the boundary is in $L^2(F, d\mu)$. Hence it determines an element of $H(l)$. For general χ this element is not represented by a smooth function on $\tilde{G}$. It depends on the choice of F.

6.1.8 *Square integrability of automorphic forms.* Let $\mathbf{c}$ be a growth condition for χ, and let $f \in A_l(\chi, \mathbf{c}, s)$. We show that the square integrability of f is determined by the Fourier terms corresponding to $P \in \mathcal{P}$, $n \in \mathbf{c}(P)$.

The automorphic form $f \in A_l(\chi, \mathbf{c}, s)$ represents an element of $H(l)$, if and only if the function $f_l : z \mapsto f(p(z))$ is square integrable on F. The decomposition of Y discussed in 3.5.5 gives a decomposition of F. As f_l is smooth on the compact set $F \setminus \bigcup_{P \in \mathcal{P}} \overline{\mathrm{pr}}^{-1} U_P(A_P)$, it is square integrable on it.

Take $P \in \mathcal{P}$. On $F \cap \overline{\mathrm{pr}}^{-1} U_P(A_P)$ the function f_l is the sum of its $\mathbf{c}$-remainder $f[\mathbf{c}, P]_l$ and the Fourier terms corresponding to $n \in \mathbf{c}(P)$. If $P \in \mathcal{P}_Y$, then $f[\mathbf{c}, P]_l$

is smooth on $F \cap \overline{\mathrm{pr}}^{-1} U_P(A_P)$, hence it is square integrable. If $P \in X^\infty$, the definition of regularity in 4.1.9 amounts to square integrability. We use that χ is unitary to see that the factor $e^{-2\pi i \nu x}$ in 4.1.9 does not matter. Fourier terms for different (P, n) are orthogonal to each other (expand $d\mu(z)$ in Iwasawa and polar coordinates to see this orthogonality). Hence $f \in H(l)$ if and only if all Fourier terms for $P \in \mathcal{P}$, $n \in \mathbf{c}(P)$ are square integrable on $F \cap \overline{\mathrm{pr}}^{-1} U_P(A_P)$.

For $P \in X^\infty$ square integrability amounts to regularity. For $P \in \mathcal{P}_Y$ regular Fourier terms are square integrable, but if $n = l$ all Fourier terms are square integrable (see 4.2.9, and use that $d\mu = 4\, du \wedge d\eta$ in polar coordinates).

Suppose the growth condition $\mathbf{c}$ is the minimal one for f, i.e., $f \notin A_l(\chi, \mathbf{c}_1, s)$ for $\mathbf{c}_1 \subset \mathbf{c}$, $\mathbf{c}_1 \neq \mathbf{c}$. Then $f \in H(l)$ if and only if the following conditions hold:

$$\begin{array}{ll} \mathbf{c}(P) \subset \{0\} & \text{if } P \in X^\infty \\ \tilde{F}_{P,0} f \in W_l^0(P, 0, s) & \text{if } P \in X^\infty \text{ and } \mathbf{c}(P) = \{0\} \\ \mathbf{c}(P) \subset \{0\} & \text{if } P \in \mathcal{P}_Y. \end{array}$$

In particular, $S_l(\chi, \mathbf{c}^{(\chi)}, s) \subset H(l)$. We check that the resolvent kernel discussed in 5.1.8 is also square integrable. Part i) of Lemma 5.1.4 implies that any Fourier term of order zero at a cusp is bounded, hence it is square integrable (remember that $\mathrm{Re}\, s > \frac{1}{2}$ in the domain of convergence). Other Poincaré series as defined in the previous chapter are in general not in $H(l)$. Square integrability occurs if $\mu_l(P, n, s) \in W_l^0(P, n, s)$ for $P \in X^\infty$.

6.1.9 *Explicit examples.* In Chapter 1 we have mentioned square integrable modular forms: the holomorphic cusp form Δ, see 1.3.2 and 1.4.2; the cuspidal Maass forms in 1.4.5; and η_r for $r \geq 0$, see 1.5.5. Except for $\eta_0 = \mathbf{1}$ these functions are cusp forms.

See 14.3.1, p. 269, for a more interesting example of a square integrable automorphic form that is not a cusp form.

6.2 The subspace for the energy norm

We define the subspace $D(\chi, l)$ of $H(\chi, l)$ by 'square integrability of the first derivatives'. This subspace turns out to be the completion of $K(l)$ with respect to another norm, the 'energy norm'. $D(\chi, l)$ will contain the domain of the selfadjoint extension of ω (to be defined in Section 6.5).

$\mathbf{E}^+$ and $\mathbf{E}^-$ are the derivatives that we use to define $D(\chi, l)$. The maps $\mathbf{E}^\pm : K(l) \to K(l \pm 2)$ are not continuous if we provide $K(l)$ with the norm $\|\cdot\|_l$, and $K(l \pm 2)$ with $\|\cdot\|_{l \pm 2}$. Hence they cannot be extended as maps $H(l) \to H(l \pm 2)$. But the identification of elements of $H(l \pm 2)$ with continuous linear forms on $K(l \pm 2)$ allows us to make sense of the condition $\mathbf{E}^\pm f \in H(l \pm 2)$.

6.2.1 *Extension of differential operators.* $\mathbf{E}^\pm : K(l) \to K(l \pm 2)$ can be extended to $H(l)$ in distribution sense: for $f \in H(l)$ we define $\mathbf{E}^\pm f$ as the antilinear form

$$\mathbf{E}^\pm f : k \mapsto -(f, \mathbf{E}^\mp k)_l$$

on $K(l \pm 2)$. This form is continuous for the norm $k \mapsto \|\mathbf{E}^{\mp}k\|_l + \|k\|_{l\pm 2}$. In this way $\mathbf{E}^{\pm}f$ makes sense as a distribution for each $f \in H(l)$.

The map $\mathbf{E}^{\pm}f : k \mapsto -(f, \mathbf{E}^{\mp}k)_l$ may happen to be continuous for $\|.\|_{l\pm 2}$. In that case the antilinear form is given by an element of $H(l \pm 2)$, which we also denote by $\mathbf{E}^{\pm}f$. We have $(\mathbf{E}^{\pm}f, k)_{l\pm 2} = -(f, \mathbf{E}^{\mp}k)_l$ for all $k \in K(l \pm 2)$.

Lemma 6.1.4 shows that if $f \in C^\infty(\tilde{G}) \cap H(l)$, and the usual derivative $\mathbf{E}^{\pm}f$ is in $H(l \pm 2)$, then $\mathbf{E}^{\pm}f$ represents $k \mapsto -(f, \mathbf{E}^{\mp}k)_l$.

6.2.2 *Definition.*

$$D(\chi, l) = D(l) = \left\{ f \in H(l) : \mathbf{E}^+f \in H(l+2) \text{ and } \mathbf{E}^-f \in H(l-2) \right\}.$$

For $f, g \in D(l)$ we define

$$\begin{aligned} (f,g)_{D,l} &= (f,g)_l + \frac{1}{8}\sum_{\pm}(\mathbf{E}^{\pm}f, \mathbf{E}^{\pm}g)_{l\pm 2} \\ \|f\|^2_{D,l} &= (f,f)_{D,l} = \|f\|_l^2 + \frac{1}{8}\sum_{\pm}\|\mathbf{E}^{\pm}f\|^2_{l\pm 2}. \end{aligned}$$

We call $\|\cdot\|_{D,l}$ the *energy norm.*

6.2.3 *Examples.* Let $\chi = 1$. Then $\mathbf{1} \in H(l)$. Moreover, $\mathbf{E}^{\pm}\mathbf{1} = 0$ implies $\mathbf{1} \in D(0)$ and $\|\mathbf{1}\|_0 = \|\mathbf{1}\|_{D,0} = \sqrt{\mathrm{vol}(\tilde{\Gamma}\backslash\mathfrak{H})}$.

All cusp forms are elements of $D(l)$; see Proposition 4.5.3. The differentiation result in 5.1.11 implies that the resolvent kernel is not an element of $D(l)$.

6.2.4. As $K(l) \subset D(l)$, the space $D(l)$ is dense in $H(l)$.

$D(l)$, provided with the scalar product $(.,.)_{D,l}$, is a Hilbert space itself. Indeed, any Cauchy sequence (f_n) in $D(l)$ with respect to $\|.\|_{D,l}$ determines $f = \lim_{n\to\infty} f_n$ in $H(l)$ with respect to $\|.\|_l$, and $f_\pm = \lim_{n\to\infty}\mathbf{E}^{\pm}f_n$ in $H(l \pm 2)$ with respect to $\|.\|_{l\pm 2}$. We check that $\mathbf{E}^{\pm}f = f_\pm$: for $k \in K(l \pm 2)$

$$(f_\pm, k)_{l\pm 2} = \lim_{n\to\infty}(\mathbf{E}^{\pm}f_n, k)_{l\pm 2} = \lim_{n\to\infty} -(f_n, \mathbf{E}^{\mp}k)_l = -(f, \mathbf{E}^{\mp}k)_l.$$

6.2.5 Lemma. *If $h \in D(l)$ satisfies $(h,k)_{D,l} = 0$ for all $k \in K(l)$, then h is represented by a real analytic function on $\tilde{G}$.*

Remark. In the next proposition we shall see that $h = 0$.

Proof. Put $\eta(\psi) = \int_{\mathfrak{H}} \psi(z)\overline{h(p(z))}\,d\mu(z)$ for $\psi \in C_c^\infty(\mathfrak{H})$. As $z \mapsto h(p(z))$ is locally square integrable on $\mathfrak{H}$, this defines a distribution η. Moreover,

$$\begin{aligned} \eta(\psi) &= \int_F \sum_{\gamma\in\bar{\Gamma}} \psi(\gamma z)\overline{h(p(\gamma\cdot z))}\,d\mu(z) \\ &= \int_F \frac{1}{2}\sum_{\gamma\in\Gamma} \chi_0(\tilde{\gamma})^{-1}e^{-il\arg(c_\gamma+d_\gamma)}\psi(\gamma z)\overline{h(p(z))}\,d\mu(z) \\ &= (k_\psi, h)_l \end{aligned}$$

with

$$k_\psi(p(z)k(\theta)) = \frac{1}{2}\sum_{\gamma\in\Gamma} \chi(\tilde\gamma)^{-1} e^{-il\arg(c_\gamma+d_\gamma)}\psi(\gamma z)e^{il\theta}.$$

This sum is a finite one for each $z \in \mathfrak{H}$, uniformly for z in compact sets in $\mathfrak{H}$. This implies that $k_\psi \in C^\infty(\tilde G)$. If $k_\psi(p(z)k(\theta)) \neq 0$, then $z \in \tilde\Gamma \cdot \operatorname{supp}\psi$. Hence $k_\psi \in K(l)$. The condition $(h,k)_{D,l} = 0$ for all $k \in K(l)$ implies $((\omega + \frac{1}{4}l^2+1)k_\psi, h)_l = 0$. As $(\omega + \frac{1}{4}l^2 + 1)k_\psi = k_\rho$, with $\rho = (-y^2\partial_x^2 - y^2\partial_y^2 + ily\partial_x)\psi$, we see that η is an eigendistribution of an elliptic differential operator on $\mathfrak{H}$ with analytic coefficients. Hence $z \mapsto \overline{h(p(z))}$ is an analytic function, see, e.g., [29], App. 4, §5, and [3], p. 207–210.

6.2.6 Proposition. *$K(l)$ is dense in $D(l)$ with respect to $\|.\|_{D,l}$.*

Proof. Suppose $h \in D(l)$ is $(.,.)_{D,l}$-orthogonal to $K(l)$. The previous lemma shows that h is represented by a real analytic function. Take, for n large, $\psi_n \in C^\infty(\tilde\Gamma\backslash\mathfrak{H}) = C^\infty(\tilde\Gamma\backslash\tilde G/\tilde K)$ in the following way:

$$\begin{array}{ll} 0 \le \psi_n \le 1 & \\ \psi_n = 1 & \text{outside } \overline{\mathrm{pr}}^{-1}\bigcup_{P\in X^\infty} U_P(n) \\ \psi_n(g_P\cdot z) = \tilde\psi(y/n) & \text{for } g_P p(z) \in \mathrm{pr}^{-1}U_P(n),\ P\in X^\infty \end{array}$$

with $\tilde\psi \in C^\infty(0,\infty)$, $\tilde\psi(y) = 1$ for $y \le 2$, $\tilde\psi(y) = 0$ for $y \ge 3$. Then $\psi_n \cdot h \in K(l)$ and $\lim_{n\to\infty}\psi_n h = h$ in $H(l)$. For $P \in X^\infty$, $y > n$ we have

$$\mathbf{E}^\pm\psi_n(g_P\cdot z) = \frac{2y}{n}\tilde\psi'(y/n) = \mathcal{O}(1).$$

As $\psi_n h \in K(l)$, we have $(\psi_n h, h)_{D,l} = 0$. So for large n

$$\sum_{P\in X^\infty}\int_{\overline{\mathrm{pr}}^{-1}U_P(n)\cap F}\left(\psi_n|h|^2 + \frac{1}{8}\sum_\pm\left(\mathbf{E}^\pm\psi_n\right)h\,\overline{\mathbf{E}^\pm h} + \frac{1}{8}\sum_\pm\psi_n\left|\mathbf{E}^\pm h\right|^2\right)d\mu$$
$$+ \int_{\text{remainder of } F}\left(|h|^2 + \frac{1}{8}\sum_\pm\left|\mathbf{E}^\pm h\right|^2\right)d\mu$$

equals zero. If h were non-zero, the integral over the remainder of F would eventually give a positive contribution, which does not decrease when n grows. So there would be a positive number τ such that for all large n

$$-\tau \ge \sum_{P\in X^\infty}\int_{\overline{\mathrm{pr}}^{-1}U_P(n)\cap F}\frac{1}{8}\sum_\pm\left(\mathbf{E}^\pm\psi_n\right)h\,\overline{\mathbf{E}^\pm h}\,d\mu.$$

h and $\mathbf{E}^\pm h$ determine elements of $L^2(F,d\mu)$. This implies the integrability of the functions $\frac{1}{8}\left(\mathbf{E}^\pm\psi_n\right)h\,\overline{\mathbf{E}^\pm h}$. As the $\mathbf{E}^\pm\psi_n$ are bounded and have pointwise limit 0 as $n \to \infty$, we obtain by dominated convergence

$$\lim_{n\to\infty} \sum_{P\in X^\infty} \int_{\overline{\mathrm{pr}}^{-1}U_P(n)\cap F} \frac{1}{8} \sum_{\pm} \left(\mathbf{E}^\pm \psi_n\right) h\, \overline{\mathbf{E}^\pm h}\, d\mu = 0.$$

This gives a contradiction.

6.2.7 Lemma. *For $f \in D(l)$, $h \in D(l \pm 2)$*

$$\left(\mathbf{E}^\pm f, h\right)_{l\pm 2} = -\left(f, \mathbf{E}^\mp h\right)_l .$$

Proof. This holds for $h \in K(l \pm 2)$ by the definition of $\mathbf{E}^\pm f$. Both sides of the equality depend continuously on h with respect to the norm $\|.\|_{D,l}$.

6.3 Fourier coefficients

The elements $f \in D(\chi, l)$ are defined by square integrability of f, $\mathbf{E}^+ f$, and $\mathbf{E}^- f$. We shall see in Lemma 6.3.7 that this implies an estimate of most Fourier terms of f at the cusps. This estimate will imply that the selfadjoint extension of ω has a compact resolvent if all cusps are regular (Theorem 6.5.5). It will also be essential in Section 8.4, in the truncated situation.

Up till now we have defined the Fourier term operator $F_{P,n}$ for C^∞-functions only, see Section 4.1. To define it in the L^2-context, we again view square integrable functions as antilinear forms on a space of test functions.

Let $X^\infty \subset \mathcal{P} \subset X$ as before. Consider $P \in \mathcal{P}$ and $n \in \mathcal{C}_P(\chi)$.

6.3.1 *Definition.* For each $\varphi \in C_c^\infty(0,\infty)$ we define

$$\left(\Theta_{P,n,l}\varphi\right)(g) = \sum_{\gamma\in\Delta_P\backslash\tilde{\Gamma}} \chi(\gamma)^{-1} h_\varphi(\gamma g),$$

with

$$\begin{cases} h_\varphi(g_P p(z) k(\theta)) = e^{2\pi i n x}\varphi(y) e^{il\theta} & \text{if } P \in X^\infty \\ h_\varphi(g_P k(\eta) a(t_u) k(\psi)) = e^{in\eta}\varphi(u) e^{il\psi} & \text{if } P \in \mathcal{P}_Y. \end{cases}$$

The sum is finite for each g, uniformly for g in compact sets. In the case $P \in \mathcal{P}_Y$ this follows from the fact that $\mathrm{supp}(h_\varphi)/\tilde{K}$ is compact and meets only finitely many different $\tilde{\Gamma}$-translates of the fundamental domain F. In the case $P \in X^\infty$ only finitely many translates of $\Delta_P \cdot F$ intersect $\mathrm{supp}(h_\varphi)/\tilde{K}$.

We see in particular that $\mathrm{pr}\,\mathrm{supp}(\Theta_{P,n,l}\varphi) = \mathrm{pr}\,\mathrm{supp}(h_\varphi)$ is compact. So $\Theta_{P,n,l} : C_c^\infty(0,\infty) \to K(l)$.

6.3.2 *Notation.* We provide $(0,\infty)$ with the measure $d\nu_P$, given by $d\nu_P(y) = y^{-2}dy$ if $P \in X^\infty$, and $d\nu_P(u) = 4\pi\, du$ if $P \in \mathcal{P}_Y$.

Remember that $\tilde{I}_P \subset (0,\infty)$ is equal to $(\tilde{A}_P, \infty)$, respectively $(0, \tilde{A}_P)$. The next lemma shows that we have chosen $d\nu_P$ in such a way that $\Theta_{P,n,l} : C_c^\infty(\tilde{I}_P) \to K(l)$ is an isometry for the $L^2(\tilde{I}_P, d\nu_p)$-norm on $C_c^\infty(\tilde{I}_P)$ and $\|.\|_l$ on $K(l)$.

6.3.3 Lemma. *Let $J \subset (0,\infty)$ be of the form (B,∞) if $P \in X^\infty$ and of the form $(0,B)$ if $P \in \mathcal{P}_Y$, with $B > 0$. The operator $\Theta_{P,n,l} : C_c^\infty(J) \to K(l)$ has a continuous extension $\Theta_{P,n,l} : L^2(J, d\nu_P) \longrightarrow H(l)$. If $J \subset \tilde{I}_P$, then $\Theta_{P,n,l}$ is an isometry.*

Proof. We use the identification of $H(l)$ with $L^2(F, d\mu)$, where F is the canonical fundamental domain that we have chosen.

Let $P \in \mathcal{P}_Y$. For each $\gamma \in \tilde{\Gamma}$:

$$\int_F |h_\varphi(\gamma p(z))|^2 \, d\mu(z) \leq \int_{\mathfrak{H}} |h_\varphi(g_P p(z))|^2 \, d\mu(z) = \|\varphi\|^2.$$

$\mathrm{supp}(h_\varphi)/\tilde{K} = g_P \tilde{K}\{\, a(t_u) : 0 \leq u \leq B \}\tilde{K}/\tilde{K}$ meets only finitely many $\tilde{\Gamma}$-translates of F.

Let $P \in X^\infty$. Now $\mathrm{supp}(h_\varphi)/\tilde{K} \cong \{\, g_P \cdot z : y \geq B \,\}$ meets only finitely many translates of $\Delta_P \cdot F$. Moreover F is contained in $\{\, g_P \cdot z : \alpha \leq x \leq \alpha + 1 \,\}$ for some $\alpha \in \mathbb{R}$. Hence

$$\int_F |h_\varphi(\gamma p(z))|^2 \, d\mu(z) \leq \int_{x=\alpha}^{\alpha+1} \int_{y=B}^{\infty} |h_\varphi(g_P p(z))|^2 \, d\mu(z) = \|\varphi\|^2.$$

In the case that $J \subset \tilde{I}_P$, there is exactly one translate meeting $\mathrm{supp}(h_\varphi)/\tilde{K}$; hence we obtain equality.

6.3.4 *Definition.* Let B and J be as in the previous lemma. We denote the adjoint map of $\Theta_{P,n,l}$ by $F_{P,n} : H(l) \longrightarrow L^2(J, d\nu_P)$.

So $F_{P,n}$ is a bounded operator with bound depending on B. If $J \subset \tilde{I}_P$, then $F_{P,n}$ is an orthogonal projection followed by an isometry. If $f \in C^\infty(\tilde{G}_{\mathcal{P}})$ represents an element of $H(l)$, then the $F_{P,n}f$ defined in Section 4.1 represent the $F_{P,n}f \in L^2(J, d\nu_P)$ as defined here.

We shall mostly use $F_{P,n}$ on $J \subset \tilde{I}_P$. But the general definition may be useful as well. In [29], Ch. XIII, §1, we see that the $\Theta_{P,0,l}$ and $F_{P,0}$ with $P \in X^\infty$ singular play a role in the study of the continuous part of the spectral decomposition of $H(l)$.

6.3.5 *Distributions.* Let J be as above. Any $h \in L^2(J, \nu_P)$ determines a distribution $C_c^\infty(J) \longrightarrow \mathbb{C} : \varphi \mapsto \int_J h\varphi \, d\nu_P$ on J. We define the derivatives of distributions on $(0,\infty)$ *with respect to $d\nu_P$*: if η is a distribution on $(0,\infty)$, then

$$\begin{aligned} \partial_y \eta &: \varphi \mapsto \eta\left(-\varphi' + \tfrac{2}{y}\varphi\right) && \text{if } P \in X^\infty \\ \partial_u \eta &: \varphi \mapsto -\eta\left(\varphi'\right) && \text{if } P \in \mathcal{P}_Y. \end{aligned}$$

If η is given by a C^∞-function, then the usual derivative determines the same distribution.

6.3.6 Lemma. *For $f \in D(l)$ we have in distribution sense on $(0,\infty)$:*

$$F_{P,n}\mathbf{E}^{\pm}f = \begin{cases} (2y\partial_y \mp 4\pi ny \pm l)\, F_{P,n}f & \text{if } P \in X^{\infty} \\ \left(2\sqrt{u^2+u}\,\partial_u \pm \frac{(n-l)/2-ul}{\sqrt{u^2+u}}\right) F_{P,n}f & \text{if } P \in \mathcal{P}_Y. \end{cases}$$

Proof. We treat the case $P \in X^{\infty}$; the other case is proved similarly.

Let $\varphi \in C_c^{\infty}(0,\infty)$, and put $\varphi_1(y) = 2y\varphi'(y) + (\pm 4\pi ny \mp l - 2)\varphi(y)$.

$$\begin{aligned} &\int_0^{\infty} F_{P,n}\mathbf{E}^{\pm}f \cdot \varphi\, d\nu_P = \left(\mathbf{E}^{\pm}f, \Theta_{P,n,l\pm 2}\bar{\varphi}\right)_{l\pm 2} \\ &= \; -\left(f, \mathbf{E}^{\mp}\Theta_{P,n,l\pm 2}\bar{\varphi}\right)_l = -\left(f, \Theta_{P,n,l}\bar{\varphi}_1\right)_l = \int_0^{\infty} F_{P,n}f \cdot (-\varphi_1)\, d\nu_P, \end{aligned}$$

6.3.7 Lemma. *Let $P \in X^{\infty}$, $n \in \mathcal{C}_P(\chi)$, $n \neq 0$. Let $J = (B,\infty)$ with $B > 0$. Put*

$$\begin{aligned} A_B(l,n) &= \begin{cases} 2\pi|n| - \frac{|l|}{2B} & \text{if } nl > 0 \\ 2\pi|n| & \text{if } nl \leq 0 \end{cases} \\ \tilde{A}_B(l,n) &= 2\pi|n|B - \frac{1}{2}l\,\mathrm{sign}(n). \end{aligned}$$

Let $f \in D(l)$. Then $y \mapsto yF_{P,n}f(y)$ represents an element of $L^2(J, d\nu_P)$. If $B > \frac{l}{4\pi n}$, then

$$\int_B^{\infty} \frac{1}{8}\sum_{\pm} \left|F_{P,n}\mathbf{E}^{\pm}f(y)\right|^2 \frac{dy}{y^2} \geq \begin{cases} A_B(l,n)^2 \int_B^{\infty} |yF_{P,n}f(y)|^2 \frac{dy}{y^2} \\ \tilde{A}_B(l,n)^2 \int_B^{\infty} |F_{P,n}f(y)|^2 \frac{dy}{y^2}. \end{cases}$$

Remark. If $ln \leq 0$ then the condition $B > \frac{l}{4\pi n}$ is automatically satisfied.

Proof. Put $h = F_{P,n}f$, $h_1 = \frac{1}{2}\sum_{\pm} F_{P,n}\mathbf{E}^{\pm}f$ and $h_2(y) = (l - 4\pi ny)h(y)$. By the previous lemma $h_2 = \frac{1}{2}F_{P,n}\mathbf{E}^{+}f - \frac{1}{2}F_{P,n}\mathbf{E}^{-}f$. So h, h_1, $h_2 \in L^2(J, d\nu_P)$. As $n \neq 0$, we obtain the first assertion of the lemma. Furthermore,

$$\begin{aligned} &\int_B^{\infty} \frac{1}{8}\sum_{\pm} \left|F_{P,n}\mathbf{E}^{\pm}f\right|^2 \frac{dy}{y^2} = \int_B^{\infty} \frac{1}{4}\left(|h_1(y)|^2 + |h_2(y)|^2\right) \frac{dy}{y^2} \\ &\geq \frac{1}{4}\int_B^{\infty} (l - 4\pi ny)^2 |h(y)|^2 \frac{dy}{y^2} \\ &\geq \begin{cases} \int_B^{\infty} \left(2\pi|n| - \frac{l\,\mathrm{sign}(n)}{2y}\right)^2 y^2|h(y)|^2 \frac{dy}{y^2} \geq A_B(l,n)^2 \int_B^{\infty} y^2|h(y)|^2 \frac{dy}{y^2} \\ \int_B^{\infty} \left(2\pi|n|y - \frac{1}{2}l\,\mathrm{sign}\, n\right)^2 |h(y)|^2 \frac{dy}{y^2} \geq \tilde{A}_B(l,n)^2 \int_B^{\infty} |h(y)|^2 \frac{dy}{y^2}. \end{cases} \end{aligned}$$

6.3.8 *Decomposition.* Consider $F_{P,n} : H(l) \to L^2(\tilde{I}_P, d\nu_P)$. Each $\Theta_{P,n,l}F_{P,n}$ is the orthogonal projection on a subspace $H_{P,n}(l)$ of $H(l)$ that is isomorphic to $L^2(\tilde{I}_P, d\nu_P)$. If $(P,n) \neq (P',n')$, then $(\Theta_{P,n,l}\varphi, \Theta_{P',n',l}\varphi') = 0$ for all $\varphi, \varphi' \in L^2(\tilde{I}_P, d\nu_p)$. So we have an orthogonal decomposition

$$H(l) = L^2\left(F \setminus \overline{\mathrm{pr}}^{-1}\left(\bigcup_P U_P(\tilde{A}_P)\right), d\mu\right) \oplus \bigoplus_P \bigoplus_n H_{P,n}(l).$$

6.4 Compactness

We consider the inclusion map $\mathrm{Id} : D(l) \to H(l)$. It is a bounded operator, if we provide $H(l)$ with the norm $\|.\|_l$, and $D(l)$ with $\|.\|_{D,l}$. If Id is a compact operator that has consequences for the automorphic spectrum. The values of $\chi(\pi_P)$ for $P \in X^\infty$ regulate this compactness.

6.4.1 Lemma. *Let $\tilde{Y}$ be a compact subset of Y. Denote by $\tilde{D}$, respectively $\tilde{H}$, the subspace of $D(l)$, respectively $H(l)$, of elements supported in $\mathrm{pr}^{-1}\tilde{Y}$, provided with the topology induced by $D(l)$, respectively $H(l)$. Then the inclusion $\tilde{D} \to \tilde{H}$ is compact.*

Proof. We reduce this to the usual Sobolev theory on a torus, see, e.g., [29], App. 4, §4. Take a fundamental domain F and put $\tilde{F} = F \cap \overline{\mathrm{pr}}^{-1}\tilde{Y}$. Choose $C \subset \mathfrak{H}$ open, $\bar{C}$ compact, such that $\bar{\tilde{F}} \subset C$. Take $\psi \in C_c^\infty(\mathfrak{H})$, $\psi|_{\tilde{F}} = 1$, $0 \le \psi \le 1$, $\mathrm{supp}\,\psi \subset C$.

C meets only a finite number of $\tilde{\Gamma}$-translates of F. This implies continuity of the map $\tilde{H} \to L^2(C, d\mu)$, obtained by identifying elements of $\tilde{H}$ with functions on $\mathfrak{H}$.

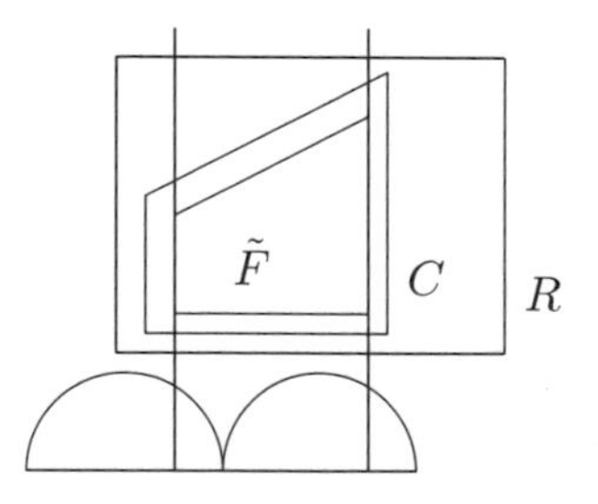

Let R be an euclidean rectangle in $\mathfrak{H}$, containing C in its interior. This rectangle may be viewed as a fundamental domain of a torus T. We map $L^2(C, d\mu)$ continuously to $L^2(T, dx \wedge dy)$ by means of $C \hookrightarrow R \to T$. This is no isometry, as the measures differ. By usual Sobolev theory there is for each $\varepsilon > 0$ a subspace $\tilde{H}(\varepsilon)$ of the first Sobolev space $H^1(T)$ on T, of finite codimension, such that

$$\|h\|_T^2 < \varepsilon \left(\|h\|_T^2 + \left\| \frac{\partial}{\partial x} h \right\|_T^2 + \left\| \frac{\partial}{\partial y} h \right\|_T^2 \right)$$

for all $h \in \tilde{H}(\varepsilon)$. The derivatives are to be understood in distribution sense; so if $f \in \tilde{D}$, then we may consider f as function on $\mathfrak{H}$, and ψf, $\psi \frac{\partial}{\partial y} f$ and $\psi \frac{\partial}{\partial x} f$ are in $L^2(C, d\mu)$. The derivatives $\frac{\partial}{\partial y}(\psi f)$ and $\frac{\partial}{\partial x}(\psi f)$ are also in $L^2(C, d\mu)$. So there is a subspace $\tilde{D}(\varepsilon)$ of finite codimension in $\tilde{D}$ such that $f \mapsto \psi f$ maps $\tilde{D}(\varepsilon)$ into $\tilde{H}(\varepsilon)$.

$$\begin{array}{ccc} \tilde{D} & \stackrel{f \mapsto \psi f}{\longrightarrow} & H^1(T) \\ \cup & & \cup \\ \tilde{D}(\varepsilon) & \longrightarrow & \tilde{H}(\varepsilon) \end{array}$$

For $f \in \tilde{D}(\varepsilon)$:

$$\begin{aligned} \|f\|_l^2 &= \|f\|_{\tilde{F}}^2 \ll_{\tilde{Y}} \|\psi f\|_T^2 \\ &< \varepsilon \left(\|\psi f\|_C^2 + \left\| \frac{\partial}{\partial x}(\psi f) \right\|_C^2 + \left\| \frac{\partial}{\partial y}(\psi f) \right\|_C^2 \right) \\ &\ll_C \varepsilon \left(\|f\|_C^2 + \|\psi_x f\|_C^2 + \left\| \frac{\partial}{\partial x} f \right\|_C^2 + \|\psi_y f\|_C^2 + \left\| \frac{\partial}{\partial y} f \right\|_C^2 \right) \\ &\ll_{l,C,\psi} \varepsilon \left(\|f\|_C^2 + \sum_{\pm} \|\mathbf{E}^{\pm} f\|_C^2 \right) \\ &\ll_{l,C,\psi} \varepsilon N \|f\|_{D,l}^2. \end{aligned}$$

6.4.2 Proposition. *Suppose that all cuspidal orbits are regular for χ. Then the inclusion* Id : $(D(l), \|.\|_{D,l}) \longrightarrow (H(l), \|.\|_l)$ *is compact.*

Remark. The regularity condition means that $\chi(\pi_P) \neq 1$ for all $P \in X^\infty$.
Proof. For each $\varepsilon > 0$, we have to produce a subspace D_ε with finite index in $D(l)$ such that $\|f\|_l < \varepsilon \|f\|_{D,l}$ for all $f \in D_\varepsilon$. We have seen in the previous lemma that on compact parts of Y this follows from the usual Sobolev theory. We use the estimates in Lemma 6.3.7 on a neighborhood of a cusp P. This is possible if the $n \in \mathcal{C}_P(\chi)$ stay away from zero; that is why we need the regularity condition.

Let $\varepsilon > 0$ be given. Choose for each $P \in X^\infty$ a number $\alpha(P)$ strictly larger than $\max\left(\tilde{A}_P, \frac{|l|/2+1/\varepsilon}{2\pi M_P}\right)$, with $M_P = \min\{\, |n| : n \in \mathcal{C}_P(\chi) \,\}$. As $\chi(\pi_P) \neq 0$, we know that $M_P > 0$. This choice of $\alpha(P)$ implies that $\tilde{A}_{\alpha(P)}(.,.)$ in Lemma 6.3.7 satisfies $\tilde{A}_{\alpha(P)}(l, M_P) > \frac{1}{\varepsilon}$. Define sets U_1 and U by $U_1 = \bigcup_{P \in X^\infty} \dot{U}_P(2\alpha(P))$,

$U = \bigcup_{P \in X^\infty} \dot{U}_P(\alpha(P))$; put $\tilde{Y} = Y \setminus U_1$. Fix a function $\psi \in C^\infty(\tilde{\Gamma}\backslash\tilde{G}/\tilde{K})$ such that $0 \leq \psi \leq 1$, $\psi = 1$ on $\mathrm{pr}^{-1}(Y \setminus U)$ and $\psi = 0$ on $\mathrm{pr}^{-1}U_1$. We apply the previous lemma to $\tilde{Y}$. As $f \mapsto \psi f$ maps $D(l)$ into $\tilde{D}$, there is a subspace D_ε of finite codimension in $D(l)$ such that

$$\|\psi f\|_l \leq \varepsilon \|\psi f\|_{D,l} \quad \text{for all } f \in D_\varepsilon.$$

For all $f \in D(l)$ we apply Lemma 6.3.7 to $f_1 = (1-\psi)f$:

$$\begin{aligned}
\|f_1\|_l^2 &= \int_{F \cap \overline{\mathrm{pr}}^{-1} U} |f_1|^2 \, d\mu \leq \sum_{P \in X^\infty} \sum_n \int_{\alpha(P)}^\infty |F_{P,n} f_1(y)|^2 \frac{dy}{y^2} \\
&\leq \sum_{P \in X^\infty} \frac{1}{\tilde{A}_{\alpha(P)}(l, M_P)^2} \int_{F \cap \overline{\mathrm{pr}}^{-1} \dot{U}_P(\alpha(P))} \frac{1}{8} \sum_\pm |\mathbf{E}^\pm f_1|^2 \, d\mu \\
&< \varepsilon^2 \int_{F \cap \overline{\mathrm{pr}}^{-1} U} \frac{1}{8} \sum_\pm |\mathbf{E}^\pm f_1|^2 \, d\mu \leq \varepsilon^2 \|f_1\|_{D,l}^2.
\end{aligned}$$

So for $f \in D_\varepsilon$

$$\|f\|_l \leq \|\psi f\|_l + \|(1-\psi) f\|_l \leq \varepsilon \left(\|\psi f\|_{D,l} + \|(1-\psi) f\|_{D,l} \right).$$

Use, for $\eta = \psi$ or $1 - \psi$,

$$\begin{aligned}
\|\mathbf{E}^\pm (\eta f)\|_{l\pm 2}^2 &= \int_F \left| \eta \mathbf{E}^\pm f + (\mathbf{E}^\pm \eta) f \right|^2 \, d\mu \\
&\leq 2 \left(\|\mathbf{E}^\pm f\|_{l\pm 2}^2 + a_\eta \|f\|_l^2 \right),
\end{aligned}$$

with $a_\eta = \sup\{ |\mathbf{E}^\pm \eta(g)|^2 : g \in \tilde{G} \}$, to obtain

$$\|f\|_l \leq \varepsilon \left(\sqrt{1 + \frac{1}{2} a_\psi} + \sqrt{1 + \frac{1}{2} a_{1-\psi}} \right) \|f\|_{D,l} \quad \text{for all } f \in D_\varepsilon.$$

One problem remains: do a_ψ and $a_{1-\psi}$ introduce a dependence on ε? If we take ψ of the form

$$\psi(g_P p(z) k(\theta)) = \tau \left(\frac{y}{\alpha(P)} \right) \qquad \text{for } y \geq \alpha(P),\ P \in X^\infty,$$

with $\tau \in C^\infty(0,\infty)$, $0 \leq \tau \leq 1$, $\tau(y) = 1$ for $y \leq 1$ and $\tau(y) = 0$ for $y \geq 2$, then

$$\mathbf{E}^\pm \psi(g_P p(z) k(\theta)) = e^{\pm 2i\theta} 2 \frac{y}{\alpha(P)} \tau' \left(\frac{y}{\alpha(P)} \right),$$

which is bounded uniformly in $\alpha(P)$. The same holds for the function $\mathbf{E}^\pm(1-\psi)$.

6.4.3 *Counterexample.* If some cuspidal orbit $P \in X^\infty$ is singular, then the inclusion $\mathrm{Id} : D(l) \to H(l)$ is bounded, but not compact.

Indeed, take f of the form $\Theta_{P,0,l}\varphi$ with $\varphi \in C_c^\infty(0,\infty)$, with support contained in $I_P = (A_P, \infty)$. Then $\|\Theta_{P,0,l}\varphi\|_l^2 = \int_{A_P}^\infty |\varphi(y)|^2 \frac{dy}{y^2}$ and

$$\frac{1}{8}\sum_{\pm} \|\mathbf{E}^\pm \Theta_{P,0,l}\varphi\|_{D,l}^2 = \int_{A_P}^\infty \left(y^2|\varphi'(y)|^2 + \frac{1}{4}l^2|\varphi(y)|^2 \right) \frac{dy}{y^2}.$$

So $f = \Theta_{P,0,l}\varphi \in K(l) \subset D(l)$ satisfies

$$\begin{aligned} \|f\|_{D,l}^2 &= \left(1 + \frac{1}{4}l^2\right)\|f\|_l^2 + \int_{A_P}^\infty |\varphi'(y)|^2\, dy \\ \|f\|_l^2 &= \int_{A_P}^\infty |\varphi(y)|^2 y^{-2} dy. \end{aligned}$$

If we take $\varphi_a(y) = \varphi(ay)$ with $a \le 1$, $\varphi \ne 0$, we obtain infinitely many linearly independent $f_a \in D(l)$ for which

$$\|f_a\|_l^2 = a\|f\|_l^2, \qquad \|f_a\|_{D,l}^2 = a\|f\|_{D,l}^2.$$

This contradicts compactness of $\mathrm{Id} : D(l) \to H(l)$.

6.5 Extension of the Casimir operator

The Casimir operator ω maps $K(l)$ into itself. It is not bounded with respect to the norms $\|\cdot\|_l$ and $\|\cdot\|_{D,l}$. So it cannot be extended to the whole of $H(\chi, l)$, or of $D(\chi, l)$. But it has a selfadjoint extension $A(\chi, l)$ in $H(\chi, l)$, with domain contained in $D(\chi, l)$. To construct this extension, it suffices to have available the results in Sections 6.1 and 6.2. Sections 6.3 and 6.4 allow us to prove that $A(\chi, l)$ has a compact resolvent if all cusps are regular for χ.

See [12], and also [31], for this approach to the selfadjoint extension of ω.

6.5.1 *Definition.* We define $\mathfrak{s}$ as the sesquilinear form on $D(l)$ given by

$$\mathfrak{s}[f,h] = \frac{1}{8}\sum_{\pm}(\mathbf{E}^\pm f, \mathbf{E}^\pm h)_{l\pm 2} - \frac{1}{4}l^2 (f,h)_l.$$

As $K(l) \subset D(l)$ is dense in $H(l)$, the form $\mathfrak{s}$ is *densely defined* in $H(l)$. It is *closed, symmetric* and *bounded from below* by $-\frac{1}{4}l^2$:

$$\mathfrak{s}[f,f] \ge -\frac{1}{4}l^2\|f\|_l^2 \quad \text{for all } f \in D(l).$$

See, e.g., [25], Ch. VI, §1, No. 1–3, for the definitions of these properties of sesquilinear forms see.

The form $\mathfrak{s}$ is continuous with respect to the norm $\|.\|_{D,l}$ on $D(l)$. In fact, $\mathfrak{s}[f,h] = (f,h)_{D,l} - \left(1+\frac{1}{4}l^2\right)(f,h)_l$.

6.5.2 *Definition.* Sesquilinear forms of this type correspond to selfadjoint operators in $H(l)$; see [25], Ch.VI, §2.1. By $A = A(\chi, l)$ we denote the unique selfadjoint operator in $H(l)$ with $\mathrm{dom}(A) \subset D(l)$, such that

$$\mathfrak{s}[f,h] = (Af,h)_l \quad \text{for all } f \in \mathrm{dom}(A),\ h \in D(l).$$

The domain of A consists of those $f \in D(l)$ for which $h \mapsto \mathfrak{s}[f,h]$ is continuous with respect to $\|.\|_l$ and hence is given by an element of $H(l)$. That element is Af.

6.5.3 *Extension of the Casimir operator.* A computation based on Lemma 6.1.4 and the relations in Section 2.2.3 shows, for $k \in K(l)$ and $h \in D(l)$,

$$\mathfrak{s}[k,h] = -\frac{1}{8}\sum_{\pm}(\mathbf{E}^{\mp}\mathbf{E}^{\pm}k,h)_l - \frac{1}{4}l^2(k,h)_l = (\omega k,h)_l.$$

So $K(l) \subset \mathrm{dom}(A)$ and $A = \omega$ on $K(l)$.

A is the *Friedrichs extension* of $\omega : K(l) \to H(l)$; see [25], Ch. VI, §2.3.

6.5.4 *Spectrum.* The spectrum of A is contained in $[-\frac{1}{4}l^2, \infty)$. At least for each $\lambda \in \mathbb{C} \setminus [-\frac{1}{4}l^2, \infty)$ the *resolvent* $\mathrm{R}(\lambda) = (A-\lambda)^{-1}$ is a bounded operator in $H(l)$. See [25], Ch. V, §3.5.

6.5.5 Theorem. *Suppose that all cusps of $\tilde{\Gamma}$ are regular for χ. Then the resolvent $\lambda \mapsto \mathrm{R}(\lambda)$ is a family of compact operators $H(l) \to H(l)$, and the spectrum of A is a discrete set. The spectrum consists of eigenvalues λ_j, $-\frac{1}{4}l^2 \le \lambda_1 < \lambda_2 < \lambda_3 < \cdots$, with $\lim_{j\to\infty}\lambda_j = \infty$. All eigenspaces $\ker(A-\lambda_j)$ have finite dimension; they are orthogonal to each other and span $H(l)$.*

Remark. This result is well known. See, e.g., [21], (10.3) on p. 99 for the case of weight 0.

Proof. Theorem 6.29 in [25], Ch. III, §6.8 shows that it suffices to prove compactness of the resolvent for only one value of λ to obtain it for all λ outside the spectrum. Furthermore, compactness implies that the spectrum is a discrete set, that each $\tilde{\lambda}$ in the spectrum is an eigenvalue of A, and that $\ker(A-\tilde{\lambda})$ has finite dimension. Apply Theorem 2.10 in *loc. cit.*, Ch. V, §2.3 to $(A-\lambda)^{-1}$ with $\lambda < -\frac{1}{4}l^2$ to obtain the orthogonality and completeness of its $\frac{1}{\lambda-\lambda_j}$-eigenspaces. As $H(l)$ is a separable Hilbert space, there are countably many eigenvalues.

We show that $\left(A - \left(-\frac{1}{4}l^2 - 1\right)\right)^{-1}$ is compact. Provide $D(l)$ with the norm $\|.\|_{D,l}$. By Proposition 6.4.2 the inclusion $\mathrm{Id} : D(l) \to H(l)$ is compact. Hence its adjoint $\mathrm{Id}^* : H(l) \to D(l)$ is compact as well; see *loc. cit.*, Ch. III, §4.2, Theorem 4.10. For $g \in H(l)$, $h \in D(l)$ we obtain

$$\mathfrak{s}[\mathrm{Id}^*g,h] \quad = \quad (\mathrm{Id}^*g,h)_{D,l} - \left(1+\frac{1}{4}l^2\right)(\mathrm{Id}^*g,h)_l$$

$$= \quad (g, \mathrm{Id}h)_l - \left(1 + \frac{1}{4}l^2\right)(\mathrm{Id}^* g, h)_l.$$

So $h \mapsto \mathfrak{s}[\mathrm{Id}^* g, h]$ is represented by $g - \left(1 + \frac{1}{4}l^2\right)\mathrm{Id}^* g \in H(l)$. This means that $\mathrm{Id}^* g \in \mathrm{dom}(A)$ and $\left(A + 1 + \frac{1}{4}l^2\right)\mathrm{Id}^* g = g$. Hence $\mathrm{Id}^* = \left(A + 1 + \frac{1}{4}l^2\right)^{-1}$. This operator is compact as an operator $H(l) \to D(l)$, hence as an operator $H(l) \to H(l)$ as well.

6.5.6 *Remark.* If some cuspidal orbit is singular for χ, then the spectrum of $A(\chi, l)$ is not discrete. The part of $H(l)$ corresponding to the continuous spectrum is described by the extension of Eisenstein series; see, e.g., [29], Ch. XIV, §13.

6.6 Relation to automorphic forms

In this book we use the selfadjoint extension $A(\chi, l)$ of the Casimir operator as a means to obtain information on automorphic forms. We now indicate the relation between both concepts by proving that for each $s \in \mathbb{C}$ the kernel of $A(\chi, l) - \frac{1}{4} + s^2$ corresponds to a subspace of $A_l(\chi, \mathbf{c}^{(\chi)}, s)$. The growth condition $\mathbf{c}^{(\chi)}$ is the minimal one for χ. This is in contrast to the previous chapter, where Poincaré series provided us with examples of automorphic forms for a growth condition that is in general not minimal.

6.6.1 *Notation.* Let $s \in \mathbb{C}$. We put $A_l^2(\chi, s) = A_l(\chi, \mathbf{c}^{(\chi)}, s) \cap H(\chi, l)$ for the unitary character χ and the real eigenvalue l. We call the elements of $A_l^2(\chi, s)$ *smooth square integrable automorphic forms.*

In the notation the growth condition is not needed. These spaces of automorphic forms are defined only for unitary characters χ and real weights l.

The Poincaré series $P_l(P, l, s)$ with $P \in \mathcal{P}_Y$ (resolvent kernel) are examples of square integrable automorphic forms that are not smooth on $\tilde{G}$. So these functions are not in $A_l^2(\chi, s)$.

6.6.2 Proposition. *Let $\lambda \in \mathbb{C}$ and let $\mathbf{c}^{(\chi)}$ be the minimal growth condition for χ. The selfadjoint extension $A = A(\chi, l)$ of ω satisfies*

$$\ker(A - \lambda) = A_l^2\left(\chi, \sqrt{\tfrac{1}{4} - \lambda}\right).$$

Remark. The choice of $\pm$ in $\sqrt{\frac{1}{4} - \lambda}$ does not matter.

Proof. First let $f \in \mathrm{dom}(A)$, $Af = \lambda f$. We have to prove that $f \in C^\infty(\tilde{G})$ and $\omega f = \lambda f$. Let $\psi \in C_c^\infty(\mathfrak{H})$, define $h_\psi(p(z)k(\theta)) = \psi(z)e^{il\theta}$. Then

$$k_\psi(g) = \sum_{\gamma \in \tilde{Z}\backslash\tilde{\Gamma}} \chi(\gamma)^{-1} h_\psi(\gamma g)$$

defines $k_\psi(g) \in K(l)$ and $\omega k_\psi = k_\eta$ with $\eta = \left(-y^2\partial_y^2 - y^2\partial_x^2 + ily\partial_x\right)\psi$. The function $z \mapsto f(p(z))$ is locally square integrable on $\mathfrak{H}$, so it gives a distribution Φ.

$$\Phi(\bar\psi) = \int_{\mathfrak{H}} f(p(z))\overline{\psi(z)}\,d\mu(z) = \int_F f(p(z))\overline{k_\psi(p(z))}\,d\mu(z) = (f, k_\psi)_l.$$

The distribution derivative $\left(-y^2\partial_y^2 - y^2\partial_x^2 + ily\partial_x\right)\Phi$ is given by

$$\begin{aligned}\left(\left(-y^2\partial_y^2 - y^2\partial_x^2 + ily\partial_x\right)\Phi\right)(\bar\psi) &= \Phi(\bar\eta) = (f, \omega k_\psi)_l \\ &= \overline{\mathfrak{s}[k_\psi, f]} = \mathfrak{s}[f, k_\psi] = (Af, k_\psi)_l = (\lambda f, k_\psi)_l = \lambda\Phi(\bar\psi).\end{aligned}$$

So Φ is an eigendistribution of an elliptic differential operator with analytic coefficients, hence it is given by a real analytic function; c.f. 2.1.4. This means that f is represented by an element of $C^\infty(\tilde G)$ (also denoted f) and that $\omega f = \lambda f$. Consider Definition 4.3.4. All conditions are satisfied, the regularity even without going over to the $\mathbf{c}$-remainder. So $f \in A_l(\chi, \mathbf{c}^{(\chi)}, s)$ with $\lambda = \frac14 - s^2$.

Conversely, let $f \in A_l(\chi, \mathbf{c}^{(\chi)}, s)$ represent an element of $H(l)$. As $\mathbf{c}^{(\chi)}(P) = \emptyset$ for all $P \in \mathcal{P}_Y$, we have $f \in C^\infty(\tilde G)$. At $P \in X^\infty$ the square integrability implies regularity. We apply Lemma 4.5.2. As $\chi \in \mathcal{X}_u$, we need no condition on the $F_{P,n}f$. We conclude that $\mathbf{E}^+ f$ and $\mathbf{E}^- f$ are regular at all $P \in \mathcal{P}$. So $\mathbf{E}^\pm f$ represent elements of $H(l \pm 2)$ that coincide with the distribution derivatives $\mathbf{E}^\pm f$. This means $f \in D(l)$ and for $k \in K(l)$:

$$\mathfrak{s}[f, k] = (f, \omega k)_l = (\omega f, k)_l = \left((\frac{1}{4} - s^2)f, k\right)_l$$

by 6.5.3 and repeated application of Lemma 6.1.4. As $k \mapsto \mathfrak{s}[f,k]$ and $k \mapsto (\frac14 - s^2)(f,k)_l$ are continuous for the norm $\|.\|_{D,l}$ on $K(l)$, the equality is valid for all $k \in D(l)$. Hence it shows that $f \in \operatorname{dom}(A)$ and $Af = (\frac14 - s^2)f$.

6.6.3 *Remarks.* All cusp forms are smooth square integrable automorphic forms. So the spaces $S_l(\chi, \mathbf{c}^{(\chi)}, s)$ are zero for $\frac14 - s^2 \notin [-\frac14 l^2, \infty)$.

For $f \in A_l(\chi, \mathbf{c}^{(\chi)}, s)$, $f \notin S_l(\chi, \mathbf{c}^{(\chi)}, s)$, the condition $f \in H(l)$ amounts to $\tilde F_{P,0} f$ being regular for all $P \in X^\infty$ with $\chi(\pi_P) = 1$. Combining the condition $\frac14 - s^2 \in [-\frac14 l^2, \infty)$ with Proposition 4.2.11, we see that this is possible only if $-\frac14 l^2 \le \frac14 - s^2 < \frac14$, and that all those $\tilde F_{P,0} f$ should be multiples of $\mu_l(P, 0, -|s|)$. So $A_l^2(\chi, s) \ne S_l(\chi, \mathbf{c}^{(\chi)}, s)$ is possible only if χ is singular at some cusp, and $\pm s \in \left((0, \frac12\sqrt{l^2+1}\right]$. (Even stronger conclusions can be drawn, see Proposition 6.7.8.)

6.6.4 Lemma. *If* $f \in \ker\left(A(\chi, l) - \lambda\right)$, *then* $\mathbf{E}^\pm f \in \ker\left(A(\chi, l \pm 2) - \lambda\right)$ *and* $\mathbf{E}^\mp \mathbf{E}^\pm f = \left(-4\lambda - l^2 \mp 2\right) f$.

Proof. Take $\lambda = \frac14 - s^2$. Proposition 6.6.2 implies that each $f \in \ker(A(\chi, l) - \lambda)$ satisfies $f \in A_l^2(\chi, s)$, with $\lambda = \frac14 - s^2$. Proposition 4.5.3 implies that $\mathbf{E}^\pm f \in A_{l\pm2}(\chi, \mathbf{c}^{(\chi)}, s)$. As $f \in D(l)$, we obtain $\mathbf{E}^\pm f \in \ker(A(\chi, l \pm 2) - \lambda)$ from Proposition 6.6.2. This reasoning, applied to $\mathbf{E}^\pm f$, gives $\mathbf{E}^\mp \mathbf{E}^\pm f \in \ker(A(\chi, l) - \lambda)$. Now use the relations in 2.2.3.

6.7 The discrete spectrum

We have seen in 6.6.3 that the spectral parameter s has to satisfy rather strong conditions if the space $A_l^2(\chi, s)$ of smooth square integrable automorphic forms is to be non-zero. The representational point of view discussed in 4.5.4 gives some more conditions. Those conditions are the subject of this section. Actually, we more or less redo the classification of irreducible unitary representations of $\mathfrak{g}$.

6.7.1 *Automorphic* $\mathfrak{g}$*-modules.* Consider a non-zero smooth square integrable automorphic form $f \in A_l^2(\chi, s)$. Proposition 6.6.2 and Lemma 6.6.4 imply that $(\mathbf{E}^+)^m(\mathbf{E}^-)^n f \in A_{l-2n+2m}^2(\chi, s)$ for all $n, m \in \mathbb{N}_{\geq 0}$. This leads to the submodule $U(\mathfrak{g}) \cdot f$ of the Lie algebra module

$$\bigoplus_{m \equiv l \bmod 2} A_m(\chi, \mathbf{c}^{(\chi)}, s).$$

$U(\mathfrak{g})$ is the enveloping algebra of $\mathfrak{g}$. It consists of all non-commutative polynomials in $\mathbf{E}^+$, $\mathbf{E}^-$ and $\mathbf{W}$. The relations $[\mathbf{E}^-, \mathbf{E}^+] = 4i\mathbf{W}$ and $[\mathbf{W}, \mathbf{E}^\pm] = \pm 2i\mathbf{E}^\mp$ in $\mathfrak{g}$, see 2.2.3, imply that each element of $U(\mathfrak{g}) \cdot f$ may be written as a linear combination of elements $(\mathbf{E}^+)^n(\mathbf{E}^-)^r\mathbf{W}^k \cdot f$. But $\mathbf{W} \cdot f = ilf$, and $\omega f_1 = (\frac{1}{4} - s^2) f_1$ for all $f_1 \in U(\mathfrak{g}) \cdot f$. This can be used to obtain a basis of $U(\mathfrak{g}) \cdot f$ that consists of elements of the form $(\mathbf{E}^+)^n \cdot f$ and $(\mathbf{E}^-)^k \cdot f$. All these basis elements are automorphic forms themselves, so $U(\mathfrak{g}) \cdot f$ does not look too bad after all.

$U(\mathfrak{g}) \cdot f$ carries a scalar product obtained from the scalar products on the $A_m^2(\chi, s) \subset H(\chi, m)$. Lemma 6.2.7 implies that $(\mathbf{E}^+ f_1, f_2) + (f_1, \mathbf{E}^- f_2) = 0$. Hence $(\mathbf{Q} f_1, f_2) + (f_1, \mathbf{Q} f_2) = 0$ for all $\mathbf{Q} \in \mathfrak{g}_r$.

The classification of $U(\mathfrak{g})$ modules with these properties has been carried out. It is closely related to the classification of irreducible unitary representations of $\tilde{G}$. We shall give the various possibilities in Proposition 6.7.3.

6.7.2 *Representations of* $\tilde{G}$. The classification of irreducible unitary representations of $G = \mathrm{SL}_2(\mathbb{R})$ is discussed in, e.g., [29], Ch. VI; consult [48] for the universal covering group $\tilde{G}$ of $\mathrm{SL}_2(\mathbb{R})$. We find three classes of representations, related to the eigenvalue λ of the Casimir operator in the representation:

$$\begin{array}{ll} \lambda \in [\frac{1}{4}, \infty) & \text{principal series,} \\ \lambda \in [\frac{1}{4} - \tau^2, \frac{1}{4}) & \text{complementary series,} \\ \lambda \in (-\infty, \frac{1}{4} - \tau^2) & \text{discrete series.} \end{array}$$

The number $\tau \in (-\frac{1}{2}, \frac{1}{2}]$ depends on the weights occurring in the representation: $2\tau \equiv l + 1 \bmod 2$. It equals $\frac{1}{2}$ for even weights, and 0 for odd weights. For our purpose the following distinction is more useful:

$$\lambda > \frac{1}{4} - \tau^2 \qquad \lambda \text{ is of } \textit{continuous series} \text{ type}$$

$$\lambda \leq \frac{1}{4} - \tau^2 \qquad \lambda \text{ is of } \textit{discrete series} \text{ type.}$$

We shall not explicitly consider these irreducible unitary representations, but start with some non-zero $f \in \ker(A(\chi, l) - \lambda)$ and construct other automorphic forms in $U(\mathfrak{g}) \cdot f$ by differentiation.

For f as indicated, we define $f_q \in \ker(A(\chi, q) - \lambda)$ for $q \equiv l \bmod 2$ by

$$\begin{aligned} f_l &= f \\ f_{q+2} &= \mathbf{E}^+ f_q \qquad \text{for } q \geq l \\ f_{q-2} &= \mathbf{E}^- f_q \qquad \text{for } q \leq l. \end{aligned}$$

Lemma 6.6.4 shows that indeed $f_q \in \ker(A(\chi, q) - \lambda)$. Let us consider the set $\mathcal{L}(f) = \{ q : q \equiv l \bmod 2,\ f_q \neq 0 \}$. Always $l \in \mathcal{L}(f)$. It turns out that the eigenvalue λ determines the form of this set.

6.7.3 Proposition. *Define $\tau = \tau(\chi)$ by $\tau \in (-\frac{1}{2}, \frac{1}{2}]$, $\chi(k(\pi)) = -e^{2\pi i \tau}$. Let $f \in \ker(A(\chi, l) - \lambda)$, $f \neq 0$. Then the eigenvalue λ and the weight l determine the form of the set $\mathcal{L}(f) = \{ q : q \equiv l \bmod 2,\ f_q \neq 0 \}$ according to the following rules:*

i) Continuous series type. *If $\lambda > \frac{1}{4} - \tau^2$, then $\mathcal{L}(f) = l + 2\mathbb{Z}$.*

ii) Holomorphic discrete series type. *If $l \equiv b \bmod 2$ and $\lambda = \frac{1}{2}b(1 - \frac{1}{2}b)$ for some $b \in (0, l]$, then $\mathcal{L}(f) = \{ q : q \equiv l \bmod 2,\ q \geq b \}$.*

iii) Antiholomorphic discrete series type. *If $l \equiv -b \bmod 2$ and $\lambda = \frac{1}{2}b(1 - \frac{1}{2}b)$ for some $b \in (0, -l]$, then $\mathcal{L}(f) = \{ q : q \equiv l \bmod 2,\ q \leq -b \}$.*

iv) Constant function. *If $l = \lambda = 0$, then $\mathcal{L}(f) = \{0\}$.*

These four cases exhaust the possibilities.

Remarks. If λ has discrete series type then $\mathbf{E}^- f_q = 0$ or $\mathbf{E}^+ f_q = 0$ occurs for some $q \in \mathcal{L}(f)$. We have noted in 2.3.7 that $\mathbf{E}^- f_q = 0$ means that f_q corresponds to a holomorphic automorphic form. Similarly, if $\mathbf{E}^+ f_q = 0$, then f_q corresponds to an antiholomorphic automorphic form on $\mathfrak{H}$.

If λ has continuous series type, $\mathbf{E}^\pm f_q = 0$ does not occur. The corresponding f_q are often called *Maass forms.*

Proof. As $\lambda \in \mathbb{R}$, we may write $\lambda = \frac{1}{4} - s^2$ with $s \in i(0, \infty) \cup [0, \infty)$. Take a non-zero $f \in \ker(A(\chi, l) - \lambda)$ and form $\mathcal{L}(f)$ as indicated above. Lemmas 6.2.7 and 6.6.4 give, for each $q \in \mathcal{L}(f)$ with $\pm(q - l) \geq 0$:

$$\begin{aligned} \|f_{q\pm 2}\|^2_{q\pm 2} &= (\mathbf{E}^\pm f_q, f_{q\pm 2})_{q\pm 2} \\ &= -(f_q, \mathbf{E}^\mp \mathbf{E}^\pm f_q)_q = 4\left(\frac{1 \pm q}{2} + s\right)\left(\frac{1 \pm q}{2} - s\right) \|f_q\|^2_q. \end{aligned}$$

This has the following consequences:

$$q,\, q+2 \in \mathcal{L}(f) \implies \left(\frac{1+q}{2}+s\right)\left(\frac{1+q}{2}-s\right) > 0,$$
$$q \in \mathcal{L}(f),\ \left(\frac{q\pm 1}{2}+s\right)\left(\frac{q\pm 1}{2}-s\right) > 0 \implies q\pm 2 \in \mathcal{L}(f).$$

The way we have constructed $\mathcal{L}(f)$ implies that there are b and c with $-\infty \leq b \leq l \leq c \leq \infty$, such that $\mathcal{L}(f) = \{\, q \equiv l \bmod 2 : b \leq q \leq c \,\}$.

The condition $\left(\frac{1+q}{2}+s\right)\left(\frac{1+q}{2}-s\right) > 0$ is satisfied for all $q \equiv l \bmod 2$ if and only if either $s \in i(0,\infty)$ or $0 \leq s < |\tau|$. This corresponds to $\lambda > \frac{1}{4} - \tau^2$. In this case $\mathcal{L}(f) = l + 2\mathbb{Z}$.

If $\mathcal{L}(f)$ has a minimum b, then $b \equiv l \bmod 2$, $b \leq l$ and

$$0 = \|f_{b-2}\|_{b-2}^2 = 4\left(\frac{1-b}{2}+s\right)\left(\frac{1-b}{2}-s\right)\|f_b\|_b^2,$$

so $\left(\frac{b-1}{2}+s\right)\left(\frac{b-1}{2}-s\right) = 0$. This implies $s = \left|\frac{b-1}{2}\right|$, $\lambda = \frac{1}{2}b(1-\frac{1}{2}b)$. Similarly, if $\mathcal{L}(f)$ has a maximum c, then $c \equiv l \bmod 2$, $c \geq l$, $s = \left|\frac{1+c}{2}\right|$, $\lambda = \frac{1}{2}(-c)(1-\frac{1}{2}(-c)) = -\frac{1}{2}c(1+\frac{1}{2}c)$.

If $\mathcal{L}(f)$ is bounded, then $b = \min \mathcal{L}(f)$ and $c = \max \mathcal{L}(f)$ satisfy $\frac{1}{2}b(1-\frac{1}{2}b) = -\frac{1}{2}c(1+\frac{1}{2}c)$, hence $b = -c$, $c \geq 0$ and $b, c \in \mathbb{Z}$. If $c > 0$, then $-c, -c+2 \in \mathcal{L}(f)$, hence $\left(\frac{1-c}{2}+\frac{c+1}{2}\right)\left(\frac{1-c}{2}-\frac{c+1}{2}\right) > 0$; this is impossible. So $b = c = l = \lambda = 0$ is the only possibility for bounded $\mathcal{L}(f)$.

In all other cases $\mathcal{L}(f)$ is bounded on only one side. We consider the case $b = \min \mathcal{L}(f)$, and leave the other case to the reader.

We have $|\tau| \leq s = \left|\frac{b-1}{2}\right|$, $b \leq l$, and $b \equiv l \bmod 2$. The product $\left(\frac{1+q}{2}+s\right)\left(\frac{1+q}{2}-s\right)$ has to be positive for all $q \equiv b \bmod 2$, $q \geq b$. Take $q = 2\tau - 1$ to see that $b > 2\tau - 1$, hence $b > 0$. If $b > 0$, then $s = \left|\frac{b-1}{2}\right|$ and $\left(\frac{1+q}{2}+s\right)\left(\frac{1+q}{2}-s\right) = \left(\frac{1+q}{2}+\frac{b-1}{2}\right)\left(\frac{1+q}{2}-\frac{b-1}{2}\right) > 0$ for all $q \geq b$, $q \equiv b \bmod 2$, so $\mathcal{L}(f) = \{\, q : q \equiv l \bmod 2,\ q \geq b \,\}$.

6.7.4 *Remarks.* This proposition implies the lower bound $\frac{1}{2}|l| - \frac{1}{4}l^2$ for the eigenvalues; this is better than the bound $-\frac{1}{4}l^2$ for $A(\chi, l)$ itself. As we have not shown that the continuous spectrum is contained in $[\frac{1}{4}, \infty)$, we keep using the lower bound $-\frac{1}{4}l^2$ for $A(\chi, l)$.

The eigenvalues $\lambda \leq \frac{1}{4} - \tau^2$ come from automorphic forms f obtained by repeated differentiation of a holomorphic or an antiholomorphic automorphic form on $\mathfrak{H}$. So these eigenvalues are related to automorphic forms that are reasonably well known. The last case in the proposition, $\mathcal{L}(f) = \{0\}$, corresponds to constant functions, so it can only occur if $\chi = 1$.

There is much less information on the automorphic forms with eigenvalue $\lambda > \frac{1}{4} - \tau^2$. In the spectral theory of automorphic forms, one often knows only distribution results averaged over the spectrum. In applications it is sometimes important to know the eigenvalues smaller than $\frac{1}{4}$, usually called *exceptional eigenvalues.* The exceptional eigenvalues $\leq \frac{1}{4} - \tau^2$ are known from the holomorphic theory. If $l \notin 1 + 2\mathbb{Z}$, the interval $(\frac{1}{4} - \tau, \frac{1}{4})$ may contain exceptional eigenvalues of continuous series type.

6.7.5 *Cuspidal eigenvalues.* We call $\frac{1}{4} - s^2 \in \mathbb{R}$ a *cuspidal eigenvalue* for $\tilde{\Gamma}$ and χ, if $S_l(\chi, \mathbf{c}^{(\chi)}, s) \neq \{0\}$ for some l satisfying $e^{\pi i l} = \chi(k(\pi))$.

The next proposition shows that eigenvalues of discrete series type tend to be cuspidal.

6.7.6 Proposition. *Let $b \geq 1$, $l \equiv \pm b \bmod 2$, $\pm l \geq b$. Then the kernel of $A(\chi, l) - \frac{1}{2}b(1 - \frac{1}{2}b)$ is equal to $S_l(\chi, \mathbf{c}^{(\chi)}, \frac{b-1}{2})$.*

Let $b > 0$, $l \equiv \pm b \bmod 2$, $\pm l \geq b$. If $P \in \mathcal{P}$, $n \in \mathcal{C}_P$ satisfy $\pm n < 0$ if $P \in X^\infty$, $\pm n < b$ if $P \in \mathcal{P}_Y$, then $F_{P,n} f = 0$ for all $f \in \ker\left(A(\chi, l) - \frac{1}{2}b(1 - \frac{1}{2}b)\right)$.

Remark. The first assertion need not hold for $0 < b < 1$. See 14.3.1 for a counterexample.

Proof. Application of $\mathbf{E}^\mp$ reduces the proof to the case $l = \pm b$.

Let $f \in \ker(A(\chi, \pm b) - \frac{1}{2}b(1 - \frac{1}{2}b))$, $f \neq 0$. Then $\mathbf{E}^\mp f = 0$. The Fourier coefficient $F_{P,0} f$ is in $L^2((A_P, \infty), y^{-2} dy)$ for each $P \in X^\infty$ with $\chi(\pi_P) = 1$. It satisfies $e^{\mp i\theta} (\mp 2iy\partial_x + 2y\partial_y \pm i\partial_\theta) F_{P,0} f(y) e^{\pm ib\theta} = 0$, hence it is a multiple of $y \mapsto y^{b/2}$. As $b \geq 1$, it can only be square integrable if it vanishes.

Let $P \in \mathcal{P}$, $n \in \mathcal{C}_P$, $n \neq 0$ if $P \in X^\infty$. Suppose $F_{P,n} f \neq 0$. As $\tilde{F}_{P,n} f$ is a multiple of $\omega_{\pm b}(P, n, \frac{b-1}{2})$, we conclude from $\mathbf{E}^\mp f = 0$ that $\mathbf{E}^\mp \omega_{\pm b}(P, n, \frac{b-1}{2}) = 0$. In Table 4.1 on p. 63 we see that this implies $\mp n < 0$ if $P \in X^\infty$, and $\pm n \geq b$ otherwise.

6.7.7 *Spectral sets.* We define

$$\begin{aligned}
\Sigma(\chi, l) &= \left\{ s \in \mathbb{C} : \operatorname{Re} s \geq 0,\ A_l^2(\chi, s) \neq \{0\} \right\} \\
\Sigma_c(\chi) = \Sigma_c(\chi, l) &= \left\{ s \in \Sigma(\chi, l) : \frac{1}{4} - s^2 > \frac{1}{4} - \tau(\chi)^2 \right\} \\
\Sigma_d(\chi, l) &= \left\{ s \in \Sigma(\chi, l) : s \geq |\tau(\chi)| \right\} \\
\Sigma^0(\chi, l) &= \left\{ s \in \Sigma(\chi, l) : S_l(\chi, \mathbf{c}^{(\chi)}, s) \neq \{0\} \right\} \\
\Sigma^e(\chi, l) &= \left\{ s \in \Sigma(\chi, l) : A_l^2(\chi, s) \neq S_l(\chi, \mathbf{c}^{(\chi)}, s) \right\} \\
\Sigma_a^b(\chi, l) &= \Sigma^b(\chi, l) \cap \Sigma_a(\chi, l) \quad \text{for } a = c, d \text{ and } b = 0, e.
\end{aligned}$$

The set $\Sigma_c(\chi, l)$ does not depend on l, see Proposition 6.7.3. Clearly $\Sigma(\chi, l) = \Sigma_d(\chi, l) \sqcup \Sigma_c(\chi)$ (disjoint union), and $\Sigma(\chi, l) = \Sigma^0(\chi, l) \cup \Sigma^e(\chi, l)$ (not necessarily disjoint). $\Sigma^e(\chi,l) \neq \emptyset$ can occur only if some $P \in X^\infty$ is singular for χ. Proposition 6.7.3 implies

$$\begin{aligned} \Sigma_c(\chi) &\subset i\mathbb{R} \cup (0, |\tau(\chi)|) \\ \Sigma_d(\chi,l) &\subset \left\{ \left|\frac{b-1}{2}\right| : 0 < b \leq |l|,\, b \equiv |l| \bmod 2 \right\} \quad \text{if } l \neq 0 \\ \Sigma_d(\chi,0) &\subset \left\{\frac{1}{2}\right\}. \end{aligned}$$

So $\Sigma_d(\chi,l)$ is a finite set. In Corollary 9.2.7 and 11.2.5 we shall see that $\Sigma_c(\chi)$ is a discrete, not necessarily finite set. In the case that all cusps are regular for the character χ, the discreteness of $\Sigma(\chi,l)$ follows from the compactness of the resolvent, see Theorem 6.5.5. In that case $\Sigma(\chi,l)$ is infinite.

In the modular case, with trivial character and weight 0 as considered in Section 1.2, we have $\Sigma_d(1,0) = \{\frac{1}{2}\}$, corresponding to the constant functions. All other smooth square integrable modular forms are of continuous series type: the cuspidal Maass forms, see 1.4.5. We have to go to other weights to obtain holomorphic cusp forms. The discussion above gives the reason from a representational point of view. (For another point of view see, e.g., Theorem 2.24 in Shimura, [53].) Section 13.3 gives more information concerning the modular case.

6.7.8 Proposition. $\Sigma^e(\chi,l) \subset (0, |\tau(\chi)|]$.

If $\tau(\chi) = \frac{1}{2}$, *then* $\frac{1}{2} \in \Sigma^e(\chi,l)$ *if and only if* $l = 0$ *and* $\chi = 1$. *In that case* $A_0^2(1,0) = \mathbb{C} \cdot \mathbf{1}$.

Proof. Let $s \in \Sigma^e(\chi,l)$. Remark 6.6.3 shows that $s > 0$. Propositions 6.7.3 and 6.7.6 imply $s \leq |\tau(\chi)|$.

Let $\tau(\chi) = \frac{1}{2}$. If $s = \frac{1}{2}$, then Proposition 6.7.3 leaves $l = 0$ as the only possibility. Any $f \in A_0^2(\chi, \frac{1}{2})$ has to satisfy $\mathbf{E}^\pm f = 0$, again by Proposition 6.7.3. Hence f is constant, and $\chi = 1$ if f is non-zero.

Chapter 7
Families of automorphic forms

Up till now we have studied individual automorphic forms. These turn out to occur in families in a natural way. For instance, the Poincaré series depends continuously on the weight l, the unitary character χ, and the spectral parameter s. The dependence on s is holomorphic. It is not clear from the series that there might be a holomorphic or meromorphic dependence on the character and the weight. In Chapters 9 and 10 we shall get meromorphy by functional analytic methods.

In Section 7.2 we state our concept of holomorphic families of automorphic forms. There are various concepts of holomorphic dependence on the parameters. One concept is pointwise holomorphy, another one is holomorphy in Hilbert space sense. In this book, it turns out to be useful to work with a mixture of these concepts.

We view the parameters as functions on a complex manifold, the parameter space, discussed in Section 7.1. As this parameter space has, in general, complex dimension larger than one, we need results on holomorphic functions in several complex variables. Most of these can be found in Hörmander, [23]; occasionally we refer to [17] of Grauert and Remmert. Families of automorphic forms need not be defined on the whole parameter space, and even if they are, we are sometimes interested solely in the behavior near a certain point. The language of sheaves is suitable for discussions of this kind.

The next chapters use mainly the concepts and results stated in Sections 7.1 and 7.2. The other sections of this chapter give proofs, and further information to which we shall refer only occasionally.

The Fourier terms of holomorphic families of automorphic forms depend on the parameters in a holomorphic way as well. For that reason we consider in Sections 7.3 and 7.4 a more general concept: 'holomorphic family of eigenfunctions'. Later on we specialize this to families of automorphic forms and to families of functions that may occur in Fourier expansions.

7.1 Parameter spaces

We use complex manifolds as parameter spaces. In this book we restrict ourselves to simply connected parameter spaces. This is a practical restriction. It implies that the equation $e^g = f$ has a solution g holomorphic on an open set Ω for each holomorphic f that is nowhere zero on Ω. The complications of topological nature that we avoid in this way may be interesting.

A still greater generalization would be to allow all complex spaces as a parameter space. That would make generalization of the results in 7.3.2 difficult and perhaps impossible.

7.1.1 *Structure sheaf.* The fact that a topological space W is a complex manifold implies that each point of W is provided with a neighborhood and a homeomorphism from this neighborhood to an open set in some $\mathbb{C}^n$. Functions on an open subset $\Omega \subset W$ are called holomorphic if their restrictions to such neighborhoods correspond to holomorphic functions on the corresponding open sets in $\mathbb{C}^n$, i.e., locally given by converging power series in the coordinates.

By $\mathcal{O}(\Omega)$ we indicate the ring of holomorphic functions on Ω, and by $\mathcal{O}(\Omega)^*$ the group of units in this ring. The sheaf $\mathcal{O} = \mathcal{O}_W$, given by $\Omega \mapsto \mathcal{O}(\Omega)$, is called the structure sheaf of W. Consult, e.g., Grauert & Remmert, [17], Ch. 1, §1.1–5, for complex manifolds and more general complex spaces. According to *loc. cit.* we should have said '$\mathcal{O}$ is the sheaf associated to the presheaf $\Omega \mapsto \mathcal{O}(\Omega)$', but for sheaves we follow the terminology of Hartshorne, [19], Ch. II, §1.

7.1.2 *Definition.* A *parameter space* for $\tilde{\Gamma}$ is a simply connected complex manifold $(W, \mathcal{O})$ provided with a holomorphic function $s \in \mathcal{O}(W)$ and a group homomorphism $\chi : \tilde{\Gamma} \to \mathcal{O}(W)^*$.

The group $\tilde{\Gamma}$ is finitely generated. So the homomorphism χ is determined by finitely many holomorphic functions that have no zeros in W.

We shall use $\lambda = \frac{1}{4} - s^2 \in \mathcal{O}(W)$ as the eigenvalue. The condition that the spectral parameter s is holomorphic on W is a further restriction of the concept of parameter space.

7.1.3 *Weights and Fourier term orders.* For each parameter space we define

$$\begin{aligned} \mathbf{wt} &= \{\, l \in \mathcal{O}(W) : e^{\pi i l} = \chi(k(\pi)) \,\} \\ \mathcal{C}_P &= \{ n \in \mathcal{O}(W) : n(w) \in \mathcal{C}_P(\chi(w)) \text{ for all } w \in W \,\} \qquad \text{for } P \in \mathcal{P}. \end{aligned}$$

The set **wt** of *weights* and the set $\mathcal{C}_P$ of *Fourier term orders at* $P \in \mathcal{P}$ do not depend on the choice of s. These subsets of $\mathcal{O}(W)$ are non-empty, as we have supposed W to be simply connected. For more general parameter spaces one should define **wt** and the $\mathcal{C}_P$ as locally constant sheaves.

7.1.4 *Trivial example.* Choose $s \in \mathbb{C}$ and $\chi \in \mathcal{X}$. Let W consist of one point t; then $\mathcal{O}(W) = \mathbb{C}$. Take $s : t \mapsto s$ and $\chi : t \mapsto \chi$ to view W as a parameter space.

In this example $\mathcal{C}_P = \mathcal{C}_P(\chi)$, as discussed in 4.1.2 and 4.1.3. Hence it does no harm to use the same notation.

7.1.5 *Other examples.* Fix $\chi \in \mathcal{X}_u$. Take $W = \mathbb{C}_{1/2} = \{\, s \in \mathbb{C} : \operatorname{Re} s > \frac{1}{2} \,\}$; put $s : s \mapsto s$ and $\chi : s \mapsto \chi$. This is the parameter space on which the family of Poincaré series $s \mapsto P_l(P, n, \chi, s)$ lives. The sets **wt** and $\mathcal{C}_P$ consist of constant functions, as χ is constant.

For the powers of the eta function, see 13.1.7, a suitable parameter space would be $W = \mathbb{C}$, with $s : r \mapsto \frac{r-1}{2}$ and $\chi : r \mapsto \chi_r$. For $P = \tilde{\Gamma}_{\mathrm{mod}} \cdot i\infty$ we see that $\mathcal{C}_P$ consists of the functions $r \mapsto \frac{r}{12} + \nu$, with $\nu \in \mathbb{Z}$. The weights are $\mathbf{wt} = \{r \mapsto r + 2k : k \in \mathbb{Z}\}$. As $\chi_{r+12} = \chi_r$ one might be inclined to take $W = \mathbb{C} \bmod 12\mathbb{Z}$ as the parameter space. Then it would be impossible to define $\mathcal{C}_P$ and $\mathbf{wt}$ globally.

See 15.6.1 for another example.

7.1.6 *Morphisms.* A morphism of parameter spaces $j : W_1 \to W$ is a holomorphic map $j : W_1 \to W$ such that $s_{W_1} = s_W \circ j$ and $\chi_{W_1}(\gamma) = \chi_W(\gamma) \circ j$ for all $\gamma \in \tilde{\Gamma}$. Composition with j also maps $\mathbf{wt}_W$ into $\mathbf{wt}_{W_1}$; similarly for the $\mathcal{C}_P$.

We shall work almost exclusively with a fixed parameter space W, and only occasionally use a morphism $j : W_1 \to W$ of parameter spaces. Then objects like $\mathcal{O}$, $\mathbf{wt}$, or $\mathcal{C}_P$ will refer to the main parameter space W. If an object is meant to be related to W_1, we attach the name of the morphism as a subscript between brackets. For example, $\mathcal{O}_{(j)}$ denotes the structure sheaf of W_1 and $\mathcal{C}_{P,(j)}$ is a subset of $\mathcal{O}_{(j)}(W_1)$.

7.2 Holomorphic families of automorphic forms and Fourier terms

In Chapter 4 we have defined automorphic forms satisfying a growth condition. In this section we extend the concepts considered in that chapter. Instead of individual automorphic forms, we study holomorphic families of automorphic forms. Taking a Fourier term becomes a map which associates a family of Fourier terms to a family of automorphic forms. In this section we give definitions and state the main results. We give proofs in the later sections of this chapter. We view families of automorphic forms as sections of a sheaf of $\mathcal{O}$-modules on W, similarly for families of cusp forms and of Fourier terms. Taking a Fourier term gives rise to a morphism of $\mathcal{O}$-modules. The same holds for the differential operators $\mathbf{E}^{\pm}$.

7.2.1 *Definition.* A *growth condition* $\mathbf{c}$ for $\mathcal{P} \subset X$ on the open subset Ω of the parameter space W is a map assigning a *finite* subset $\mathbf{c}(P)$ of $\mathcal{C}_P \subset \mathcal{O}(W)$ to each $P \in \mathcal{P}$; if $P \in X^\infty$ the set $\mathbf{c}(P)$ has to contain at least those $n \in \mathcal{C}_P$ for which $\mathrm{Re}\, n$ is zero somewhere in Ω. This implies that $\mathbf{c}$ determines a growth condition $\mathbf{c}(w)$ suitable for $\chi(w)$ for each $w \in \Omega$; see 4.3.1.

For a large parameter space W it may happen that there are no growth conditions on the whole of W. For instance, consider the parameter space $W = \mathbb{C}$, $s : r \mapsto \frac{r-1}{2}$, $\chi : r \mapsto \chi_r$, mentioned in 7.1.5. For $P = \tilde{\Gamma}_{\mathrm{mod}} \cdot i\infty$ the set $\mathcal{C}_P$ consists of the functions $r \mapsto \frac{r}{12} + \nu$, with $\nu \in \mathbb{Z}$; all these functions have real part equal to zero somewhere in W. The set $\mathbf{c}(P)$ should be finite, so there are no global growth conditions on this parameter space.

7.2.2 *Notations.* For growth conditions $\mathbf{c}$ and $\mathbf{d}$ on $\Omega \subset W$, we define $\mathbf{c} \subset \mathbf{d}$ to mean $\mathbf{c}(P) \subset \mathbf{d}(P)$ for all $P \in \mathcal{P}$. Take in this situation $\Omega_{\mathrm{c}} \supset \Omega$ maximal such that $\mathbf{c}$ is a growth condition on Ω_{c}, and similarly Ω_{d} for $\mathbf{d}$. Then $\Omega_{\mathrm{d}} \supset \Omega_{\mathrm{c}}$.

Define $\mathbf{n}$ by $\mathbf{n}(P) = \emptyset$ for all $P \in \mathcal{P}$. This does not define a growth condition $\mathbf{n}$ on $\Omega \subset W$ if there are $P \in X^\infty$ and $n \in \mathcal{C}_P$ for which $\mathrm{Re}\, n$ has a zero somewhere in Ω. If $\mathbf{n}$ is a growth condition we call it *the empty growth condition.*

7.2.3 *Holomorphic families of automorphic forms.* Let $\Omega \subset W$ be open. We have to decide under what conditions we call a function f on $\Omega \times \tilde{G}_{\mathcal{P}}$ a *holomorphic family of automorphic forms* on Ω *for the growth condition* $\mathbf{c}$. It seems sensible to impose at least the following conditions:

i) $f \in C^\infty(\Omega \times \tilde{G}_{\mathcal{P}})$.

ii) $f(w) : g \mapsto f(w; g)$ is in $A_{l(w)}(\chi(w), \mathbf{c}(w), s(w))$ for each $w \in \Omega$.

iii) $f_g : w \mapsto f(w; g)$ is holomorphic on Ω for each $g \in \tilde{G}_{\mathcal{P}}$.

The regularity condition iii) in Definition 4.3.4 should also be generalized. Let us stay as close as possible to Definition 4.3.4 of individual automorphic forms.

Let $P \in \mathcal{P}$ and $n \in \mathcal{C}_P$. The *Fourier term* $\tilde{F}_{P,n} f \in C^\infty(\Omega \times \mathrm{pr}^{-1} \dot{U}_P(A_P))$ is defined by $\left(\tilde{F}_{P,n} f\right)(w; g) = \left(\tilde{F}_{P,n(w)} f(w)\right)(g)$. As a Fourier term is given by integration over a compact set, we have holomorphy of $w \mapsto \tilde{F}_{P,n} f(w; g)$ for each $g \in \mathrm{pr}^{-1} \dot{U}_P(A_P)$. We define $f[\mathbf{c}, P] = f - \sum_{n \in \mathbf{c}(P)} \tilde{F}_{P,n} f$; this is a C^∞-function on $\Omega \times \mathrm{pr}^{-1} \dot{U}_P(A_P)$.

It seems sensible to impose the following additional conditions on f:

iv) For each $P \in \mathcal{P}_Y$ the function $f[\mathbf{c}, P]$ can be extended to give a C^∞-function on $\Omega \times U_P(A_P)$.

v) For $P \in X^\infty$ we take some $n_P \in \mathcal{C}_P$.

$$F_{w,P}(z) = e^{-2\pi i n_P(w) \,\mathrm{Re}(z)} f[\mathbf{c}, P](w, g_P p(z))$$

defines a square integrable function on $S_{P,A_P} = \{\, z \in \mathfrak{H} : 0 \leq x < 1,\, y > A_P \,\}$ with respect to $d\mu(z) = y^{-2} dx \wedge dy$. The condition is that $w \mapsto F_{w,P}$ is an L^2-holomorphic function $\Omega \to L^2(S_{P,A_P}, d\mu)$ for each $P \in X^\infty$.

The L^2-holomorphy means that near each $\tilde{w} \in \Omega$ the $F_{w,P}$ are given by a series

$$\sum w_1^{n_1} \cdots w_N^{n_N} h_{n_1,\ldots,n_N}$$

converging in the L^2-norm; $w_1, \ldots, w_N$ are complex coordinates with $w_j(\tilde{w}) = 0$.

In the case $P \in \mathcal{P}_Y$, the function $w \mapsto f[\mathbf{c}, P](w, g_0)$ is holomorphic for g_0 with $\mathrm{pr}(g_0) = P$ as well. Indeed, the functions $\partial_{\overline{w_j}} f[\mathbf{c}, P]$ are elements of $C^\infty(\Omega \times \mathrm{pr}^{-1} U_P(A_P))$, and vanish on the dense subset $\Omega \times \mathrm{pr}^{-1} \dot{U}_P(A_P))$. Hence the derivatives $\partial_{\overline{w_j}} f[\mathbf{c}, P]$ are identically zero.

7.2.4 *Example: the family* $r \mapsto \eta_r$. Take for the modular group $\tilde\Gamma_{\text{mod}}$ the parameter space $W = \{ r \in \mathbb{C} : \operatorname{Re} r > 0 \}$, $\chi : r \mapsto \chi_r$, $s : r \mapsto \frac{r-1}{2}$. Take $\mathcal{P} = \{P\}$, $P = \tilde\Gamma_{\text{mod}} \cdot i\infty$. The empty growth condition $\mathbf{n}$ is suitable on W. We consider the family $f : r \mapsto \eta_r$ given by $\eta_r(p(z)k(\theta)) = y^{r/2} e^{2r \log \eta(z)} e^{ir\theta}$, with η the eta-function of Dedekind. Conditions i)–iii) above are clearly satisfied, condition iv) is not relevant. For condition v), note that $f[\mathbf{n}, P] = f$. We have to check that $H(r) : z \mapsto e^{-\pi i r x/6} \eta_r(p(z)$ is L^2-holomorphic on $S_{P,a} = \{ z \in \mathfrak{H} : y > a \}$ for some $a \geq 1$. The norm in $L^2(S_{P,2})$ of this function, given by

$$\|H(r)\|^2 = \int_a^\infty \int_{-1/2}^{1/2} y^{\operatorname{Re} r} e^{-\pi y \operatorname{Re} r/6} \left| \prod_{m=1}^{\infty} (1 - e^{2\pi i m z})^{2r} \right|^2 dx \, \frac{dy}{y^2},$$

is bounded for r in compact sets in W. As H is pointwise holomorphic in r, all scalar products with compactly supported functions on $S_{P,a}$ are holomorphic in r. We obtain the L^2-holomorphy from the fact that weak holomorphy implies holomorphy in norm. (See, e.g., [29], App. 5, §1; in Lemma 7.4.3 we have given a version of this result that discusses L^2-holomorphy of families of more than one variable.)

7.2.5 *Example: Poincaré series.* In 7.5.3 we shall see that the Poincaré series defined in Proposition 5.1.6 gives a holomorphic family of automorphic forms with s as the parameter.

7.2.6 *Definition.* Let $\mathbf{c}$ be a growth condition on an open subset Ω_0 of W; let $l \in \mathbf{wt}$. For each open $\Omega \subset \Omega_0$, we define $\mathcal{A}_l(\mathbf{c}; \Omega)$ as the linear space of all holomorphic families of automorphic forms on Ω for the growth condition $\mathbf{c}$. If $\Omega_1 \subset \Omega$ we have the restriction map $\mathcal{A}_l(\mathbf{c}; \Omega) \to \mathcal{A}_l(\mathbf{c}; \Omega_1)$. One can check that $\Omega \mapsto \mathcal{A}_l(\mathbf{c}; \Omega)$ defines a *sheaf* $\mathcal{A}_l(\mathbf{c})$ on Ω_0. Multiplication by a holomorphic function on Ω maps $\mathcal{A}_l(\mathbf{c}; \Omega)$ into itself. This gives $\mathcal{A}_l(\mathbf{c})$ the structure of an $\mathcal{O}$-module on Ω_0.

7.2.7 *Definition.* $\mathcal{S}_l(\mathbf{c})$ is the $\mathcal{O}$-submodule of $\mathcal{A}_l(\mathbf{c})$ of holomorphic families of *cusp forms* for the growth condition $\mathbf{c}$. It is defined by

$$\mathcal{S}_l(\mathbf{c}; \Omega) = \left\{ f \in \mathcal{A}_l(\mathbf{c}; \Omega) : f(w) \in S_{l(w)}(\chi(w), \mathbf{c}(w), s(w)) \text{ for all } w \in \Omega \right\}.$$

7.2.8 *Definition.* Let $P \in \mathcal{P}$, $n \in \mathcal{C}_P$, $l \in \mathbf{wt}$. Put $\tilde G_P = \tilde G \setminus \{P\}$. Note that $\tilde G_P = \tilde G$ if $P \in X^\infty$.

The $\mathcal{O}$-module $\mathcal{W}_l(P, n)$ of Fourier terms is determined by the condition that $\mathcal{W}_l(P, n; \Omega)$ consists of the $f \in C^\infty(\Omega \times \tilde G_P))$ that satisfy

i) $f(w) : g \mapsto f(w; g)$ is in $W_{l(w)}(P, n(w), s(w))$ for each $w \in \Omega$.

ii) $w \mapsto f(w; g)$ is in $\mathcal{O}(\Omega)$ for each $w \in \tilde G_P$.

The condition $f(w) \in W^0_{l(w)}(P, n(w), s(w))$ for all $w \in \Omega$ determines a submodule $\mathcal{W}^0_l(P,n)$ of $\mathcal{W}_l(P,n)$.

7.2.9 Proposition. *Let* $\mathbf{c}$ *be a growth condition on* $\Omega \subset W$, *and let* $l \in \mathbf{wt}$.

i) Differentiation. *The operators* $\mathbf{E}^+$ *and* $\mathbf{E}^-$ *define morphisms of* $\mathcal{O}$*-modules* $\mathbf{E}^\pm : \mathcal{A}_l(\mathbf{c}) \to \mathcal{A}_{l\pm 2}(\mathbf{c})$ *and* $\mathbf{E}^\pm : \mathcal{S}_l(\mathbf{c}) \to \mathcal{S}_{l\pm 2}(\mathbf{c})$.

ii) Divisibility. *Let* $f \in \mathcal{A}_l(\mathbf{c}; \Omega_1)$, *respectively* $\mathcal{S}_l(\mathbf{c}; \Omega_1)$. *Let* $\psi \in \mathcal{O}(\Omega_1)$, ψ *not identically zero on* Ω_1. *Suppose that for each* $g \in \tilde{G}_{\mathcal{P}}$ *the function* $w \mapsto \frac{1}{\psi(w)} f(w; g)$ *extends as a holomorphic function on* Ω_1. *Then there exists* $h \in \mathcal{A}_l(\mathbf{c}; \Omega_1)$, *respectively* $\mathcal{S}_l(\mathbf{c}; \Omega_1)$, *such that* $f = \psi \cdot h$.

iii) Fourier terms. *For each* $P \in \mathcal{P}$, $n \in \mathcal{C}_P$, *there is a morphism of* $\mathcal{O}$*-modules* $\tilde{F}_{P,n} : \mathcal{A}_l(\mathbf{c}) \to \mathcal{W}_l(P,n)$. *It is given by* $\left(\tilde{F}_{P,n} f\right)(w) = \tilde{F}_{P,n(w)} f(w)$, *for* $f \in \mathcal{A}_l(\mathbf{c}; \Omega)$.

If $n \notin \mathbf{c}(P)$, *then* $\tilde{F}_{P,n} \mathcal{A}_l(\mathbf{c}) \subset \mathcal{W}^0_l(P,n)$. *If* $n \in \mathbf{c}(P)$, *then* $\tilde{F}_{P,n} \mathcal{S}_l(\mathbf{c}) = \{0\}$.

Proof. See 7.5.1, 7.5.2, and 7.7.2.

7.2.10 *Dependence on the growth condition.* If $\mathbf{c} \subset \mathbf{d}$ for growth conditions $\mathbf{c}$ and $\mathbf{d}$ on Ω, then $\mathcal{S}_l(\mathbf{d}) \subset \mathcal{S}_l(\mathbf{c}) \subset \mathcal{A}_l(\mathbf{c}) \subset \mathcal{A}_l(\mathbf{d})$. Furthermore, $\mathcal{S}_l(\mathbf{n}) = \mathcal{A}_l(\mathbf{n})$ and $\bigcap_{\mathbf{d} \supset \mathbf{c}} \mathcal{S}_l(\mathbf{d}) = \{0\}$. For the last fact, note that the $\tilde{F}_{P,n}$ describe the Fourier terms on the set $\mathrm{pr}^{-1} \dot{U}_P(A_P)$. If an analytic function vanishes on only one $\mathrm{pr}^{-1} \dot{U}_P(A_P)$, it vanishes on $\tilde{G}_{\mathcal{P}}$. (We can restrict the intersection to a collection of growth conditions for which $\bigcup_{\mathbf{d}} \mathbf{d}(Q) = \mathcal{C}_Q$ for one $Q \in \mathcal{P}$.)

7.2.11 Proposition. *Let* $P \in \mathcal{P}$, $n \in \mathcal{C}_P$, $l \in \mathbf{wt}$.

i) $\mathcal{W}_l(P,n)$ *is a free* $\mathcal{O}$*-module of rank* 2.

ii) $\mathcal{W}^0_l(P,n)$ *is a free* $\mathcal{O}$*-module of rank* 1 *on* $\Omega \subset W$ *in the following cases:*

- *a)* $P \in \mathcal{P}_Y$.
- *b)* $P \in X^\infty$, $n(w) \neq 0$ *for all* $w \in \Omega$.
- *c)* $P \in X^\infty$, n *vanishes identically on* Ω, *and* $\mathrm{Re}\, s(w) \neq 0$ *for all* $w \in \Omega$.

iii) The operators $\mathbf{E}^+$ *and* $\mathbf{E}^-$ *define morphisms of* $\mathcal{O}$*-modules* $\mathbf{E}^\pm : \mathcal{W}_l(P,n) \to \mathcal{W}_{l\pm 2}(P,n)$. *If* P, n *and* Ω *satisfy one of the conditions a)–c) in ii), then* $\mathbf{E}^\pm \mathcal{W}^0_l(P,n) \subset \mathcal{W}^0_{l\pm 2}(P,n)$ *on* Ω.

Proof. See 7.6.2, 7.6.7, and 7.7.1. In Section 7.6 we shall give explicit sections that form $\mathcal{O}$-bases.

7.3 Families of eigenfunctions

The division property in Proposition 7.2.9 looks trivial. Nevertheless, a careful proof takes some work. That is one of the things we do in the Sections 7.3–7.5. This being done, we shall use the divisibility result freely in the next chapters, sometimes without stating it explicitly.

In this section we start with families that are more general than families of automorphic forms.

We keep fixed a parameter space (W, s, χ).

7.3.1 *Definition.* Let $\Omega \subset W$ and $U \subset \tilde{G}$ be non-empty open sets, with $U\tilde{K} = U$, and let $l \in \mathbf{wt}$. A *holomorphic family of eigenfunctions* f on $\Omega \times U$ of weight l is an element $f \in C^\infty(\Omega \times U)$ such that

i) $f(w; gk(\theta)) = f(w; g)e^{il(w)\theta}$ for all $w \in \Omega$, $g \in U$, $\theta \in \mathbb{R}$,

ii) $\omega f(w) = (\frac{1}{4} - s(w)^2)f(w)$ on U for each $w \in \Omega$,

iii) $w \mapsto f(w; g)$ is in $\mathcal{O}(\Omega)$ for each $g \in U$.

Often we shall just say *family of eigenfunctions;* the holomorphy in w will be understood. The operator of which these families are eigenfunctions is the Casimir operator.

A holomorphic family of automorphic forms, as considered in the previous section, is a family of eigenfunctions on $\Omega \times \tilde{G}_{\mathcal{P}}$. Elements of $\mathcal{W}_l(Q, n; \Omega)$ are also families of eigenfunctions on $\Omega \times \tilde{G}_Q$.

7.3.2 *Analyticity.* A family of eigenfunctions is a real analytic function. To see this, we consider a small Ω on which we have complex coordinates $w_1, \ldots, w_N$. Let $\hat{U}$ be the image of U in $\mathfrak{H} \cong \tilde{G}/\tilde{K}$. A family of eigenfunctions f on $\Omega \times U$ determines a function F on $\Omega \times \hat{U}$ by $F(w, z) = f(w; p(z))$; this function is annihilated by the elliptic differential operator

$$-\partial_{w_1}\partial_{\overline{w_1}} - \cdots - \partial_{w_N}\partial_{\overline{w_N}} - y^2\partial_x^2 - y^2\partial_y^2 + il(w)\partial_x - \frac{1}{4} + s(w)^2,$$

hence it is real analytic on $\Omega \times \hat{U}$.

This also shows that the C^∞-condition in Definition 7.3.1 can be weakened. We formulate this as a lemma:

7.3.3 Lemma. *Let* $w_1, \ldots, w_N$ *be complex coordinates on an open set* $\Omega \subset W$, *and let* U *be open in* $\tilde{G}$. *Take* $\hat{U}$ *as above. If the distribution* Φ *on* $\Omega \times \hat{U}$ *satisfies*

i) $\left(-y^2\partial_x^2 - y^2\partial_y^2 + il(w)\partial_x\right)\Phi = (\frac{1}{4} - s(w)^2)\Phi$ *on* $\Omega \times \hat{U}$,

ii) $\partial_{\overline{w_j}}\Phi = 0$ *for* $j = 1, \ldots, N$, *then Φ is given by a real analytic function and*

$$f(w; p(z)k(\theta)) = \Phi(w, z)e^{il(w)\theta}$$

defines a family of eigenfunctions on $\Omega \times U$.

7.3.4 *Operations on families of eigenfunctions.*

- *Restriction* to a smaller Ω or to a smaller U sends families of eigenfunctions to families of eigenfunctions. This also holds for composition with a morphism of parameter spaces.
- *Multiplication* by an element of $\mathcal{O}(\Omega)$ sends families of eigenfunctions to families of eigenfunctions.
- *Differentiation* by $\mathbf{E}^-$ or $\mathbf{E}^+$ sends families of eigenfunctions of weight l to families of weight $l \mp 2$.

The conditions defining a family of eigenfunctions are local: If $\Omega = \bigcup_i \Omega_i$, then a family of eigenfunctions f on $\Omega \times U$ is determined by its restrictions f_i to the $\Omega_i \times U$. Any collection $\{f_i\}$ of families of eigenfunctions on the $\Omega_i \times U$ determines a family of eigenfunctions on $\Omega \times U$, provided $f_i = f_j$ on the intersections $(\Omega_i \cap \Omega_j) \times U$.

7.3.5 *Definition.* Let $U \subset \tilde{G}$ and $l \in \mathbf{wt}$ as above. For each open $\Omega \subset W$, we define $\mathcal{F}_l(U; \Omega)$ to be the linear space of all holomorphic families of eigenfunctions on $\Omega \times U$. If $\Omega_1 \subset \Omega$, restriction gives a map $\mathcal{F}_l(U; \Omega) \to \mathcal{F}_l(U; \Omega_1)$. Clearly $\Omega \mapsto \mathcal{F}_l(U; \Omega)$ defines a *sheaf* $\mathcal{F}_l(U)$ on W. Multiplication by a holomorphic function on Ω maps $\mathcal{F}_l(U; \Omega)$ into itself. This gives $\mathcal{F}_l(U)$ the structure of an *$\mathcal{O}$-module* on W.

The $\mathcal{O}$-module $\mathcal{F}_l(U)$ is torsion-free: if for $w \in W$ the stalks $\psi \in \mathcal{O}_w$ and $f \in \mathcal{F}_l(U)_w$ satisfy $\psi f = 0$ in $\mathcal{F}_l(U)_w$, then $\psi = 0$ or $f = 0$.

One could vary the U in $\mathcal{F}_l(U)$, and consider $\mathcal{F}_l$ as a sheaf on $W \times Y$. In the context of this book I refrain from taking this point of view.

7.3.6 *Zero set.* For $\psi \in \mathcal{O}(\Omega)$ the zero set is $N(\psi) = \{ w \in \Omega : \psi(w) = 0 \}$. This is of course an analytic set (locally the intersection of the zero sets of finitely many holomorphic functions). If $\psi \neq 0$, then $\Omega \smallsetminus N(\psi)$ is open and dense in Ω.

We define the *zero set* $N(f)$ of $f \in \mathcal{F}_l(U; \Omega)$ as the intersection $\bigcap_{g \in U} N(f_g)$, with $f_g : w \mapsto f(w; g)$. This set $N(f)$ is also analytic; see [17], Ch. 5, §6.1.

The next three lemmas concern divisibility in $\mathcal{F}_l(U)$.

7.3.7 Lemma. *Let Ω be open in W, $\psi \in \mathcal{O}(\Omega)$, $\psi \neq 0$, $\Omega_1 = \Omega \smallsetminus N(\psi)$. Let U be open in $\tilde{G}$. Let $f \in \mathcal{F}_l(U; \Omega_1)$, and suppose that for each $g \in U$ there exists $f_g \in \mathcal{O}(\Omega)$ such that $f_g(w) = f(w; g)$ for $w \in \Omega_1$. Then $f(w; g) = f_g(w)$ defines f as an element of $\mathcal{F}_l(U; \Omega)$.*

Postcard
Stamp
Sender:
Birkhäuser Verlag AG
Marketing
P.O. Box 133
CH–4010 Basel/Switzerland

Remark. The sole role of ψ in this lemma is to determine $\Omega \setminus \Omega_1$. In particular, the set $\Omega \setminus \Omega_1$ cannot be too wild.

Proof. The assertion is local on Ω, so we take $a \in N(\psi)$ and replace Ω by a smaller neighborhood if necessary.

We can assume that $\Omega \subset \mathbb{C}^N$ with complex coordinates $w_1, \ldots, w_N$ vanishing at a, and we can assume that $\psi(w_1, 0, \ldots, 0) = w_1^m \tau(w_1)$ with τ holomorphic at $w_1 = 0$, $\tau(0) \neq 0$ (see [17], Ch. 2, §1.3). So there exists $\delta > 0$ such that $\{ (w_1, 0, \ldots, 0) : |w_1| = \delta \}$ is a compact subset of Ω_1. By *loc. cit.* Ch. 2, §1.2 we can assume that ψ is a Weierstrass polynomial: $\psi(w_1, w_2, \ldots, w_N) = w_1^m + \sum_{j=0}^{m-1} a_j(w_2, \ldots, w_N) w_1^j$ with a_j holomorphic on a neighborhood of 0 in $\mathbb{C}^{N-1}$ and $a_j(0) = 0$ for $0 \leq j \leq m-1$. From *loc. cit.* Ch. 2, §3.5, Supplement, we conclude that there exists $\varepsilon > 0$ such that $\{ (w_1, w_2, \ldots, w_N) : |w_1| = \delta, |w_j| < \varepsilon$ for $j = 2, \ldots, N \}$ is contained in Ω_1. So if $|w_1| < \delta$ and $|w_j| < \varepsilon$ for $j = 2, \ldots, N$, we find

$$f_g(w_1, \ldots, w_N) = \frac{1}{2\pi i} \int_{|\zeta|=\delta} \frac{1}{\zeta - w_1} f(\zeta, w_2, \ldots, w_N; g)\, d\zeta.$$

Differentiate under the integral to see that $(w, g) \mapsto f_g(w)$ is a C^∞-function at each point of $\{a\} \times U$. The validity of conditions i) and ii) in Definition 7.3.1 extends from $\Omega_1 \times U$ to $\Omega \times U$ by continuity.

7.3.8 Lemma. *Let $f \in \mathcal{F}_l(U; \Omega)$, let $\psi \in \mathcal{O}(\Omega)$, $\psi \neq 0$, and suppose that for each $g \in \mathrm{pr}^{-1} U$ there exists $h_g \in \mathcal{O}(\Omega)$ such that $f(w; g) = \psi(w) h_g(w)$ for all $w \in \Omega$. Then $h(w; g) = h_g(w)$ defines $h \in \mathcal{F}_l(U; \Omega)$.*

Proof. Clear from the previous lemma.

In this lemma we have assumed the existence of the h_g. We assume less in the next lemma.

7.3.9 Lemma. *Let $f \in \mathcal{F}_l(U; \Omega)$, let $a \in \Omega$, and $\psi \in \mathcal{O}(\Omega)$. Suppose that ψ is irreducible in the stalk $\mathcal{O}_a$ and that $f(w; g) = 0$ for all $w \in N(\psi)$ and all $g \in \mathrm{pr}^{-1} U$. Then there exists a holomorphic family of eigenfunctions $h \in \mathcal{F}_l(U; \tilde{\Omega})$ on an open neighborhood $\tilde{\Omega} \subset \Omega$ of a, such that $f = \psi h$ on $\tilde{\Omega} \times U$.*

Remarks. If W has complex dimension 1, then this is trivial. Let w be a coordinate near a. Then ψ may be supposed to be $\psi = w - w(a)$ on a small enough neighborhood $\tilde{\Omega}$ of a in Ω. For each $g \in U$, the function f_g is divisible by ψ on $\tilde{\Omega}$; take $h_g = \frac{1}{\psi} f_g$, and apply the previous lemma.

The higher dimensional case is more complicated. In the stalk $\mathcal{O}_a$ each f_g is divisible by ψ, but the corresponding $h_g = \frac{1}{\psi} f_g$ might be represented by holomorphic functions on different neighborhoods of a, the intersection of which one might fear to be $\{a\}$.

Proof. The sheaf of ideals $\mathcal{I}$ of $N(\psi)$ is a coherent $\mathcal{O}$-module, see [17], Ch. 4, §2. By Ch. 4, §1.5 of *loc. cit.* this sheaf $\mathcal{I}$ is the radical of the sheaf $\mathcal{O} \cdot \psi$. The irreducibility of ψ implies $\mathcal{I}_a = \mathcal{O}_a \cdot \psi$. So $\mathcal{I}_w = \mathcal{O}_w \cdot \psi$ for all w in some neighborhood $\tilde{\Omega} \subset \Omega$ of a, see *loc. cit.*, Annex, §3.1. Hence $\tilde{\Omega}$ has a covering by open sets Ω_j and each $b \in \mathcal{I}(\tilde{\Omega})$ is of the form $b = g_j\psi$ on Ω_j for some $g_j \in \mathcal{O}(\Omega_j)$. As $\mathcal{O}$ has no zero divisors, $g_i = g_j$ on $\Omega_i \cap \Omega_j$, hence $b = g\psi$ on $\tilde{\Omega}$ with $g \in \mathcal{O}(\tilde{\Omega})$. We conclude that $\mathcal{I}(\tilde{\Omega}) = \mathcal{O}(\tilde{\Omega}) \cdot \psi$. Apply this to the $w \mapsto f(w;g)$ to obtain $h_g \in \mathcal{O}(\tilde{\Omega})$ such that $h_g(w) = \frac{1}{\psi(w)} f(w;g)$ for $w \in \tilde{\Omega} \setminus N(\psi)$. The previous lemma completes the proof.

7.3.10 *Divisibility.* These three lemmas state that the $\mathcal{F}_l(U)$ has the divisibility property that we want to prove for $\mathcal{A}_l(\mathbf{c})$ and $\mathcal{S}_l(\mathbf{c})$. These $\mathcal{O}$-modules are submodules of $\mathcal{F}_l(\tilde{G}_{\mathcal{P}})$. It will suffice to show that if a section f of $\mathcal{A}_l(\mathbf{c})$ is divisible in $\mathcal{F}_l(\tilde{G}_{\mathcal{P}})$ by a holomorphic function ψ, then the quotient $\frac{1}{\psi} f$ is still a section of $\mathcal{A}_l(\mathbf{c})$. We proceed a bit more generally, and define an eigenfunction module as a submodule of an $\mathcal{F}_l(U)$ with this divisibility property.

7.3.11 *Definition.* Let l and U be as above. We call an $\mathcal{O}$-submodule $\mathcal{H}$ of $\mathcal{F}_l(U)$ an *eigenfunction module* if it has the following property for each $w \in W$: *if $f \in \mathcal{F}_l(U)_w$ and $h \in \mathcal{H}_w$ satisfy $h = \psi f$ for some non-zero $\psi \in \mathcal{O}_w$, then $f \in \mathcal{H}_w$.*

The Lemmas 7.3.7–7.3.9 imply immediately:

7.3.12 Proposition. *Let $\mathcal{H}$ be an eigenfunction module in $\mathcal{F}_l(U)$. Let $\Omega \subset W$ be open, let $\psi \in \mathcal{O}(\Omega)$ such that $\Omega_1 = \Omega \setminus N(\psi)$ is dense in Ω.*

i) Let $f \in \mathcal{H}(\Omega_1)$. Suppose that for each $g \in U$ there exists $f_g \in \mathcal{O}(\Omega)$ such that $f_g(w) = f(w;g)$ for $w \in \Omega_1$. Then $h(w;g) = f_g(w)$ defines $h \in \mathcal{H}(\Omega)$, with $h|_{\Omega_1} = f$.

ii) Let $f \in \mathcal{H}(\Omega)$. Suppose that for each $g \in \mathrm{pr}^{-1}(U)$ there exists $h_g \in \mathcal{O}(\Omega)$ such that $f(w;g) = \psi(w)h_g(w)$ for all $w \in \Omega$. Then $h(w;g) = h_g(w)$ defines $h \in \mathcal{H}(\Omega)$ such that $f = \psi h$ on Ω.

iii) Let $f \in \mathcal{H}(\Omega)$, let $a \in \Omega$. Suppose that ψ is irreducible in the stalk $\mathcal{O}_a$, and that $f(w) = 0$ for all $w \in N(\psi)$. Then $f = \psi h$ for some $h \in \mathcal{H}(\tilde{\Omega})$ for some open neighborhood $\tilde{\Omega}$ of a in Ω.

Remark. We shall prove that this proposition applies to $\mathcal{A}_l(\mathbf{c})$ and $\mathcal{S}_l(\mathbf{c})$ in $\mathcal{F}_l(\tilde{G}_{\mathcal{P}})$, and $\mathcal{W}_l(P,n)$ and $\mathcal{W}_l^0(P,n)$ in $\mathcal{F}_l(\tilde{G}_P)$, by showing that these four $\mathcal{O}$-modules are eigenfunction modules.

7.3.13 Lemma.

i) If $\mathcal{C}$ is a collection of eigenfunction modules in $\mathcal{F}_l(U)$, then $\bigcap_{\mathcal{H}\in\mathcal{C}} \mathcal{H}$ is an eigenfunction module as well.

ii) Let $p : \mathcal{H} \to \mathcal{N}$ be a morphism of $\mathcal{O}$-modules. If $\mathcal{H}$ is an eigenfunction module, then its $\mathcal{O}$-submodule $\ker(p)$ *is an eigenfunction module as well.*

Remark. It turns out that all eigenfunction modules we shall need can be obtained by taking intersections and kernels of morphisms.

Proof. Part i) is easily checked.

Put $\mathcal{K} = \ker(p)$. So $\mathcal{K} \subset \mathcal{H} \subset \mathcal{F}_l(U)$ for some l and U. Let $w \in W$. Consider $h \in \mathcal{K}_w$, $\psi \in \mathcal{O}_w$, $\psi \neq 0$, $f \in \mathcal{F}_l(U)_w$ with $h = \psi f$. Then $f \in \mathcal{H}_w$. As $\mathcal{F}_l(U)$ is torsion-free, we conclude $pf = 0$ from $0 = ph = p(\psi f) = \psi pf$. So $f \in \mathcal{K}_w$.

7.3.14 *More functoriality.* If $j : W_1 \to W$ is a morphism of parameter spaces, then composition maps $\mathcal{F}_l(U;\Omega)$ to $\mathcal{F}_{l\circ j}(U; j^{-1}\Omega)$. This gives a morphism of sheaves $j^\sharp : \mathcal{F}_l(U) \to j_*\mathcal{F}_{l\circ j}(U)_{(j)}$. (If $\mathcal{G}$ is a sheaf on W_1, then $j_*\mathcal{G}$ is the sheaf $\Omega \mapsto \mathcal{G}(j^{-1}\Omega)$ on W. See [19], p. 65 and p. 72.)

If U_1 is an open subset of U, then restriction gives an injective morphism of $\mathcal{O}$-modules $\mathcal{F}_l(U) \to \mathcal{F}_l(U_1)$.

7.4 Families of eigenfunctions with automorphic transformation behavior

In this section we turn our attention to families of χ-l-equivariant eigenfunctions. We work above open subsets of Y, and generalize the condition of regularity.

Up till now we have discussed only holomorphic families of eigenfunctions. In 1.4.4 we have seen that we have to discuss meromorphic families as well. This we do in 7.4.10–7.4.15.

7.4.1 *Families of eigenfunctions with automorphic transformation behavior.* Let V be an open subset of Y and $l \in \mathbf{wt}$. It is easy to check that for each $\gamma \in \tilde{\Gamma}$,

$$(q_\gamma f)(w;g) = f(w;\gamma g) - \chi(\gamma;w) f(w;g)$$

defines a morphism of $\mathcal{O}$-modules $q_\gamma : \mathcal{F}_l(\mathrm{pr}^{-1}V) \to \mathcal{F}_l(\mathrm{pr}^{-1}V)$. We define

$$\mathcal{A}_l[V] = \bigcap_{\gamma \in \tilde{\Gamma}} \ker q_\gamma.$$

Part i) of Lemma 7.3.13 shows that $\mathcal{A}_l[V]$ is an eigenfunction module. As $\tilde{\Gamma}$ is finitely generated, the intersection is in fact a finite one.

If $V = Y$ or $V = Y_\mathcal{P}$, then $\mathcal{A}_l[V]$ is a — in general very large — module of families of automorphic forms, without any growth condition.

7.4.2 *Regularity.* Let $\Omega \subset W$, $P \in \mathcal{P}$, $V \subset Y$, $P \notin V$, such that $\dot{U}_P(a) \subset V$ for some $a \in I_P$. We define the property *regular at* P for sections $f \in \mathcal{A}_l[V](\Omega)$:

i) If $P \in \mathcal{P}_Y$ we call $f \in \mathcal{A}_l[V](\Omega)$ regular at P if it is the restriction to $\Omega \times V$ of a section in $\mathcal{A}_l[V \cup \{P\}](\Omega)$.

ii) Let $P \in X^\infty$, take some $n_P \in \mathcal{C}_P$. To f we associate the family of functions

$$F_{w,P}(z) = e^{-2\pi i n_P(w)x} f(w; g_P p(z))$$

on $S_{P,b} = \{ z \in \mathfrak{H} : 0 \leq x < 1,\, y > b \}$. We call f regular at P if this gives a L^2-holomorphic family $w \mapsto F_{w,P} : \Omega \to L^2(S_{P,b}, d\mu)$ for some $b > a$.

These conditions generalize iv) and v) in Definition 7.2.3.

If the L^2-holomorphy holds for one $b > a$, then it holds for all such b. Indeed, a section in $\mathcal{A}_l[V](\Omega)$ determines an L^2-holomorphic family $w \mapsto F_{w,P}$ in each L^2-space on $\{ z \in \mathfrak{H} : 0 \leq x < 1,\, b_1 \leq y \leq b_2 \}$, with $a < b_1 < b_2$. This follows from the next lemma, applied with K equal to the image of $C_c^\infty(S_{P,b})$ and $f_w : k \mapsto (F_{w,P}, k)$. We may replace Ω by a relatively compact neighborhood of the point at which we want to prove the L^2-holomorphy.

7.4.3 Lemma. *Let H be a Hilbert space with a dense subspace K, let Ω be open in $\mathbb{C}^N$. Suppose that for each $w \in \Omega$ there is a semilinear form $k \mapsto f_w(k)$ on K such that*

i) $w \mapsto f_w(k)$ is holomorphic on Ω for each $k \in K$.

ii) $|f_w(k)| \leq C\|k\|$ for all $w \in \Omega$, $k \in K$, for some $C \geq 0$.

Then there is a holomorphic map $F : \Omega \to H$ such that $f_w(k) = (F(w), k)$ for all $k \in K$.

Remarks. This is the well known result that weak holomorphy implies holomorphy in norm. See, e.g., [29], App. 5, §1.

Condition ii) may be weakened to hold only on compact subsets of Ω, with C depending on the subset.

Proof. For each $w \in \Omega$ condition ii) implies that $f_w(k) = (F(w), k)$ for some $F(w) \in H$ with $\|F(w)\| \leq C$.

It is sufficient to prove holomorphy of F in Hilbert space sense at $w = 0 \in \Omega \subset \mathbb{C}^N$. For $\delta > 0$ small enough, we define for $k \in K$, $n_1, \ldots, w_N \geq 0$

$$a(n_1, \ldots, n_N)(k) = \frac{1}{(2\pi i)^N} \int_{|w_1|=\delta} \cdots \int_{|w_N|=\delta} \frac{1}{w_1^{n_1+1} \cdots w_N^{n_N+1}} f_w(k)\, dw_1 \cdots dw_N.$$

So

$$f_w(k) = \sum_{n_1, \ldots, n_N \geq 0} a(n_1, \ldots, n_N)(k) w_1^{n_1} \cdots w_N^{n_N}$$

for $|w_j| < \delta$; see, e.g., [23], Theorem 2.2.6. Condition ii) implies the inequalities $|a(n_1, \ldots, n_N)(k) \leq C\delta^{-n_1 - \cdots - n_N} \|k\|$. So there are $A(n_1, \ldots, n_N) \in H$ with

$\|A(n_1,\ldots,n_N)\| \leq C \cdot \delta^{-n_1-\cdots-n_N}$ and $a(n_1,\ldots,n_N)(k) = (A(n_1,\ldots,n_N),k)$. The series

$$\sum_{n_1,\ldots,n_N\geq 0} w_1^{n_1}\cdots w_N^{n_N} A(n_1,\ldots,n_N)$$

converges in H for $|w_j| < \delta$, and represents $F(w)$.

7.4.4 *Properties.* Regularity is preserved under addition and under multiplication by elements of $\mathcal{O}(\Omega)$. Moreover, regularity is a local property on W: if every $w \in \Omega$ has an open neighborhood Ω_w such that the restriction of f to $\Omega_w \times \mathrm{pr}^{-1}V$ is regular at P, then $f \in \mathcal{A}_l[V](\Omega)$ is regular at P.

7.4.5 Lemma. *Let $\Omega \subset W$, $V \subset Y$, $P \in \mathcal{P}$, $P \notin V$, $\dot{U}_P(a) \subset V$ for some $a \in I_P$. Suppose $h \in \mathcal{A}_l[V](\Omega)$ is regular at P, and is of the form $h = \psi f$ with $\psi \in \mathcal{O}(\Omega)$, $f \in \mathcal{F}_l(\mathrm{pr}^{-1}V;\Omega)$. Then f is regular at P.*

Remark. $\mathcal{A}_l[V]$ is an eigenfunction module. So $f \in \mathcal{A}_l[V](\Omega)$, and the property of regularity makes sense.

Proof. As regularity is a local property, it suffices to consider $w_0 \in W$, a regular $h \in \mathcal{A}_l[V]_{w_0}$, $\psi \in \mathcal{O}_{w_0}$, and $f \in \mathcal{F}_l(\mathrm{pr}^{-1}V)_{w_0}$, with $\psi \neq 0$ and $h - \psi f$, and to show that f is regular. If $\psi(w_0) \neq 0$ this is clear. If not, then we treat irreducible factors of ψ in $\mathcal{O}_{w_0}$ one by one. This reduces the proof to the case that ψ is irreducible in $\mathcal{O}_{w_0}$.

First consider $P \in \mathcal{P}_Y$. Put $V_1 = V \cup \{P\}$. As h is regular we can represent it by the restriction to $\Omega \times \mathrm{pr}^{-1}V$ of some $h \in \mathcal{A}_l[V_1](\Omega)$ for some open neighborhood Ω of w_0. Let $g_1 \in \mathrm{pr}^{-1}\{P\}$. As $h(w;g_1) = \lim_{g\to g_1,\, \mathrm{pr}g\in V} h(w;g)$, the function $w \mapsto h(w;g_1)$ vanishes on $N(\psi)$. Now we apply part iii) of Proposition 7.3.12 to finish the proof in the interior case.

Let $P \in X^\infty$. Take coordinates $w_1,\ldots,w_N$ on a neighborhood of w_0 in the same way as in the proof of Lemma 7.3.7. There we have seen how to adapt ψ and to choose $\delta > 0$ and $\varepsilon > 0$ to obtain for $g \in \mathrm{pr}^{-1}\dot{U}(b)$, $|w_1| < \delta$, and $|w_j| < \varepsilon$, $j = 2,\ldots,N$:

$$\begin{aligned} &f(w_1,\ldots,w_N;g) \\ &\quad = \frac{1}{2\pi i}\int_{|\zeta|=\delta} (\zeta - w_1)^{-1}\psi(\zeta,w_2,\ldots,w_N)^{-1} h(\zeta,w_2,\ldots,w_N;g)\,d\zeta. \end{aligned}$$

Let $\tau \in C_c^0(S_{P,b})$. Then for $(w_1,\ldots,w_N)$ as indicated

$$\begin{aligned} &\int_{S_{P,b}} f(w_1,\ldots,w_N;g_P p(z))\overline{\tau(z)}\,d\mu(z) \\ &\quad = \frac{1}{2\pi i}\int_{|\zeta|=\delta} (\zeta - w_1)^{-1}\psi(\zeta,w_2,\ldots,w_N)^{-1} \\ &\qquad\qquad \cdot \int_{S_{P,b}} h(\zeta,w_2,\ldots,w_N;p(z))\overline{\tau(z)}\,d\mu(z)\,d\zeta \end{aligned}$$

$$= \frac{1}{2\pi i}\int_{|\zeta|=\delta} (\zeta - w_1)^{-1}\psi(\zeta, w_2, \ldots, w_N)^{-1}\left(H_{(\zeta,w_2,\ldots,w_N)}, \tau\right) d\zeta,$$

where $w \mapsto H_w$ is the L^2-holomorphic family corresponding to $h = \psi f$. In particular $w \mapsto (H_w, \tau)$ is holomorphic, and bounded for $|w_1| \le \delta$, $|w_j| \le \varepsilon$, $j = 2, \ldots, N$, by (constant) $\cdot\ \|\tau\|$. By Lemma 7.4.3 this shows that for $|w_1| < \frac{1}{2}\delta$, $|w_j| < \varepsilon$, $j = 2, \ldots, N$, the family f determines $F_{(w_1,\ldots,w_N),P} \in L^2(S_{P,b}, d\mu)$, and that $w \mapsto F_{w,P}$ is L^2-holomorphic.

7.4.6 *Families of eigenfunctions above subsets of X.* Let $V_1 \subset Y$ and $\Omega \subset W$ be open, $P \in \mathcal{P}$. For $P \in \mathcal{P}_Y$, restriction gives a natural identification between $\mathcal{A}_l[V_1 \cup \{P\}](\Omega)$ and the space of sections in $\mathcal{A}_l[V_1](\Omega)$ that are regular at P. So it seems sensible to define the sheaf $\mathcal{A}_l[V]$ for any open $V \subset X$ by

$$\mathcal{A}_l[V](\Omega) = \{\, f \in \mathcal{A}_l[V \cap Y](\Omega) : f \text{ is regular at } P \text{ for all } P \in V \cap X^\infty \,\}.$$

7.4.7 Proposition. *Let V be open in X. The sheaf $\mathcal{A}_l[V]$ is an eigenfunction module.*

Proof. $V \setminus Y$ is finite. Apply Lemma 7.4.5 to each $P \in V \cap X^\infty$.

7.4.8 *Taking a Fourier term.* Let $P \in \mathcal{P}$, $n \in \mathcal{C}_P$, $l \in \mathbf{wt}$.

$$\left(\tilde{F}_{P,n} f\right)(w; g) = \begin{cases} \int_0^1 e^{-2\pi i n(w)x} f(w; g_P n(x) g_P^{-1} g)\, dx & \text{if } P \in X^\infty \\ \int_0^\pi e^{-in(w)\eta} f(w; g_P k(\eta) g_P^{-1} g)\, \frac{d\eta}{\pi} & \text{if } P \in \mathcal{P}_Y \end{cases}$$

defines a morphism of $\mathcal{O}$-modules $\tilde{F}_{P,n} : \mathcal{A}_l[\dot{U}_P(A_P)] \to \mathcal{F}_l(U)$, with $\dot{U}_P(A_P)$ as defined in 3.5.3, and $U = g_P \tilde{N}\{\, a(y) : y > A_P \,\}\tilde{K}$ if $P \in X^\infty$, and $U = g_P \tilde{K}\{\, a(t_u) : 0 < u < A_P \,\}\tilde{K}$ if $P \in \mathcal{P}_Y$. Indeed, the integral preserves the conditions in Definition 7.3.1, and commutes with multiplication by a holomorphic function.

If f is a section of $\mathcal{A}_l[\dot{U}_P(A_P)]$, then $\tilde{F}_{P,n} f(w; \gamma g) = \chi(\gamma; w)\tilde{F}_{P,n} f(w; g)$ for those $\gamma \in \tilde{\Gamma}$ and $g \in \tilde{G}$ for which both g and γg are in U. We extend $\tilde{F}_{P,n} f$ in a $\tilde{\Gamma}$-equivariant way to $\tilde{\Gamma} U$, and obtain an element of $\mathcal{A}_l[\dot{U}_P(A_P)]$. In this way we can consider $\tilde{F}_{P,n}$ as a morphism $\mathcal{A}_l[\dot{U}_P(A_P)] \to \mathcal{A}_l[\dot{U}_P(A_P)]$.

If $X \supset V \supset \dot{U}_P(A_P)$, we first take the restriction $\mathcal{A}_l[V] \to \mathcal{A}_l[\dot{U}_P(A_P)]$, and then apply $\tilde{F}_{P,n}$. We also use $\tilde{F}_{P,n}$ to denote the resulting morphism of $\mathcal{O}$-modules $\mathcal{A}_l[V] \to \mathcal{A}_l[\dot{U}_P(A_P)]$.

In 7.6.5 we shall indicate how to view $\tilde{F}_{P,n}$ as a morphism $\mathcal{A}_l[Y_{\mathcal{P}}] \to \mathcal{W}_l(P, n)$.

7.4.9 Lemma. $\tilde{F}_{P,n}\mathcal{A}_l[U_P(A_P)] \subset \mathcal{A}_l[U_P(A_P)]$ *for each* $P \in \mathcal{P}$.

Proof. Clear if $P \in \mathcal{P}_Y$.

For $P \in X^\infty$ suppose that $w \mapsto F_{w,P}$ is the L^2-holomorphic family on Ω with values in $L^2(S_{P,b}, d\mu)$ such that

$$f(w; g_P p(z)k(\theta)) = e^{2\pi i n_P(w)x} F_{w,P}(z) e^{il(w)\theta}$$

is a section of $\mathcal{A}_l(U_P(b))$. Note that $\nu = n - n_P : \Omega \to \mathbb{Z}$ is constant. Then

$$\tilde{F}_{P,n} f(w; g_P p(z)k(\theta)) = e^{2\pi i n_P(w)x} G_w(z) e^{il(w)\theta},$$

where $G_w = J F_{w,P}$:

$$Jh(z) = \int_0^1 e^{-2\pi i x' \nu} h(z + x')\, dx'.$$

We have extended $F_{w,P}$ and G_w to $\{ z \in \mathfrak{H} : y > b \}$ by invariance under $z \mapsto z+1$. The integration operator J is bounded with respect to the L^2-norm, hence it preserves L^2-holomorphy.

7.4.10 *Meromorphic functions.* Let $\mathcal{M}$ be the sheaf of meromorphic functions on W. So $f \in \mathcal{M}(\Omega)$ is locally given as the quotient $\frac{g}{h}$ with g and h sections of $\mathcal{O}$, $h \neq 0$. See, e.g., [17], Ch. 6, §3.

Each $f \in \mathcal{M}(\Omega)$ has a zero set $N(f)$ and a polar set $\mathrm{Pol}(f)$; in *loc. cit.* a definition in terms of sheaves is given. On a complex manifold all sections of $\mathcal{O}$ are functions; this leads to the following description:

$$\begin{aligned} \mathrm{Pol}(f) &= \left\{ w \in \Omega : \text{if } f = \frac{g}{h} \text{ on } \Omega_1 \ni w,\ g, h \in \mathcal{O}(\Omega_1),\ \text{then } h(w) = 0 \right\}, \\ N(f) &= \left\{ w \in \Omega : \text{if } f = \frac{g}{h} \text{ on } \Omega_1 \ni w,\ g, h \in \mathcal{O}(\Omega_1),\ \text{then } g(w) = 0 \right\}. \end{aligned}$$

$\mathrm{Pol}(f)$ and $N(f)$ are analytic subsets of Ω. If $f \neq 0$ the sets $\Omega \smallsetminus \mathrm{Pol}(f)$ and $\Omega \smallsetminus N(f)$ are open and dense in Ω.

Let $w \in \Omega \subset W$ and $f \in \mathcal{M}(\Omega)$. The local ring $\mathcal{O}_w$ is factorial, as W is a complex manifold; see [17], Ch. 2, §2.1. So the representation $f = \frac{g}{h}$ with $g, h \in \mathcal{O}(\Omega_1)$, $w \in \Omega_1 \subset \Omega$, may be arranged such that g and h are relatively prime in $\mathcal{O}_w$. Then g and h are relatively prime in $\mathcal{O}_u$ for all u in a neighborhood of w; see [23], Theorem 6.2.3.

7.4.11 *Meromorphic sections of an $\mathcal{O}$-module.* For each $\mathcal{O}$-module $\mathcal{H}$ on W there exists the sheaf $\mathcal{M} \otimes_{\mathcal{O}} \mathcal{H}$ associated to the presheaf $\Omega \mapsto \mathcal{M}(\Omega) \otimes_{\mathcal{O}(\Omega)} \mathcal{H}(\Omega)$. If $f \in (\mathcal{M} \otimes_{\mathcal{O}} \mathcal{H})(\Omega)$, then each point $a \in \Omega$ has a neighborhood Ω_a on which f is represented as $f = \frac{1}{\psi} h$ with ψ a non-zero element of $\mathcal{O}(\Omega_a)$, and $h \in \mathcal{H}(\Omega_a)$.

We call the sections of the sheaves $\mathcal{M} \otimes_{\mathcal{O}} \mathcal{F}_l(U)$ *meromorphic families of eigenfunctions.* Let $f \in (\mathcal{M} \otimes_{\mathcal{O}} \mathcal{F}_l(U))(\Omega)$. Then each $a \in \Omega$ has a neighborhood Ω_a on which $f = \frac{1}{\psi} h$ with $\psi \in \mathcal{O}(\Omega_a)$, $\psi \neq 0$, $h \in \mathcal{F}_l(U; \Omega)$. This means that for each $g \in U$ there is a meromorphic function $w \mapsto f(w;g)$ on Ω. Moreover, the f_g have locally a common denominator.

Lemma 7.3.8 has a convenient consequence: a meromorphic family of eigenfunctions f on $\Omega \times U$ for which all maps $w \mapsto f(w;g)$ are holomorphic, is a holomorphic family.

7.4.12 *Eigenfunction modules.* We can characterize an eigenfunction module as a submodule $\mathcal{H}$ of some $\mathcal{F}_l(U)$ satisfying

$$\mathcal{H}_w = \mathcal{F}_l(U)_w \cap (\mathcal{M} \otimes_{\mathcal{O}} \mathcal{H})_w$$

for all $w \in W$.

7.4.13 *Polar set and zero set.* Let $f \in (\mathcal{M} \otimes_{\mathcal{O}} \mathcal{F}_l(U))(\Omega)$.

The *zero set* $N(f)$ of f is the set of those $a \in \Omega$ for which $a \in N(\psi f)$ for all $\psi \in \mathcal{O}_a$ satisfying $\psi f \in \mathcal{F}_l(U)_a$. The *polar set* $\mathrm{Pol}(f)$ is the set of all $w \in \Omega$ at which f is not holomorphic. By holomorphic at w we mean that its restriction to some neighborhood of w is a holomorphic family of cigenfunctions.

If $\psi \in \mathcal{O}(\Omega_2)$, $\psi \neq 0$, such that ψf is a holomorphic family on $\Omega_2 \times U$, then $\mathrm{Pol}(f) \cap \Omega_2 \subset N(\psi)$.

7.4.14 Lemma. *Let f be a meromorphic family of eigenfunctions on $\Omega \times U$.*

i) $N(f)$ and $\mathrm{Pol}(f)$ are analytic sets.

ii) Let $a \in \mathrm{Pol}(f)$. There exists a non-zero $\psi \in \mathcal{O}(\Omega_0)$ on a neighborhood Ω_0 of a contained in Ω, such that ψf extends to $\Omega_0 \times U$ as a holomorphic family of eigenfunctions and such that there are $w \in N(\psi)$ satisfying $(\psi f)(w) \neq 0$.

Proof. As $N(f)$ is locally the intersection of analytic sets $N(\psi f)$, it is an analytic set itself; see [17], Ch. 5, §6.1.

The existence of a neighborhood Ω_0 of a and of $\psi \in \mathcal{O}(\Omega_0)$, $\psi \neq 0$, such that ψf extends as a holomorphic family on $\Omega_0 \times U$, follows from the definition of a meromorphic family.

Write ψ as the product of its irreducible factors $\psi_1, \ldots, \psi_k$ in the factorial ring $\mathcal{O}_a$. We may take Ω_0 such that all $\psi_j \in \mathcal{O}(\Omega_0)$. If $N(\psi_j) \subset N(\psi f)$, then $\frac{\psi}{\psi_j} f$ is holomorphic on Ω_0 by Lemma 7.3.9. We remove this factor ψ_j from ψ. Repeating this we arrive at the situation $N(\psi) \not\subset N(\psi f)$.

Clearly $\mathrm{Pol}(f) \subset N(\psi)$. Consider one of the factors ψ_j that are left in ψ. As $N(\psi_j) \not\subset N(\psi f)$, there is a $g \in \mathrm{pr}^{-1} U$ such that $\varphi_g : w \mapsto (\psi f)(w;g)$ and ψ_j are relatively prime in $\mathcal{O}_a$, hence in all $\mathcal{O}_w$ with $w \in \Omega_0$, if we take Ω_0 sufficiently small. So f itself cannot be holomorphic on $N(\psi_j) \cap \Omega_0$. Hence $\mathrm{Pol}(f) \cap \Omega_0 = N(\psi) \cap \Omega_0$. This shows that $\mathrm{Pol}(f)$ is an analytic set.

7.4.15 *Complex dimension one.* In the case that the parameter space W has complex dimension one, we apply Lemma 7.3.9 to show that each meromorphic family of eigenfunctions has a well defined order at each point of its domain. Indeed, consider such a family f at the point w_0. Let q be a local coordinate on W at w_0, with $q(w_0) = 0$. There is a unique order $m \in \mathbb{Z}$ such that $w \mapsto q(w)^{-m} f(w)$ is holomorphic and non-zero at w_0.

For parameter spaces of higher dimension the situation is much more complicated; see Section 12.2.

7.5 Families of automorphic forms

The results obtained up till now enable us to identify $\mathcal{A}_l(\mathbf{c})$ and $\mathcal{S}_l(\mathbf{c})$ as eigenfunction modules contained in $\mathcal{A}_l[Y_\mathcal{P}]$. This implies the divisibility statement in Proposition 7.2.9.

7.5.1 *Redefinition of* $\mathcal{A}_l(\mathbf{c})$. Let $l \in \mathbf{wt}$, and let $\mathbf{c}$ be a growth condition on $\Omega \subset W$. For each $P \in \mathcal{P}$ and $n \in \mathcal{C}_P$ there is the morphism of $\mathcal{O}$-modules

$$\tilde{F}_{P,n} : \mathcal{A}_l[Y_\mathcal{P}] \longrightarrow \mathcal{A}_l[\dot{U}_P(A_P)] \bmod \mathcal{A}_l[U_P(A_P)].$$

Consider the intersection

$$\bigcap_{P \in \mathcal{P}} \ker \left(\mathrm{Id} - \sum_{n \in \mathbf{c}(P)} \tilde{F}_{P,n} \right).$$

This is an eigenfunction module; see Lemma 7.3.13. Up to the Fourier terms indicated by $\mathbf{c}$, its sections are regular at each $P \in \mathcal{P}$. It turns out to be the $\mathcal{O}$-module $\mathcal{A}_l(\mathbf{c})$, defined in Definition 7.2.6. This means that the divisibility in part ii) of Proposition 7.2.9 follows from Proposition 7.3.12.

7.5.2 *Redefinition of* $\mathcal{S}_l(\mathbf{c})$. It is also clear that

$$\mathcal{S}_l(\mathbf{c}) = \bigcap_{P \in \mathcal{P}} \bigcap_{n \in \mathbf{c}(P)} \ker \left(\tilde{F}_{P,n} : \mathcal{A}_l(\mathbf{c}) \longrightarrow \mathcal{A}_l[\dot{U}_P(A_P)] \right).$$

Hence $\mathcal{S}_l(\mathbf{c})$ is an eigenfunction module, and the divisibility in Proposition 7.2.9 is clear for $\mathcal{S}_l(\mathbf{c})$.

7.5.3 *Poincaré series.* Take $W = \mathbb{C}_{1/2} = \{ s \in \mathbb{C} : \mathrm{Re}\, s > \frac{1}{2} \}$ as in 7.1.5, with $\chi \in \mathcal{X}_u$ fixed, $l \in \mathbb{R}$ a weight for χ, $P \in \mathcal{P}$, $n \in \mathcal{C}_P(\chi) = \mathcal{C}_P$.

$$f : (s, g) \mapsto P_l(P, n; \chi, s; g)$$

defines $f \in \mathcal{A}_l[Y_P](\mathbb{C}_{1/2})$, with $Y_P = Y \setminus \{P\}$, and $\tilde{G}_P = \mathrm{pr}^{-1} Y_P$. To see the pointwise holomorphy, we do not try to differentiate under the sum defining the Poincaré series, but proceed weakly. f is continuous on $\mathbb{C}_{1/2} \times \tilde{G}_P$, hence it defines a distribution: $\hat{f} : C_c^\infty(\mathbb{C}_{1/2} \times \mathfrak{H}_P) \to \mathbb{C}$ given by

$$\psi \mapsto \int_{\mathbb{C}_{1/2} \times \mathfrak{H}_P} f(s; p(z)) \psi(s, z) \frac{ds \wedge d\bar{s}}{-2i} \, d\mu(z)$$

with $\mathfrak{H}_P = \mathfrak{H} \setminus \overline{\mathrm{pr}}^{-1}\{P\}$. This integral is easily interchanged with the sum in Proposition 5.1.6. Application of Lemma 7.3.3 shows that f is a holomorphic family of eigenfunctions.

We take the growth condition $\mathbf{c}$ minimal for this situation: $\mathbf{c}(Q)$ contains n if $Q = P$, and 0 if $Q \in X^\infty$ is singular for χ (i.e., $\chi(\pi_Q) = 1$). To show that $f \in \mathcal{A}_l(\mathbf{c}; \mathbb{C}_{1/2})$, we have to consider regularity at $\{P\} \cup X^\infty$.

If $P \in \mathcal{P}_Y$, then the terms in the Poincaré series with $\gamma \cdot z_P = z_P$ are not necessarily smooth at the points of $p(z_P)\tilde{K}$. If we ignore these terms, the remaining sum is pointwise holomorphic at all $g \in \mathrm{pr}^{-1}\{P\}$, and defines $\tilde{f} \in \mathcal{A}_l[U_P(A_P)]$. The terms that we have ignored contribute only to $\tilde{F}_{P,n} f$. In Lemma 7.4.9 we see that $(\mathrm{Id} - \tilde{F}_{P,n})\tilde{f}$ is regular at P.

At $Q \in X^\infty$ we put $\tilde{f}$ equal to f if $P \neq Q$, and if $P = Q$ to $f - \mu_l(P,n,s)$. In Lemma 5.1.4 we see that $\tilde{f}$ is bounded on $\mathrm{pr}^{-1}\dot{U}_Q(A_Q)$ uniformly for s in compact sets. As $n \in \mathbb{R}$, the exponential factor in the definition of regularity in 7.4.2 does not matter for the square integrability. The pointwise holomorphy and the uniform bound for s in compact sets are sufficient to obtain L^2-holomorphy with the help of Lemma 7.4.3. The term $\mu_l(P,n,s)$ that is omitted contributes solely to the Fourier term of order n at $Q = P$. So we may proceed as in the case $P \in \mathcal{P}_Y$.

7.5.4 *Functoriality in W.* Similarly to what we remarked in 7.3.14, every morphism of parameter spaces $j : W_1 \to W$ gives rise (by composition) to morphisms of sheaves $\mathcal{A}_l(\mathbf{c}) \to j_* \mathcal{A}_l(\mathbf{c})_{(j)}$ and $\mathcal{S}_l(\mathbf{c}) \to j_* \mathcal{S}_l(\mathbf{c})_{(j)}$.

7.6 Families of Fourier terms

We identify the $\mathcal{O}$-modules $\mathcal{W}_l(P,n)$ with submodules of $\mathcal{A}_l[\dot{U}_P(A_P)]$, and prove the statements i) and ii) in Proposition 7.2.11.

We shall see that the standard sections ω_l and μ_l from Section 4.2 suffice to obtain a basis of $\mathcal{M} \otimes_{\mathcal{O}} \mathcal{W}_l(P,n)$. If one wants holomorphic bases, one has to consider many cases. We carry this out from 7.6.12 on. The reader may want to skip that material at first reading.

Fix $P \in \mathcal{P}$, $l \in \mathbf{wt}$, and $n \in \mathcal{C}_P$.

7.6.1 *Redefinition of $\mathcal{W}_l(P,n)$.* To see that $\mathcal{W}_l(P,n)$ is an eigenfunction module, we characterize it as an intersection of such modules.

We define the subgroup M of $\tilde{G}$ and the group homomorphism $\nu : M \to \mathcal{O}(W)^*$ by

$$\begin{aligned} M &= \tilde{N}, \quad \nu(n(x); w) = e^{2\pi i n(w) x} && \text{if } P \in X^\infty \\ M &= \tilde{K}, \quad \nu(k(\eta); w) = e^{i n(w) \eta} && \text{if } P \in \mathcal{P}_Y. \end{aligned}$$

Take $b \in [0,\infty)$ if $P \in X^\infty$, and $b \in (0,\infty]$ if $P \in \mathcal{P}_Y$, and consider $U_b \subset \tilde{G}$ of the form $M\{\, a(y) : y > b \,\}\tilde{K}$ if $P \in X^\infty$, and $M\{\, a(t_u) : 0 < u < b \,\}\tilde{K}$ if $P \in \mathcal{P}_Y$.

For each $m \in M$ we have a morphism of $\mathcal{O}$-modules $p_{n,m} : \mathcal{F}_l(U_b) \to \mathcal{F}_l(U_b)$ given by $(p_{n,m}f)(w;g) = f(g_P m g_P^{-1} g; w) - \nu(m;w) f(g;w)$. We define

$$\mathcal{W}_l^{(b)}(P,n) = \bigcap_{m \in M} \ker p_{n,m}.$$

Lemma 7.3.13 tells us that $\mathcal{W}_l^{(b)}(P,n)$ is an eigenfunction module. It is easy to check that $\mathcal{W}_l(P,n) = \mathcal{W}_l^{(b)}(P,n)$, with $b=0$ if $P \in X^\infty$, and $b = \infty$ if $p \in \mathcal{P}_Y$.

7.6.2 *A basis.* Consider $P \in \mathcal{P}$, $n \in \mathcal{C}_P$ and fix $a \in (0,b)$ if $P \in \mathcal{P}_Y$, $a \in (b,\infty)$ if $P \in X^\infty$. At this point we identify $W_{l(w)}(P, n(w), s(w))$ with the corresponding functions in y, respectively u, on $(0,\infty)$. For each $w \in W$ there is a unique basis α, β of $W_{l(w)}(P,n(w),s(w))$ such that $\alpha(a) = 1$, $\alpha'(a) = 0$, $\beta(a) = 0$ and $\beta'(a) = 1$. These solutions are holomorphic in the parameters $n(w)$, $l(w)$ and $s(w)$. This gives $\alpha_{P,n}$, $\beta_{P,n} \in \mathcal{W}_l^{(b)}(P,n;W)$.

Each section $f \in \mathcal{W}_l^{(b)}(P,n;\Omega)$ determines holomorphic functions on Ω by differentiation and evaluation at a. If $P \in X^\infty$ these are the functions $w \mapsto f(w; g_P a(a))$ and $w \mapsto \partial_y f(w; g_P a(y))|_{y=a}$, and if $P \in \mathcal{P}_Y$ the functions $w \mapsto f(w; g_P a(t_a))$ and $w \mapsto \partial_u f(w; g_P a(t_u))|_{u=a}$. This expresses f as a linear combination of $\alpha_{P,n}$ and $\beta_{P,n}$ with coefficients in $\mathcal{O}(\Omega)$. In this way

$$\mathcal{W}_l^{(b)}(P,n) = \mathcal{O} \cdot \alpha_{P,n} \oplus \mathcal{O} \cdot \beta_{P,n} \cong \mathcal{O}^2.$$

In particular, $\mathcal{W}_l^{(b)}(P,n)$ and $\mathcal{W}_l(P,n)$ are isomorphic as $\mathcal{O}$-modules for each b. We have also proved part i) of Proposition 7.2.11.

The isomorphism between $\mathcal{W}_l(P,n)$ and $\mathcal{W}_l^{(b)}(P,n)$ with $b \in (0,\infty)$, is canonical; it is obtained by restriction. The basis α, β of $\mathcal{W}_l$ is rather uncanonical. It is more useful to distinguish sections of $\mathcal{W}_l$ by their asymptotic behavior, as we have done in Section 4.2. By $\omega_l(P,n;w;g) = \omega_{l(w)}(P,n(w),s(w);g)$, and similarly for μ_l, we obtain explicit sections of $\mathcal{W}_l(P,n)$. The pointwise holomorphy follows easily from known results concerning the special function involved.

7.6.3 *Case $P \in \mathcal{P}_Y$.* The difference $l-n$ is constant, hence $p = \frac{1}{4}|n-l|$, occurring in the definition of $\omega_l(P,n,s)$ in 4.2.9, is constant as well. $\omega_l(P,n)$ is an element of $\mathcal{W}_l(P,n;W)$ for each parameter space W.

On $\{ w \in W : s(w) \notin -\frac{1}{2}\mathbb{N} \}$ we have the section $\mu_l(P,n)$. One may check explicitly that the multiple $w \mapsto \Gamma(1+s(w))^{-1} \mu_l(P,n;w)$ extends to give an element of $\mathcal{W}_l(P,n;W)$. Hence $\mu_l(P,n) \in (\mathcal{M} \otimes_{\mathcal{O}} \mathcal{W}_l(P,n))(W)$.

By $\tilde{\mu}_l(P,n) \in (\mathcal{M} \otimes_{\mathcal{O}} \mathcal{W}_l(P,n))(W)$ we denote the section determined by $\tilde{\mu}_l(P,n;w;g) = \mu_{l(w)}(P,n(w),-s(w);g)$.

7.6.4 *Case $P \in X^\infty$.* If $\Omega = \{ w \in W : \operatorname{Re} n(w) \neq 0 \}$ is not empty, there is $\omega_l(P,n) \in \mathcal{W}_l(P,n;\Omega)$.

As before, $\mu_l(P,n,s)$ gives a section $\mu_l(P,n)$ in $\mathcal{M} \otimes_{\mathcal{O}} \mathcal{W}_l(P,n)(W)$. Multiplication by $\Gamma(1+2s)^{-1}$ turns it into a holomorphic section. We form $\tilde{\mu}_l(P,n)$ by replacing s by $-s$.

Remark. To distinguish between μ_l and $\tilde{\mu}_l$ we need to know s as a function on W, and not only the eigenvalue $\frac{1}{4}-s^2$. We have built the distinction between μ_l and $\tilde{\mu}_l$ into our definition of a parameter space.

7.6.5 *Further identification.* Take $b = A_P$. On the one hand, $\mathcal{W}_l^{(A_P)}(P,n)$ is isomorphic to $\mathcal{W}_l(P,n)$. On the other hand, we can extend sections of $\mathcal{W}_l^{(A_P)}(P,n)$ to $\mathrm{pr}^{-1}\dot{U}_P(A_P)$ by χ-equivariance, obtaining sections of $\mathcal{A}_l[\dot{U}_P(A_P)]$. Hence $\mathcal{W}_l(P,n)$ is canonically isomorphic to a submodule of $\mathcal{A}_l[\dot{U}_P(A_P)]$. The image of $\tilde{F}_{P,n}$: $\mathcal{A}_l[Y_{\mathcal{P}}] \to \mathcal{A}_l[\dot{U}_P(A_P)]$ is contained in the submodule isomorphic to $\mathcal{W}_l(P,n)$. We can view $\tilde{F}_{P,n}$ as a morphism $\mathcal{A}_l[Y_{\mathcal{P}}] \to \mathcal{W}_l(P,n)$.

7.6.6 *Regular Fourier terms.* Consider $\mathcal{W}_l(P,n)$ as a submodule of $\mathcal{A}_l[\dot{U}_P(A_P)]$. Define $\mathcal{W}_l^0(P,n) = \mathcal{W}_l(P,n) \cap \mathcal{A}_l[U_P(A_P)]$. By Proposition 7.4.7 and Lemma 7.3.13 this is an eigenfunction module.

7.6.7 *Proof of part ii) of Proposition 7.2.11.* We show that the following sections form a basis of $\mathcal{W}_l^0(P,n)$ on $\Omega \subset W$ (provided the indicated set Ω is non-empty and open):

$$\begin{array}{lll}
\omega_l(P,n) & \text{on } \Omega = W & \text{if } P \in \mathcal{P}_Y \\
\omega_l(P,n) & \text{on } \Omega = \{\, w \in W : n(w) \neq 0 \,\} & \text{if } P \in X^\infty \\
\tilde{\mu}_l(P,0) & \text{on } \Omega = \{\, w \in w : n(w) = 0,\ \mathrm{Re}\, s(w) > 0 \,\} & \text{if } P \in X^\infty \\
\mu_l(P,0) & \text{on } \Omega = \{\, w \in W : n(w) = 0,\ \mathrm{Re}\, s(w) < 0 \,\} & \text{if } P \in X^\infty.
\end{array}$$

For $P \in \mathcal{P}_Y$, inspection of the definition in 4.2.9 shows that $\omega_l(P,n)$ is regular at P. For $P \in X^\infty$, the asymptotic behavior gives a bound for the L^2-norm of $\omega_l(P,n)$, $\tilde{\mu}_l(P,n)$, and $\mu_l(P,n)$ on S_{P,A_P}, uniform on compact sets in Ω. The pointwise holomorphy gives weak holomorphy in $L^2(S_{P,A_P}, d\mu)$: integrate against test functions in $C_c^0(S_{P,A_P})$. Lemma 7.4.3 gives holomorphy in L^2-norm.

Consider $f \in \mathcal{W}_l^0(P,n;\Omega_1)$ with $\Omega_1 \subset \Omega$. Let η be the element of $\mathcal{W}_l^0(P,n;\Omega)$ indicated above. Proposition 4.2.11 shows that $f(w) = c(w)\eta(w)$ with $c(w) \in \mathbb{C}$ for each $w \in \Omega$. Let $a \in \Omega$. There are a neighborhood Ω_a and $g \in \dot{U}_P(A_P)$ such that $\eta(w;g) \neq 0$ for all $w \in \Omega_a$; see the asymptotic behavior of η as $y \to \infty$, respectively $u \downarrow 0$. So $c(w) = \frac{f(w;g)}{\eta(w;g)}$ for $w \in \Omega_a$. This shows that $c \in \mathcal{O}(\Omega)$.

7.6.8 *Wronskian.* If $f \in \mathcal{W}_l(P,n;W)$ is kept fixed, then $g \mapsto \mathrm{Wr}(g,f)$ defines a morphism of $\mathcal{O}$-modules $\mathcal{W}_l(P,n) \to \mathcal{O}$.

If we take η as in the proof of the previous proposition, then $\mathcal{W}_l^0(P,n)$ is the kernel of the morphism $g \mapsto \mathrm{Wr}(g,\eta)$.

7.6.9 *Fourier coefficients of Poincaré series.* The Fourier terms $\tilde{F}_{Q,m}\dot{P}_l(P,n,\chi)$ are holomorphic sections over $\mathbb{C}_{1/2}$ of $\mathcal{W}_l(Q,m)$ for all $Q \in \mathcal{P}$ and $m \in \mathcal{C}_Q$; if $m \notin \mathbf{c}(Q)$

they are sections in $\mathcal{W}_l^0(Q, m; \mathbb{C}_{1/2})$. The $c_l(Q, m; P, n; \chi) : s \mapsto c_l(Q, m; P, n; \chi, s)$ discussed in Proposition 5.2.1 are holomorphic on $\mathbb{C}_{1/2}$. To see this note that $\dot{P}_l(P, n, \chi) - \delta_{P,Q} w_P \mu_l(P, n)$ is regular at Q, and apply 7.6.8, and Lemma 7.4.9.

Consider $P, Q \in X^\infty$. This is the situation discussed in Proposition 5.2.9. If $l \in (-1, 1]$ the holomorphic function $s \mapsto G_l(Q, m; P, n; s)$ has no zeros in $\mathbb{C}_{1/2}$. Hence the series

$$\sum_c \mathrm{S}_\chi(Q, m; P, n; c) J(Q, m; P, n; s; c)$$

is holomorphic in $s \in \mathbb{C}_{1/2}$.

7.6.10 *Functoriality in W.* We have defined the eigenfunction modules $\mathcal{W}_l(P, n)$ for each parameter space W. If $j : W_1 \to W$ is a morphism of parameter spaces, then the restriction of $j^\sharp$ (see 7.3.14) gives a morphism of sheaves $\mathcal{W}_l(P, n) \to j_* \mathcal{W}_l(P, n)_{(j)}$. Under this morphism the explicit sections $\omega_l(P, n)$ and $\mu_l(P, n)$ correspond to $\omega_l(P, n)_{(j)}$ and $\mu_l(P, n)_{(j)}$. The same holds for $\alpha_{P,n}$ and $\beta_{P,n}$, and for the explicit sections discussed below.

7.6.11 *Explicit basis, meromorphic case.* As $\mathcal{W}_l(P, n) \cong \mathcal{O}^2$, the dimension of the vector space $(\mathcal{M} \otimes_{\mathcal{O}} \mathcal{W}_l(P, n))(W)$ over $\mathcal{M}(W)$ is equal to 2.

The explicit expressions for the Wronskians in Section 4.2 imply that $\mu_l(P, n)$, $\tilde{\mu}_l(P, n)$ is a basis, unless s is constant on W with value in $\frac{1}{2}\mathbb{Z}$.

Similarly, $\mu_l(P, n)$, $\omega_l(P, n)$ is a basis, provided $\mathrm{Wr}(\mu_l(P, n), \omega_l(P, n))$ does not vanish identically on W, and, in the case $P \in X^\infty$, the function $\mathrm{Re}\, n$ has no zeros on W.

7.6.12 *Basis, holomorphic case.* It is more difficult to obtain explicit $\mathcal{O}$-bases of $\mathcal{W}_l(P, n)$ itself. Of course one could use $\alpha_{P,n}$ and $\beta_{P,n}$ with, e.g., $a = A_P$, but this is not very canonical.

In the remainder of this section we discuss explicit bases that we shall need later on, in the study of singularities of Poincaré families in Chapters 11 and 12.

Let $w_0 \in W$. We look for explicit holomorphic sections f and g of $\mathcal{W}_l(P, n)$ on a neighborhood of w_0 that form an $\mathcal{O}$-basis of $\mathcal{W}_l(P, n)$ on this neighborhood. By explicit we mean that we want to express f and g in $\omega_l(P, n)$ and $\mu_l(P, n)$ with known functions (e.g., gamma functions and exponentials) in the coefficients.

In the applications we have in mind, it suffices to do this under the assumptions $\chi(w_0) \in \mathcal{X}_u$ and $\mathrm{Re}\, s(w_0) \geq 0$. We use the notation $l_0 = l(w_0)$, $n_0 = n(w_0)$ and $s_0 = s(w_0)$. Note that n_0, $l_0 \in \mathbb{R}$.

We work in the stalk at w_0. It is sufficient to exhibit two $\mathcal{M}_{w_0}$-linear combinations f and g of $\mu_l(P, n)$ and $\omega_l(P, n)$ or $\tilde{\mu}_l(P, n)$, and check that

i) (holomorphy) f, $g \in \mathcal{W}_l(P, n)_{w_0}$,

ii) (linear independence) $\mathrm{Wr}(f, g)(w_0) \neq 0$.

7.6.13 Lemma. Regular case. *Suppose* $n_0 \neq 0$ *or* $P \in \mathcal{P}_Y$. *Take* $g = \omega_l(P,n)$, *and take* f *in the following way:*

i) *Case* $P \in X^\infty$. *Put* $\varepsilon = \mathrm{sign}(n_0)$.

   If $0 \leq s_0 \leq (\varepsilon l_0 - 1)/2$ *and* $s_0 \equiv (\varepsilon l_0 - 1)/2 \bmod 1$, *take* $f = \hat{\omega}_l(P,n)$, *given by* $\hat{\omega}_l(P,n;w;g) = \hat{\omega}_{l(w)}(P,n(w),s(w);g)$, *see 4.2.8.*

   Take $f = \mu_l(P,n)$ *otherwise.*

ii) *Case* $P \in \mathcal{P}_Y$. *Take* $\varepsilon, \zeta \in \{1,-1\}$ *such that* $\varepsilon n_0 \geq \varepsilon l_0$ *and* $\zeta(n_0 + l_0) \geq 0$; *put* $p = \varepsilon(n_0 - l_0)/4$ *and* $q = \zeta(n+l)/4$.

   If $0 \leq s_0 \leq q(w_0) - p - \frac{1}{2}$ *and* $s_0 \equiv q(w_0) - p - \frac{1}{2} \bmod 1$, *then take* $f = \hat{\omega}_l(P,n,\zeta)$, *given by* $\hat{\omega}_l(P,n,\zeta;w;g) = \hat{\omega}_{l(w)}(P,n(w),\zeta,s(w);g)$, *see 4.2.9.*

   Take $f = \mu_l(P,n)$ *otherwise.*

Then f *and* g *form an* $\mathcal{O}_{w_0}$*-basis of* $\mathcal{W}_l(P,n)_{w_0}$.

Remark. In part ii) the function $n - l$ is constant on W. The choice of ε does not matter if $n = l$.

Proof. g and $\mu_l(P,n)$ are holomorphic at w_0. So the only problem in choosing $f = \mu_l(P,n)$ may be the vanishing of the Wronskian $\mathrm{Wr}_0 = \mathrm{Wr}(\mu_l(P,n), \omega_l(P,n))$ at w_0.

Let $P \in X^\infty$. Then

$$\mathrm{Wr}_0 = (4\pi\varepsilon n_0)^{-s_0+1/2}\Gamma(2s_0+1)\Gamma(\frac{1}{2} + s_0 - \frac{1}{2}\varepsilon l_0)^{-1}.$$

The Wronskian Wr_0 vanishes precisely under the conditions in i). The definitions in 4.2.8 imply that $\hat{\omega}_l(P,n) \in \mathcal{W}_l(P,n)_{w_0}$. In the special case, the non-vanishing of $\mathrm{Wr}(f,g)$ at w_0 is also clear from 4.2.8.

Let $P \in \mathcal{P}_Y$. Put $q_0 = q(w_0)$. In 4.2.9 we see that $\mathrm{Wr}_0(w_0) = (2p)!\Gamma(1+2s_0)$ $\Gamma(\frac{1}{2} + s_0 + p + q_0)^{-1}\Gamma(\frac{1}{2} + s_0 + p - q_0)^{-1}$. Solely the factor $\Gamma(\frac{1}{2} + s_0 + p - q_0)$ may give a zero for $\mathrm{Re}\, s_0 \geq 0$. This happens if and only if the conditions in ii) are satisfied. So only under these conditions the choice $f = \mu_l(P,n)$ is not right. In 4.2.9 we see that $\hat{\omega}_l(P,n,\zeta)$ is in $\mathcal{W}_l(P,n)_{w_0}$, and that the relevant Wronskian does not vanish at w_0 under these conditions.

7.6.14 Lemma. New families. *Let* $P \in X^\infty$, $n \in \mathcal{C}_P$, $l \in$ **wt**.

i) *There exists* $\lambda_l(P,n) \in (\mathcal{M} \otimes_{\mathcal{O}} \mathcal{W}_l(P,n))(W)$ *such that* $2s\lambda_l(P,n)$ *is equal to* $\mu_l(P,n) - \tilde{\mu}_l(P,n)$, *and* $\mathrm{Wr}(\lambda_l(P,n), \tilde{\mu}_l(P,n)) = 1$.

$\lambda_l(P,n)$ is holomorphic at each $w \in W$ for which $s(w) = 0$. If $s(w) = n(w) = 0$, then $\lambda_l(P,n;w;g_P a(y)) = y^{1/2}\log y$.

ii) Let $b \in \mathbb{N}$. Put

$$w_l^b(n,s) = (4\pi n)^b \left(\tfrac{1-2s-l}{2}\right)_b \frac{\Gamma(-2s)}{\Gamma(2s)}.$$

There exists $\nu_l^b(P,n) \in (\mathcal{M} \otimes \mathcal{O}\mathcal{W}_l(P,n))(W)$ that is holomorphic near all points $w \in W$ with $s(w) = \frac{b}{2}$, and satisfies

$$\mathrm{Wr}(\mu_l(P,n), \nu_l^b(P,n)) = 2s,$$

and if $s(w) \notin \frac{1}{2}\mathbb{Z}$

$$\nu_l^b(P,n;w) = \tilde{\mu}_l(P,n;w) + w_{l(w)}^b(n(w),s(w))\mu_l(P,n;w).$$

Proof. The values at $w \in W$ of the sections $\mu_l(P,n)$ and $\tilde{\mu}_l(P,n)$ depend only on $s(w)$, $l(w)$ and $n(w)$. So for $\mathcal{W}_l(P,n)$ we could use $\mathbb{C}^3$ as a parameter space. First we can consider $(s,l,n) \mapsto \mu_l(P,n,s)$ as defined on $\mathbb{C}^3$, and afterwards on W with help of composition with the holomorphic map $\varphi = (s,l,n) : W \to \mathbb{C}^3$. It suffices to prove the lemma on $\mathbb{C}^3$. The results in the previous sections stay valid, as long as we do not use χ.

$\lambda(P;s,l,n) = \frac{1}{2s}\left(\mu_l(P,n,s) - \mu_l(P,n,-s)\right)$ defines $\lambda(P)$ as a meromorphic family of eigenfunctions on $\mathbb{C}^3$. It is holomorphic at all points with $|\mathrm{Re}\, s| < \frac{1}{2}$, $s \neq 0$, and the Wronskian $\mathrm{Wr}(\lambda(P;s,l,n), \mu_l(P,n,-s))$ is equal to 1. Apply Lemma 7.3.9 to $(s,l,n) \mapsto 2s\lambda(P;s,l,n)$ and the function $(s,n,l) \mapsto s$ to see that $\lambda(P)$ is holomorphic at all points with $s = 0$. Define $\lambda_l(P,n;w) = \lambda(P;s(w),l(w),n(w))$.

Let $b \in \mathbb{N}$. Then $(s,l,n) \mapsto (s - \frac{b}{2})\mu_l(P,n,-s)$ is holomorphic at points with $s = \frac{b}{2}$, and

$$\lim_{s \to b/2} (s - \tfrac{b}{2})\mu_l(P,n,-s) = \frac{1}{2}\frac{(-1)^b}{(b-1)!\, b!}\left(\tfrac{1-b-l}{2}\right)_b (4\pi n)^b\, \mu_l(P,n,\tfrac{b}{2}).$$

As $(s,l,n) \mapsto \mu_l(P,n,s)$ and $(s,l,n) \mapsto (s - \frac{b}{2})w_l^b(n,s)$ are holomorphic near $s = \frac{b}{2}$, and

$$\lim_{s \to b/2} (s - \tfrac{b}{2})w_l^b(n,s) = -\frac{1}{2}\frac{(-1)^b}{(b-1)!\, b!}\left(\tfrac{1-b-l}{2}\right)_b (4\pi n)^b,$$

we see that $\nu^b(P;s,l,n) = \mu_l(P,n,-s) + w_l^b(n,s)\mu_l(P,n,s)$ defines a holomorphic section $\nu^b(P)$ on a neighborhood of $s = \frac{b}{2}$, with the Wronskian $-2s$ with $\mu_l(P,n,s)$.

Define $\nu_l^b(P,n) : w \mapsto \nu^b(P;s(w),l(w),n(w))$.

7.6.15 Lemma. Singular case. *Let $P \in X^\infty$, and $n_0 = 0$.*

i) If $s_0 = 0$, take $g = \tilde{\mu}_l(P,n)$ and $f = \lambda_l(P,n)$.

ii) If $s_0 = \frac{b}{2} > 0$, $b \in \mathbb{N}$ take $f = \mu_l(P,n)$ and $g = \nu_l^b(P,n)$.

iii) In all other cases take $f = \mu_l(P,n)$ and $g = \tilde{\mu}_l(P,n)$.

Then f and g form an $\mathcal{O}_{w_0}$-basis of $\mathcal{W}_l(P,n)_{w_0}$.

Proof. There may occur two kind of difficulties. The first is the vanishing of the Wronskian $\mathrm{Wr}(\mu_l(P,n), \tilde{\mu}_l(P,n))$ at w_0; this happens if $s_0 = 0$. The other difficulty occurs if $\tilde{\mu}_l(P,n)$ is not holomorphic at w_0; that may happen if $s_0 \in \frac{1}{2}\mathbb{N}$. In both cases the previous lemma shows that the choice of f and g is suitable.

7.7 Differentiation

We prove the differentiation results in part i) of Proposition 7.2.9, and part iii) of Proposition 7.2.11. This is easy for families of Fourier terms, as we have explicit formulas for the basis elements. For families of automorphic forms, a bit more care is needed.

7.7.1 *Fourier terms.* The statement concerning $\mathbf{E}^\pm : \mathcal{W}_l(P,n) \to \mathcal{W}_{l\pm 2}(P,n)$ is easily checked.

We have given a basis of $\mathcal{W}_l^0(P,n)$ over $\Omega \subset W$ in the cases a)–c) in part ii) of Proposition 7.2.11, see 7.6.7. The differentiation results in Table 4.1 on p. 63 show that $\mathbf{E}^\pm\omega_l(P,n)$ is a holomorphic multiple of $\omega_{l\pm 2}(P,n)$. We proceed similarly in the case with $n = 0$. This completes the proof of Proposition 7.2.11.

7.7.2 *Proof of part i) of Proposition 7.2.9.* It is straightforward to check that $\mathbf{E}^+$ and $\mathbf{E}^-$ give morphisms of $\mathcal{O}$-modules $\mathbf{E}^\pm : \mathcal{A}_l[V] \to \mathcal{A}_{l\pm 2}[V]$ for each open $V \subset Y$. This means in particular that they preserve regularity at $P \in \mathcal{P}_Y$. The only fact left to prove is preservation of regularity at $P \in X^\infty$.

It is possible to construct examples where regularity is not preserved under differentiation. (Use elements of $\mathcal{W}_l(P,n)$ with $P \in X^\infty$ and $\mathrm{Re}\, n = 0$ somewhere on W.) To get part i) of Proposition 7.2.9, we need to proof regularity at $P \in X^\infty$ of $\mathbf{E}^\pm f[\mathbf{c}, P]$, for $f \in \mathcal{A}_l[\dot{U}_P(A_P)](\Omega)$ such that $f[\mathbf{c}, P]$ is regular at P.

Several reductions are possible. It is sufficient to prove regularity on a neighborhood Ω of a point $w_0 \in W$; this neighborhood may be as small as we want. Furthermore, we may enlarge $\mathbf{c}(P)$ by adding finitely many $n \in \mathcal{C}_P$. Let $w \mapsto F_{w,P}$ be the holomorphic family with values in $L^2(S_{P,A_P}, d\mu)$ given by $F_{w,P}(z) = e^{-2\pi i n_P(w)x} f[\mathbf{c}, P](g_P p(z))$, with $n_P \in \mathcal{C}_P$ fixed, and denote the family corresponding to the derivative by $F^\pm_{w,P}(z) = e^{-2\pi i n_P(w)x}\mathbf{E}^\pm f[\mathbf{c}, P](g_P p(z))$. It suffices to show L^2-holomorphy of $w \mapsto F^\pm_{w,P}$ on $S_{P,b}$ for some large $b > A_P$. We apply Lemma 7.4.3. The pointwise holomorphy is clear. So it suffices to show that the norm $\|F^\pm_{w,P}\|_b$ in $L^2(S_{P,b}, d\mu)$ is bounded.

As in Lemma 4.3.7, we have

$$F_{w,P}(z) = \sum_n c_n(w) e^{2\pi i(n(w)-n_P(w))x} \omega_l(P,n;w;g_P a(y)),$$

with n running through $\mathcal{C}_P \smallsetminus \mathbf{c}(P)$. We can assume that $|\operatorname{Re} n(w)| > N$ for all such n and all $w \in \Omega$, for some large N. The family $F^{\pm}_{w,P}(z)$ is given by its Fourier expansion. The operators $\mathbf{E}^{\pm}$ and the $\tilde{F}_{P,n}$ commute. So we obtain

$$F^{\pm}_{w,P}(z) = \sum_n c_n(w) t^{\pm}_n(w) e^{2\pi i(n(w)-n_P(w))x} \omega_{l\pm 2}(P,n;w;g_P a(y)).$$

The explicit form of the $t^{\pm}_n(w)$ can be found in Table 4.1 on p. 63. We may arrange Ω such that the $t^{\pm}_n(w)$ are bounded in $w \in \Omega$, uniformly in n. For the norms we obtain the expressions $\|F_{w,P}\|^2_{A_P} = \sum_n |c_n(w)|^2 I_l(n,w,A_P)$, and $\|F^{\pm}_{w,P}\|^2_b = \sum_n |c_n(w) t^{\pm}_n(w)|^2 I_{l\pm 2}(n,w,b)$, with

$$I_l(n,w,b) = \int_b^\infty \left|\omega_{l(w)}(P,n;w;g_P a(y))\right|^2 y^{-2} dy.$$

To conclude the boundedness of $\|F^{\pm}_{w,P}\|_b$ on Ω from the boundedness of $\|F_{w,P}\|_{A_P}$, it suffices to show that $I_{l\pm 2}(n,w,b)/I_l(n,w,A_P)$ is bounded, uniformly in $w \in \Omega$ and in n with $|\operatorname{Re} n(w)| > N$.

The asymptotic behavior of the Whittaker function $W_{\cdot,\cdot}$ mentioned in 4.2.8 is uniform on compact sets in the space of the parameters; this can be arranged by taking Ω relatively compact. So there exists $b_0 \geq A_P$ such that for all $b \geq b_0$ and all $w \in \Omega$

$$\begin{aligned} I_l(n,w,b) &\geq \frac{1}{2} \int_b^\infty |4\pi n(w) y|^{-|\operatorname{Re} l(w)|} e^{-\pi |\operatorname{Im} l(w)|/2} e^{-4\pi |\operatorname{Re} n(w)| y}\, dy, \\ I_{l\pm 2}(n,w,b) &\leq 2 \int_b^\infty |4\pi n(w) y|^{|\operatorname{Re} l(w)|+2} e^{\pi |\operatorname{Im} l(w)|/2} e^{-4\pi |\operatorname{Re} n(w)| y}\, dy. \end{aligned}$$

Take $\zeta \in \left(0, \frac{1}{2}N\right)$. We choose $b \geq b_0$ and $c_0 > 0$ such that $y^{|\operatorname{Re} l(w)|+2} \leq c_0 e^{4\pi\zeta y}$ for all $y \geq b$. There is also a constant C_1 to take care of the factors $|4\pi n(w)|^{|\operatorname{Re} l(w)|+2}$ and $e^{\pi |\operatorname{Im} l(w)|/2}$. We obtain

$$\begin{aligned} I_l(n,w,b) &\geq \frac{1}{2C_1 c_0} \int_b^\infty e^{4\pi(-\zeta - |\operatorname{Re} n(w)|)y}\, dy = \frac{1}{8\pi C_1 c_0} \frac{e^{-4\pi(\zeta + |\operatorname{Re} n(w)|)b}}{\zeta + |\operatorname{Re} n(w)|}, \\ I_{l\pm 2}(n,w,b) &\leq 2C_1 c_0 \int_b^\infty e^{4\pi(\zeta - |\operatorname{Re} n(w)|)y}\, dy = \frac{C_1 c_0}{2\pi} \frac{e^{4\pi(\zeta - |\operatorname{Re} n(w)|)b}}{|\operatorname{Re} n(w)| - \zeta}, \\ \frac{I_{l\pm 2}(n,w,b)}{I_l(n,w,A_P)} &\leq \frac{I_{l\pm 2}(n,w,b)}{I_l(n,w,b)} \leq 4(C_1 c_0)^2 \left(1 + \frac{2\zeta}{N - \zeta}\right) e^{8\pi\zeta b}. \end{aligned}$$

Remark. I would prefer a proof that would not explicitly use the asymptotic properties of Whittaker functions.

Chapter 8
Transformation and truncation

In Chapters 8–10 we shall prove the existence of meromorphic families of automorphic forms, in particular the meromorphic continuation of Poincaré series. We have explained the method in Section 1.6. In this chapter we generalize the contents of 1.6.1–1.6.2.

In Section 8.2 we construct a transformation function that will enable us to identify the Hilbert spaces $H(\chi, l)$ for various unitary characters χ and weights l. The resulting identification is not canonical; several choices have to be made. But we take care to arrange the transformation in such a way that the terms in the Fourier series expansions near the points $P \in \mathcal{P}$ are not jumbled.

The space denoted H_r in Section 1.6 is equal to $H(\chi_r, r)$ in the notation of Chapter 6, with the character χ_r of $\tilde{\Gamma}_{\text{mod}}$ as defined in 13.1.3. In Section 1.6 we have used the transformation function to relate all $H(\chi_r, r)$ to $H(1, 0)$. In the general setting we leave more freedom. The group of characters $\mathcal{X}$ may have more than one component, and even within one component we want to be able to put the center of perturbation at whatever pair $(\chi_0, l_0) \in \mathcal{X}_u \times \mathbb{R}$ we want.

In Section 8.3 we define a closed subspace ${}^aH(\chi_0, l_0)$ of the Hilbert space $H(\chi_0, l_0)$, consisting of truncated functions. In Section 1.6 we have truncated the zero order Fourier term at the cusp. Here we truncate at each $P \in \mathcal{P}$ a finite number of Fourier terms, specified by a fixed growth condition $\mathbf{c}$. In Section 8.4 we consider a subspace ${}^aD(\chi_0, l_0)$ of ${}^aH(\chi_0, l_0)$ that is closed for the energy norm, similar to what we did in Chapter 6. In the next chapter we shall study a selfadjoint holomorphic family of operators in the fixed Hilbert space ${}^aH(\chi_0, l_0)$. That family will be constructed in the same way as we obtained $A(\chi, l)$ (the extension of the Casimir operator) in Section 6.5.

We aim at the construction of families of automorphic forms. In Proposition 8.3.6 we shall see that sections of a sheaf $\mathcal{A}_l(\mathbf{c})$ of families of automorphic forms give rise to functions from the parameter space to ${}^aH(\chi_0, l_0)$ that are holomorphic in L^2-sense. Lemma 4.5.2 suggests that these families might have their values in ${}^aD(\chi_0, l_0)$. In Section 8.5 we shall give a condition under which this is the case.

8.1 Parameter space

In Chapters 8–10 we aim at the meromorphic continuation of Poincaré series, jointly in the character and the spectral parameter. In general the character group $\mathcal{X}$ is not simply connected, hence $\mathcal{X} \times \mathbb{C}$ is not allowed as a parameter space. Lemma 3.4.9 suggests to parametrize $\mathcal{X}$ by $(\chi_0, \varphi) \mapsto \chi_0 \cdot \mathbf{exp}(\varphi)$ with $\chi_0 \in \mathcal{X}_0$ and $\varphi \in \mathcal{V}$. We want to construct families that are meromorphic in (φ, s) for each $\chi_0 \in \mathcal{X}_0$. It may be useful not to work on the whole of $\mathcal{V}$, but on a linear variety in it. For instance, in Section 15.5 we restrict our attention to Eisenstein series of weight 0. This is accomplished by working on a two-dimensional linear subspace of the three-dimensional space $\mathcal{V}$ of homomorphisms $\tilde{\Gamma}_{\mathrm{com}} \to \mathbb{C}$.

These considerations lead us to take a slightly more general parameter space in this chapter and the next one. We allow any unitary character $\chi_0 \in \mathcal{X}_u$ as the central point for the perturbation, and restrict φ to a linear subspace of $\mathcal{V}$. The point of view is local: we study families on some neighborhood of $\varphi = 0$. For suitable growth conditions this neighborhood can be quite large.

8.1.1 *Notation.* By V_r we denote an $\mathbb{R}$-linear subspace of the space $\mathcal{V}_r$ of group homomorphism $\tilde{\Gamma} \to \mathbb{R}$. Put $V = \mathbb{C} \otimes_{\mathbb{R}} V_r \subset \mathcal{V} = \hom(\tilde{\Gamma}, \mathbb{C})$. We fix a unitary character $\chi_0 \in \mathcal{X}_u$ and use the parameter space $V \times \mathbb{C}$ with $\chi(\varphi, s) = \chi_0 \cdot \mathbf{exp}(\varphi)$, and $s(\varphi, s) = s$. Often we shall write $\chi(\varphi)$ instead of $\chi(\varphi, s)$.

8.1.2 *Comment.* As we shall use L^2-methods, we need a Hilbert space to work in. This is accomplished by using $\chi_0 \in \mathcal{X}_u$ as the center of perturbation, and not a general, possibly non-unitary, character in $\mathcal{X}$.

8.1.3 *Weights and Fourier term orders.* Fix $l_0 \in \mathbb{R}$ such that $\chi_0(\zeta) = e^{\pi i l_0}$. Then **wt** consists of the functions $(\varphi, s) \mapsto l_0 + \frac{\varphi(\zeta)}{\pi} + m$ with $m \in 2\mathbb{Z}$; we use $l : (\varphi, s) \mapsto l_0 + \frac{\varphi(\zeta)}{\pi}$.

For $P \in \mathcal{P}$ put

$$N_P = \begin{cases} \{\, \nu \in \mathbb{R} : e^{2\pi i \nu} = \chi_0(\pi_P) \,\} & \text{if } P \in X^\infty \\ \{\, \nu \in \mathbb{R} : e^{\pi i \nu / v_P} = \chi_0(\varepsilon_P) \,\} & \text{if } P \in \mathcal{P}_Y. \end{cases}$$

The set N_P is the set $\mathcal{C}_P$ introduced in Section 4.1 for the character χ_0. Here we use another notation for it to avoid confusion with $\mathcal{C}_P \subset \mathcal{O}(V \times \mathbb{C})$, as defined in 7.1.3. The set $\mathcal{C}_P$ consists of the functions

$$n_\nu \,:\, (\varphi, s) \mapsto \begin{cases} \nu + \frac{1}{2\pi}\varphi(\pi_P) & \text{if } P \in X^\infty \\ \nu + \frac{1}{\pi} v_P \varphi(\varepsilon_P) & \text{if } P \in \mathcal{P}_Y, \end{cases}$$

with ν running through N_P.

Often we write $n_\nu(\varphi)$, respectively $l(\varphi)$, instead of $n_\nu(\varphi, s)$, respectively $l(\varphi, s)$.

Giving a growth condition $\mathbf{c}$ corresponds to giving finite subsets $\mathbf{c}[P]$ of the N_P: take $\mathbf{c}[P] = \{\, \nu \in N_P : n_\nu \in \mathbf{c}(P) \,\}$. We shall impose the *condition* that $0 \in \mathbf{c}[P]$ for all $P \in X^\infty$ for which $0 \in N_P$. This is the same condition as in 4.3.1.

8.2 Transformation

In 1.6.1 we have discussed a transformation function for the modular case. The construction of a similar function in this section works for all cofinite discrete groups $\tilde{\Gamma}$. It does not depend on 'known' functions like $\log \eta$.

A transformation function has appeared in 7.4.2: to define L^2-holomorphy near $P \in X^\infty$ we replaced the function $f \in C^\infty(\Omega \times \mathrm{pr}^{-1}\dot{U}_P(b))$ by a family $w \mapsto F_{w,P}$ on Ω of elements of $C^\infty(\mathrm{pr}^{-1}\dot{U}_P(b))$ that are π_P-invariant ($\dot{U}_P(b)$ is a neighborhood of P in Y; see 3.5.3). In this way we could formulate the L^2-holomorphy in the fixed Hilbert space $L^2(S_{P,b}, d\mu)$. This transformation was accomplished as a multiplication operator by the *transformation function* $(w, g_P p(z)) \mapsto e^{-2\pi i n_P(w)x}$. The choice of $n_P \in \mathcal{C}_P$ does not matter for the holomorphy condition.

The transformation function to be constructed in this section is global on Y. On the sets $\mathrm{pr}^{-1}\dot{U}_P(A_P)$ with $P \in X^\infty$ it will have the same form $(w, g_P p(z)) \mapsto e^{-2\pi i n_P(w)x}$ as in 7.4.2.

8.2.1 Lemma. *There exists a $\mathbb{C}$-linear map $t : \mathcal{V} \to C^\infty(\tilde{G}) : \varphi \mapsto t_\varphi$ such that*

i) t_φ is real-valued for $\varphi \in \mathcal{V}_r$.

ii) $t_\varphi(\gamma g k(\theta)) = \varphi(\gamma) + t_\varphi(g) + \frac{\varphi(\zeta)}{\pi}\theta$ for $\gamma \in \tilde{\Gamma}$, $g \in \tilde{G}$, $\theta \in \mathbb{R}$. ($\zeta = k(\pi)$ generates the center $\tilde{Z}$ of $\tilde{\Gamma}$.)

iii) If $P \in X^\infty$, then $t_\varphi(g_P p(z)k(\theta)) = \varphi(\pi_P)x + \frac{\varphi(\zeta)}{\pi}\theta$ on $\mathrm{pr}^{-1}\dot{U}_P(A_P)$;
if $P \in \mathcal{P}_Y$, then $t_\varphi(g_P k(\eta)a(t_u)k(\psi)) = \frac{\varphi(\zeta)}{\pi}(\eta + \psi)$ on $\mathrm{pr}^{-1}\dot{U}_P(A_P)$.

Remarks. t is far from unique. For instance, we may add to t_φ any real-valued element of $C^\infty(\tilde{\Gamma}\backslash\tilde{G}/\tilde{K})$ whose support does not meet the $U_P(A_P)$.

Condition ii) ensures that e^{-it_φ} has the desired transformation behavior. Condition i) is useful in a Hilbert space context, as it implies $|e^{it_\varphi}| = 1$ for $\varphi \in \mathcal{V}_r$. Condition iii) says that e^{it_φ} has a very simple form near the points of $\mathcal{P}$. This will be important when we shall deal with Fourier expansions.

One might want to impose a differential equation on the t_φ. But if one does so it becomes hard to satisfy condition iii).

Proof. It is sufficient to construct real-valued t_φ satisfying ii) and iii) for φ in an $\mathbb{R}$-basis of $\mathcal{V}_r$. So we consider $\varphi \in \mathcal{V}_r$, and construct t_φ.

First we construct $f \in C^\infty(\tilde{G})$, real valued, such that $f(\gamma g) = \varphi(\gamma) + f(g)$ for all $\gamma \in \tilde{\Gamma}$. Take $\psi \in C_c^\infty(\tilde{G})$ with $\psi \geq 0$, $\int_{\tilde{G}} \psi(g)\, dg = 1$. Let F be a fundamental domain of $\tilde{\Gamma}$ in $\mathfrak{H}$. Put

$$f_0(p(z)k(\theta)) = \begin{cases} 1 & \text{if } z \in F,\ \theta \in [0, \pi) \\ 0 & \text{otherwise.} \end{cases}$$

So $\sum_{\gamma\in\tilde\Gamma} f_0(\gamma g) = 1$ for all g except those for which $g\tilde K$ corresponds to a boundary point of the fundamental domain $F \subset \mathfrak{H} \cong \tilde G/\tilde K$. The sum is locally finite. Consider $f_1 = f_0 * \psi : g \mapsto \int_{\tilde G} f_0(gg_1^{-1})\psi(g_1)\, dg_1$. So $f_1 \in C^\infty(\tilde G)$. For $g_1 \in \operatorname{supp}\psi$ and $g \in \tilde G$ we have $\sum_{\gamma\in\tilde\Gamma} f_0(\gamma g g_1^{-1}) = 1$, except for g_1 in a set of measure 0. So $\sum_{\gamma\in\tilde\Gamma} f_1(\gamma g) = 1$. This is again a locally finite sum. Put $f(g) = -\sum_{\gamma\in\tilde\Gamma} \varphi(\gamma) f_1(\gamma g)$. Note that $f \in C^\infty(\tilde G)$, and that it satisfies $f(\gamma g) = \varphi(\gamma) + f(g)$ for all $\gamma \in \tilde\Gamma$; in particular $f(g\zeta) = f(\zeta g) = \varphi(\zeta) + f(g)$. Put

$$f_2(g) = \int_{\mathbb{R}/\pi\mathbb{Z}} \left(f(gk(\eta)) - \frac{\varphi(\zeta)}{\pi}\eta \right) \frac{d\eta}{\pi}.$$

$f_2 \in C^\infty(\tilde G)$ is real valued and satisfies ii).

If $h_1, \cdots, h_N$ are real valued C^∞-functions satisfying $h_j(\gamma g k(\theta)) = \varphi(\gamma) + h_j(g) + \frac{\varphi(\zeta)}{\pi}\theta$ and $\psi_1, \cdots, \psi_N \in C^\infty(\tilde\Gamma\backslash\tilde G)$ are real valued with $\sum_{n=1}^N \psi_n = 1$, then $\sum_{n=1}^N \psi_n h_n$ is also real valued and satisfies ii). The h_n need satisfy ii) only on $\mathrm{pr}^{-1}\operatorname{supp}\psi_n$.

Take for each $P \in \mathcal{P}$ a real valued function $\psi_P \in C^\infty(\tilde\Gamma\backslash\tilde G/\tilde K)$, such that $\psi_P = 1$ on the set $\mathrm{pr}^{-1}(U_P(A_P))$ and $\mathrm{pr}(\operatorname{supp}(\psi_P)) \subset U_P(\tilde A_P)$, and put $\psi_* = 1 - \sum_{P\in\mathcal{P}} \psi_P$. Take $h_* = f_2$. Let h_P be given by the formulas in ii) and iii) on $\mathrm{pr}^{-1}\dot U_P(\tilde A_P)$ and equal to 0 elsewhere. Then $\psi_* h_* + \sum_{P\in\mathcal{P}} \psi_P h_P$ may be taken as t_φ. The smoothness at points of $g_P\tilde K$ for $P \in \mathcal{P}_Y$ follows from the fact that in polar coordinates $(\eta, u, \psi) \mapsto \eta + \psi$ is smooth on $\tilde G$, see 2.2.6.

8.2.2 *Remark.* $t_\varphi(p(z)k(\theta)) = t_\varphi(p(z)) + \varphi(\zeta)\theta/\pi$. So the function ωt_φ is a function on $\tilde G/\tilde K \cong \mathfrak{H}$. From ii) in the lemma it follows that $\omega t_\varphi \in C^\infty(\tilde\Gamma\backslash\tilde G/\tilde K)$, and from iii) that $\omega t_\varphi \in C_c^\infty(\tilde\Gamma\backslash\tilde G/\tilde K)$. (Explicit differentiation shows that ωt_φ vanishes on the sets $\mathrm{pr}^{-1}\dot U_P(A_P)$.)

8.2.3 *Derivatives.* We define $b_\varphi^\pm = i\mathbf{E}^\pm t_\varphi$. The $b_\varphi^\pm \in C^\infty(\tilde\Gamma\backslash\tilde G)$ are functions of weight ± 2.

$$\begin{aligned}
\overline{b_\varphi^\pm} &= -b_{\bar\varphi}^\mp \\
b_\varphi^\pm(g_P p(z)k(\theta)) &= \pm\left(-2\varphi(\pi_P)y + \frac{\varphi(\zeta)}{\pi}\right)e^{\pm 2i\theta} \quad \text{if } P \in X^\infty,\ y > A_P \\
b_\varphi^\pm(g_P k(\eta)a(t_u)k(\psi)) &= \mp\frac{\varphi(\zeta)}{\pi}\sqrt{\frac{u}{u+1}}\, e^{\pm 2i\psi} \quad \text{if } P \in \mathcal{P}_Y,\ 0 < u < A_P.
\end{aligned}$$

8.2.4 *Transformation function.* We shall use e^{-it_φ} as the transformation function. It is an element of $C^\infty(\tilde G)$, vanishes nowhere, and satisfies

$$\begin{aligned}
e^{-it_\varphi}(\gamma g k(\theta)) &= \mathbf{exp}\,(\varphi)(\gamma)^{-1} e^{-it_\varphi}(g) e^{-i\varphi(\zeta)/\pi} \\
e^{-it_\varphi}(g_P p(z)k(\theta)) &= e^{-i\varphi(\pi_P)x - i\varphi(\zeta)\theta/\pi} \quad \text{if } P \in X^\infty,\ y > A_P \\
e^{-it_\varphi}(g_P k(\eta)a(t_u)k(\psi)) &= e^{-i(\eta+\psi)\varphi(\zeta)/\pi} \quad \text{if } P \in \mathcal{P}_Y,\ 0 < u < A_P.
\end{aligned}$$

8.2.5 *Transformation.* Let V be as in 8.1.1. For $\varphi \in V$ the map

$$C^\infty(\tilde{G}_\mathcal{P}) \longrightarrow C^\infty(\tilde{G}_\mathcal{P}) \,:\, f \mapsto e^{-it_\varphi} f$$

gives a bijective correspondence between $(\chi_0 \cdot \mathbf{exp}\,(\varphi))$-$\left(l_0 + \frac{\varphi(\zeta)}{\pi}\right)$-equivariant and χ_0-l_0-equivariant functions.

On the sets $\mathrm{pr}^{-1}\dot{U}_P(A_P)$ there is a term-by-term correspondence between the Fourier expansions:

$$\tilde{F}_{P,\nu}\left(e^{-it_\varphi} f\right) = e^{-it_\varphi} \tilde{F}_{P,n_\nu} f,$$

with $\nu \in N_P$, and n_ν as in 8.1.3.

8.2.6 *Differential operator.* Under this transformation the differential operator $\omega - \frac{1}{4} + s^2$ corresponds to $M(\varphi, s) = e^{-it_\varphi} \circ (\omega - \frac{1}{4} + s^2) \circ e^{it_\varphi}$.

We use $\mathbf{E}^\pm(\varphi) = e^{-it_\varphi} \circ \mathbf{E}^\pm \circ e^{it_\varphi} = \mathbf{E}^\pm + b_\varphi^\pm$. On the functions of weight l_0:

$$\begin{aligned} M(\varphi, s) &= -\frac{1}{4}\mathbf{E}^\pm(\varphi)\mathbf{E}^\mp(\varphi) - \frac{1}{4}(l_0 + \frac{\varphi(\zeta)}{\pi})^2 \pm \frac{1}{2}(l_0 + \frac{\varphi(\zeta)}{\pi}) - \frac{1}{4} + s^2 \\ &= \omega - \frac{1}{4} + s^2 \\ &\quad - \frac{1}{4} b_\varphi^- \mathbf{E}^+ - \frac{1}{4} b_\varphi^+ \mathbf{E}^- + i(\omega t_\varphi) - \frac{1}{2\pi} l_0 \varphi(\zeta) \\ &\quad - \frac{1}{4} b_\varphi^+ b_\varphi^- - \frac{1}{4\pi^2}\varphi(\zeta)^2. \end{aligned}$$

$M(\varphi, s)$ is a perturbation of $\omega - \frac{1}{4} + s^2$. It is polynomial of degree two in the coordinates of φ.

$M(\varphi, s)$ is an elliptic differential operator with coefficients in $C^\infty(\tilde{G})$, but these smooth coefficients are not real analytic on $\tilde{G}$. So outside the $\mathrm{pr}^{-1}\dot{U}_P(A_P)$ the eigenfunctions of $M(\varphi, s)$ are smooth, but not necessarily real analytic.

8.2.7 *Families of automorphic forms.* Let Ω be an open subset of $V \times \mathbb{C}$, let $m : (\varphi, s) \mapsto l + \varphi(\zeta)/\pi$ and f a family of automorphic forms in $\mathcal{A}_m(\mathbf{c}; \Omega)$. Then $F(\varphi, s) = e^{-it_\varphi} f(\varphi, s)$ determines $F \in C^\infty(\Omega \times \tilde{G}_\mathcal{P})$ satisfying $F(w; \gamma g k(\theta)) = \chi_0(\gamma) F(w; g) e^{il_0\theta}$ for all $\gamma \in \tilde{\Gamma}$, and $M(\varphi, s) F(\varphi, s) = 0$, pointwise holomorphic in $w \in \Omega$, such that $w \mapsto F(w)[\mathbf{c}, P]$ determines an L^2-holomorphic family on Ω with values in $L^2(S_{P,A_P}, d\mu)$ for each $P \in X^\infty$ (see 7.2.3, and 7.4.2), and such that $F(w)[\mathbf{c}, P]$ has a C^∞-extension to a neighborhood of $\Omega \times \mathrm{pr}^{-1}\{P\}$ for each $P \in \mathcal{P}_Y$. We define the $\mathbf{c}$-remainder by $F[\mathbf{c}, P](g) = F(g) -$ Fourier terms of order $\nu \in \mathbf{c}[P])$ for $g \in \mathrm{pr}^{-1}\dot{U}_P(A_P)$, $P \in \mathcal{P}$; we consider it as a function on $\Omega \times \mathrm{pr}^{-1}\dot{U}_P(A_P)$. This correspondence is bijective. We do not introduce a separate notation for the set of these transformed elements of $\mathcal{A}_m(\mathbf{c}; \Omega)$.

8.3 Truncation

In 1.6.2 we have considered a subspace ${}^aH$ of $H_0 = H(\chi_0, 0) \cong L^2(F)$. Its elements have a Fourier coefficient of order zero that vanishes above $y = a$. In this section we consider truncation in a more general context.

The problem that we avoid by working in the subspace ${}^aH$ of $H(\chi_0, l_0)$ can be seen in 6.4.3 and Theorem 6.5.5: if some of the cuspidal orbits $P \in X^\infty$ are singular for χ_0, then the selfadjoint extension in $H(l_0)$ of the Casimir operator ω does not have a compact resolvent. It is the Fourier term at P of order 0 that causes the trouble. The idea from [12] is to throw away the offending Fourier term. Here we even omit a finite number of Fourier terms. In this way we can avoid 'offending Fourier terms' not only for χ_0, but for all $\chi_0 \cdot \mathbf{exp}\,(\varphi)$ with $\varphi \in U$ for a large $U \subset V$, if we might wish so.

We consider a growth condition $\mathbf{c}$ such that $0 \notin N_P \smallsetminus \mathbf{c}[P]$ for all $P \in X^\infty$.

8.3.1 *Truncation points.* For each $P \in \mathcal{P}$, we fix a real number $a(P) \in I_P$; if $P \in X^\infty$, this number has to satisfy the additional condition

$$a(P) > \frac{|l_0|}{2\pi|\nu|} \quad \text{for all } \nu \in N_P \smallsetminus \mathbf{c}[P] \text{ with } l_0\nu > 0.$$

Put ${}^aY = X \smallsetminus \bigcup_{P\in\mathcal{P}} U_P(a(P))$, ${}^aI_P = (A_P, a(P)]$ if $P \in X^\infty$, ${}^aI_P = [a(P), A_P)$ if $P \in \mathcal{P}_Y$, and ${}^a\hat{I}_P = I_P \smallsetminus {}^aI_P$. So I_P is the interval on which the

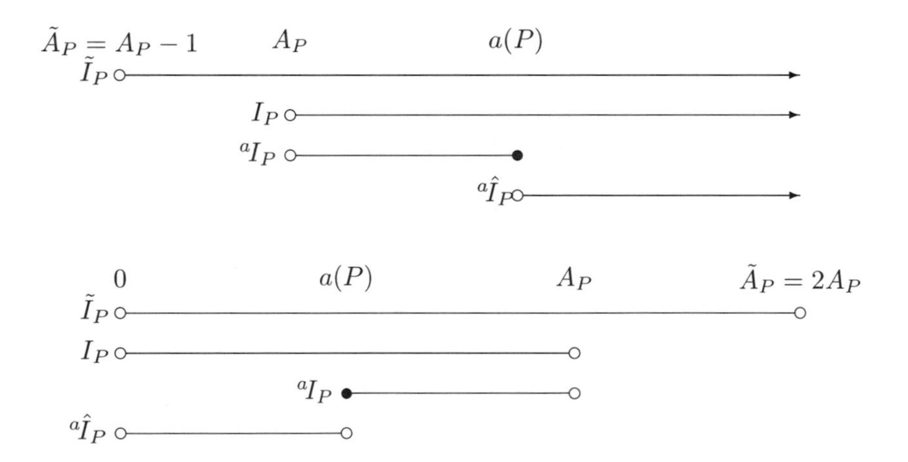

Figure 8.1 The intervals $\tilde{I}_P$, I_P, ${}^aI_P$ and ${}^a\hat{I}_P$, for $P \in X^\infty$ (top) and $P \in \mathcal{P}_Y$ (bottom).

Fourier expansion is transformed term by term. On $\tilde{I}_P \smallsetminus I_P$ we have carried out the gluing in the proof of Lemma 8.2.1. The truncation will be carried out at $a(P)$. (See Figure 8.1.)

8.3.2 *Truncation.* For a χ_0-l_0-equivariant function $f \in C^\infty(\tilde{G}_\mathcal{P})$ we define ${}^{(a)}f : \tilde{G}_\mathcal{P} \to \mathbb{C}$ to be the χ_0-l_0-equivariant function satisfying

$$\begin{aligned} {}^{(a)}f &= f \quad \text{on } \mathrm{pr}^{-1}{}^aY \\ {}^{(a)}f &= f[\mathbf{c}, P] \quad \text{on } \mathrm{pr}^{-1}\dot{U}_P(a(P)) \text{ for each } P \in \mathcal{P}. \end{aligned}$$

Of course this truncation may be defined for more general functions than those in $C^\infty(\tilde{G}_\mathcal{P})$. We use it for $f \in H(l_0)$ too. In the orthogonal decomposition

$$H(l_0) \cong L^2\left(F \cap \overline{\mathrm{pr}}^{-1}{}^aY, d\mu\right) \oplus \bigoplus_{P \in \mathcal{P}} \bigoplus_{\nu \in N_P} L^2({}^aI_P, d\nu_P),$$

it amounts to omitting the $L^2({}^aI_P, d\nu_P)$ with $\nu \in \mathbf{c}[P]$.

The truncation depends on the growth condition $\mathbf{c}$. This is not visible in the notation.

8.3.3 *Definition.* We put

$$\begin{aligned} {}^aH(l_0) &= \text{the image of } H(l_0) \text{ under } h \mapsto {}^{(a)}h \\ {}^aK(l_0) &= {}^aH(l_0) \cap K(l_0) \\ {}^{(a)}K(l_0) &= \text{the image of } K(l_0) \text{ under } h \mapsto {}^{(a)}h. \end{aligned}$$

Often we shall use an abbreviated notation: ${}^aH = {}^aH(l_0)$, ${}^aK = {}^aK(l_0)$, and ${}^{(a)}K = {}^{(a)}K(l_0)$. In the notation we cannot see that these objects depend on $\mathbf{c}$ as well.

Truncation may destroy differentiability, hence ${}^{(a)}K \not\subset K(l_0)$. The space ${}^{(a)}K$ is dense in ${}^aH$, as $h \mapsto {}^{(a)}h$ is continuous and $K(l_0)$ is dense in $H(l_0)$. Even ${}^aK$ is dense in ${}^aH$; this may be seen by considering separately $L^2\left(F \cap \overline{\mathrm{pr}}^{-1}{}^aY, d\mu\right)$ and $L^2\left(F \cap \overline{\mathrm{pr}}^{-1}U_P(a(P)), d\mu\right)$ for $P \in \mathcal{P}$. In each case $h \in L^2$ may be approximated by $k \in K(l_0)$ with support modulo $\tilde{\Gamma} \times \tilde{K}$ in the interior of the corresponding subset of F. If $h \in {}^aH$, then removal of the Fourier coefficients of k corresponding to $n \in \mathbf{c}[P]$ improves the approximation in $L^2(F \cap \overline{\mathrm{pr}}^{-1}U_P(a(P)))$ and brings k into ${}^aK$.

8.3.4 *Hilbert subspace.* ${}^aH(l_0)$ may be characterized as the closed subspace of $H(l_0)$ orthogonal to all $\Theta_{P,\nu,l_0}\psi$ with $P \in \mathcal{P}$, $\nu \in \mathbf{c}[P]$, $\psi \in C_c^\infty({}^a\hat{I}_P)$ (see 6.3.1 for the definition of the theta series $\Theta_{P,\nu,l_0}\psi$). So ${}^aH(l_0)$ is a Hilbert subspace of $H(l_0)$.

If there are no singular cusps for χ_0, then we may take $\mathbf{c} = \mathbf{n}$. In that case ${}^aH(l_0) = H(l_0)$. The space ${}^aH$ in 1.6.2 is obtained by taking $\tilde{\Gamma} = \tilde{\Gamma}_{\mathrm{mod}}$, $\chi_0 = 1$, $l_0 = 0$, $P = \tilde{\Gamma}_{\mathrm{mod}} \cdot i\infty$, $\mathcal{P} = \{P\}$, and $\mathbf{c}[P] = \{0\}$.

8.3.5 *Automorphic forms.* We consider families of automorphic forms on $\Omega \subset V \times \mathbb{C}$, and we transform and truncate them. This gives holomorphic maps $\Omega \to {}^aH$ characterized in the next proposition.

Actually, we do not restrict ourselves to families on a subset of $V \times \mathbb{C}$, but consider, more generally, families on a parameter space W that allows a morphism of parameter spaces $j : W \to V \times \mathbb{C}$. Recall the notation convention in 7.1.6: a subscript (j) denotes objects living on W. The growth condition $\mathbf{c}_{(j)}$ is defined by $\mathbf{c}_{(j)}(P) = \{ \nu \circ j : \nu \in \mathbf{c}(P) \}$.

8.3.6 Proposition. *Let $j : W \to V \times \mathbb{C}$ be a morphism of parameter spaces. Define $m \in \mathbf{wt}_{(j)}$ by $m = l \circ j : w \mapsto l_0 + \varphi(j(w))/\pi$. Let $\Omega \subset W$ be open. For each $f \in \mathcal{A}_m(\mathbf{c}_{(j)}; \Omega)$ there exists $F_f : \Omega \to {}^aH(l_0)$ given by $F_f(w) = {}^{(a)}\big(e^{-it_{\varphi(j(w))}} f(w)\big)$. This gives a bijection from $\mathcal{A}_m(\mathbf{c}_{(j)}; \Omega)$ onto the space of $F : \Omega \to {}^aH$ satisfying*

i) $F : \Omega \to {}^aH$ is L^2-holomorphic,

ii) $\left(F(w), M\left(\overline{\varphi(w)}, \overline{s(w)}\right) k\right)_{l_0} = 0$ for all $w \in \Omega$ and all $k \in {}^aK$.

Remark. This characterizes ${}^{(a)}\big(e^{-it_\varphi} A_{l_0+\varphi(\zeta)/\pi}(\chi_0 \cdot \mathbf{exp}(\varphi), \mathbf{c}_1, s)\big)$ inside ${}^aH(l_0)$ in the case that W consists of one point. Here $\mathbf{c}_1(P) = \mathbf{c}[P] + \frac{\varphi(\pi_P)}{2\pi}$ if $P \in X^\infty$ and $\mathbf{c}_1(P) = \mathbf{c}[P] + v_P \varphi(\varepsilon_P)/\pi$ otherwise.

Proof. We have defined $\mathcal{A}_m(\mathbf{c}_{(j)})$ in such a way that we get L^2-holomorphy on neighborhoods of $P \in X^\infty$. On compact sets in the fundamental domain, the pointwise holomorphy gives L^2-holomorphy by Lemma 7.4.3. So condition i) for F_f is clear. As $M(\varphi, s)^a K \subset {}^aK$ for all $\varphi \in V$, $s \in \mathbb{C}$, we obtain

$$\left(F_f(w), M(\overline{\varphi(w)}, \overline{s(w)})k\right)_{l_0} = \left(e^{-it_{\varphi(w)}} f(w), M(\overline{\varphi(w)}, \overline{s(w)})k\right)_{l_0}$$

for all $k \in {}^aK$. Use $M(\bar\varphi, \bar s) = e^{-it_{\bar\varphi}} \circ (\omega - \frac{1}{4} + \bar s^2) \circ e^{it_{\bar\varphi}}$ and Lemma 6.1.4 to obtain ii).

To prove the converse, consider $F : \Omega \to {}^aH$ satisfying i) and ii). We shall consider the distribution Φ on $\Omega \times \overline{\mathrm{pr}}^{-1} Y_\mathcal{P} \subset W \times \mathfrak{H}$ corresponding to the family F. We suppose Ω to be small enough to admit complex coordinates $w_1, \ldots, w_N$.

To define the distribution Φ we consider $\kappa \in C_c^\infty(\Omega \times \mathfrak{H})$. Define the function $\tilde\kappa \in C^\infty(\Omega \times \tilde G)$ by $\tilde\kappa(w; p(z)k(\theta)) = \kappa(w; z) e^{il_0\theta}$. Then

$$k_\kappa(w; g) = \sum_{\gamma \in \tilde Z \backslash \tilde\Gamma} \chi_0(\gamma)^{-1} \overline{\tilde\kappa(w; \gamma g)}$$

defines $k_\kappa \in C^\infty(\Omega \times \tilde G)$ with compact support in $\Omega \times (\tilde\Gamma \backslash \tilde G / \tilde K)$ such that $k_\kappa(w; .) \in K(l_0)$ for each $w \in \Omega$. This gives a C^∞-map $w \mapsto k_\kappa(w; \cdot)$ on Ω with values in $H(l_0)$. So $\Phi(\kappa) = \int_\Omega (F(w), k_\kappa(w))_{l_0} d\,\mathrm{vol}(w)$, with $d\,\mathrm{vol}(w) = (-2i)^{-N} dw_1 \wedge d\overline{w_1} \wedge \cdots \wedge dw_N \wedge d\overline{w_N}$, is well defined. If we restrict the support of κ to a compact set in $\Omega \times \mathfrak{H}$, then $\kappa \mapsto \Phi(\kappa)$ is continuous in the sup-norm. So $\kappa \mapsto \Phi(\kappa)$ is a distribution.

The sum defining k_κ is locally finite, hence $k_{\partial_{\overline{w_j}}\kappa} = \partial_{w_j} k_\kappa$. This pointwise derivative represents the derivative in L^2-sense. Hence $\partial_{\overline{w_j}}\Phi(\kappa) = -\Phi\left(\partial_{\overline{w_j}}\kappa\right) = -\int_\Omega \left(F(w), \partial_{w_j} k_\kappa\right)_{l_0} \, d\,\mathrm{vol}(w) = \int_\Omega \left(\partial_{\overline{w_j}} F(w), k_\kappa\right)_{l_0} \, d\,\mathrm{vol}(w) = 0$, see condition i). There is a differential operator $D(w)$ on $\Omega \times \mathfrak{H}$ such that $M\left(\overline{\varphi(w)}, \overline{s(w)}\right) k_\kappa(w) = k_{D(w)\kappa}(w)$. Its transpose $D(w)^t$ is the operator $M(\varphi(w), s(w))$, acting on the functions on $\Omega \times \mathfrak{H}$.

Suppose $\mathrm{supp}(\kappa) \subset \Omega \times \text{interior}\left(\overline{\mathrm{pr}}^{-1\,a}Y\right)$. Then $k_\kappa(w) \in {}^aK$ for all $w \in \Omega$. Condition ii) implies that Φ is annihilated by $D(w)^t$ on $\Omega \times \text{interior}\left(\overline{\mathrm{pr}}^{-1\,a}Y\right)$. Add the sum of the $-\partial_{w_j}\partial_{\overline{w_j}}$ to $D(w)^t$ to get an elliptic differential operator. As in 7.3.2, we conclude that Φ is given by a C^∞-function on $\Omega \times \text{interior}\left(\overline{\mathrm{pr}}^{-1\,a}Y\right)$ (not real analytic this time, as the coefficients of $M(\varphi, s)$ are only smooth). On $\Omega \times \overline{\mathrm{pr}}^{-1} U_P(a(P))$, we proceed similarly. From $\mathrm{supp}(\kappa) \subset \Omega \times \overline{\mathrm{pr}}^{-1} U_P(a(P))$ does not necessarily follow that $k_\kappa(w) \in {}^aK$; but removal of the Fourier terms corresponding to $\nu \in \mathbf{c}[P]$ does not change $(F(w), k_\kappa(w))_{l_0}$.

We have shown that Φ is given by a smooth function outside the truncation sets $\Omega \times \overline{\mathrm{pr}}^{-1}\left(\text{boundary}(U_P(a(P)))\right)$, satisfying $\partial_{\overline{w_j}}\Phi = 0$, and $D(w)^t\Phi = 0$; remember that $D(w)^t$ is nothing else than $M(\varphi(w), s(w))$ written down as a differential operator on $\mathfrak{H}$.

It now makes sense to form Fourier terms of Φ. The reasoning above shows that $\Phi[\mathbf{c}, P]$ is smooth even at the truncation sets.

Consider $\nu \in \mathbf{c}[P]$. The Fourier term of Φ of order (P, ν) is an element of $\mathcal{W}_m(P, n_\nu; \Omega)$ on the interior of $\Omega \times \left(\overline{\mathrm{pr}}^{-1} U_P(A_P) \smallsetminus \overline{\mathrm{pr}}^{-1} U_P(a(P))\right)$. This we use to extend it as a function on $\Omega \times \overline{\mathrm{pr}}^{-1} \dot{U}_P(A_P)$ annihilated by $M(\varphi(.), s(.))$. Doing this for all $\nu \in \mathbf{c}[P]$ for all $P \in \mathcal{P}$, we obtain $\tilde{\Phi} \in C^\infty(\Omega \times \overline{\mathrm{pr}}^{-1} Y_{\mathcal{P}})$, holomorphic in $w \in \Omega$; for each $w \in \Omega$ it is an eigenfunction of $M(\varphi(w), s(w))$. Further, $\tilde{F}(w; p(z)k(\theta)) = \tilde{\Phi}(w, z)e^{il_0\theta}$ defines $\tilde{F}$ on $\Omega \times \tilde{G}_{\mathcal{P}}$ such that ${}^{(a)}\tilde{F}(w) = F(w)$. The properties of $\tilde{\Phi}$ imply that $\tilde{F}(w) = e^{-it_{\varphi(w)}} f(w)$ for some $f \in \mathcal{F}_m(Y_{\mathcal{P}}; \Omega)$. At each $P \in \mathcal{P}$, the $\mathbf{c}$-remainder $f[\mathbf{c}_{(j)}, P]$ is regular, as it corresponds to $F(w)$ on $\Omega \times \mathrm{pr}^{-1} \dot{U}_P(a(P))$.

8.4 The subspace for the energy norm

In Section 6.2 we have used the space $D(l_0) \subset H(l_0)$ to get the selfadjoint extension A of ω on $H(l_0)$. Here we consider the analogous space ${}^aD(l_0)$ in ${}^aH(l_0)$. Proposition 8.4.5 will give conditions under which reasonably smooth functions in ${}^aH(l_0)$ are elements of ${}^aD(l_0)$.

The important advantage of the truncated set-up is the compactness of the inclusion ${}^aD(l_0) \to {}^aH(l_0)$ (Lemma 8.4.8). Complications, however, are caused by the fact that ${}^{(a)}K(l_0) \not\subset K(l_0)$.

Lemma 8.4.11 is rather technical but quite important. It gives an estimate for the difference $\mathbf{E}^\pm(\varphi) - \mathbf{E}^\pm$ considered as an operator ${}^aD(l_0) \to {}^aH(l_0 \pm 2)$. We need this lemma in the application of analytic perturbation theory in the next chapter.

8.4.1 *Definition.* ${}^aD(l_0) = {}^aH(l_0) \cap D(l_0)$. So ${}^aD(l_0)$ consists of those $f \in {}^aH(l_0)$ for which the distribution derivatives $\mathbf{E}^{\pm} f$ are square integrable.

If $f \in {}^aD(l_0)$, then $\mathbf{E}^{\pm} f \in H(l_0 \pm 2)$. So for $P \in \mathcal{P}$, $\nu \in \mathbf{c}[P]$ and $\psi \in C_c^\infty({}^a\hat{I}_P)$:

$$(\mathbf{E}^{\pm} f, \Theta_{P,\nu,l_0\pm 2}\psi)_{l_0\pm 2} = -(f, \Theta_{P,\nu,l_0}\psi_{\pm})_{l_0} = 0,$$

with $\psi_{\pm} \in C_c^\infty({}^a\hat{I}_P)$ given by $\psi_{\pm}(y) = 2y\psi'(y) + (\pm(4\pi\nu y - l_0) - 2)\,\psi(y)$ if $P \in X^\infty$, and $\psi_{\pm}(u) = 2\sqrt{u^2+u}\psi'(u) \pm (-\nu + (l_0 \pm 2)(2u+1))\,\psi(u)/2\sqrt{u^2+u}$ if $P \in \mathcal{P}_Y$, compare Lemma 6.3.6. This shows that $\mathbf{E}^{\pm\,a}D(l_0)$ is contained in ${}^aH(l_0 \pm 2)$.

On ${}^aD(l_0)$ we put the scalar product $(.,.)_{D,l_0}$ and the energy norm $\|.\|_{D,l_0}$ of $D(l_0)$. For this norm, ${}^aD(l_0)$ is a Hilbert subspace of $D(l_0)$. Note that ${}^aK(l_0) \subset {}^aD(l_0)$.

8.4.2 *Example.* If $\mathbf{c} \neq \mathbf{n}$ it is easy to obtain elements of ${}^aH(l_0)$ that are not in ${}^aD(l_0)$. For example, take $\nu \in \mathbf{c}[P]$, $P \in \mathcal{P}$. Choose $\psi \in C_c^\infty(I_P)$. If ψ vanishes on a neighborhood of $a(P)$, then ${}^{(a)}\Theta_{P,\nu,l_0}\psi \in {}^aK(l_0) \subset {}^aD(l_0)$. For general $\psi \in C_c^\infty(I_P)$ consider $F_{P,\nu}\mathbf{E}^{\pm\,(a)}\Theta_{P,\nu,l_0}\psi$ as the distribution on I_P sending $\eta \in C_c^\infty(I_P)$ to

$$\begin{aligned}
&(\mathbf{E}^{\pm\,(a)}\Theta_{P,\nu,l_0}\psi, \Theta_{P,\nu,l_0\pm 2}\bar{\eta})_{l_0\pm 2}\\
&= -({}^{(a)}\Theta_{P,\nu,l_0}\psi, \mathbf{E}^{\mp}\Theta_{P,\nu,l_0\pm 2}\bar{\eta})_{l_0} = -\left({}^{(a)}\Theta_{P,\nu,l_0}\psi, \mathbf{E}^{\mp}\Theta_{P,\nu,l_0\pm 2}\bar{\eta}\right)_{l_0}\\
&= \begin{cases} \frac{-2\psi(a(P))\eta(a(P))}{a(P)} + \int_{A_P}^{a(P)} \left(2y\psi'(y) \pm (-4\pi\nu y + l)\psi(y)\right)\eta(y)\frac{dy}{y^2} \\ \qquad\qquad\qquad\qquad \text{if } P \in X^\infty \\ 8\pi\sqrt{a(P)^2 + a(P)}\psi(a(P))\eta(a(P)) + \\ \int_{a(P)}^{A_P}\left(2\sqrt{u^2+u}\psi'(u) \pm \frac{\nu - l(1+2u)}{2\sqrt{u^2+u}}\psi(u)\right)\eta(u)4\pi\,du \\ \qquad\qquad\qquad\qquad \text{if } P \in \mathcal{P}_Y, \end{cases}
\end{aligned}$$

compare Lemma 6.3.6. This distribution is in $L^2(I_P, d\nu_P)$ if and only if $\psi(a(P)) = 0$. So ${}^{(a)}\Theta_{P,\nu,l_0}\psi \notin {}^aD(l_0)$ if and only if $\psi(a(P)) \neq 0$.

We may carry out this reasoning in a bit more complicated situation and obtain the next lemma. (See Figure 8.1 on p. 138 for I_P, ${}^aI_P$, and ${}^a\hat{I}_P$.)

8.4.3 Lemma. *Let $f \in {}^aD$, $P \in \mathcal{P}$, $\nu \in \mathbf{c}[P]$. Suppose that the restriction of $F_{P,\nu}f$ to ${}^aI_P$ is represented by a C^∞-function on* interior$({}^aI_P)$. *Then $F_{P,\nu}f$ has a continuous extension to I_P with value 0 on ${}^a\hat{I}_P$, and $F_{P,\nu}\mathbf{E}^{\pm}f$ is represented by the function on $I_P \smallsetminus \{a(P)\}$ given by*

$$\begin{cases} (2y\partial_y \mp 4\pi\nu y \pm l_0)\,F_{P,\nu}f & \text{if } P \in X^\infty \\ \left(2\sqrt{u^2+u}\,\partial_u \pm \frac{(\nu - l_0) - 2ul_0}{\sqrt{u^2+u}}\right)F_{P,\nu}f & \text{if } P \in \mathcal{P}_Y. \end{cases}$$

Proof. Let $\varphi_1 \in C^\infty({}^aI_P \smallsetminus \{a(P)\})$ be given by the formula in the lemma. By Lemma 6.3.6, it represents $F_{P,\nu}\mathbf{E}^\pm f$ on ${}^aI_P \smallsetminus \{a(P)\}$. As $C_c^\infty({}^aI_P \smallsetminus \{a(P)\})$ is dense in $L^2({}^aI_P, d\nu_P)$, the extended function φ_1, zero on $\{a(P)\} \cup {}^a\hat{I}_P$, represents $F_{P,\nu}\mathbf{E}^\pm f$. In particular it is in $L^2(I_P, d\nu_P)$.

For $b \in {}^aI_P \smallsetminus \{a(P)\}$ put $I(b) = (A_P, b)$ if $P \in X^\infty$ and $I(b) = (b, A_P)$ otherwise. For $\eta \in L^2(I_P, d\nu_P)$ we have

$$\int_{{}^aI_P} \eta\overline{\varphi_1}\, d\nu_P = \lim_{b \to a(P)} \int_{I(b)} \eta\overline{\varphi_1}\, d\nu_P,$$

with b varying in the interior of ${}^aI_P$.

Consider $\eta \in C_c^\infty(I_P)$. Partial integration shows that

$$\begin{aligned}(\mathbf{E}^\pm f, \Theta_{P,\nu,l_0\pm 2}\bar\eta)_{l_0\pm 2} &= -\int_{I_P} F_{P,\nu}f \cdot \eta_\pm\, d\nu_P \\ &= \lim_{b\to a(P)} \left(\int_{I(b)} \varphi_1 \cdot \eta\, d\nu_P + \begin{cases} -\frac{2}{b} F_{P,\nu}f(b)\eta(b) & \text{if } P \in X^\infty \\ 8\pi\sqrt{b^2+b}F_{P,\nu}f(b)\eta(b) & \text{if } P \in \mathcal{P}_Y \end{cases} \right).\end{aligned}$$

The limit of the integral exists. Hence $F_{P,\nu}f$ has a continuous extension to ${}^aI_P$. The value at $a(P)$ can be nothing but 0, otherwise $\eta \mapsto (\mathbf{E}^\pm f, \Theta_{P,\nu,l_0}\bar\eta)_{l_0\pm 2}$ would not be given by an element of $L^2(I_P, d\nu_P)$.

8.4.4. Let us consider the converse situation: $f \in C^\infty(G_{\mathcal{P}})$, ${}^{(a)}f \in H(l_0)$, ${}^{(a)}\mathbf{E}^\pm f \in H(l_0 \pm 2)$, $F_{P,\nu}f(a(P)) = 0$ for all $P \in \mathcal{P}$, $\nu \in \mathbf{c}[P]$. The question is whether this suffices to conclude that ${}^{(a)}f \in {}^aD(l_0)$.

That amounts to the question whether ${}^a\mathbf{E}^+ f$ represents $\mathbf{E}^{\pm(a)}f$. So we want to know whether

$$\left({}^{(a)}\mathbf{E}^\pm f, k\right)_{l_0\pm 2} + \left({}^{(a)}f, \mathbf{E}^\mp k\right)_{l_0} = 0 \quad \text{for all } k \in K(l_0 \pm 2).$$

Consider first $k \in K(l_0 \pm 2)$ with ${}^{(a)}k = k$, and hence also ${}^{(a)}\mathbf{E}^\mp k = \mathbf{E}^\mp k$. Take $b(P) \in {}^a\hat{I}_P$ for each $P \in \mathcal{P}$. Denote $F_b = F \smallsetminus \bigcup_{P\in\mathcal{P}} \overline{\mathrm{pr}}^{-1}(U_P(b(P)))$. For each such choice of b's,

$$\begin{aligned}&\left({}^{(a)}\mathbf{E}^\pm f, k\right)_{l_0\pm 2} + \left({}^{(a)}f, \mathbf{E}^\mp k\right)_{l_0} \\ &\quad = \int_{F_b} \left(\mathbf{E}^\pm f(p(z))\overline{k(p(z))} + f(p(z))\overline{\mathbf{E}^\mp k(p(z))}\right) d\mu(z).\end{aligned}$$

In the same way as in the proof of Lemma 6.1.4, this turns out to be equal to the integral over the boundary of F_b of the 1-form $-2f(p(z))\overline{k(p(z))\frac{dx\mp idy}{y}}$. The contributions of most of the boundary components cancel each other. We are left with a component for each $P \in \mathcal{P}$, its image in X encircles P once. For $P \in X^\infty$, we take $b(P)$ large, such that the corresponding path is outside the support of k. The corresponding contributions vanish.

Consider $P \in \mathcal{P}_Y$. The Fourier terms of order (P,ν) with $\nu \in \mathbf{c}[P]$ are zero for k. So we replace f by ${}^{(a)}f$, without changing the value of the integral. We obtain

$$-2\int_{\eta=0}^{\pi} {}^{(a)}f(g_P k(\eta) a(t_{b(P)}))\overline{k(g_P k(\eta) a(t_{b(P)}))}\frac{dz_{\eta,b(P)}}{y_{\eta,b(P)}},$$

with $z_{\eta,b} = g_P k(\eta) \cdot it_b$ and $y_{\eta,b} = \operatorname{Im} z_{\eta,b}$. Note that $\lim_{b\downarrow 0} t_b = 1$. A small computations shows that

$$\frac{dz_{\eta,b}}{y_{\eta,b}} = \frac{\cos\eta + it_b \sin\eta}{\cos\eta - it_b\sin\eta}\, d\eta.$$

If ${}^{(a)}f$ is smooth at $p(z_P)$, then the integral has limit 0 as $b \downarrow 0$. If ${}^{(a)}f$ is singular at the points above $P \in \mathcal{P}_Y$, the limit of the integral need not to vanish.

So we have to add a continuity condition on ${}^{(a)}f$ to the assumptions we started with, if we want to be sure that ${}^{(a)}\mathbf{E}^{\pm} f$ represents $\mathbf{E}^{\pm\,(a)}f$.

A general $k \in K(l_0 \pm 2)$ can be expressed as $k = k_a + \sum_{P,\nu} \Theta_{P,\nu,l_0\pm 2}\psi_{P,\nu}$, with $k_a \in K(l_0 \pm 2)$ satisfying $k_a = {}^{(a)}k_a$, and $\psi_{P,\nu} \in C^\infty(I_P)$ vanishing on a neighborhood of A_P. If $P \in X^\infty$ then $\psi_{P,\nu} \in C_c^\infty(I_P)$. We still have to check the desired equality for the $\psi_{P,\nu}$; it amounts to $\int_{{}^aI_P} F_{P,\nu}\mathbf{E}^{\pm} f \cdot \overline{\psi_{P,\nu}}\, d\nu_P + \int_{{}^aI_P} F_{P,\nu} f \cdot \overline{(\psi_{P,\nu})_{\pm}}\, d\nu_P = 0$, with $(\psi_{P,\nu})_{\pm}$ defined as in 8.4.1. Partial integration shows that this equality follows from the assumption $F_{P,\nu} f(a(P)) = 0$.

This reasoning and Lemma 8.4.3 together yield the following proposition.

8.4.5 Proposition. *Let $f \in C^\infty(\tilde{G}_{\mathcal{P}})$ be χ_0-l_0-equivariant, and suppose that ${}^{(a)}f \in {}^aH(l_0)$, that ${}^{(a)}f$ is C^∞ on a neighborhood of $\mathrm{pr}^{-1}\{P\}$ for each $P \in \mathcal{P}$, and that ${}^{(a)}\mathbf{E}^{\pm} f \in {}^aH(l_0 \pm 2)$.*

Then ${}^{(a)}f \in {}^aD(l_0)$ if and only if $F_{P,\nu} f(a(P)) = 0$ for all $P \in \mathcal{P}$, $\nu \in \mathbf{c}[P]$. If this holds, then $\mathbf{E}^{\pm\,(a)}f$ is represented by ${}^{(a)}\mathbf{E}^{\pm} f$.

Remark. f itself need not be smooth at the points above $P \in \mathcal{P}_Y$, but this non-smoothness is allowed to involve only the Fourier terms specified by the growth condition $\mathbf{c}$.

8.4.6 *Examples.* We apply this proposition to $f \in A_{l_0}(\chi_0, \mathbf{c}, s)$. We know that ${}^{(a)}\mathbf{E}^{\pm} f \in {}^aH(l_0 \pm 2)$, see Proposition 4.5.3. So all $f \in S_{l_0}(\chi_0, \mathbf{c}, s)$ give ${}^{(a)}f = f \in {}^aD(l_0)$.

If one accepts the meromorphic continuation of the Eisenstein series in the modular case, as given by the Fourier series expansion in 1.4.4, one may obtain examples from this family of non-cuspidal modular forms. Denote by $E_0(s)$ the modular form on $\tilde{G}$ corresponding to the Eisenstein series $E(s)$ on $\mathfrak{H}$. The proposition implies that ${}^{(a)}E_0(s) \in {}^aD$ if and only if $a^{2s} = -\Lambda(2s)/\Lambda(2s+1)$, with

$\Lambda(u) = \pi^{-u/2}\Gamma(u/2)\zeta(u)$. Consider $s = it$, $t > 0$. The equation becomes

$$t \log a + \arg \Lambda(1 + 2it) \in \pi(\frac{1}{2} + \mathbb{Z}),$$

for a continuous choice of $\arg \Lambda$. One may check that $\arg \Lambda(1+2it) = \frac{1}{2} t \log t + \mathcal{O}(t)$ as $t \to \infty$; use Stirling's formula and a reasoning as in 9.4 of [58] for $\arg \zeta(1+2it)$. This implies that for each $a \geq 5$ there are infinitely many t such that ${}^{(a)}E_0(it) \in {}^aD$.

8.4.7 Lemma. *${}^aK(l_0)$ is dense in ${}^aD(l_0)$ with respect to $\|.\|_{D,l_0}$.*

Proof. Suppose $f \in {}^aD(l_0)$ is $(.,.)_{D,l_0}$-orthogonal to ${}^aK(l_0)$. Then

$$(f, (\omega + 1 + \frac{1}{4} l_0^2) k)_{l_0} = 0 \qquad \text{for all } k \in {}^aK(l_0).$$

As $\omega + 1 + \frac{1}{4} l_0^2 = M(0, s_0)$, with $\frac{1}{4} - s_0^2 = -1 - \frac{1}{4} l_0^2$, Proposition 8.3.6, applied to $W = \{(0, s_0)\} \subset V \times \mathbb{C}$, shows that $f = {}^{(a)}h$ for some $h \in A_{l_0}(\chi_0, \mathbf{c}, s_0)$. Proposition 8.4.5 implies that $\mathbf{E}^{\pm} f$ is represented by ${}^{(a)}\mathbf{E}^{\pm} h$.

For each $\nu \in \mathbf{c}[P]$, $P \in \mathcal{P}$, we multiply the Fourier coefficient of order (P, ν) of $\mathbf{E}^{\pm} h$ by $\psi \in C^{\infty}(I_P)$, $\psi = 1$ on a neighborhood of ${}^aI_P$, but $\psi = 0$ on ${}^a\hat{I}_P \setminus$ (a neighborhood of $a(P)$). Thus we obtain $h_{\pm} \in D(l_0 \pm 2)$. As ${}^{(a)}\mathbf{E}^{\pm} h = \mathbf{E}^{\pm} f \in {}^aH(l_0 \pm 2)$ and $f \in {}^aH(l_0)$:

$$\begin{aligned}
(\mathbf{E}^{\pm} f, \mathbf{E}^{\pm} f)_{l_0 \pm 2} &= (\mathbf{E}^{\pm} f, {}^{(a)}\mathbf{E}^{\pm} h)_{l_0 \pm 2} = (\mathbf{E}^{\pm} f, h_{\pm})_{l_0 \pm 2} \\
&= -(f, \mathbf{E}^{\mp} h_{\pm})_{l_0} \qquad \text{(Lemma 6.2.7)} \\
&= -(f, \mathbf{E}^{\mp}\mathbf{E}^{\pm} h)_{l_0} = -(f, (-4\omega - l_0^2 \mp 2l_0) h)_{l_0} \\
&= (-4 \pm 2l_0)(f, h)_{l_0} = (-4 \pm 2l_0) \|f\|_{l_0}^2.
\end{aligned}$$

We obtain $\|f\|_{D,l_0}^2 = \frac{1}{8} \sum_{\pm} (-4 \pm 2l_0) \|f\|_{l_0}^2 + \|f\|_{l_0}^2 = 0$.

8.4.8 Lemma. *The inclusion ${}^aD(l_0) \to {}^aH(l_0)$ is compact.*

Proof. Given $\varepsilon > 0$, choose $\alpha(P) > a(P)$ for $P \in X^{\infty}$ such that $\alpha(P) > \frac{|l_0| + 2/\sqrt{\varepsilon}}{4\pi|\nu|}$ for all $\nu \in N_P \setminus \mathbf{c}[P]$. Take $U_1 = \bigcup_{P \in X^{\infty}} U_P(2\alpha(P))$, $\tilde{Y} = Y \setminus U_1$, $Y = \tilde{Y} \cup U$, $U = \bigcup_{P \in X^{\infty}} U_P(\alpha(P))$. Lemma 6.3.7 implies for $f \in {}^aD$:

$$\int_{F \cap \overline{\mathrm{pr}}^{-1} U} |f|^2 \, d\mu \leq \varepsilon \int_{F \cap \overline{\mathrm{pr}}^{-1} U} \left(|f|^2 + \frac{1}{8} \sum_{\pm} |\mathbf{E}^{\pm} f|^2 \right) d\mu.$$

We multiply a given $f \in {}^aD$ by $\psi \in C^{\infty}(\tilde{\Gamma} \backslash \tilde{G} / \tilde{K})$ such that $\psi = 1$ on $\mathrm{pr}^{-1} Y \setminus U$ and $\psi = 0$ on $\mathrm{pr}^{-1} U_1$. We apply Lemma 6.4.1 with $\tilde{Y}$ as given here. Then $\psi \cdot f \in \tilde{D}$.

We conclude that for all f in a subspace of finite codimension in ${}^aD$:

$$\begin{aligned}\int_{F\cap\overline{\mathrm{pr}}^{-1}Y\smallsetminus U}|f|^2\,d\mu &\leq \int_{F\cap\overline{\mathrm{pr}}^{-1}\tilde{Y}}|f|^2\,d\mu\\ &\leq \varepsilon\int_{F\cap\overline{\mathrm{pr}}^{-1}\tilde{Y}}\left(|f|^2+\frac{1}{8}\sum_{\pm}|\mathbf{E}^{\pm}f|^2\right)d\mu.\end{aligned}$$

8.4.9 *Perturbation of* $\mathbf{E}^{\pm}$. The operators $\mathbf{E}^{\pm}: {}^aD(l_0)\to{}^aH(l_0\pm 2)$ are continuous if we provide ${}^aD(l_0)$ with the norm $\|\cdot\|_{D,l_0}$, and ${}^aH(l_0\pm 2)$ with $\|\cdot\|_{l_0}$. For $\varphi\in V$, we may view $\mathbf{E}^{\pm}(\varphi)=e^{-it_\varphi}\circ\mathbf{E}^{\pm}\circ e^{it_\varphi}=\mathbf{E}^{\pm}+b_\varphi^{\pm}$ as a perturbation of $\mathbf{E}^{\pm}$. The question is whether $\mathbf{E}^{\pm}(\varphi)$ gives a continuous operator ${}^aD(l_0)\to{}^aH(l_0\pm 2)$. This is a question concerning the operator 'multiplication by $b_\varphi^{\pm}$'. Lemma 8.4.11 gives a positive answer, and expresses the norm of this operator in terms of two seminorms on V.

8.4.10 *Definition.* $\|\varphi\|_c=\max_{P\in X^\infty}|\varphi(\pi_P)|$ defines a seminorm on V. It does not depend on the truncation data, and satisfies $\|\bar\varphi\|_c=\|\varphi\|_c$.

8.4.11 Lemma.

i) Let $\varphi\in V$. The map $f\mapsto b_\varphi^{\pm}f$ sends ${}^aD(l_0)$ to ${}^aH(l_0\pm 2)$.

ii) Put

$$\xi=\min_{P\in X^\infty,\,\nu\in N_P\smallsetminus\mathbf{c}[P]}A_{a(P)}(l_0,\nu),$$

with $A_B(.,.)$ as in Lemma 6.3.7. There are a number $n_1\geq 0$ and a seminorm $\|.\|_b$ on V satisfying $\|\bar\varphi\|_b=\|\varphi\|_b$, such that for each $\varepsilon>0$

$$\begin{aligned}\|b_\varphi^{\pm}f\|_{l_0\pm 2}^2 &\leq \left(\|\varphi\|_b^2+\left(n_1+\frac{\sqrt{n_1}}{\varepsilon\xi}\right)\|\varphi\|_c^2\right)\|f\|_{l_0}^2\\ &\quad+\left(\frac{4}{\xi^2}+\frac{4\varepsilon\sqrt{n_1}}{\xi}\right)\|\varphi\|_c^2\|f\|_{D,l_0}^2\end{aligned}$$

for all $f\in{}^aD(l_0)$, $\varphi\in V$.

Remarks. In the proof we shall see that the seminorm $\|.\|_b$ depends on many choices, but that $\|.\|_c$ and n_1 do not.

$A_B(l_0,\nu)\sim 2\pi|\nu|$ as $|\nu|\to\infty$ for fixed l_0. So ξ becomes a large quantity, if we sufficiently enlarge the $\mathbf{c}[P]$ for $P\in X^\infty$. Then the influence of $\|f\|_{D,l_0}$ on the estimate is small. In the application of analytic perturbation theory, the factor in front of $\|f\|_{D,l_0}$ will determine the region in V on which we have a holomorphic family of operators. If we want meromorphic families of automorphic forms on a large region, we should take the $\mathbf{c}[P]$, with $P\in X^\infty$, large. If our aim is only local, it is more sensible to take $\mathbf{c}$ minimal.

Proof. From 8.2.3 it is clear that $b^\pm_\varphi f$ is χ_0-$(l_0 \pm 2)$-equivariant. From Lemma 6.3.7 we obtain the square integrability of $b^\pm_\varphi f$.

We take $\tilde{Y}$ and U as in the proof of Lemma 8.4.8, but with $\alpha(P) = a(P)$. Clearly

$$\|b^\pm_\varphi f\|^2_{l_0 \pm 2} \leq \int_{F\cap\overline{\mathrm{pr}}^{-1}\tilde{Y}} |b^\pm_\varphi f|^2 \, d\mu + \int_{F\cap\overline{\mathrm{pr}}^{-1}U} |b^\pm_\varphi f|^2 \, d\mu.$$

Define $\|\varphi\|_b = \sup_{\pm,\, z\in\overline{\mathrm{pr}}^{-1}\tilde{Y}} \left|b^\pm_\varphi(p(z))\right|$. The integral over $F \cap \overline{\mathrm{pr}}^{-1}\tilde{Y}$ is bounded by $\|\varphi\|^2_b \|f\|^2_{l_0}$. On $\overline{\mathrm{pr}}^{-1}U$ we define f_1 and f_2 by

$$\begin{aligned} f_1(g_P p(z)) &= \mp 2\varphi(\pi_P) y f(g_P p(z)) \\ f_2(g_P p(z)) &= \pm\pi^{-1}\varphi(\zeta) f(g_P p(z)), \end{aligned}$$

for $P \in X^\infty$ and $g_P p(z) \in \overline{\mathrm{pr}}^{-1}\dot{U}_P(a(P))$, see 8.2.3 for the relation with $b^\pm_\varphi$. We use Lemma 3.4.4 and Lemma 6.3.7 to obtain for the integrals over $F \cap \overline{\mathrm{pr}}^{-1}U$

$$\begin{aligned} \left(\int \left|b^\pm_\varphi f\right|^2 d\mu\right)^{1/2} &\leq \left(\int |f_1|^2 \, d\mu\right)^{1/2} + \left(\int |f_2|^2 \, d\mu\right)^{1/2} \\ &\leq \frac{2\|\varphi\|_c}{\xi} \left(\int \frac{1}{8} \sum_\pm \left|\mathbf{E}^\pm f\right|^2 d\mu\right)^{1/2} + \beta\|\varphi\|_c \left(\int |f|^2 \, d\mu\right)^{1/2} \\ &\leq \|\varphi\|_c \left(\frac{2}{\xi}\|f\|_{D,l_0} + \beta\|f\|_{l_0}\right), \end{aligned}$$

with $\beta = 2\pi\,|X^\infty|\,\mathrm{vol}(\tilde{\Gamma}\backslash\mathfrak{H})^{-1}$. Use $xy \leq \frac{1}{4\varepsilon}x^2 + \varepsilon y^2$ for $x, y \geq 0$ to obtain

$$\|b^\pm_\varphi f\|^2_{l_0\pm 2} \leq \left(\|\varphi\|^2_b + \left(\beta^2 + \frac{\beta}{\varepsilon\xi}\right)\|\varphi\|^2_c\right)\|f\|^2_{l_0} + \left(\frac{4}{\xi^2} + \frac{4\beta\varepsilon}{\xi}\right)\|\varphi\|^2_c \|f\|^2_{D,l_0}.$$

Take $n_1 = \beta^2$.

8.5 Families of automorphic forms

Proposition 8.3.6 gives a bijection $f \mapsto F_f$ between families of automorphic forms and families of elements of ${}^aH(l_0)$. Under what conditions is $F_f(w) \in {}^aD(l_0)$? Could F_f be a holomorphic family with respect to the energy norm $\|.\|_{D,l_0}$ as well? Our tool is Proposition 8.4.5, but before we can apply it, we have to relate $\mathbf{E}^\pm$ and $\mathbf{E}^\pm(\varphi)$.

8.5.1 *Differentiation of families of automorphic forms.* We use the notations of Proposition 8.3.6 and consider $f \in \mathcal{A}_m(\mathbf{c}_{(j)}; \Omega)$. We have already seen that $\mathbf{E}^\pm f \in \mathcal{A}_{m\pm 2}(\mathbf{c}_{(j)}; \Omega)$. Let

$$F(w) = {}^{(a)}\!\left(e^{-it_{\varphi(j(w))}} f(w)\right), \quad F_\pm(w) = {}^{(a)}\!\left(e^{-it_{\varphi(j(w))}} \mathbf{E}^\pm f(w)\right).$$

Proposition 8.3.6 shows that $F : \Omega \to {}^aH(l_0) : w \mapsto F(w)$ and $F_\pm : \Omega \to {}^aH(l_0 \pm 2) : w \mapsto F_\pm(w)$ are L^2-holomorphic.

8.5.2 Lemma. *In the notations given above:*

$$\begin{aligned} &\{\, w \in \Omega : F(w) \in {}^aD(l_0) \,\} \\ &\quad = \;\{\, w \in \Omega : F_{P,\nu} f(w; a(P)) = 0 \text{ for all } \nu \in \mathbf{c}[P],\ P \in \mathcal{P} \,\}. \end{aligned}$$

Proof. We fix $w \in \Omega$, and write f, F, $F_\pm$, and φ instead of $f(w)$, $F(w)$, $F_\pm(w)$, and $\varphi(j(w))$.

We want to apply Proposition 8.4.5 to $g = e^{-it_\varphi} f$. Clearly ${}^{(a)}g = F \in {}^aH(l_0)$. The transformation preserves regularity. So ${}^{(a)}g$ is smooth above $P \in \mathcal{P}_Y$. The problem is to show that ${}^{(a)}\mathbf{E}^\pm g \in {}^aH(l_0 \pm 2)$. Once we have done that the lemma follows.

$$\begin{aligned} {}^{(a)}\mathbf{E}^\pm g &= {}^{(a)}\mathbf{E}^\pm (e^{-it_\varphi} f) \\ &= {}^{(a)}\mathbf{E}^\pm(\varphi)\left(e^{-it_\varphi} f\right) - {}^{(a)}\left(b_\varphi^\pm e^{-it_\varphi} f\right) \\ &= {}^{(a)}\left(e^{-it_\varphi}\mathbf{E}^\pm f\right) - b_\varphi^\pm \cdot {}^{(a)}\left(e^{-it_\varphi} f\right) \\ &= F_\pm - b_\varphi^\pm \cdot F \end{aligned}$$

So the question is whether $b_\varphi^\pm F \in {}^aH(l_0 \pm 2)$.

The square integrability of $b_\varphi^\pm F$ is clear outside the sets $F \cap \overline{\mathrm{pr}}^{-1}\dot{U}_P(A_P)$ with $P \in X^\infty$. Change the $F_{P,n_\nu} f$ with $\nu \in \mathbf{c}[P]$, $P \in X^\infty$, smoothly on I_P such that they vanish on ${}^a\hat{I}_P$ and on a neighborhood of $a(P)$. This results in a function $\tilde{f}$, and in the corresponding function $\tilde{F} = {}^{(a)}\left(e^{-it_\varphi}\tilde{f}\right)$. Proposition 8.4.5, applied to $e^{-it_\varphi}\tilde{f}$, shows that $\tilde{F} \in {}^aD(l_0)$. Hence $b_\varphi^\pm \tilde{F} \in {}^aH(l_0 \pm 2)$, see Lemma 8.4.11. The functions $\tilde{F}$ and F coincide outside $\bigcup_{P \in X^\infty}\left(\mathrm{pr}^{-1}U_P(A_P) \smallsetminus \mathrm{pr}^{-1}U_P(a(P))\right)$. So $b_\varphi^\pm\left(F - \tilde{F}\right) \in {}^aH(l_0 \pm 2)$ as well.

Now we have got $b_\varphi^\pm F \in {}^aH(l_0 \pm 2)$, and apply Proposition 8.4.5 to complete the proof.

For the implication from left to right we have $F \in {}^aD(l_0)$ directly, and do not need the introduction of $\tilde{f}$ and $\tilde{F}$.

8.5.3 *Cusp forms.* In particular, if $f \in \mathcal{S}_m(\mathbf{c}_{(j)}; \Omega)$, then $F(w) \in {}^aD(l_0)$ for all $w \in \Omega$.

8.5.4 Lemma. *Use the notations given above. The family F is holomorphic $\Omega \to {}^aD(l_0)$ with respect to the norm $\|.\|_{D,l_0}$, if and only if $F_{P,n_\nu} f(w; a(P)) = 0$ for all $w \in \Omega$, $\nu \in \mathbf{c}[P]$, $P \in \mathcal{P}$.*

Proof. By the previous lemma it suffices to prove that the $\|.\|_{D,l_0}$-holomorphy of F follows from the fact that $F(w) \in {}^aD(l_0)$ for all $w \in \Omega$.

By Proposition 8.3.6, the families $F : \Omega \to {}^aH(l_0)$ and $F_\pm : \Omega \to {}^aH(l_0 \pm 2)$ are L^2-holomorphic. Proposition 8.4.5 implies that $F_\pm(w) - b^\pm_{\varphi(j(w))} F(w)$ represents $\mathbf{E}^\pm F(w)$. So if we take Ω small enough to have $\|F(w)\|_{l_0}$, $\|F_\pm(w)\|_{l_0 \pm 2}$, $\|\varphi(j(w))\|_c$, and $\|\varphi(j(w))\|_d$ bounded for all $w \in \Omega$, then we obtain boundedness of $\|F(w)\|_{D,l_0}$ as well; see Lemma 8.4.11. For $k \in {}^aK(l_0)$ consider

$$\begin{aligned} w \mapsto\ & (F(w), k)_{D,l_0} \\ =\ & \frac{1}{8} \sum \Big((F_\pm(w), \mathbf{E}^\pm k)_{l_0 \pm 2} - (F(w), b^\mp_{\overline{\varphi(j(w))}} \mathbf{E}^\pm k)_{l_0} \Big) + (F(w), k)_{l_0}. \end{aligned}$$

If this is holomorphic for all $k \in {}^aK(l_0)$, we obtain the lemma by application of Lemma 7.4.3. For $w \mapsto (F(w), b^\mp_{\overline{\varphi(j(w))}} k)_{l_0}$ the holomorphy is not immediately clear. But $w \mapsto F(w)$ is holomorphic $\Omega \to {}^aH(l_0)$. For fixed $k_1 \in {}^aK(l_0 \pm 2)$ the map $\Omega \to {}^aK(l_0) : w \mapsto b^\mp_{\overline{\varphi(j(w))}} k_1$ is antilinear in the coordinates of $\varphi(j(w))$ with respect to some basis of V_r. As the morphism of parameter spaces $j : W \to V \times \mathbb{C}$ is a holomorphic map, the coordinates of $\varphi(j(w))$ depend on w in a holomorphic way. This gives the holomorphy of $w \mapsto (F(w), b^\mp_{\overline{\varphi(j(w))}} \mathbf{E}^\pm k)_{l_0}$.

8.5.5 *Remark.* These results reinforce the message of Proposition 8.3.6: study families of functions in ${}^aH(l_0)$ or ${}^aD(l_0)$ to obtain information on families of automorphic forms.

Chapter 9
Pseudo Casimir operator

In this chapter we consider — in a more general context — the operator ${}^aA$ discussed in 1.6.2, and its resolvent. We call ${}^aA$ the pseudo Casimir operator, as it is the generalization of the pseudo Laplacian of Colin de Verdière, [12]. We use the resolvent to prove a central result of this book, Theorem 9.4.1. This theorem gives the existence of meromorphic families of automorphic forms that satisfy certain conditions on their Fourier terms. The theorem implies that — generically on an open neighborhood in $\mathcal{V} \times \mathbb{C}$ of any point in $\mathcal{V}_r \times \mathbb{C}$ — there are as many families of automorphic forms as the Maass-Selberg relation allows, see Section 9.5. In the context of Section 1.6, Theorem 9.4.1 amounts to the construction of ${}^a\tilde{E}(r, s)$ in 1.6.4.

The pseudo Casimir operator ${}^aA$ depends on a parameter φ in $\mathcal{V} = \hom(\tilde{\Gamma}, \mathbb{C})$ (or on the weight r in the notations of Section 1.6). So it is actually a family of operators, which extends the differential operator $M(\varphi, \frac{1}{2})$, defined in 8.2.6. We obtain this extension by the same procedure we used in Section 6.5 to get the selfadjoint operator $A(\chi, l)$ extending ω. When we discuss in Section 9.1 the sesquilinear form we need, we have to take into account the dependence on φ. We obtain a family of sesquilinear forms that has nice properties on a neighborhood of 0 in $\mathcal{V}$. Section 9.2 gives the construction of ${}^aA$, and the relation between its eigenfunctions and automorphic forms.

The proof of Theorem 9.4.1 uses the resolvent of the family ${}^aA$. In 1.6.4, we could just refer to Kato's book [25] for the properties of the resolvent as a function of (r, s). But here the parameter $r \in \mathbb{C}$ is replaced by $\varphi \in \mathcal{V}$, and the dimension of $\mathcal{V}$ may very well be larger than one. Section 9.3 shows that the resolvent of ${}^aA$ has the properties we shall need in Section 9.4.

I emphasize that this chapter is largely built on Colin de Verdière's approach of the meromorphic continuation of the Eisenstein series, [12].

We keep fixed a unitary character χ_0, a suitable real weight l_0 for χ_0, a linear subspace V of $\mathcal{V}$, a set of exceptional points $\mathcal{P}$, a growth condition $\mathbf{c}$, and truncation points $a(P)$, as in the previous chapter.

9.1 Sesquilinear form

We obtained the selfadjoint extension $A(\chi, l)$ of the Casimir operator $\omega : K(l) \to K(l)$ with help of the sesquilinear form $\mathfrak{s}$ on $D(l)$ introduced in Section 6.5. In this chapter we associate a family of operators to a family $\varphi \mapsto \mathfrak{s}(\varphi)$ of sesquilinear forms in ${}^aD(l_0)$, corresponding to $\varphi \mapsto M(\varphi, \frac{1}{2})$ with $\varphi \in V$. Remember that under the transformation introduced in the previous chapter, the differential operator $M(\varphi, s)$ corresponds to $\omega - \frac{1}{2} + s^2$ acting in the $\chi_0 \cdot \mathbf{exp}(\varphi)$-$l(\varphi)$-equivariant functions.

We shall show that there is a neighborhood V_0 of 0 in V on which $\mathfrak{s}(\varphi)$ is relatively bounded with respect to $\|.\|_{D,l_0}$; this is a condition under which the family of operators associated to $\mathfrak{s}$ has nice properties.

9.1.1 *Definition.* We define for $\varphi \in V$, and $f, g \in {}^aD(l_0)$,

$$\mathfrak{s}(\varphi)[f,g] = \frac{1}{8}\sum_{\pm} \left(\mathbf{E}^{\pm}(\varphi)f, \mathbf{E}^{\pm}(\bar{\varphi})g\right)_{l_0 \pm 2} - \frac{1}{4}\left(l_0 + \frac{\varphi(\zeta)}{\pi}\right)^2 (f,g)_{l_0}.$$

$\mathfrak{s}(\varphi)$ is a sesquilinear form on ${}^aD = {}^aD(l_0)$.

9.1.2 *Casimir operator.* For $\varphi \in V$, $k \in {}^aK(l_0)$, and $f \in {}^aD(l_0)$:

$$(f, M(\bar{\varphi}, \bar{s})k)_{l_0} = \mathfrak{s}(\varphi)[f,k] - (\frac{1}{4} - s^2)(f,k)_{l_0}.$$

$M(\varphi, s)$ is the differential operator corresponding to $\omega - \frac{1}{4} + s^2$ under the transformation discussed in the previous chapter.

To see that $(\mathbf{E}^{\pm}(\varphi)f, \mathbf{E}^{\pm}(\bar{\varphi})k)_{l_0 \pm 2} = -(f, \mathbf{E}^{\mp}(\bar{\varphi})\mathbf{E}^{\pm}(\bar{\varphi})k)_{l_0}$, we use the fact that $\mathbf{E}^{\pm}(\bar{\varphi})k \in K(l_0 \pm 2)$, and apply 6.2.1 and 8.2.3.

9.1.3 *Decomposition.* If $f, g \in {}^aD$, then $\varphi \mapsto \mathfrak{s}(\varphi)[f,g]$ is a quadratic polynomial in the coordinates of φ with respect to any basis of V. The decomposition into homogeneous parts is $\mathfrak{s}(\varphi) = \mathfrak{s}_0 + \mathfrak{s}_1(\varphi) + \mathfrak{s}_2(\varphi)$ with

$$\begin{aligned}
\mathfrak{s}_0[f,g] &= \frac{1}{8}\sum_{\pm}(\mathbf{E}^{\pm}f, \mathbf{E}^{\pm}g)_{l_0\pm 2} - \frac{1}{4}l_0^2(f,g)_{l_0} \\
\mathfrak{s}_1(\varphi)[f,g] &= \frac{1}{8}\sum_{\pm}\Big((\mathbf{E}^{\pm}f, b^{\pm}_{\bar{\varphi}}g)_{l_0\pm 2} + (b^{\pm}_{\varphi}f, \mathbf{E}^{\pm}g)_{l_0\pm 2}\Big) - \frac{l_0}{2\pi}\varphi(\zeta)(f,g)_{l_0} \\
\mathfrak{s}_2(\varphi)[f,g] &= \frac{1}{8}\sum_{\pm}(b^{\pm}_{\varphi}f, b^{\pm}_{\bar{\varphi}}g)_{l_0\pm 2} - \frac{1}{4\pi^2}\varphi(\zeta)^2(f,g)_{l_0}.
\end{aligned}$$

9.1.4 *Remarks.* $\mathfrak{s}_0$ is the restriction to ${}^aD$ of the form $\mathfrak{s}$ in Section 6.5.

$\mathfrak{s}_0$ is a symmetric closed sesquilinear form in ${}^aH$, densely defined, and bounded from below by $-\frac{1}{4}l_0^2\,\|.\|_{l_0}^2$. On its domain ${}^aD(l_0)$, it is relatively bounded with respect to $\|.\|_{D,l_0}$; in fact, $\mathfrak{s}_0 = \|.\|_{D,l_0}^2 - \left(1 + \frac{1}{4}l_0^2\right)\|.\|_{l_0}^2$.

A form $\mathfrak{t}$ is *relatively bounded* with respect to a sectorial form $\mathfrak{t}_1$ if $\mathrm{dom}(\mathfrak{t}) \subset \mathrm{dom}(\mathfrak{t}_1)$, and if there are $a, b \geq 0$ such that $|\mathfrak{t}[\mathfrak{u},\mathfrak{u}]| \leq \mathfrak{a}|\mathfrak{t}_1[\mathfrak{u},\mathfrak{u}]| + \mathfrak{b}\|\mathfrak{u}\|^2$ for all $u \in \mathrm{dom}(\mathfrak{t})$; see Kato, [25], Ch. VI, Remark 1.32.

Sectoriality generalizes the concept *bounded from below.* The values $\mathfrak{t}_1[\mathfrak{f},\mathfrak{f}]$, with $f \in \mathrm{dom}(\mathfrak{t}_1)$, of a *sectorial* sesquilinear form $\mathfrak{t}_1$ are situated in a sector of the form $\{\, z \in \mathbb{C} : |\arg(z - \gamma)| \leq \theta \,\}$, with $\gamma \in \mathbb{R}$ and $0 \leq \theta < \pi/2$; see [25], Ch. VI, §1.2.

9.1.5 Lemma. *Let $\varphi \in V$.*

i) The adjoint form $\mathfrak{s}(\varphi)^*[f,g] = \overline{\mathfrak{s}(\varphi)[g,f]}$ is given by $\mathfrak{s}(\varphi)^* = \mathfrak{s}(\bar\varphi)$.

ii) $\mathfrak{s}_1(\varphi)$ and $\mathfrak{s}_2(\varphi)$ are relatively bounded with respect to $\|.\|^2_{D,l_0}$; there are $a_{j,q}, b_{j,q} \geq 0$, with $j \in \{1,2\}$, $q \in \{b,c\}$, such that for all $\varphi \in V$, $f \in {}^aD$:

$$
\begin{aligned}
&|\mathfrak{s}_1(\varphi)[f,f]| \\
&\quad \leq \; (a_{1,b}\|\varphi\|_b + a_{1,c}\|\varphi\|_c)\|f\|^2_{l_0} + (b_{1,b}\|\varphi\|_b + b_{1,c}\|\varphi\|_c)\|f\|^2_{D,l_0} \\
&|\mathfrak{s}_2(\varphi)[f,f]| \\
&\quad \leq \; (a_{2,b}\|\varphi\|^2_b + a_{2,c}\|\varphi\|^2_c)\|f\|^2_{l_0} + b_{2,c}\|\varphi\|^2_c\|f\|^2_{D,l_0},
\end{aligned}
$$

with the seminorms $\|\cdot\|_b$ and $\|\cdot\|_c$ as defined in 8.4.10 and Lemma 8.4.11. Take n_1 and ξ as in Lemma 8.4.11, and let $\varepsilon > 0$. Then the $a_{j,q}$ and $b_{j,q}$ may be chosen such that

$$
\begin{aligned}
b_{1,b} &= 4\varepsilon^2, \\
b_{1,c} &= 4/\xi + 2\sqrt{n_1}\varepsilon + 4\varepsilon^2\sqrt{n_1 + \sqrt{n_1/\varepsilon\xi}}, \\
b_{2,c} &= 2\xi^{-2} + \varepsilon\sqrt{n_1}/\xi.
\end{aligned}
$$

Remark. The relative boundedness with respect to $\mathfrak{s}_0$ is equivalent to relative boundedness with respect to $\|.\|^2_{D,l_0}$; see [25], Ch. VI, §1.6.

Proof. The first part follows directly from the definition. From Lemma 8.4.11 we obtain, with $\alpha = n_1 + \frac{\sqrt{n_1}}{\varepsilon\xi}$ and $\beta = \frac{4}{\xi^2} + \frac{4\varepsilon\sqrt{n_1}}{\xi}$,

$$
\begin{aligned}
&|\mathfrak{s}_1(\varphi)[f,f]| \\
&\leq \; 2\sum_{\pm}\frac{1}{8}\|\mathbf{E}^\pm f\|_{l_0\pm 2}\sqrt{(\|\varphi\|^2_b + \alpha\|\varphi\|^2_c)\,\|f\|^2_{l_0} + \beta\|\varphi\|^2_c\|f\|^2_{D,l_0}} + |l_0|\frac{|\varphi(\zeta)|}{2\pi}\|f\|^2_{l_0} \\
&\leq \; \frac{\sqrt{n_1}}{2\pi}|l_0|\|\varphi\|_c\|f\|^2_{l_0} + 2\left(\|\varphi\|_b + \sqrt{\alpha}\|\varphi\|_c\right)\|f\|_{l_0}\|f\|_{D,l_0} + 2\sqrt{\beta}\|\varphi\|_c\|f\|^2_{D,l_0}.
\end{aligned}
$$

Lemma 3.4.4 shows how to express $|\varphi(\zeta)|$ in terms of $\|\varphi\|_c$. Use $xy \leq \frac{1}{8\varepsilon^2}x^2 + 2\varepsilon^2 y^2$ for $x, y \geq 0$ to obtain the estimate of $\mathfrak{s}_1(\varphi)[f,f]$ with

$$
\begin{aligned}
a_{1,b} &= 1/4\varepsilon^2 \\
a_{1,c} &= |X^\infty|\,\mathrm{vol}(\tilde\Gamma\backslash\mathfrak{H})^{-1} + \sqrt{\alpha}/4\varepsilon^2 \\
b_{1,b} &= 4\varepsilon^2 \\
b_{1,c} &= 4\sqrt{\alpha}\varepsilon^2 + 2\sqrt{\beta} \leq 4\varepsilon^2\sqrt{n_1 + \sqrt{n_1/\varepsilon\xi}} + 4/\xi + 2\varepsilon\sqrt{n_1}.
\end{aligned}
$$

Finally

$$\begin{aligned} \mathfrak{s}_2(\varphi)[f,f] &\le \frac{1}{8}\sum_{\pm} \|b^{\pm}_{\varphi} f\|_{l_0\pm 2}\|b^{\pm}_{\bar\varphi} f\|_{l_0\pm 2} + \frac{|\varphi(\zeta)|^2}{4\pi^2}\|f\|^2_{l_0} \\ &\le \frac{1}{4}\Big(\left(\|\varphi\|^2_b + \alpha\|\varphi\|^2_c\right)\|f\|^2_{l_0} + \beta\|\varphi\|^2_c\|f\|^2_{D,l_0}\Big) + \frac{|\varphi(\zeta)|^2}{4\pi^2}\|f\|^2_{l_0} \end{aligned}$$

gives

$$\begin{aligned} a_{2,b} &= 1/4 \\ a_{2,c} &= n_1/4 + \sqrt{n_1}/4\varepsilon\xi + |X^\infty|^2 \operatorname{vol}(\tilde\Gamma\backslash\mathfrak{H})^{-2} \\ b_{2,c} &= \xi^{-2} + \varepsilon\sqrt{n_1}/\xi. \end{aligned}$$

9.1.6 Lemma. *There is an open subset $V_0 = V_0(l_0,\chi_0,\mathbf{c})$ in V such that:*

i) $V_0(l_0,\chi_0,\mathbf{c})$ is invariant under $\varphi\mapsto\bar\varphi$ and $\varphi\mapsto t\varphi$ with $t\in\mathbb{C}$, $|t|\le 1$.

ii) $\mathfrak{s}(\varphi)$ is a closed sectorial sesquilinear form on ${}^aD$ for all $\varphi\in V_0(l_0,\chi_0,\mathbf{c})$.

iii) $V_0(-l_0,\chi_0^{-1},-\mathbf{c}) = V_0(l_0,\chi_0,\mathbf{c})$.

If $\mathcal{P}$ and the $a(P)$ have been fixed, we can choose $V_0(l_0,\chi_0,\mathbf{c})$ arbitrarily large, provided we take the $\mathbf{c}[P]$ with $P\in X^\infty$ large enough.

Remarks. In 9.1.4 we have indicated the meaning of 'sectorial'.

Assertion ii) is the main one. It says that $\mathfrak{s}$ behaves nicely on V_0. Assertion iii) is convenient when we apply the Maass-Selberg relation.

Proof. Theorem 1.33 in [25], Ch. VI, §1.6, gives a sufficient condition for $\mathfrak{s}(\varphi)$ to be a closed sesquilinear form:

$$|\mathfrak{s}_1(\varphi)[f,f] + \mathfrak{s}_2(\varphi)[f,f]| \le A\|f\|^2_{l_0} + B\|f\|^2_{D,l_0} \quad \text{for all } f\in {}^aD$$

for some constants A and B with $B<1$. By the previous lemma, this is the case if

$$b_{1,b}\|\varphi\|_b + b_{1,c}\|\varphi\|_c + b_{2,c}\|\varphi\|^2_c < 1.$$

So we can easily arrange V_0 as desired.

9.1.7 *Remark.* The family $\varphi\mapsto\mathfrak{s}(\varphi)$ is a *holomorphic family of type (a)* on V_0 in the sense of Kato, [25], Ch. VII, §4.2, except for the fact that in *loc. cit.* the parameter runs through an open set in $\mathbb{C}$, whereas the dimension of V_0 may be larger than one.

9.1.8 *How large can V_0 be?* Consider the modular case with $P=\tilde\Gamma_{\mathrm{mod}}$, $\mathcal{P}=\{P\}$. The seminorm $\|.\|_b$ in Lemma 8.4.11 is of the form $\|r\alpha\|_b = |r|N_b$; see 13.1.7 for the definition of α. The positive number N_b is not known explicitly (it depends on choices we have made). The other seminorm is $\|r\alpha\|_c = \frac{\pi}{6}|r|$.

First take $\mathbf{c} = \mathbf{n}$, $l_0 = r_0 \in (0, 6]$, and let χ_{r_0} play the role of the center of perturbation χ_0. The condition $a(P) > \frac{|l_0|}{2\pi|\nu|}$ for $\nu \in \mathcal{C}_P$, $l_0\nu > 0$ in 8.3.1 amounts to $a(P) > \frac{6}{\pi}$. But we have to take $a(P) > A_P = 2$ anyhow. We obtain $\xi = \xi(r_0) = \left(\frac{\pi}{6} - \frac{1}{2a(P)}\right) r_0$. In the proof of Lemma 8.4.11 we have chosen $n_1^2 = 2\pi|X^\infty|/\operatorname{vol}(\Gamma\backslash\mathfrak{H})$; so $n_1 = \sqrt{6}$. If

$$\left(4\varepsilon^2 N_b + \frac{\pi}{6}\left(\frac{4}{\xi} + 2\sqrt{6}\varepsilon + 4\varepsilon^2\sqrt{6 + \sqrt{6/\varepsilon\xi}}\right)\right)|r| + \frac{\pi^2}{36}\left(\frac{2}{\xi^2} + \frac{\varepsilon\sqrt{6}}{\xi}\right)|r|^2 < 1,$$

then $r\alpha$ is an element of V_0. Hence V_0 contains each open disk around r_0 with radius strictly smaller than $\left(\frac{1}{2}\sqrt{6} - 1\right) r_0$. Indeed, the terms $\frac{4\pi}{6\xi}|r|$ and $\frac{2\pi^2}{36\xi^2}|r|^2$ in the inequality are the main ones. The other terms can be handled by making $\varepsilon\xi$ small. To come near to $|r| = \left(\frac{1}{2}\sqrt{6} - 1\right) r_0$ we have to choose $a(P)$ sufficiently large.

Another case is $l_0 = r_0 = 0$, and $\mathbf{c}[P] = \{\, \nu \in \mathbb{Z} : |\nu| < N \,\}$ with $N > 0$. We may take any $a(P) > 2$, and find $\xi = 2\pi N$. In this case, V_0 can be chosen as a disk around $r_0 = 0$ with radius proportional to N.

In 15.5.2 we discuss an example where $V_0(l_0, \chi_0, \mathbf{c})$ is unbounded even for a small growth condition $\mathbf{c}$.

9.1.9 *Comparison at real points.* Let $\varphi \in V_r = V \cap \mathcal{V}_r$. The map $U_\varphi : f \mapsto e^{it_\varphi} f$ gives a unitary isomorphism of $H(\chi_0, l_0) \to H(\chi_0 \cdot \mathbf{exp}(\varphi), l_0 + \varphi(\zeta)/\pi)$.

For $\varphi \in V_r$ we have the following relation between theta series

$$e^{it_\varphi}\Theta_{P,\nu,l_0,\chi_0}\psi = \Theta_{P,n_1,l_1,\chi_1}\psi,$$

with $n_1 = n_\nu(\varphi)$, $l_1 = l_0 + \varphi(\zeta)/\pi$, and $\chi_1 = \chi_0 \cdot \mathbf{exp}(\varphi)$. (See 6.3.1.) This implies that $U_\varphi{}^a H(\chi_0, l_0) = {}^a H(\chi_1, l_1)$ as long as $\mathbf{c}$ is a growth condition for χ_1. (The additional condition in 8.3.1 on the truncation points $a(P)$ with $P \in X^\infty$ does not matter here.)

As $U_\varphi \circ \mathbf{E}^\pm(\varphi) = \mathbf{E}^\pm \circ U_\varphi$ and $b_\varphi^\pm {}^a D(l_0) \subset H(l_0 \pm 2)$, we obtain $U_\varphi{}^a D(\chi_0, l_0) \subset {}^a D(\chi_1, l_1)$. We can reverse the roles of χ_0 and $\chi_1 = \chi_0 \cdot \mathbf{exp}(\varphi)$ and use $U_{-\varphi} = U_\varphi^{-1}$ to obtain equality.

This gives, for $f, g \in {}^a D(\chi_0, l_0)$,

$$\left(U_\varphi f, U_\varphi g\right)_{D,l_1} = \mathfrak{s}(\varphi)[f,g] + \left(1 + \frac{1}{4}l_1^2\right)(f,g)_{l_0},$$

hence $(f,g) \mapsto \mathfrak{s}(\varphi)[U_{-\varphi} f, U_{-\varphi} g]$ is a restriction of the sesquilinear form introduced in Section 6.5, but taken in the weight $l_1 = l_0 + \varphi(\zeta)/\pi$.

In particular, for $\mathbf{c} = \mathbf{n}$ the family $\varphi \mapsto \mathfrak{s}(\varphi)$ on $\{\varphi \in V_r : n_\nu(\varphi) \neq 0$ for all $\nu \in \mathcal{C}_P$, $P \in X^\infty\}$ describes the sesquilinear forms in the various subspaces $D(l_0 + \varphi(\zeta)/\pi)$; c.f. Section 6.5.

9.2 Pseudo Casimir operator

In Section 6.5 we used the form $\mathfrak{s}$ to define the selfadjoint extension of the Casimir operator $\omega : K(l) \to K(l)$ in $H(l)$. Here we do the same for $\mathfrak{s}(\varphi)$. This leads to an operator ${}^aA(\varphi)$ in ${}^aH(l_0)$ extending $M(\varphi, \frac{1}{2}) : {}^aK(l_0) \to {}^aK(l_0)$. If the weight is zero this is the pseudo Laplacian of Colin de Verdière, [12].

This works well for $\varphi \in V_0$. We obtain a family $\varphi \mapsto {}^aA(\varphi)$ of operators in ${}^aH(l_0)$ with compact resolvent. We have to pay a price for the truncation. We may view the extension $A(\chi_0, l_0)$ of the Casimir operator ω as given by the action of ω on the distributions. This is no longer true for the pseudo Casimir operator. In distribution sense, ${}^aA(\varphi)$ differs from $M(\varphi, \frac{1}{2})$ by an operator that is described in Proposition 9.2.5.

9.2.1 *Definition.* Let $\varphi \in V_0$. We define the *pseudo Casimir operator* ${}^aA(\varphi)$: $\operatorname{dom} {}^aA(\varphi) \to {}^aH$ as the operator in ${}^aH$ associated to $\mathfrak{s}(\varphi)$; see [25], Ch. VI, Theorem 2.1. So $\operatorname{dom} {}^aA(\varphi) = \{ u \in {}^aD : (\exists w \in {}^aH)\, (\forall v \in {}^aD)\; \mathfrak{s}(\varphi)[u,v] = (w,v)_{l_0} \}$. If the w in the equality $\mathfrak{s}(\varphi)[u,v] = (w,v)_{l_0}$ exists, it is unique, and then ${}^aA(\varphi)u = w$. Lemma 8.4.7 implies that we can replace $\forall v \in {}^aD$ by $\forall v \in {}^aK$.

The adjoint is ${}^aA(\varphi)^* = {}^aA(\bar{\varphi})$. In particular, if $\varphi \in V_r \cap V_0$ the operator ${}^aA(\varphi)$ is selfadjoint and bounded from below:

$$({}^aA(\varphi)f, f)_{l_0} \geq -\frac{1}{4}\left(l_0 + \frac{\varphi(\zeta)}{\pi}\right)^2 \|f\|^2_{l_0}.$$

9.2.2 Proposition. *The operator ${}^aA(\varphi)$ has compact resolvent in ${}^aH$ for each $\varphi \in V_0(l_0, \varphi_0, \mathbf{c})$.*

Remark. The compactness of the resolvent is an important advantage of working in ${}^aH$ instead of $H(l_0)$. The set of eigenvalues of each ${}^aA(\varphi)$ is discrete.

Proof. Theorem 3.4 in Ch. VI of [25] shows that having a compact resolvent is a rather stable property. Apply it to the sesquilinear forms $(.,.)_{D,l_0} = \left(\frac{1}{4}l_0^2 + 1\right) (.,.)_{l_0} + \mathfrak{s}(0)$ and $\mathfrak{s}(\varphi) = (.,.)_{D,l_0} + \left(\mathfrak{s}_1(\varphi) + \mathfrak{s}_2(\varphi) - \left(\frac{1}{4}l_0^2 + 1\right)(.,.)_{l_0}\right)$ to see that ${}^aA(\varphi)$ has compact resolvent if the resolvent of ${}^aA(0) + \left(\frac{1}{4}l_0^2 + 1\right) \mathrm{Id}$ is compact. By *loc. cit.*, Ch. III, Theorem 6.29 it is sufficient to show that ${}^aA(0) + \left(\frac{1}{4}l_0^2 + 1\right) \mathrm{Id}$ has a compact inverse.

The adjoint $\mathrm{Id}^* : {}^aH \to {}^aD$ of the compact inclusion ${}^aD \to {}^aH$ is compact as well, see Lemma 8.4.8 and [25], Ch. III, §4.2, Theorem 4.10. We show that $\left({}^aA(0) + \left(\frac{1}{4}l_0^2 + 1\right)\mathrm{Id}\right)\mathrm{Id}^*$ is the indentity on ${}^aH$. Let $g \in {}^aH$. For each $h \in {}^aD$:

$$\left(\left({}^aA(0) + \left(\frac{1}{4}l_0^2 + 1\right)\mathrm{Id}\right)\mathrm{Id}^* g, h\right)_{l_0} = (\mathrm{Id}^* g, h)_{D,l_0} = (g, \mathrm{Id}h)_{l_0} = (g,h)_{l_0}.$$

9.2.3 Lemma. *${}^aK \subset \operatorname{dom} {}^aA(\varphi)$ and ${}^aA(\varphi) = M(\varphi, \frac{1}{2})$ on ${}^aK$.*

Proof. Let $k \in {}^aK$, $v \in {}^aD$. Lemma 6.2.7 implies

$$\begin{aligned}(M(\varphi, \frac{1}{2})k, v)_{l_0} &= \frac{1}{8}\sum_{\pm}(\mathbf{E}^{\pm}(\varphi)k, \mathbf{E}^{\pm}(\bar{\varphi})v)_{l_0 \pm 2} - \frac{1}{4}\left(l_0 + \frac{\varphi(\zeta)}{\pi}\right)^2 (k, v)_{l_0} \\ &= \mathfrak{s}(\varphi)[k, v].\end{aligned}$$

9.2.4 *Distributions.* Let $f \in \operatorname{dom} {}^aA(\varphi)$. Then ${}^aA(\varphi)f \in {}^aH \subset H$ defines a linear form $k \mapsto (k, {}^aA(\varphi)f)_{l_0}$ on $K(l_0)$, which we view as a distribution on $\mathfrak{H}$. Similarly, $k \mapsto (k, f)_{l_0}$ corresponds to a distribution. We denote this distribution by f. The distribution derivative $M(\varphi, \frac{1}{2})f : K(l_0) \to \mathbb{C} : k \mapsto (M(\bar{\varphi}, \frac{1}{2})k, f)_{l_0}$ coincides with the distribution ${}^aA(\varphi)f$ on $\overline{\mathrm{pr}}^{-1}(\text{interior } {}^a\!Y)$ and on all open sets $\overline{\mathrm{pr}}^{-1}\dot{U}_P(a(P))$. This follows from the previous lemma. As $f \in {}^aD$, we can ignore Fourier terms specified by the growth condition.

9.2.5 Proposition. *Let $f \in \operatorname{dom} {}^aA(\varphi)$. For all $P \in \mathcal{P}$, $\nu \in \mathbf{c}[P]$ there are $d_{P,\nu}(f) \in \mathbb{C}$ such that for all $k \in K(l_0)$*

$$({}^aA(\varphi)f, k)_{l_0} = (f, M(\bar{\varphi}, \frac{1}{2})k)_{l_0} + \sum_{P \in \mathcal{P},\, \nu \in \mathbf{c}[P]} d_{P,\nu}(f)\overline{F_{P,\nu}k(a(P))}.$$

If $F_{P,\nu}f$ is represented by a C^∞-function on ${}^aI_P$, then

$$d_{P,\nu}(f) = \begin{cases} -\left(F_{P,\nu}f\right)'(a(P)) & \text{if } P \in X^\infty \\ 4\pi\left(a(P)^2 + a(P)\right)\left(F_{P,\nu}f\right)'(a(P)) & \text{if } P \in \mathcal{P}_Y. \end{cases}$$

Remark. Differentiability on ${}^aI_P$ is to be understood as differentiability on the interior, and as one-sided differentiability at the boundary point $a(P) \in {}^aI_P$. In particular, all derivatives of $F_{P,\nu}f \in C^\infty({}^aI_P)$ are bounded functions on ${}^aI_P$.

$F_{P,\nu}f$ vanishes on ${}^a\hat{I}_P$. So $F_{P,\nu}f$ need not be an element of $C^\infty(I_P)$.

Proof. The first result holds for $k \in {}^aK$, as we have seen above. It is sufficient to consider $k = \Theta_{P,\nu,l_0}\psi$ with $P \in \mathcal{P}$, $\nu \in C_P$, $\psi \in C_c^\infty(I_P)$. Put

$$\delta(\psi) = ({}^aA(\varphi)f, k)_{l_0} - (f, M(\bar{\varphi}, \frac{1}{2})k)_{l_0}.$$

Note that $\mathbf{E}^{\pm}f \in {}^aH(l_0 \pm 2)$. We obtain

$$\begin{aligned}\delta(\psi) = \int_{{}^aI_P} \Bigg(& F_{P,\nu}{}^aA(\varphi)f \cdot \bar{\psi} - \frac{1}{8}\sum_{\pm} F_{P,\nu}\mathbf{E}^{\pm}(\varphi)f \cdot \overline{e^{\pm}(\bar{\varphi})\psi} \\ & + \frac{1}{4}\left(l_0 + \frac{\varphi(\zeta)}{\pi}\right)^2 F_{P,\nu}f \cdot \bar{\psi}\Bigg)\, d\nu_P,\end{aligned}$$

with $e^{\pm}(\varphi)\psi \in C^\infty(I_P)$ such that $\mathbf{E}^{\pm}(\varphi)\Theta_{P,\nu,l_0}\psi = \Theta_{P,\nu,l_0\pm 2}e^{\pm}(\varphi)\psi$; see 6.3.1 and 6.3.4. As $F_{P,\nu}{}^aA(\varphi)f$, $F_{P,\nu}\mathbf{E}^{\pm}(\varphi)f$ and $F_{P,\nu}f$ are square integrable on I_P and vanish outside ${}^aI_P$, the distribution $\psi \mapsto \overline{\delta(\psi)}$ on I_P is an element of the first Sobolev space (i.e., it is continuous for the norm $\psi \mapsto \|\psi\| + \|\psi'\|$; see, e.g., [29], App. 4, §1). Its support is contained in ${}^aI_P$. But if $\operatorname{supp}\psi$ is contained in the interior of ${}^aI_P$, then $k = \Theta_{P,\nu,l_0}\psi \in {}^aK$, hence $\delta(\psi) = 0$. So $\psi \mapsto \overline{\delta(\psi)}$ has support inside $\{a(P)\}$. This is possible only if it is a multiple of the δ-distribution at $a(P)$; see [26], Ch. XI, §4, Theorem 4.1. This shows that $({}^aA(\varphi)f, k)_{l_0} - (f, M(\bar\varphi, \frac{1}{2})k)_{l_0}$ has the form indicated in the proposition.

Let $F_{P,\nu}f \in C^\infty({}^aI_P)$. We want to rewrite the expression for $\delta(\psi)$ as $\delta(\psi) = \int_{{}^aI_P} g\overline{\psi}\, d\nu_P + C \cdot \overline{\psi(a(P))}$ for some $g \in L^2({}^aI_P, d\nu_P)$. Then the integral has to be zero, and C equals $d_{P,\nu}(f)$.

The terms with $F_{P,\nu}{}^aA(\varphi)f$ and $\frac{1}{4}\left(l_0 + \frac{\varphi(\zeta)}{\pi}\right)^2 F_{P,\nu}f$ already have the form $\int_{{}^aI_P} g\overline{\psi}\, d\nu_P$. The functions $F_{P,\nu}\mathbf{E}^{\pm}(\varphi)f$ are in $C^\infty({}^aI_P)$, so their derivatives are bounded on ${}^aI_P$. We have, modulo terms of the form $\int_{{}^aI_P} g\overline{\psi}\, d\nu_P$:

$$\int_{{}^aI_P} \left(-\frac{1}{8}\right) \sum_{\pm} F_{P,\nu}\mathbf{E}^{\pm}(\varphi)f \cdot \overline{e^{\pm}(\varphi)\psi}\, d\nu_P$$

$$\equiv -\frac{1}{8}\sum_{\pm}\int_{{}^aI_P} F_{P,\nu}\mathbf{E}^{\pm}(\varphi)f \cdot \left\{ \begin{matrix} 2y\,\overline{\psi'(y)} \\ 2\sqrt{u^2+u}\,\overline{\psi'(u)} \end{matrix} \right\} \cdot d\nu_P$$

$$\equiv \left\{ \begin{array}{l} -\frac{1}{4}\sum_{\pm} \frac{1}{a(P)} \left(F_{P,\nu}\mathbf{E}^{\pm}(\varphi)f\right)'(a(P)) \cdot \overline{\psi(a(P))} \\ \pi \sum_{\pm} \sqrt{a(P)^2 + a(P)}\left(F_{P,\nu}\mathbf{E}^{\pm}(\varphi)f\right)'(a(P)) \cdot \overline{\psi(a(P))}. \end{array} \right.$$

Lemma 8.4.3, with ν and l_0 adapted, gives the form of $F_{P\nu}\mathbf{E}^{\pm}(\varphi)f$. We work out the case $P \in X^\infty$, and leave the other case to the reader. Write $a(P) = a$.

$$\begin{aligned} &\left(F_{P,\nu}\mathbf{E}^{\pm}(\varphi)f\right)'(a) \\ &= 2a\left(F_{P,\nu}f\right)'(a) \mp (4\pi\nu a - l_0)F_{P,\nu}f(a) \pm \left(-2\varphi(\pi_P)a + \frac{\varphi(\zeta)}{\pi}\right) F_{P,\nu}f(a) \\ &= 2a\left(F_{P,\nu}f\right)'(a). \end{aligned}$$

We have used $F_{P,\nu}f(a) = 0$, see Lemma 8.4.3. This leads to the desired value of $d_{P,\nu}f$.

9.2.6 Proposition. *Let* $(\varphi, s) \in V_0 \times \mathbb{C}$. *Then* $f \mapsto {}^{(a)}\left(e^{-it_\varphi}f\right)$ *gives a bijection from* $\{f \in A_{l_0+\varphi(\zeta)/\pi}(\chi_0 \cdot \mathbf{exp}(\varphi), \mathbf{c}, s)\colon F_{P,n}f(a(P)) = 0$ *for all* $P \in \mathcal{P}$, $n \in \mathbf{c}(P)(\varphi)\}$ *onto* $\ker\left({}^aA(\varphi) - \frac{1}{4} + s^2\right)$.

Remark. We have seen 8.4.6 that Eisenstein series for the modular group correspond to elements of ${}^aD(\dot{0})$ for an infinite sequence of values of the spectral parameter. This proposition tells us that those functions are eigenfunctions of the pseudo Casimir operator.

Proof. Proposition 8.3.6 and Lemma 8.5.2 show that the map $f \mapsto {}^{(a)}\left(e^{-it_\varphi} f\right)$ gives a bijection onto $\{ h \in {}^aD : (h, M(\bar\varphi, \bar s)k)_{l_0} = 0 \text{ for all } k \in {}^aK \}$. To see that this set is equal to $\ker\left({}^aA(\varphi) - \frac14 - s^2\right)$, note that 9.1.2 gives $\mathfrak{s}(\varphi)[h,k] - \left(\left(\frac14 - s^2\right) h, k\right)_{l_0} = (h, M(\bar\varphi, \bar s)k)_{l_0}$ for all $h \in {}^aD$, $k \in {}^aK$.

9.2.7 Corollary. *Let $\varphi \in V_0$. The set of $s \in \mathbb{C}$ such that the space of cusp forms $S_{l_0+\varphi(\zeta)/\pi}(\chi_0 \cdot \mathbf{exp}\,(\varphi), \mathbf{c}, s)$ is not trivial, is discrete in $\mathbb{C}$.*

Remark. This gives the discreteness of $\Sigma^0(\chi, l)$ not only for unitary χ and real l, but for nearby (χ, l) as well.

Proof. Use the compactness in Proposition 9.2.2.

9.2.8 *Gluing.* The treatment of the pseudo Casimir operator in this book is local on $\mathcal{V}_r$. We use it in Section 10.2 to obtain Poincaré families that are meromorphic in several variables. There we glue the local results of this chapter on sets that we call cells of continuation. In [5]–[7] the gluing was done for the family of pseudo Casimir operators itself. As that may be useful in other situations, we indicate the principle.

For $\tilde\chi_0 \in \mathcal{X}_u$, $\mathbf{c}$ a growth condition suitable for $\tilde\chi_0$, and V_r an $\mathbb{R}$-linear subspace of $\mathcal{V}_r$, let J be the connected component of 0 of the set of $\varphi_0 \in V_r$ such that $\mathbf{c}$ is suitable for $\tilde\chi_0 \cdot \mathbf{exp}\,(\varphi_0)$. For each $\varphi_0 \in J$ we may take $\chi_0 = \tilde\chi_0 \cdot \mathbf{exp}\,(\varphi_0)$, and obtain a holomorphic family of pseudo Casimir operators $\varphi \mapsto {}^aA_{\varphi_0}(\varphi - \varphi_0)$ in the space ${}^aH_{\varphi_0}$. This family is defined on an open neighborhood $\varphi_0 + V_0(l(\chi_0), \chi_0, \mathbf{c})$ of φ_0 in V. For a neighboring point $\varphi_1 \in J$ such that its neighborhood intersects that of φ_0, the families ${}^aA_{\varphi_1}$ and ${}^aA_{\varphi_0}$ correspond to each other on the intersection, by means of the unitary map $U_{\varphi_1 - \varphi_0}$ discussed in 9.1.9. This one sees by checking that the sesquilinear forms agree.

Thus we may identify all ${}^aH_{\varphi_0}$ to ${}^aH = {}^aH_0$, and glue the various families ${}^aA_{\varphi_0}$ to one family ${}^aA$ in ${}^aH$, holomorphic on a neighborhood of J in V. This family has the properties discussed in this section. If $\dim_{\mathbb{C}} V = 1$, the eigenvalues of ${}^aA(\varphi)$ depend on $\varphi \in J$ in an analytic way, see [25], Ch. VII, Theorem 3.9 and Remark 4.22.

9.2.9 *Untruncated case.* Suppose $\mathbf{c} = \mathbf{n}$. Take J as above. All $P \in X^\infty$ are regular for $\chi_0 \cdot \mathbf{exp}\,(\varphi_0)$ with $\varphi_0 \in J$. Under the transformation U_φ discussed in 9.1.9 the form $\mathfrak{s}(\varphi)$ corresponds to the form $\mathfrak{s}$ used to define $A(\tilde\chi_0 \cdot \mathbf{exp}\,(\varphi), l_0 + \varphi(\zeta)/\pi)$. So ${}^aA(\varphi)$ is $U_\varphi \circ A(\chi_0 \cdot \mathbf{exp}\,(\varphi), l_0 + \varphi(\zeta)/\pi) \circ U_{-\varphi}$.

9.3 Meromorphy of the resolvent

The pseudo Casimir operator ${}^aA(\varphi)$ depends on a parameter φ that varies in a space that has in general complex dimension larger than one. So we cannot directly consider ${}^aA(.)$ as a holomorphic family of operators in the sense of Kato, [25], Chap. VII. Instead, we use the resolvent $({}^aA(\varphi)-\lambda)^{-1}$. We prove that the resolvent is meromorphic in φ and λ jointly. Most of the results we need are somewhere in [25]; we prefer to work it out here in order to get some idea of what is going on.

We keep fixed χ_0, l_0, $\mathcal{P}$, $a(P)$ for $P \in \mathcal{P}$, and truncation data as before. In this section we do not use s as the spectral parameter, but the eigenvalue $\lambda = \frac{1}{4} - s^2$. We use the subscript λ if we consider objects as a function of λ.

9.3.1 *Resolvent.* Let $\varphi \in V_0$. The *resolvent set* ${}^a\mathsf{P}_\lambda(\varphi)$ consists of those $\lambda \in \mathbb{C}$ for which the *resolvent* $\mathrm{R}_\lambda(\varphi,\lambda) = ({}^aA(\varphi)-\lambda)^{-1}$ exists as a bounded operator on ${}^aH$. (The subscript λ in ${}^a\mathsf{P}_\lambda(\varphi)$ is just a symbol. It is not an instance of the variable λ.) Proposition 9.2.2 implies that $\mathbb{C} \setminus {}^a\mathsf{P}_\lambda(\varphi)$ is a discrete set, the *spectrum* of ${}^aA(\varphi)$.

We put ${}^a\mathsf{P}_\lambda = \{ (\varphi,\lambda) \in V_0 \times \mathbb{C} : \lambda \in {}^a\mathsf{P}_\lambda(\varphi) \}$.

Later on, when s will be the spectral parameter, we shall use ${}^a\mathsf{P}(\varphi) = \{ s \in \mathbb{C} : \frac{1}{4} - s^2 \in {}^a\mathsf{P}_\lambda(\varphi) \}$, ${}^a\mathsf{P} = \{ (\varphi, s) : (\varphi, \frac{1}{4}-s^2) \in {}^a\mathsf{P}_\lambda \}$ and $\mathrm{R}(\varphi,s) = \mathrm{R}_\lambda(\varphi, \frac{1}{4}-s^2)$.

9.3.2 Proposition. *${}^a\mathsf{P}_\lambda$ is a dense open subset of $V_0 \times \mathbb{C}$; its complement $(V_0 \times \mathbb{C}) \setminus {}^a\mathsf{P}_\lambda$ is a complex analytic set.*

If $(\varphi,\lambda) \notin {}^a\mathsf{P}_\lambda$ and $\varphi \in V_0 \cap \mathcal{V}_r$, then $\lambda \in \mathbb{R}$ and $\lambda \geq -\frac{1}{4}\left(l_0 + \varphi(\zeta)/\pi\right)^2$.

$(\varphi,\lambda) \mapsto \mathrm{R}_\lambda(\varphi,\lambda)$ is a bounded-holomorphic family of operators ${}^aH \to {}^aD$ on ${}^a\mathsf{P}_\lambda$ and a bounded-meromorphic family of operators ${}^aH \to {}^aD$ on $V_0 \times \mathbb{C}$.

Remarks. By *bounded-holomorphy* of R_λ on ${}^a\mathsf{P}_\lambda$ we mean that for each $(\varphi_0, \lambda_0) \in {}^a\mathsf{P}_\lambda$ there is, on a neighborhood of (φ_0,λ_0), a power series representation of $\mathrm{R}_\lambda(\varphi,\lambda)$ in $\lambda - \lambda_0$ and the coordinates of $\varphi - \varphi_0$ with bounded operators ${}^aH \to {}^aD$ as coefficients. The series has to converge with respect to the operator norm. See [25], Ch. VII, §1.1. By the family R_λ being *bounded-meromorphic* on $V_0 \times \mathbb{C}$, we mean that for each $w_0 = (\varphi_0,\lambda_0) \in V_0 \times \mathbb{C}$ there is a non-zero holomorphic function ψ on a neighborhood Ω of w_0 such that $\psi \cdot \mathrm{R}_\lambda : (\varphi,\lambda) \mapsto \psi(\varphi,\lambda)\mathrm{R}_\lambda(\varphi,\lambda)$ extends as a bounded-holomorphic family on Ω.

In terms of s instead of λ we obtain: ${}^a\mathsf{P}$ *is dense in $V_0 \times \mathbb{C}$; it is the complement of a complex analytic set.*

If $(\varphi, s) \in (V_0 \times \mathbb{C}) \setminus {}^a\mathsf{P}$, and $\varphi \in V_0 \cap \mathcal{V}_r$, then $s \in \mathbb{R} \cup i\mathbb{R}$ and $\frac{1}{4} - s^2 \geq -\frac{1}{4}\left(l_0 + \varphi(\zeta)/\pi\right)^2$.

$(\varphi, s) \mapsto \mathrm{R}(\varphi, s)$ is a bounded-meromorphic family of operators ${}^aH \to {}^aD$; it is bounded-holomorphic on ${}^a\mathsf{P}$.

The *proof* of Proposition 9.3.2 is completed in 9.3.8. It is prepared in a sequence of lemmas. Lemma 9.3.7 is interesting in its own right. It gives information on how $\ker({}^aA(\varphi)-\lambda)$ behaves locally on $V \times \mathbb{C}$.

9.3.3 Lemma. *There exists a self-adjoint operator* H *in* ${}^aH$, *with* $\mathrm{dom}(\mathrm{H}) = {}^aD$, *such that* $(f,g)_{D,l_0} = (\mathrm{H}f, \mathrm{H}g)_{l_0}$ *for all* $f, g \in {}^aD$.

H *has an inverse* H^{-1} *which is bounded as an operator* ${}^aH \to {}^aD$.

Remark. $\mathrm{H}^{-2} = \mathrm{Id}^*$, the compact operator in the proof of Lemma 9.2.2. Indeed, for $f, g \in {}^aH$:

$$\left(\mathrm{H}^{-2}f, g\right)_{D,l_0} = \left(\mathrm{H}^{-1}f, \mathrm{H}g\right)_{l_0} = \left(\mathrm{H}\mathrm{H}^{-1}f, g\right)_{l_0} = (f,g)_{l_0}.$$

Proof. Apply [25], Ch. VI, §2.6, Theorem 2.23 to the form $(.,.)_{D,l_0}$ on ${}^aD$ to obtain H. (In *loc. cit.* our H is called $H^{1/2}$.) As $\mathrm{H} \geq 1$, it has a bounded inverse $\mathrm{H}^{-1} : {}^aH \to {}^aH$. Clearly $\mathrm{H}^{-1}({}^aH) = {}^aD$ and $\|\mathrm{H}^{-1}f\|_{D,l_0} = \|f\|_{l_0}$ for all $f \in {}^aH$.

9.3.4 Lemma.

i) For each $\varphi \in V_0$ *there exists a bounded operator* $T(\varphi) : {}^aH \to {}^aH$ *such that* $(T(\varphi)f, g)_{l_0} = \mathfrak{s}(\varphi)[\mathrm{H}^{-1}f, \mathrm{H}^{-1}g]$ *for all* $f, g \in {}^aH$.

$T(\varphi)$ is a polynomial in the coordinates of φ with coefficients in the algebra of bounded operators ${}^aH \to {}^aH$.

ii) Let $\varphi \in V_0$, $\lambda \in \mathbb{C}$. *Suppose that* $T(\varphi) - \lambda\mathrm{H}^{-2}$ *has a bounded inverse* $Q(\varphi, \lambda)$: ${}^aH \to {}^aH$. *Then* $(\varphi, \lambda) \in {}^a\mathfrak{P}_\lambda$, *and* $\mathrm{R}_\lambda(\varphi, \lambda) = \mathrm{H}^{-1}Q(\varphi,\lambda)\mathrm{H}^{-1}$.

Remark. The operator H is used to associate a family of *bounded* operators $\varphi \mapsto T(\varphi)$ to $\mathfrak{s}(\cdot)$. These are easier to handle than the unbounded operators ${}^aA(\varphi)$.

Proof. $(f,g) \mapsto \mathfrak{s}(\varphi)[\mathrm{H}^{-1}f, \mathrm{H}^{-1}g]$ is a bounded sesquilinear form on ${}^aH$, hence it is given by a bounded operator $T(\varphi)$. For all $f, g \in {}^aH$ and $\varphi \in V_0$

$$\begin{aligned}(T(\varphi)f, g)_{l_0} &= \frac{1}{8}\sum_{\pm}\left((\mathbf{E}^{\pm} + b_{\varphi}^{\pm})\mathrm{H}^{-1}f, (\mathbf{E}^{\pm} + b_{\bar{\varphi}}^{\pm})\mathrm{H}^{-1}g\right)_{l_0\pm 2} \\ &\quad - \frac{1}{4}\left(l_0 + \frac{\varphi(\zeta)}{\pi}\right)^2 (\mathrm{H}^{-1}f, \mathrm{H}^{-1}g)_{l_0},\end{aligned}$$

hence the map $V_0 \to \mathbb{C} : \varphi \mapsto (T(\varphi)f, g)_{l_0}$ is a polynomial of degree 2 in the coordinates of φ. Each coefficient of this polynomial may be obtained as a linear combination of the values at some elements of V_0, hence it corresponds to a bounded operator too.

Suppose $T(\varphi) - \lambda\mathrm{H}^{-2}$ has a bounded inverse $Q(\varphi, \lambda) : {}^aH \to {}^aH$. Then $\mathrm{H}^{-1}Q(\varphi,\lambda)\mathrm{H}^{-1}$ is an injective bounded operator ${}^aH \to {}^aD$. For $f \in {}^aH$, $h \in {}^aD$

$$\begin{aligned}\mathfrak{s}(\varphi)[\mathrm{H}^{-1}Q(\varphi,\lambda)\mathrm{H}^{-1}f, h] &= (T(\varphi)Q(\varphi,\lambda)\mathrm{H}^{-1}f, \mathrm{H}h)_{l_0} \\ &= (f + \lambda\mathrm{H}^{-1}Q(\varphi,\lambda)\mathrm{H}^{-1}f, h)_{l_0}.\end{aligned}$$

So $\mathrm{H}^{-1}Q(\varphi,\lambda)\mathrm{H}^{-1}f \in \mathrm{dom}\, {}^aA(\varphi)$ and $\left({}^aA(\varphi) - \lambda\right)\mathrm{H}^{-1}Q(\varphi,\lambda)\mathrm{H}^{-1}f = f$.

9.3.5 Lemma. *Let $\varphi_0 \in V_0$. Take $\lambda_0 < -1-\frac{1}{4}l_0^2-a_{1,b}\|\varphi_0\|_b-a_{1,c}\|\varphi_0\|_c-a_{2,b}\|\varphi_0\|_b^2-a_{2,c}\|\varphi_0\|_c^2$, with the $a_{j,q}$ as in Lemma 9.1.5. Then there is a neighborhood Ω of (φ_0,λ_0) in $V_0\times\mathbb{C}$ such that $T(\varphi)-\lambda\mathrm{H}^{-2}$ has a bounded inverse $Q(\varphi,\lambda) : {}^aH \to {}^aH$ for all $(\varphi,\lambda)\in\Omega$. Moreover, Q is bounded-holomorphic on Ω.*

Proof. Lemma 9.1.5 implies for $f \in {}^aH$:

$$\begin{aligned}
\operatorname{Re}\left((T(\varphi_0)-\lambda_0\mathrm{H}^{-2})f,f\right)_{l_0} &= \operatorname{Re}\mathfrak{s}(\varphi_0)[\mathrm{H}^{-1}f,\mathrm{H}^{-1}f]-\lambda_0\|\mathrm{H}^{-1}f\|_{l_0}^2\\
&= \|\mathrm{H}^{-1}f\|_{D,l_0}^2-\left(1+\frac{1}{4}l_0^2\right)\|\mathrm{H}^{-1}f\|_{l_0}^2+\operatorname{Re}\mathfrak{s}_1(\varphi_0)[\mathrm{H}^{-1}f,\mathrm{H}^{-1}f]\\
&\quad +\operatorname{Re}\mathfrak{s}_2(\varphi_0)[\mathrm{H}^{-1}f,\mathrm{H}^{-1}f]-\lambda_0\|\mathrm{H}^{-1}f\|_{l_0}^2\\
&\geq \left(1-b_{1,b}\|\varphi_0\|_b-b_{1,c}\|\varphi_0\|_c-b_{2,c}\|\varphi_0\|_c^2\right)\|\mathrm{H}^{-1}f\|_{D,l_0}^2\\
&\quad +\Bigl(-1-\frac{1}{4}l_0^2-a_{1,b}\|\varphi_0\|_b-a_{1,c}\|\varphi_0\|_c\\
&\qquad -a_{2,b}\|\varphi_0\|_b^2-a_{2,c}\|\varphi_0\|_c^2-\lambda_0\Bigr)\|\mathrm{H}^{-1}f\|_{l_0}^2\\
&\geq \left(1-b_{1,b}\|\varphi_0\|_b-b_{1,c}\|\varphi_0\|_c-b_{2,c}\|\varphi_0\|_c^2\right)\|f\|_{l_0}^2.
\end{aligned}$$

As $\varphi_0\in V_0$, we have $1-b_{1,b}\|\varphi_0\|_b-b_{1,c}\|\varphi_0\|_c-b_{2,c}\|\varphi_0\|_c^2>0$. Hence $T(\varphi_0)-\lambda_0\mathrm{H}^{-2}$ has a bounded inverse in ${}^aH$; let us call it Q_0.

Consider $C(\varphi,\lambda)=\left(T(\varphi)-T(\varphi_0)+(\lambda_0-\lambda)\mathrm{H}^{-2}\right)Q_0$; it is a bounded operator ${}^aH\to{}^aH$ for all $\varphi\in V_0$, $\lambda\in\mathbb{C}$. It is a polynomial in the coordinates of $\varphi-\varphi_0$ and $\lambda-\lambda_0$ with bounded operators ${}^aH\to{}^aH$ as coefficients, and $C(\varphi_0,\lambda_0)=0$. So $\mathrm{Id}+C(\varphi,\lambda)$ has a bounded inverse, bounded-holomorphic in (φ,λ) on a neighborhood Ω of (φ_0,λ_0), and $Q_0(\mathrm{Id}+C(\varphi,\lambda))^{-1}$ is the inverse of $T(\varphi)-\lambda\mathrm{H}^{-2}$. □

9.3.6 Lemma. *${}^a\mathrm{P}_\lambda$ is open and dense in $V_0\times\mathbb{C}$, and the resolvent R_λ is bounded-holomorphic on ${}^a\mathrm{P}_\lambda$ as a family of operators ${}^aH\to{}^aD$.*

Proof. As $\mathbb{C}\setminus{}^a\mathrm{P}_\lambda(\varphi)$ is discrete in $\mathbb{C}$ for each $\varphi\in V_0$, the set ${}^a\mathrm{P}_\lambda$ is dense in $V_0\times\mathbb{C}$. As soon as we have proved the bounded-holomorphy of R_λ, it is clear too that ${}^a\mathrm{P}_\lambda$ is open.

Take $\varphi_0\in V_0$ and choose λ_0 satisfying the condition in Lemma 9.3.5. By the two preceding lemmas, (φ_0,λ_0) has a neighborhood Ω contained in ${}^a\mathrm{P}_\lambda$, such that $\mathrm{R}_\lambda(\varphi,\lambda)=\mathrm{H}^{-1}Q(\varphi,\lambda)\mathrm{H}^{-1}$ for all $(\varphi,\lambda)\in\Omega$, with Q bounded-holomorphic ${}^aH\to{}^aH$ on Ω. Each coefficient X in the power series expansion of Q at (φ_0,λ_0) gives a coefficient $\mathrm{H}^{-1}X\mathrm{H}^{-1}$ in a power series expansion of R_λ at (φ_0,λ_0), converging in the norm for operators ${}^aH\to{}^aD$. So the resolvent R_λ is bounded-holomorphic ${}^aH\to{}^aD$ at (φ_0,λ_0).

Consider $\lambda_1\in\mathbb{C}$, $\lambda_1\neq\lambda_0$, such that $(\varphi_0,\lambda_1)\in{}^a\mathrm{P}_\lambda$. Theorem 3.12 in [25], Ch. IV, §3.3 states that $\mathrm{R}_\lambda(\varphi,\lambda)=\Phi(\lambda-\lambda_0,\mathrm{R}_\lambda(\varphi,\lambda_0))$ for all $(\varphi,\lambda)\in{}^a\mathrm{P}_\lambda$, where $\Phi:(\eta,B)\mapsto\Phi(\eta,B)$ is defined on an open set in the product of $\mathbb{C}$ and the algebra of bounded operators in ^aH. This shows that R_λ is bounded-holomorphic at (φ_0, λ_1) as a family of operators $^aH \to {}^aH$. To get bounded-holomorphy $^aH \to {}^aD$ we use the *resolvent equation*

$$\mathrm{R}_\lambda(\varphi, \lambda) = \mathrm{R}_\lambda(\varphi, \lambda_0)\,(\mathrm{Id} + (\lambda - \lambda_0)\mathrm{R}_\lambda(\varphi, \lambda))\,.$$

The factor $\mathrm{R}_\lambda(\varphi, \lambda_0)$ is bounded-holomorphic $^aH \to {}^aD$ in φ. The other factor is a bounded holomorphic family $^aH \to {}^aH$.

9.3.7 Lemma. *Let $\varphi_0 \in V_0$, $\lambda_0 \in \mathbb{C}$ such that $(\varphi_0, \lambda_0) \notin {}^a\mathrm{P}_\lambda$. Then there exist the following objects:*

- *a bounded-holomorphic family P of operators $^aH \to {}^aD$ on a neighborhood $\Omega_0 \subset V_0$ of φ_0;*
- *a bounded-holomorphic family U of operators $^aH \to {}^aH$ on the same neighborhood Ω_0; the family of inverses $\varphi \mapsto U(\varphi)^{-1}$ exists as a bounded-holomorphic family of operators $^aH \to {}^aH$ on Ω_0;*
- *a finite dimensional subspace $N(\varphi_0, \lambda_0)$ of ^aD;*

with the properties

i) *$P(\varphi)^2 = P(\varphi)$ for all $\varphi \in \Omega_0$,*

ii) *$P(\varphi)^aH = U(\varphi)N(\varphi_0, \lambda_0)$, and $U(\varphi)P(\varphi_0) = P(\varphi)U(\varphi)$ for all $\varphi \in \Omega_0$,*

iii) *$(\varphi, \lambda) \mapsto \mathrm{R}_\lambda(\varphi, \lambda)(\mathrm{Id} - P(\varphi))$ is a bounded-holomorphic family of operators $^aH \to {}^aD$ on $\Omega_0 \times W$, for some neighborhood W of λ_0,*

iv) *$U(\varphi)N(\varphi_0, \lambda_0) \subset \mathrm{dom}\, {}^aA(\varphi)$ for all $\varphi \in \Omega_0$,*

v) *$\tilde{A}(\varphi) = U(\varphi)^{-1\,a}A(\varphi)U(\varphi)$ determines a linear operator in $N(\varphi_0, \lambda_0)$ that depends on φ in a holomorphic way.*

vi) *If $(\varphi, \lambda) \in \Omega_0 \times W$, then $\ker\,({}^aA(\varphi) - \lambda)^m \subset U(\varphi)N(\varphi_0, \lambda_0)$ for each $m \geq 0$.*

Remark. Compare [25], Ch. VII, §1.3, Theorem 1.7, for the ideas behind this lemma.

Proof. Fix a positively oriented contour C in $^a\mathrm{P}_\lambda(\varphi_0)$ encircling λ_0 once, but not encircling any other points of $\mathbb{C} \setminus {}^a\mathrm{P}_\lambda(\varphi_0)$. Then $C \subset {}^a\mathrm{P}_\lambda(\varphi)$ holds for all φ in a sufficiently small neighborhood Ω_0 of φ_0 in V_0; c.f. *loc. cit.*, Ch. IV, Theorems 2.14 in §2.4 and 3.1 in §3.1. Take the open region enclosed by C as the neighborhood W of λ_0.

Define for $\varphi \in \Omega_0$

$$P(\varphi) = \frac{-1}{2\pi i} \int_C \mathrm{R}_\lambda(\varphi, \lambda)\, d\lambda.$$

This is a bounded operator ${}^aH \to {}^aD$ and a compact operator ${}^aH \to {}^aH$. The proof that $P(\varphi)$ satisfies property i) may be found in *loc. cit.*, Ch. III, §6.4, proof of Theorem 6.17 and Ch. I, §5.3. We give it here for the convenience of the reader. We fix $\varphi \in \Omega_0$, and in this proof we omit φ from the notation. We assume that C is a circle, and we adapt the coordinate λ such that C is given by $|\lambda| = \delta$.

Define for $n \in \mathbb{Z}$

$$B_n = \frac{-1}{2\pi i} \int_C \lambda^{-n-1} \mathrm{R}_\lambda(\lambda)\, d\lambda.$$

This defines the B_n as bounded operators ${}^aH \to {}^aD$, with $\|B_n\| = \mathcal{O}(\delta^{-n})$. Note that $P(\varphi) = B_{-1}$. The resolvent equation implies

$$B_{-1}^2 = \frac{-1}{2\pi i} \int_C \frac{-1}{2\pi i} \int_C \frac{1}{\lambda - \mu} \Big(\mathrm{R}_\lambda(\lambda) - \mathrm{R}_\lambda(\mu) \Big)\, d\lambda\, d\mu.$$

As $(\mu, \lambda) \mapsto \frac{1}{\lambda - \mu} (\mathrm{R}_\lambda(\lambda) - \mathrm{R}_\lambda(\mu))$ is holomorphic on a neighborhood of $C \times C$ in $\mathbb{C}^2$, we can replace the cycle of integration by $C_0 \times C$, with C_0 given by $|\mu| = \delta_0$, where δ_0 is slightly larger than δ. We obtain

$$\begin{aligned} B_{-1}^2 &= \frac{-1}{2\pi i} \int_{C_0} \frac{-1}{2\pi i} \int_C \frac{1}{\lambda - \mu} \Big(\mathrm{R}_\lambda(\lambda) - \mathrm{R}_\lambda(\mu) \Big)\, d\lambda\, d\mu \\ &= \frac{-1}{2\pi i} \int_{C_0} \frac{-1}{\mu} \frac{-1}{2\pi i} \int_C \sum_{n=0}^\infty \left(\frac{\lambda}{\mu} \right)^n \Big(\mathrm{R}_\lambda(\lambda) - \mathrm{R}_\lambda(\mu) \Big)\, d\lambda\, d\mu \\ &= \frac{1}{2\pi i} \int_{C_0} \sum_{n=0}^\infty \mu^{-n-1} B_{-1-n}\, d\mu \\ &= \sum_{n=0}^\infty \frac{1}{2\pi i} \int_{C_0} \mu^{-n-1}\, d\mu\, B_{-1-n} \;=\; B_{-1}. \end{aligned}$$

This shows that $P(\varphi)$ is a projection operator.

As $C \subset {}^a\mathrm{P}_\lambda(\varphi)$, the bounded-holomorphy of R_λ on ${}^a\mathrm{P}_\lambda$ implies that $\varphi \mapsto P(\varphi)$ is a bounded-holomorphic family of operators ${}^aH \to {}^aD$.

Define for $\varphi \in \Omega_0$:

$$U(\varphi) = P(\varphi)P(\varphi_0) + (\mathrm{Id} - P(\varphi))(\mathrm{Id} - P(\varphi_0)).$$

U is bounded-holomorphic ${}^aH \to {}^aH$ and $U(\varphi_0) = \mathrm{Id}$. So $U(\varphi)^{-1}$ exists and $\varphi \mapsto U(\varphi)^{-1}$ is bounded-holomorphic ${}^aH \to {}^aH$, if we take Ω_0 sufficiently small (use the series $\sum_{n=0}^\infty X^n$ with $X = U(\varphi_0) - U(\varphi)$). We have used the method in [2], Supplement, §2, Theorem 3 to construct the family U, and not the method in [25], Ch. II, §4.2 and Ch. VII, §1.3, proof of Theorem 1.7. The latter construction does not work immediately in the higher dimensional case.

Note that

$$U(\varphi)P(\varphi_0) = P(\varphi)P(\varphi_0) = P(\varphi)U(\varphi).$$

We put $N(\varphi_0, \lambda_0) = P(\varphi_0)^a H$. Then $U(\varphi)N(\varphi_0, \lambda_0) = P(\varphi)U(\varphi)^a H = P(\varphi)^a H$; this is property ii).

We define $\tilde{A}(\varphi) = U(\varphi)^{-1}\, {}^a A(\varphi)U(\varphi)P(\varphi_0)$. It is bounded-holomorphic as a family of operators ${}^a H \to {}^a H$. To see this, we write it as $U(\varphi)^{-1}\left({}^a A(\varphi)P(\varphi)\right)U(\varphi)$; we shall show that ${}^a A(\varphi)P(\varphi)$ is bounded-holomorphic ${}^a H \to {}^a D$.

The compactness of the projection operators $P(\varphi) : {}^a H \to {}^a H$ implies that the spaces $P(\varphi)^a H$ have finite dimension. The relation $U(\varphi)P(\varphi_0) = P(\varphi)U(\varphi)$ implies that $U(\varphi) : N(\varphi_0, \lambda_0) \to P(\varphi)^a H$ is an isomorphism.

We have $\mathfrak{s}(\varphi)[\mathrm{R}_\lambda(\varphi, \lambda)f, h] = (f, h)_{l_0} + (\lambda \mathrm{R}_\lambda(\varphi, \lambda)f, h)_{l_0}$ for $\varphi \in \Omega_0$, $\lambda \in {}^a\mathcal{P}_\lambda(\varphi)$, $f \in {}^a H$, $h \in {}^a D$. Hence

$$\mathfrak{s}(\varphi)[P(\varphi)f, h] = \left(\frac{-1}{2\pi i}\int_C \lambda \mathrm{R}_\lambda(\varphi, \lambda)\, d\lambda \cdot f, h\right)_{l_0}.$$

This shows that $P(\varphi)^a H = U(\varphi)N(\varphi_0, \lambda_0) \subset \operatorname{dom} {}^a A(\varphi)$. This is property iv). We also conclude that

$$ {}^a A(\varphi)P(\varphi) = \frac{-1}{2\pi i}\int_C \lambda \mathrm{R}_\lambda(\varphi, \lambda)\, d\lambda \tag{9.1}$$

is bounded-holomorphic ${}^a H \to {}^a D$ for $\varphi \in \Omega_0$. So

$$\tilde{A}(\varphi) = U(\varphi)^{-1 a} A(\varphi)U(\varphi)P(\varphi_0) = U(\varphi)^{-1}\left({}^a A(\varphi)P(\varphi)\right)U(\varphi)$$

defines a bounded-holomorphic family of operators $N(\varphi_0, \lambda_0) \to {}^a H$.

We show that $P(\varphi_0)\tilde{A}(\varphi) = \tilde{A}(\varphi)$ by proving the equality $P(\varphi)^a A(\varphi)P(\varphi) = {}^a A(\varphi)P(\varphi)$. For $\lambda \in {}^a\mathcal{P}_\lambda(\varphi)$:

$$\begin{aligned}
\mathrm{R}_\lambda(\varphi, \lambda)^a A(\varphi)P(\varphi) &= P(\varphi) + \lambda \mathrm{R}_\lambda(\varphi, \lambda)P(\varphi), \\
P(\varphi)^a A(\varphi)P(\varphi) &= \frac{-1}{2\pi i}\int_C \left(P(\varphi) + \lambda \mathrm{R}_\lambda(\varphi, \lambda)P(\varphi)\right) d\lambda \\
&= {}^a A(\varphi)P(\varphi),
\end{aligned}$$

see Equation (9.1). So $\tilde{A}(\varphi)$ gives a holomorphic family of operators in the space $N(\varphi_0, \lambda_0)$; this is property v).

Consider $f_1, \ldots, f_m \in \operatorname{dom} {}^a A(\varphi)$, $f_0 = 0$, with $({}^a A(\varphi) - \lambda)f_j = f_{j-1}$ for $j = 1, \ldots, m$. For $\lambda_1 \in {}^a\mathcal{P}_\lambda(\varphi)$, $1 \le j \le m$:

$$\mathrm{R}_\lambda(\varphi, \lambda_1)f_j = \frac{1}{\lambda - \lambda_1} f_j - \frac{1}{\lambda - \lambda_1}\mathrm{R}_\lambda(\varphi, \lambda_1)f_{j-1},$$

hence

$$P(\varphi)f_j = f_j + \frac{-1}{2\pi i}\int_C \frac{1}{\lambda_1 - \lambda}\mathrm{R}_\lambda(\varphi, \lambda_1)f_{j-1}\, d\lambda_1.$$

For $j = 1$ we obtain $f_1 \in P(\varphi)^a H = U(\varphi)N(\varphi_0, \lambda_0)$. We proceed inductively to show that $f_j \in P(\varphi)^a H$. It suffices to show that $\frac{-1}{2\pi i}\int_C \frac{1}{\lambda_1-\lambda_0}\mathrm{R}_\lambda(\varphi,\lambda_1)\,d\lambda_1\,P(\varphi) = 0$. The induction hypothesis $f_{j-1} \in P(\varphi)^a H$ gives $(P(\varphi) - \mathrm{Id})f_j = 0$. This implies that vi) holds.

Note that

$$X(\varphi,\mu) = \frac{-1}{2\pi i}\int_C \frac{1}{\lambda_1-\mu}\mathrm{R}_\lambda(\varphi,\lambda_1)\,d\lambda_1$$

is holomorphic on $\Omega_0 \times W$. For $(\varphi,\mu) \in (\Omega_0 \times W) \cap {}^a\mathrm{P}_\lambda$ the resolvent equation gives

$$\begin{aligned} X(\varphi,\mu) &= -\mathrm{R}_\lambda(\varphi,\mu) + \frac{-1}{2\pi i}\int_C \frac{1}{\mu-\lambda_1}\Big(\mathrm{R}_\lambda(\varphi,\mu) - \mathrm{R}_\lambda(\varphi,\lambda_1)\Big)\,d\lambda_1 \\ &= -\mathrm{R}_\lambda(\varphi,\mu) + \frac{-1}{2\pi i}\mathrm{R}_\lambda(\varphi,\mu)\mathrm{R}_\lambda(\varphi,\lambda_1)\,d\lambda_1 \\ &= \mathrm{R}_\lambda(\varphi,\mu)\,(P(\varphi) - \mathrm{Id})\,. \end{aligned}$$

In this way we have obtained iii). The equality $X(\varphi,\mu)P(\varphi) = 0$ extends to $(\varphi,\mu) \in \Omega_0 \times W$ by holomorphy. This concludes the proof of vi).

9.3.8 *Proof* of Proposition 9.3.2. The statement for real φ follows from the bound in 9.2.1. The resolvent is bounded-holomorphic on ${}^a\mathrm{P}_\lambda$ by Lemma 9.3.6.

For the bounded-meromorphy of the resolvent as a family of operators ${}^aH \to {}^aD$ on a neighborhood of a point $(\varphi_0,\lambda_0) \in (V_0 \times \mathbb{C}) \setminus {}^a\mathrm{P}_\lambda$, we need consider only $(\varphi,\lambda) \mapsto \mathrm{R}_\lambda(\varphi,\lambda)P(\varphi)$, see iii) in the previous lemma.

For $(\varphi,\lambda) \in {}^a\mathrm{P}_\lambda \cap (\Omega_0 \times W)$ we have

$$\begin{aligned} &U(\varphi)^{-1}\mathrm{R}_\lambda(\varphi,\lambda)U(\varphi)\cdot\Big(\tilde{A}(\varphi) - \lambda\Big)\,P(\varphi_0) \\ &= U(\varphi)^{-1}\mathrm{R}_\lambda(\varphi,\lambda)\Big(A(\varphi) - \lambda\Big)U(\varphi)P(\varphi_0) = P(\varphi_0). \end{aligned}$$

Hence $U(\varphi)^{-1}\mathrm{R}_\lambda(\varphi,\lambda)P(\varphi)U(\varphi) = U(\varphi)^{-1}\mathrm{R}_\lambda(\varphi,\lambda)U(\varphi)P(\varphi_0)$ is the inverse of $\tilde{A}(\varphi) - \lambda$ on the finite dimensional space $N(\varphi_0,\lambda_0)$.

As $\tilde{A}(\varphi) : N(\varphi_0,\lambda_0) \to N(\varphi_0,\lambda_0)$ depends on φ in a holomorphic way, the determinant $\psi(\varphi,\lambda) = \det\Big(\tilde{A}(\varphi) - \lambda\Big)$ is a holomorphic function on $\Omega_0 \times W$ that is non-zero on ${}^a\mathrm{P}_\lambda$. So $\tilde{Q}(\varphi,\lambda) = \psi(\varphi,\lambda)\Big(\tilde{A}(\varphi) - \lambda\Big)^{-1}$ extends as a holomorphic family of linear operators in $N(\varphi_0,\lambda_0)$. For $(\varphi,\lambda) \in (\Omega_0 \times W) \cap {}^a\mathrm{P}_\lambda$:

$$\begin{aligned} \psi(\varphi,\lambda)\mathrm{R}_\lambda(\varphi,\lambda)P(\varphi) &= \psi(\varphi,\lambda)U(\varphi)\Big(\tilde{A}(\varphi) - \lambda\Big)^{-1}U(\varphi)^{-1}P(\varphi) \\ &= U(\varphi)\tilde{Q}(\varphi,\lambda)U(\varphi)^{-1}P(\varphi). \end{aligned}$$

$\psi \cdot \mathrm{R}_\lambda \cdot P$ extends as a bounded-holomorphic family of operators ${}^aH \to {}^aH$ on $\Omega_0 \times W$. From

$$\begin{aligned} U(\varphi)\tilde{Q}(\varphi,\lambda)U(\varphi)^{-1}P(\varphi) &= U(\varphi)P(\varphi_0)\tilde{Q}(\varphi,\lambda)U(\varphi)^{-1}P(\varphi) \\ &= P(\varphi)U(\varphi)\tilde{Q}(\varphi,\lambda)U(\varphi)^{-1}P(\varphi), \end{aligned}$$

follows that it is even bounded-holomorphic as a family ${}^aH \to {}^aD$. (The leftmost $P(\varphi)$ is bounded-holomorphic as a family of operators ${}^aH \to {}^aD$.)

Near (φ_0, λ_0) the set ${}^a\mathsf{P}_\lambda$ is described by $\det(\tilde{A}(\varphi) - \lambda) \neq 0$. So its complement is a complex analytic set.

9.4 Meromorphic families of automorphic forms

We are ready to prove Theorem 9.4.1, which is a central result. In the context of Section 1.6, this theorem gives the existence of ${}^a\tilde{E}(r,s)$, see 1.6.4. In the general situation, it gives the existence of many meromorphic families of automorphic forms on $V_0 \times \mathbb{C}$.

We use the the resolvent family $\mathrm{R}(.,.)$ in the way indicated in 1.6.4. In the previous section, we have seen that the resolvent forms a bounded meromorphic family of operators in ${}^aH$.

We return to the use of s as the spectral parameter. The relation $(\varphi, s) \mapsto (\varphi, \lambda) = (\varphi, \frac{1}{4} - s^2)$ connects the spaces $V_0 \times \mathbb{C}$ here and in the previous section.

9.4.1 Theorem. *Let $\Omega \subset V_0 \times \mathbb{C}$ be open; denote by $m : (\varphi, s) \mapsto l_0 + \frac{\varphi(\zeta)}{\pi}$ the weight with value l_0 at φ_0. For each collection $\rho = (\rho_{P,n})_{n \in \mathbf{c}[P],\, P \in \mathcal{P}}$ of holomorphic functions on Ω there exists a unique meromorphic family of automorphic forms $e_\rho \in (\mathcal{M} \otimes_{\mathcal{O}} \mathcal{A}_m(\mathbf{c}))(\Omega)$ such that*

$$F_{P,n} e_\rho(a(P)) = \rho_{P,n} \quad \text{for all } n \in \mathbf{c}(P),\ P \in \mathcal{P}.$$

This family is holomorphic on $\Omega \cap {}^a\mathsf{P}$.

Let $(\varphi_0, s_0) \in \Omega$, $\varphi_0 \in V_r$. If $(F_{P,n} e_\rho)'(a(P))$ is holomorphic at (φ_0, s_0) for all $n \in \mathbf{c}(P)$, $P \in \mathcal{P}$, then e_ρ is holomorphic at (φ_0, s_0).

Remarks. The set ${}^a\mathsf{P}$ is open and dense in $V_0 \times \mathbb{C}$, see Lemma 9.3.6. It intersects each line $\{\varphi\} \times \mathbb{C}$ discretely.

We may take $\Omega = V_0 \times \mathbb{C}$, and $V = \mathcal{V} = \hom(\tilde{\Gamma}, \mathbb{C})$. In that way we obtain the existence of many meromorphic families of automorphic forms on the rather general parameter space $V_0 \times \mathbb{C} \subset \mathcal{V} \times \mathbb{C}$. In particular, the $\mathcal{M}(V_0 \times \mathbb{C})$-dimension of $(\mathcal{M} \otimes_{\mathcal{O}} \mathcal{A}_m(\mathbf{c}))(V_0 \times \mathbb{C})$ is at least $|\mathbf{c}|$. Enlarging $\mathbf{c}$ enlarges V_0 sufficiently to cover $\mathcal{X}$ by sets of the form $\chi_0 \cdot \mathbf{exp}(V_0)$.

From a suitable choice of $\mathbf{c}$ and the $\rho_{P,n}$ we shall obtain the meromorphic continuation of Poincaré series in φ and s jointly, see the next chapter.

Near real points φ_0 the holomorphy of e_ρ is implied by the holomorphy of its Fourier terms that are specified by $\mathbf{c}$. I do not know whether this or a similar result holds for non-real φ.

Theorem 9.4.1 has the following truncated version:

9.4.2 Proposition. *Let $\Omega \subset V_0 \times \mathbb{C}$ and $\rho = (\rho_{P,\nu})_{\nu \in \mathbf{c}[P],\, P \in \mathcal{P}}$ a collection of holomorphic functions on Ω. Then there exists a unique L^2-meromorphic map $E_\rho : \Omega \to {}^aH(l_0)$ such that*

i) $(E_\rho(\varphi, s), M(\bar\varphi, \bar s)k)_{l_0} = 0$ for all $(\varphi, s) \in \Omega$, $k \in {}^aK(l_0)$.

ii) Let $P \in \mathcal{P}$, $\nu \in \mathbf{c}[P]$. The corresponding Fourier term has the form $\tilde F_{P,\nu} E_\rho = \rho_{P,\nu}\alpha_{P,n_\nu} + \sigma_{P,\nu}\beta_{P,n_\nu}$ on ${}^aI_P$ with $\sigma_{P,\nu} \in \mathcal{M}(\Omega)$.

iii) E_ρ is L^2-holomorphic on $\Omega \cap {}^a\mathsf{P}$.

For $w_0 = (\varphi_0, s_0) \in \Omega$, $\varphi_0 \in V_r$, the holomorphy of the $\sigma_{P,\nu}$ at w_0 implies the L^2-holomorphy of E_ρ at w_0.

Remarks. α_{P,n_ν} and β_{P,n_ν} have been introduced in 7.6.2. They are determined by their values and derivatives at $a(P)$.

The orthogonality of E_ρ to $M(\bar\varphi, \bar s)^aK$ implies that $F_{P,\nu}E_\rho$ is given by a linear combination of α_{P,n_ν} and β_{P,n_ν} on ${}^aI_P$. Property ii) states that the extension of this function to I_P has value $\rho_{P,\nu}$ at $a(P)$. Proposition 8.3.6 shows that this proposition implies Theorem 9.4.1.

9.4.3 *Remarks on the proof.* We give the proof of Proposition 9.4.2 in 9.4.4–9.4.14.

In 1.6.4, we have explained the main idea of the proof. The role of $L(r) - \frac{1}{4} + s^2$ there is played here by $M(\varphi, s)$. The E_ρ to be constructed here correspond to ${}^a\tilde E(r, s)$ in 1.6.4. Property ii) in the proposition amounts to $F_{P,\nu}E_\rho(\varphi, s; a(P)) = \rho_{P,n_\nu}(\varphi, s)$ for all $P \in \mathcal{P}$, $\nu \in \mathbf{c}[P]$. This replaces the property $F_0{}^a\tilde E(r, s; a) = 1$.

We construct E_ρ in the form $E_\rho(\varphi, s) = {}^{(a)}h(\varphi, s) - G(\varphi, s)$, with families h and G. The h in 1.6.4 corresponds to ${}^{(a)}h$ here.

We first give h, pointwise holomorphic on Ω and satisfying

$$F_{P,\nu}h(\varphi, s; a(P)) = \rho_{P,\nu}(\varphi, s)$$

for $P \in \mathcal{P}$, $\nu \in \mathbf{c}[P]$. The relation between ${}^{(a)}h$ and G is given by property i). For h sufficiently nice, this amounts to

$$\left({}^aA(\varphi) - \frac{1}{4} + s^2\right) G(\varphi, s) = {}^{(a)}(M(\varphi, s)h(\varphi, s))\,.$$

So we can get hold of $G(\varphi, s)$ by means of the resolvent, and iii) is fulfilled automatically.

Here we are interested also in the relation between holomorphy of E_ρ and of the Fourier terms specified by $\mathbf{c}$. We did not consider that aspect in Section 1.6.

9.4.4 *Construction of h.* For each $P \in \mathcal{P}$, choose $\tau_P \in C^\infty(I_P)$ such that $0 \le \tau_P \le 1$, $\tau_P = 0$ on a neighborhood of A_P and $\tau_P = 1$ on ${}^a\hat I_P$ and on a neighborhood of $a(P)$ in I_P. (See Figure 8.1 on p. 138 for the intervals I_P, ${}^a\hat I_P$, $\ldots$.) Put

$$\theta_{P,\nu}(\varphi, s) = \rho_{P,\nu}(\varphi, s) \cdot \tau_P \cdot \alpha_{P,n_\nu}(\varphi, s) \quad \text{for } (\varphi, s) \in \Omega,\ \nu \in \mathbf{c}[P].$$

Define $h(\varphi, s)$ by

$$\begin{aligned} h(\varphi,s) &= 0 && \text{on } \mathrm{pr}^{-1}\left(Y \setminus \bigcup_{P\in\mathcal{P}} U_P(A_P)\right), \\ F_{P,\nu}h(\varphi,s) &= 0 && \text{for } P\in\mathcal{P},\ \nu\in N_P\setminus \mathbf{c}[P] \\ F_{P,\nu}h(\varphi,s) &= \theta_{P,\nu}(\varphi,s) && \text{for } P\in\mathcal{P},\ \nu\in \mathbf{c}[P]. \end{aligned}$$

This gives h as a family of functions satisfying property ii) in Proposition 9.4.2 on a neighborhood of the sets $\mathrm{pr}^{-1}U_P(a(P))$. Moreover, $M(\varphi,s)h(\varphi,s)=0$ on these sets. By construction, h is pointwise holomorphic on Ω, and $M(\varphi,s)h(\varphi,s)$ is an element of ${}^aK$ for all $(\varphi,s)\in\Omega$. The map $\Omega\to{}^aH : (\varphi,s)\mapsto M(\varphi,s)h(\varphi,s)$ is L^2-holomorphic, and so is $(\varphi,s)\mapsto {}^{(a)}h(\varphi,s)$. The Fourier coefficient $F_{P,\nu}{}^{(a)}h(\varphi,s)$ is represented by $\rho_{P,\nu}(\varphi,s)\alpha_{P,n_\nu}(\varphi,s)$ on a neighborhood of $a(P)$ in ${}^aI_P$.

Lemma 6.1.4 and the definition of $M(.,.)$ give, for each $k\in{}^aK$,

$$\left({}^{(a)}h(\varphi,s), M(\bar\varphi,\bar s)k\right)_{l_0} = (h(\varphi,s), M(\bar\varphi,\bar s)k)_{l_0} = (M(\varphi,s)h(\varphi,s),k)_{l_0}.$$

9.4.5 *Construction of G.* Property i) in the proposition means that we want to solve the equation $M(\varphi,s)E_\rho(\varphi,s)=0$ in distribution sense. We have already obtained h that satisfies property ii). We try to solve $M(\varphi,s)G = M(\varphi,s)h(\varphi,s)$ in another way. The resolvent enables us to do this in L^2-sense.

$$G(\varphi,s) = \mathrm{R}(\varphi,s)\left(M(\varphi,s)h(\varphi,s)\right)$$

defines a meromorphic map $G:\Omega\to{}^aD$, holomorphic on $\Omega\cap{}^a\mathsf{P}$ (see Proposition 9.3.2). This map satisfies

$$\left({}^aA(\varphi) - \frac{1}{4} + s^2\right) G(s,\varphi) = M(\varphi,s)h(\varphi,s) \quad \text{for each } (\varphi,s)\in\Omega\cap{}^a\mathsf{P}.$$

Let $k\in{}^aK$. On the one hand, $\left(\left({}^aA(\varphi)-\frac14+s^2\right)G(\varphi,s),k\right)_{l_0} = \mathfrak{s}(\varphi)[G(\varphi,s),k] - \left(\frac14 - s^2\right)(G(\varphi,s),k)_{l_0} = (G(\varphi,s), M(\bar\varphi,\bar s)k)_{l_0}$. On the other hand, this equals $(M(\varphi,s)h(\varphi,s),k)_{l_0} = (h, M(\bar\varphi,\bar s)k)_{l_0} = \left({}^{(a)}h(\varphi,s), M(\bar\varphi,\bar s)k\right)_{l_0}$. In this way we obtain

$$\left(G(\varphi,s) - {}^{(a)}h(\varphi,s), M(\bar\varphi,\bar s)k\right)_{l_0} = 0$$

for all $k\in{}^aK$, $(\varphi,s)\in\Omega\cap{}^a\mathsf{P}$.

9.4.6 *Construction of E_ρ.* Now $E_\rho = {}^{(a)}h - G$ defines an L^2-meromorphic map $\Omega\to{}^aH$ satisfying i) and iii). On the interior of ${}^aI_P$ the Fourier coefficient $F_{P,\nu}E_\rho$ satisfies weakly

$$\left(l_{P,n_\nu}(\varphi) - \frac14 + s^2\right) F_{P,\nu}E_\rho = 0$$

(see 4.1.2 and 4.1.3 for l_{P,n_ν}), hence it is given by a linear combination of α_{P,n_ν} and β_{P,n_ν}. Lemma 8.4.3 shows that $G(\varphi,s)$ can provide only a multiple of β_{P,n_ν}. So ${}^{(a)}h$ gives us the value at $a(P)$ prescribed by property ii).

9.4.7 *Uniqueness.* Let F satisfy the properties i)–iii). Property ii) and Proposition 9.2.5 imply that $(F-E_\rho)(\varphi,s) \in \ker\left({}^aA(\varphi)-\frac14+s^2\right)$ for $(\varphi,s)\in\Omega\cap{}^a\mathsf{P}$, hence $F=E_\rho$ on $\Omega\cap{}^a\mathsf{P}$, hence on Ω.

9.4.8 *Singularities of E_ρ not reflected in the $\sigma_{P,\nu}$.* The Fourier coefficients $\sigma_{P,\nu}$ introduced in ii) are holomorphic on $\Omega\cap{}^a\mathsf{P}$. Suppose $w_0\in\mathrm{Pol}(E_\rho)$, and suppose that all $\sigma_{P,\nu}$ are holomorphic at w_0. Write $E_\rho=\frac{1}{\psi}L_\psi$ with ψ a non-zero element of $\mathcal{O}(\Omega_1)$ such that $L_\psi:\Omega_1\to{}^aH$ is L^2-holomorphic on some neighborhood Ω_1 of w_0. Choose ψ minimal with respect to divisibility in the stalk $\mathcal{O}_{w_0}$. Apply Lemma 7.4.14 to the family of eigenfunctions corresponding to E_ρ. This shows that there is an element $g\in\tilde G_{\mathcal{P}}$ such that $L_\psi(w_1;g)\neq0$ for some point $w_1=(\varphi_1,s_1)\in N(\psi)$. The holomorphy of the $\rho_{P,\nu}$ and the $\sigma_{P,\nu}$ implies that

$$F_{P,\nu}L_\psi(w_1)=\psi(w_1)\Big(\rho_{P,\nu}(w_1)\alpha_{P,\nu}(w_1)+\sigma_{P,\nu}(w_1)\beta_{P,\nu}(w_1)\Big)$$

vanishes on ${}^aI_P$. Furthermore,

$$L_\psi(w_1)=\psi(w_1)^{(a)}h(w_1)-(\psi G)(w_1)=-(\psi G)(w_1).$$

We arrange that $\psi G:\Omega_1\to{}^aD$ is holomorphic by taking Ω_1 sufficiently small. For each $q\in{}^aD$:

$$\begin{aligned}
\mathfrak{s}(\varphi_1)[(\psi G)(w_1),q] &= \lim_{s\to s_1}\mathfrak{s}(\varphi_1)[(\psi G)(\varphi_1,s),q]\\
&= \lim_{s\to s_1}\psi(\varphi_1,s)\left({}^aA(\varphi_1)G(\varphi_1,s),q\right)_{l_0}\\
&= \lim_{s\to s_1}\psi(\varphi_1,s)\left(M(\varphi_1,s)h(\varphi_1,s)+(\tfrac14-s^2)G(\varphi,s),q\right)_{l_0}\\
&= 0+\left(\frac14-s_1^2\right)\left((\psi G)(w_1),q\right)_{l_0}. \qquad (9.2)
\end{aligned}$$

So $L_\psi(w_1)=-(\psi G)(w_1)$ is an element of the kernel of ${}^aA(\varphi_1)-\frac14+s_1^2$. We have proved the next result:

9.4.9 Lemma. *Let $w_0=(\varphi_0,s_0)\in\mathrm{Pol}(E_\rho)$, but $\sigma_{P,\nu}\in\mathcal{O}_{w_0}$ for all $P\in\mathcal{P}$, $\nu\in\mathbf{c}[P]$. Take $\psi\in\mathcal{O}_{w_0}$, $\psi\neq0$, minimal with respect to divisibility in $\mathcal{O}_{w_0}$, such that ψE_ρ is holomorphic at w_0. Then for each $w_1=(\varphi_1,s_1)\in N(\psi)$:*

$$(\psi E_\rho)(w_1)\in\ker\left({}^aA(\varphi_1)-\frac14+s_1^2\right)$$

$$F_{P,\nu}(\psi E_\rho)(w_1)=0 \text{ on } {}^aI_P \text{ for all } P\in\mathcal{P},\ \nu\in\mathbf{c}[P].$$

Moreover, there are $w_1\in N(\psi)$ such that $(\psi E_\rho)(w_1)\neq0$.

Remark. Here, and in the next lemma, we do not suppose that $\varphi_0 \in V_r$. Only when combining these lemmas to prove the holomorphy statement in the proposition, we shall take φ_0 to be real.

9.4.10 Lemma. *Suppose $f \in \ker({}^aA(\varphi_1) - \frac{1}{4} + s_1^2)$ for some $(\varphi_1, s_1) \in V_0 \times \mathbb{C}$ satisfies $F_{P,\nu} f = 0$ on ${}^aI_P$ for all $P \in \mathcal{P}$, $\nu \in \mathbf{c}[P]$. Then*

$$(f, E_\rho(\bar\varphi_1, \bar s))_{l_0} = 0 \qquad \text{for all } \bar s \in {}^a\mathsf{P}(\bar\varphi_1).$$

Proof. The vanishing of the $F_{P,\nu} f$ implies

$$\begin{aligned}(f, E_\rho(\bar\varphi_1, \bar s))_{l_0} &= -(f, G(\bar\varphi_1, \bar s))_{l_0} = -\Big(\mathrm{R}(\varphi_1, s) f, M(\bar\varphi_1, \bar s) h(\bar\varphi_1, \bar s)\Big)_{l_0} \\ &= \frac{1}{s_1^2 - s^2}\Big(f, M(\bar\varphi_1, \bar s) h(\bar\varphi_1, \bar s)\Big)_{l_0} = 0.\end{aligned}$$

9.4.11 *Holomorphy at real points.* Suppose $w_0 = (\varphi_0, s_0) \in \mathrm{Pol}(E_\rho)$, but $\sigma_{P,\nu} \in \mathcal{O}_{w_0}$ for all $P \in \mathcal{P}$, $\nu \in \mathbf{c}[P]$. Now we suppose $\varphi_0 \in V_r$ as well. We have to show that this leads to a contradiction. First we show that the assumptions imply $\frac{1}{4} - s_0^2 \in \mathbb{R}$. Secondly we apply Lemma 9.4.10 to get a contradiction. We shall use the Lemmas 9.4.12 and 9.4.13; these contain general results from the theory of several complex variables.

Take $\psi \in \mathcal{O}_{w_0}$ as in Lemma 9.4.9. A holomorphic factor of ψ that has no zeros in a neighborhood of w_0 does not matter, and $N(\psi) \subset (V_0 \times \mathbb{C}) \smallsetminus {}^a\mathsf{P}$ intersects $\{\varphi_0\} \times \mathbb{C}$ discretely. So we may assume that ψ has Weierstrass form in $s - s_0$, i.e., $\psi(\varphi, s) = (s - s_0)^k + \sum_{j=0}^{k-1} a_j(\varphi)(s - s_0)^j$ with $k \geq 1$, the a_j holomorphic on a neighborhood of φ_0, and $a_j(\varphi_0) = 0$; see [23], Corollary 6.1.2, or [17], Ch. 2, §1.2.

In each neighborhood of w_0 there exists a point $w_1 = (\varphi_1, s_1) \in N(\psi)$ such that $(\psi E_\rho)(w_1) \neq 0$. Apply Lemma 9.4.12 to $\tau(z, t) = \psi(\varphi_0 + z(\varphi_1 - \varphi_0), s_0 + t)$. Use the resulting d and η to define a holomorphic family

$$f : u \mapsto (\psi E_\rho)(\varphi_0 + u^d(\varphi_1 - \varphi_0), s_0 + \eta(u))$$

on a neighborhood of 0 in $\mathbb{C}$ with values in ${}^aD$. The family f is not identically zero, hence there exists $m \geq 0$ such that $f_0 = \lim_{u \to 0} u^{-m} f(u)$ is a non-zero element of ${}^aD$. Like in Equation (9.2) on p. 170, we obtain

$$\mathfrak{s}(\varphi_0 + u^d(\varphi_1 - \varphi_0))[u^{-m} f(u), q] = \left(\frac{1}{4} - (s_0 + \eta(u))^2\right)\left(u^{-m} f(u), q\right)_{l_0}$$

for all $q \in {}^aD$. Take the limit as $u \to 0$ to see that f_0 is an element of the kernel of ${}^aA(\varphi_0) - \frac{1}{4} + s_0^2$. As ${}^aA(\varphi_0)$ is a selfadjoint operator, this implies $\frac{1}{4} - s_0^2 \in \mathbb{R}$.

Suppose there exists $w_1 = (\varphi_1, s_1) \in N(\psi)$ with $(\psi E_\rho)(w_1) \neq 0$ and $\varphi_1 \in V_r$. Then $s_1 \in \mathbb{R} \cup i\mathbb{R}$ (selfadjointness). We obtain from Lemma 9.4.10, with $\overline{s_1} = \pm s_1$,

$$\begin{aligned} \|(\psi E_\rho)(w_1)\|_{l_0}^2 &= ((\psi E_\rho)(w_1), (\psi E_\rho)(\overline{\varphi_1}, \overline{\pm s_1}))_{l_0} \\ &= \lim_{s \to \pm s_1} \Big((\psi E_\rho)(w_1), \psi(\varphi_1, \bar{s}) E_\rho(\bar{\varphi}_1, \bar{s}) \Big)_{l_0} \\ &= \lim_{s \to \pm s_1} \overline{\psi(\varphi_1, \bar{s})} \Big((\psi E_\rho)(w_1), E_\rho(\varphi_1, \bar{s}) \Big)_{l_0} \\ &= \lim_{s \to \pm s_1} \overline{\psi(\varphi_1, \bar{s})} \cdot 0 = 0. \end{aligned}$$

So we are left with the situation that $(\psi E_\rho)(\varphi_1, s_1) = 0$ for all $(\varphi_1, s_1) \in N(\psi) \cap (V_r \times \mathbb{C})$. There exists $g \in \tilde{G}_{\mathcal{P}}$ such that $l = (\psi E_\rho)(g)$ is not identically zero on $N(\psi)$. So ψ has an irreducible factor ψ_1 in $\mathcal{O}_{w_0}$ not dividing l. As l vanishes on $(V_r \times \mathbb{C}) \cap N(\psi_1)$, Lemma 9.4.13 shows that this is impossible. Hence for real φ_0, the condition $w_0 \in \mathrm{Pol}(E_\rho)$ implies that some $\sigma_{P,\nu}$ has to be singular at w_0.

9.4.12 Lemma. *Let τ be of the form*

$$\tau(z,t) = t^k + \sum_{j=0}^{k-1} a_j(z) t^j$$

with $k \geq 1$, and the a_j holomorphic on a neighborhood of 0 in $\mathbb{C}$, $a_j(0) = 0$. Then there are $d \in \mathbb{N}$ and η holomorphic at $0 \in \mathbb{C}$ such that $\eta(0) = 0$, and $\tau(u^d, \eta(u)) = 0$ for all u in a neighborhood of 0 in $\mathbb{C}$.

Remark. This result is often called the theorem of Puiseux. See §5 of [1] for a discussion in the context of formal power series.

In the Lemmas 12.1.5 and 12.1.10 we shall return to this result.

Proof. Use [14], Chap. III, §1.4–6, especially p. 132, 133. (In the second line of formula (21) one should replace x^{qh} by $x^{q(h-1)+1}$.) Essential in the reasoning is the symmetric role of z and t.

9.4.13 Lemma. *Let $\psi, h \in \mathcal{O}(\Omega)$ with $\Omega \subset \mathbb{C}^{N+1}$ a neighborhood of 0. Suppose that ψ is irreducible in $\mathcal{O}_0$ and that $h = 0$ on $(\mathbb{R}^N \times \mathbb{C}) \cap N(\psi)$. Then ψ divides h in $\mathcal{O}_0$.*

Proof. We use the fact that if a function f is holomorphic on an open neighborhood B of 0 in $\mathbb{C}^m$ and vanishes on $B \cap \mathbb{R}^m$, then $f = 0$ on a neighborhood of 0. To see this, note that the property of vanishing on $B \cap \mathbb{R}^m$ is inherited by the derivatives $\partial_{z_j} f$, where $z_j = x_j + iy_j$ for $1 \leq j \leq m$ are coordinates on $\mathbb{C}^m$. Indeed, for $b \in B \cap \mathbb{R}^m$ we have $\partial_{x_j} f(b) = 0$, hence $\partial_{z_j} f(b) = -\partial_{\bar{z}_j} f(b) = 0$.

Consider ψ and h as in the lemma. By the irreducibility of ψ it suffices to show that ψ and h cannot be relatively prime at 0.

Suppose h does not vanish along the line $\mathbb{C} \cdot (0, \ldots, 0, 1)$. Suppose h and ψ are relatively prime at (φ_0, λ_0). We may assume that ψ and h are Weierstrass polynomials in z_{N+1}; c.f. [23], Corollary 6.1.2, or [17], Ch. 2, §1.2. We proceed as in the proof of Theorem 6.2.3 in [23]: ψ and h are relatively prime in $K[z_{N+1}]$, with K the quotient field of the ring of germs at 0 of holomorphic functions on $\mathbb{C}^N$. This means that

$$l(z^{(N)}) = f_1(z^{(N)}, z_{N+1})\psi(z^{(N)}, z_{N+1}) + f_2(z^{(N)}, z_{N+1})h(z^{(N)}, z_{N+1})$$

on a sufficiently small neighborhood $\Omega_1 = W \times U$ of $(0,0)$, with l holomorphic on $W \subset \mathbb{C}^N$ and not identically zero, and f_1 and f_2 holomorphic on Ω_1, polynomial in z_{N+1}. By shrinking W, we may ensure that for each $z^{(N)} \in W$ there is an element $z_{N+1} \in U$ such that $(z^{(N)}, z_{N+1}) \in N(\psi)$; see [17], Ch. 2, §3.5. In particular, for each real $z^{(N)} \in W \cap \mathbb{R}^N$ we obtain $l(z^{(N)}) = 0$. So l vanishes on W after all, and ψ and h cannot be relatively prime at 0.

h might vanish along the line $\{(0, \ldots, 0, z_{N+1})\}$. If ψ and h are relatively prime at 0, they are so on a neighborhood ([23], Theorem 6.2.3), which we may suppose to be of the form $W \times U$. It might happen that h does not vanish along $\{z^{(N)}\} \times \mathbb{C}$ for other real points $z^{(N)} \in W \cap \mathbb{R}^N$. In that case, the reasoning above leads to a contradiction. If h vanishes at all points of $(W \cap \mathbb{R}^N) \times \mathbb{C}$ it vanishes on all real points of a neighborhood of 0, and hence is zero.

9.4.14. Lemma 9.4.13 finishes the proof of Proposition 9.4.2, and hence the proof of Theorem 9.4.1 as well.

We reformulate the Lemmas 9.4.9 and 9.4.10 in terms of families of automorphic forms.

9.4.15 Proposition. *Let $\Omega \subset V_0 \times \mathbb{C}$ be open, let $\rho_{P,n} \in \mathcal{O}(\Omega)$ for all $P \in \mathcal{P}$, $n \in \mathbf{c}(P)$. Take e_ρ as in Theorem 9.4.1. Suppose $w_0 \in \mathrm{Pol}(e_\rho)$ and $(F_{P,n}e_\rho)'(a(P)) \in \mathcal{O}_{w_0}$ for all $n \in \mathbf{c}(P)$, $P \in \mathcal{P}$. Then there exists $\psi \in \mathcal{O}(\Omega_0)$ for some open neighborhood Ω_0 of w_0 inside Ω such that ψe_ρ is holomorphic at w_0, but does not vanish identically along $N(\psi)$. For each $(\varphi_1, s_1) \in N(\psi)$,*

$$(\psi e_\rho)(\varphi_1, s_1) \in S_{l_0 + \varphi(\zeta)/\pi}(\chi_0 \cdot \mathbf{exp}(\varphi_1), \mathbf{c}, s_1).$$

9.4.16 Proposition. *Let $\Omega \subset V_0 \times \mathbb{C}$ be open, let $\rho_{P,n} \in \mathcal{O}(\Omega)$ for all $P \in \mathcal{P}$, $n \in \mathbf{c}(P)$, and let $(\varphi, s_1) \in \Omega$. Suppose $f \in S_{l_0+\varphi_1(\zeta)/\pi}(\chi_0 \cdot \mathbf{exp}(\varphi_1), \mathbf{c}, s_1)$. Then*

$$\int_{\tilde{\Gamma}\backslash\tilde{G}/\tilde{K}} f \cdot \overline{e_\rho(\bar{\varphi}_1, \bar{s})}\, dg = 0$$

for all $\bar{s} \in \mathbb{C}$ such that $(\bar{\varphi}_1, \bar{s}) \in \Omega \setminus \mathrm{Pol}(e_\rho)$.

Remark. This integral is not an ordinary Lebesgue integral if $e_\rho(\bar\varphi_1, \bar s)$ grows too fast at some $P \in \mathcal{P}$. On $\overline{\mathrm{pr}}^{-1}({}^aY) \cap F$ it is a Lebesgue integral. On $\overline{\mathrm{pr}}^{-1}(U_P(a(P)) \cap F$ it is given by

$$\sum_{n \in \mathcal{C}_P \smallsetminus \mathbf{c}(P)} \int_{{}^a\hat I_P} F_{P,n(\varphi_1)} f \cdot \overline{F_{P,n(\bar\varphi_1)} e_\rho(\bar\varphi_1, \bar s)}\, d\nu_P.$$

Proof. Under this interpretation the integral equals

$$\left(e^{-it_{\varphi_1}} f, E_\rho(\bar\varphi_1, \bar s)\right)_{l_0} = \left({}^{(a)}\left(e^{-it_{\varphi_1}} f\right), E_\rho(\bar\varphi_1, \bar s)\right)_{l_0}.$$

Use Proposition 9.2.6 to see that Lemma 9.4.10 can be applied to the truncated function ${}^{(a)}(e^{-it_{\varphi_1}} f)$. The integral vanishes for $\bar s \in {}^a\mathsf{P}(\bar\varphi_1)$. By the L^2-meromorphy of E_ρ, this extends to all $\bar s$ such that $(\bar\varphi_1, \bar s) \notin \mathrm{Pol}(e_\rho)$.

9.5 Dimension results for spaces of automorphic forms

Our main use for Theorem 9.4.1 will be in the proof of the meromorphic continuation of Poincaré series in character and spectral parameter jointly. That is the subject of the next chapter. Here we show that the theorem gives us a supply of families of automorphic forms that is sufficient to obtain some results on the dimensions of spaces of automorphic forms.

9.5.1 Proposition. *Let $\Omega \subset V_0 \times \mathbb{C}$ be open, $m : (\varphi, s) \mapsto l_0 + \varphi(\zeta)/\pi$. Then $\mathcal{M} \otimes_{\mathcal{O}} \mathcal{A}_m(\mathbf{c}; \Omega)$ has dimension $|\mathbf{c}|$ as a vector space over $\mathcal{M}(\Omega)$. It has a basis consisting of sections e_ρ with the ρ chosen in $\mathcal{O}(\Omega)^{|\mathbf{c}|}$.*

Remark. For any parameter space there are locally morphisms of parameter spaces $j : W \to \mathcal{V} \times \mathbb{C}$, with image contained in $V_0 \times \mathbb{C}$ if the growth condition has been chosen suitably. This proposition gives a lower bound on the $\mathcal{M}_{(j)}$-dimensions of sheaves of automorphic forms on W. We do not obtain an equality in the general case, as there may be non-zero families of cusp forms. (Think about the case that $j(W)$ is contained in $(V_0 \times \mathbb{C}) \smallsetminus {}^a\mathsf{P}$.)
Proof. Clearly $\mathcal{M} \otimes_{\mathcal{O}} \mathcal{A}_m(\mathbf{c}; \Omega)$ is a vector space over the field $\mathcal{M}(\Omega)$ and Theorem 9.4.1 enables us to give $|\mathbf{c}|$ linearly independent elements. Take for each $P \in \mathcal{P}$, $n \in \mathbf{c}(P)$ the collection $\rho(P, n)$ given by Kronecker deltas:

$$\rho(P,n)_{Q,m} = \delta_{P,Q}\delta_{n,m} \quad \text{for } Q \in \mathcal{P},\ m \in \mathbf{c}(Q).$$

Evaluation at $w = (\varphi, s) \in \Omega \cap {}^a\mathsf{P}$ gives $|\mathbf{c}|$ elements of $A_{m(w)}(\chi, \mathbf{c}(w), s)$ that are $\mathbb{C}$-linearly independent. (We write $\chi = \chi_0 \cdot \mathbf{exp}\,(\varphi)$.) Hence

$$\dim_{\mathbb{C}} A_{m(w)}(\chi, \mathbf{c}(w), s) \geq |\mathbf{c}| + \dim_{\mathbb{C}} S_{m(w)}(\chi, \mathbf{c}(w), s),$$

and in the same way

$$\dim_{\mathbb{C}} A_{-m(w)}(\chi^{-1}, -\mathbf{c}(w), s) \geq |\mathbf{c}| + \dim_{\mathbb{C}} S_{-m(w)}(\chi^{-1}, -\mathbf{c}(w), s).$$

The Maass-Selberg relation implies that we have equality, c.f. 4.6.6. Proposition 9.2.6 shows that $w = (\varphi, s) \in {}^a\mathsf{P}$ implies $S_{m(w)}(\chi, \mathbf{c}(w), s) = \{0\}$.

Consider $f \in \mathcal{M} \otimes_{\mathcal{O}} \mathcal{A}_m(\mathbf{c}; \Omega)$. Subtracting a linear combination of the $e_{\rho(P,n)}$, we arrange $F_{P,n}f(a(P))$ to be identically zero for all $P \in \mathcal{P}$, $n \in \mathbf{c}(P)$. So $F_{P,n}f(w; a(P)) = 0$ for each $w \in \Omega \cap {}^a\mathsf{P}$. Hence if we express $f(w)$ in the basis of $A_{m(w)}(\chi_0 \cdot \mathbf{exp}\,(\varphi), \mathbf{c}(w), s)$ given above, it turns out to be the zero linear combination. Hence f vanishes on $\Omega \cap {}^a\mathsf{P}$, and $f = 0$.

9.5.2 *Remark.* We have also shown for $w = (\varphi, s) \in (V_0 \times \mathbb{C}) \cap {}^a\mathsf{P}$:

i) $S_{m(w)}(\chi_0 \cdot \mathbf{exp}\,(\varphi), \mathbf{c}(w), s) = \{0\}$,

ii) $f \mapsto (F_{P,n}f(a(P)))_{P\in\mathcal{P},\, n\in\mathbf{c}(P)}$ gives a bijective linear map

$$A_{m(w)}(\chi_0 \cdot \mathbf{exp}\,(\varphi), \mathbf{c}(w), s) \to \mathbb{C}^{|\mathbf{c}|}.$$

9.5.3 *Restriction to a vertical line.* The statements of Proposition 9.5.1 are true on more general subsets of $V_0 \times \mathbb{C}$. We consider one-dimensional sets of the form $\{\varphi\} \times \mathbb{C}$ with $\varphi \in V_0$. Let $U \subset \mathbb{C}$ be open. For each $s \in U \cap {}^a\mathsf{P}(\varphi)$ the space $A_{m(\varphi)}(\chi_0 \cdot \mathbf{exp}\,(\varphi), \mathbf{c}(\varphi), s)$ has dimension $|\mathbf{c}|$. As in the proof of Proposition 9.5.1 the values of the $e_{\rho(P,n)}$ form a basis.

Consider the parameter space $W = \mathbb{C}$, $\chi : s \mapsto \chi$, $s : s \mapsto s$. Take $\mathbf{d} = \mathbf{c}(\varphi)$, and put $q = m(\varphi)$. The $\mathcal{M}(U)$-linear space $\mathcal{A}_q(\mathbf{d}; U)$ has dimension at least $|\mathbf{c}|$. As in the proof of Proposition 9.5.1 we conclude that this dimension is exactly $|\mathbf{c}|$.

9.5.4 Lemma. *Let $\varphi \in V_0$, $s \in \mathbb{C}$. Suppose that at least one of the following conditions holds:*

i) $\varphi \in V_r$.

ii) $S_{m(\varphi)}(\chi_0 \cdot \mathbf{exp}\,(\varphi), \mathbf{c}(\varphi), s) = \{0\}$.

Then $\dim_{\mathbb{C}} A_{m(\varphi)}(\chi_0 \cdot \mathbf{exp}\,(\varphi), \mathbf{c}(\varphi), s) = |\mathbf{c}| + \dim_{\mathbb{C}} S_{m(w)}(\chi_0 \cdot \mathbf{exp}\,(\varphi), \mathbf{c}(\varphi), s)$.

Proof. The parameter space W in 9.5.3 has dimension one. Consider $s_0 \in W$. The local ring $\mathcal{O}_{s_0} = (\mathcal{O}_W)_{s_0}$ is a principal ideal domain. $\mathcal{A}_q(\mathbf{d})_{s_0}$ is a torsion-free $(\mathcal{O}_W)_{s_0}$-module. Any finitely generated $(\mathcal{O}_W)_{s_0}$-submodule $\mathcal{U}$ of $\mathcal{A}_q(\mathbf{d})_{s_0}$ is of the form $\mathcal{U} = \bigoplus_{j=1}^{t} (\mathcal{O}_W)_{s_0} \cdot f_j$, see [24], 3.8, p. 181. From 9.5.3 it follows that $t = \dim_{\mathcal{M}_{s_0}} \mathcal{M}_{s_0} \otimes \mathcal{U} \leq |\mathbf{d}|$ for all such $\mathcal{U}$, and also that $t = |\mathbf{d}|$ for a suitably chosen $\mathcal{U}_1$. There exists $\varphi \in \mathcal{O}_{s_0}$, $\varphi \neq 0$, such that $\varphi\mathcal{A}_q(\mathbf{d})_{s_0} \subset \mathcal{U}_1$. As $\dim_{\mathbb{C}} \mathcal{O}_{s_0}/\varphi\mathcal{O}_{s_0} < \infty$, the $\mathbb{C}$-dimension of $\mathcal{A}(\mathbf{d})_{s_0}/\mathcal{U}_1$ is finite. Hence $\mathcal{A}_q(\mathbf{d})_{s_0}$ itself is finitely generated, and $\mathcal{A}_q(\mathbf{d})_{s_0} = \bigoplus_{j=1}^{|\mathbf{d}|} \mathcal{O}_{s_0} \cdot f_j$, with $f_j \in \mathcal{A}_q(\mathbf{d})_{s_0}$.

Let f be a $\mathbb{C}$-linear combination $f = \sum_{j=1}^{|\mathbf{d}|} c_j \cdot f_j$ vanishing at s_0. By Lemma 7.3.9, the section $s \mapsto \frac{1}{s-s_0} f(s)$ is in $\mathcal{A}_q(\mathbf{d})_{s_0}$. So there are $\psi_j \in \mathcal{O}_{s_0}$ such that $c_j = (s - s_0)\psi_j(s)$. This shows that all c_j vanish and that $A_q(\chi, \mathbf{d}, s_0)$ has $\mathbb{C}$-dimension at least $|\mathbf{d}|$.

By 4.6.6, it is sufficient to show that $\dim_{\mathbb{C}} (A_q(\chi, \mathbf{d}, s_0)/S_q(\chi, \mathbf{d}, s_0)) \geq |\mathbf{d}|$. Suppose that the $\mathbb{C}$-linear combination $f = \sum_{j=1}^{|\mathbf{d}|} c_j \cdot f_j$ has the cuspidal value $f(s_0) \in S_q(\chi, \mathbf{d}, s_0)$. If $S_{m(\varphi)}(\chi_0 \cdot \mathbf{exp}(\varphi), \mathbf{c}(\varphi), s) = \{0\}$, then $f(s_0) = 0$. If $\varphi \in \mathcal{V}_r$, we use $\rho_{P,\nu} = F_{P,\nu} f(a(P))$ to express f as the restriction of some e_ρ and apply Proposition 9.4.16 to see that $(f(s_0), f(s))_{l_0} = 0$ for all s near s_0. Hence $f(s_0)$ vanishes. This means that the $f_j(s_0)$ give linearly independent elements of $A_q(\chi, \mathbf{d}, s_0)/S_q(\chi, \mathbf{d}, s_0)$.

9.5.5 *Remarks.* It is essential that the parameter space W has complex dimension one. Otherwise the stalk $\mathcal{O}_{s_0}$ is not a principal ideal domain.

This proof shows that under the conditions of the lemma, each automorphic form in $A_{m(\varphi)}(\chi_0 \cdot \mathbf{exp}(\varphi), \mathbf{c}(\varphi), s)$ is the value of a family of automorphic forms on a parameter space of the form $\{\chi\} \times \Omega$ with $\Omega \subset \mathbb{C}$ open: enlarge the growth condition till condition ii) in Lemma 9.5.4 holds.

If $\varphi \in \mathcal{V}_r$ the proof shows that there are sections $f_1, \ldots, f_{|\mathbf{d}|}$ of $\mathcal{A}_q(\mathbf{d}; \Omega)$ for some neighborhood Ω of s_0 in $\mathbb{C}$ such that

$$A_q(\chi, \mathbf{d}, s_0) = S_q(\chi, \mathbf{d}, s_0) \oplus \bigoplus_{j=1}^{|\mathbf{d}|} \mathbb{C} \cdot f_j(s_0).$$

9.5.6 Proposition. *Let $\chi \in \mathcal{X}$, $s \in \mathbb{C}$, $\mathbf{d}$ a growth condition for χ and q a weight for χ. Then* $\dim_{\mathbb{C}} S_q(\chi, \mathbf{d}, s) < \infty$.

Proof. Take $\chi_0 \in \mathcal{X}_r$ and $V_r \subset \mathcal{V}_r$ such that $\chi = \chi_0 \cdot \mathbf{exp}(\varphi)$ for some $\varphi \in V = \mathbb{C} \otimes_{\mathbb{R}} V_r$. Take a growth condition $\mathbf{c}$ on $V \times \mathbb{C}$ such that $\mathbf{d} \subset \mathbf{c}(\varphi, s)$ and $(\varphi, s) \in V_0(\chi_0, \mathbf{c})$. This is possible, see Lemma 9.1.6. Take $l(\varphi, s) = q - \varphi(\zeta)/\pi$. Propositions 9.2.6 and 9.2.2 imply

$$\dim_{\mathbb{C}} S_q(\chi, \mathbf{c}(\varphi, s), s) \leq \dim_{\mathbb{C}} \ker({}^a A(\varphi) - \frac{1}{4} + s^2) < \infty.$$

As $S_q(\chi, \mathbf{c}(\varphi, s), s)$ is a subspace of $S_q(\chi, \mathbf{d}, s)$ determined by finitely many linear conditions, the proposition follows.

9.5.7 Proposition. *Let χ, s, $\mathbf{d}$ and q be as in the previous proposition. There are growth conditions $\mathbf{d}_1 \supset \mathbf{d}$ such that* $\dim_{\mathbb{C}} A_q(\chi, \mathbf{d}_1, s) = |\mathbf{d}_1| + \dim_{\mathbb{C}} S_q(\chi, \mathbf{d}_1, s)$. *If χ is unitary, this holds for $\mathbf{d}_1 = \mathbf{d}$.*

Proof. If $\chi \in \mathcal{X}_u$, we apply Lemma 9.5.4 with $\chi_0 = \chi$. Otherwise take $\mathbf{d}_1 \supset \mathbf{d}$ large enough to have $\mathbf{d}_1 = \mathbf{c}(\varphi)$, with $\mathbf{c}$ as in Lemma 9.5.4, and also $S_q(\chi, \mathbf{d}_1, s) = \{0\}$ (see 4.3.10).

Chapter 10
Meromorphic continuation of Poincaré series

In Section 1.6 we have sketched the proof of the meromorphic continuation in two variables of the Eisenstein series in the modular case. Most of this proof we have carried out for the general situation. Now we come to the last step, indicated in 1.6.5. This step amounts to adjusting and gluing the families of automorphic forms obtained in the previous chapter. In 1.6.5 we normalized one Fourier term of $\tilde{E}(r,s)$ by prescribing the value 1 at a. We have brought this term into the form $\mu_r^0(r,s) + (\text{constant}) \cdot \mu_r^0(r,-s)$ by multiplying it with a suitable meromorphic factor. We have used a functional analytic argument to show that this factor does not vanish identically.

Here we have not one, but a finite number of Fourier terms, specified by the growth condition **c**. The transition will consist of a linear transformation with meromorphic coefficients. The families e_ρ obtained in Theorem 9.4.1 have Fourier terms with a simple form when expressed in the basis $\alpha_{P,n}$, $\beta_{P,n}$. This basis is characterized by value and derivative at the truncation points $a(P)$. This amounts to the same as the condition in 1.6.5. Our aim is to obtain a similar existence result with respect to the more canonical basis of $\mathcal{W}_l(P,n)$ used in Proposition 5.2.1. That will imply the meromorphic continuation of the Poincaré series introduced in Section 5.1.

In Section 1.6 we were content to obtain the meromorphic extension of the Eisenstein series in (r,s) for r in a neighborhood of 0. Here we try to obtain the continuation on regions that are as large as possible. These regions are determined by the cells of continuation, as discussed in Section 10.1.

In Section 10.2 we prove the meromorphic continuation of Poincaré series. The conditions on the Fourier terms uniquely determine these meromorphic families of automorphic forms. This leads to the well known functional equation for $s \mapsto -s$, discussed in Section 10.3.

10.1 Cells of continuation

In the modular case, discussed in Section 1.6, we used the basis $\mu_r^0(r,s)$, $\mu_r^0(r,-s)$ for the space of Fourier terms of order $\frac{r}{12}$. All other Fourier terms of $E(r,s)$ are multiples of $\omega_r^\nu(r,s)$ with $\nu \in \mathbb{Z}$. Consider $\nu = 1$. The definition in 1.5.8 shows that $(r,s) \mapsto \omega_r^1(r,s;y)$ has no meromorphic extension to a neighborhood of the line $\{-12\} \times \mathbb{C}$ in the (r,s)-plane. Similarly, the term with $\nu = -1$ gives a problem at $r = 12$. This means that the bases used to characterize $E(r,s)$ by its Fourier terms cannot be used near $r = \pm 12$. We can expect a meromorphic continuation of the Eisenstein series only to a set $U \times \mathbb{C}$, with U a neighborhood of $(-12, 12)$ in $\mathbb{C}$. In principle one could let U protrude out of the set $\{ r \in \mathbb{C} : |\operatorname{Re} r| < 12 \}$, provided one avoids $r = 12$ and $r = -12$. But if one does so, some $\omega_r^\nu(r, s)$ are no longer quickly decreasing. That would mean that the Fourier term corresponding to the growth condition $\mathbf{c}_0[P] = \{0\}$ no longer determines the growth behavior. We call the interval $(-12, 12)$ the *cell of continuation.* As long as $\operatorname{Re} s$ stays in this interval, we may hope that $E(r, s)$ has a sensible meromorphic continuation.

If we consider a Poincaré series given in 1.5.9, with $0 < r < 12$, then the basis $\mu_r^\nu(r, s)$, $\omega_r^\nu(r, s)$ is suitable for all Fourier terms. The presence of $\omega_r^0(r, s)$ implies that meromorphic extension at $r = 0$ is not possible. Here $(0, 12)$ is the suitable cell of continuation.

In this section we consider cells of continuation for a general discrete cofinite group $\tilde{\Gamma}$.

10.1.1 *Parameter space.* The Poincaré series defines a continuous map $(\chi, s, g) \mapsto P_l(P, n, \chi, s; g)$ on $\mathcal{X}_u \times \mathbb{C}_{1/2} \times (\tilde{G} \setminus \operatorname{pr}^{-1}\{P\})$; here l and n vary as functions of χ. The obvious parameter space in this context would be $\mathcal{X} \times \mathbb{C}$, but this space is not necessarily simply connected. So we fix $\chi_1 \in \mathcal{X}_0$, and use the parameter space

$$W = \mathcal{V} \times \mathbb{C}, \quad \chi : (\varphi, s) \mapsto \chi_1 \cdot \mathbf{exp}\,(\varphi), \quad s : (\varphi, s) \mapsto s.$$

We fix $l \in \mathbf{wt}(W)$; it has the form $l : (\varphi, s) \mapsto l_1 + \varphi(\zeta)/\pi$ with $e^{\pi i l_1} = \chi_1(\zeta)$. If $P \in X^\infty$, the elements of $\mathcal{C}_P$ are $n_\nu : (\varphi, s) \mapsto \frac{1}{2\pi}\varphi(\pi_P) + \nu$. If $P \in \mathcal{P}_Y$, and $n \in \mathcal{C}_P$, the functions $l - n$ are constant on W.

10.1.2 *Null cell.* Put $J(0) = \{ \varphi \in \mathcal{V}_r : -2\pi < \varphi(\pi_P) < 2\pi \text{ for all } P \in X^\infty \}$. We call it the null cell.

As $\mathcal{X}_u = \mathcal{X}_0 \cdot \mathbf{exp}\,(\mathcal{V}_r)$, it suffices to study the Poincaré families $(\varphi, s) \mapsto P_l(P, n, \chi_1 \cdot \mathbf{exp}\,(\varphi), s)$ for a set of representatives χ_1 of $\mathcal{X}_0 / (\mathcal{X}_0 \cap \mathbf{exp}\,(\mathcal{V}))$. The domain of these families will be a subset of $U \times \mathbb{C}$, with $U = \{ \varphi \in \mathcal{V} : \operatorname{Re} \varphi \in J(0) \}$. The group $\mathcal{X}_0$ is trivial for the modular group and the other example groups in Chapters 14–15. So we take $\chi_1 = 1$ in these cases. See 3.4.7 for an example where $\mathcal{X}_0 \neq \{1\}$.

We use the following $\mathcal{M}$-basis of $\mathcal{M} \otimes_{\mathcal{O}} \mathcal{W}_l(P, n)$ on $J(0)$:

$$\begin{cases} \mu_l(P, n_0), \quad \tilde{\mu}_l(P, n_0) & \text{if } P \in X^\infty,\ n = n_0 \\ \mu_l(P, n), \quad \omega_l(P, n) & \text{otherwise.} \end{cases}$$

This is the basis used in Proposition 5.2.1 to describe the Fourier coefficients of the Poincaré series $P_{l_1 + \varphi(\zeta)/\pi}(P, n(\varphi), \chi_1 \cdot \mathbf{exp}\,(\varphi), s)$ for those φ that satisfy $n(\varphi) \in \mathbb{Z}$ if $n \in \mathcal{C}_P$, $P \in X^\infty$.

10.1.3 *Cells of continuation.* For $\varphi_0 \in J(0)$ we define $J(\varphi_0)$ as the connected component of φ_0 in

$$\{ \varphi \in \mathcal{V}_r : \forall P \in X^\infty \forall n \in \mathcal{C}_P(\mathcal{V}) : n(\varphi_0) \neq 0 \Rightarrow n(\varphi) \neq 0 \}.$$

We call $J(\varphi_0)$ the *cell of continuation of* φ_0. Note that $\varphi_0 \in J(\varphi_0)$, and that $J(0)$ is the cell of continuation of $0 \in \mathcal{V}_r$.

$\varphi_0, \varphi_1 \in \mathcal{V}_r$ may determine the same cell of continuation. We call

$$J^0(\varphi_0) = \{\, \varphi_1 \in \mathcal{V}_r : J(\varphi_1) = J(\varphi_0) \,\}$$

the *defining part* of $J(\varphi_0)$.

10.1.4 *Examples.* We parametrize $\mathcal{V}$ for the modular case by $r \mapsto r\alpha$ (see 13.1.7). As $\alpha(n(1)) = \frac{\pi}{6}$, the null cell corresponds to the interval $(-12, 12)$ on the r-line. If $r_0 \in (-12, 12)$, then

$$\begin{array}{lll} J(r_0) = (-12,0) & J^0(r_0) = (-12,0) & \text{if } -12 < r_0 < 0 \\ J(0) = (-12,12) & J^0(0) = \{0\} & \\ J(r_0) = (0,12) & J^0(r_0) = (0,12) & \text{if } 0 < r_0 < 12. \end{array}$$

See 14.5.1 and 15.1.12 for more complicated examples. If there are no cuspidal orbits at all, then $\mathcal{V}_r$ is the sole cell of continuation.

10.1.5 *Explicit description.* The cells of continuation are determined by the map

$$\psi : \mathcal{V}_r \longrightarrow \mathbb{R}^p \, : \, \varphi \mapsto \left(\frac{1}{2\pi} \varphi(\pi_P) \right)_{P \in X^\infty} ;$$

$p = |X^\infty|$ is the number of cuspidal orbits. We have defined $J(0)$ as $\psi^{-1}\left((-1,1)^p \right)$.

Let $P \in X^\infty$; then $\mathcal{C}_P = \{\, n_\nu : \nu \in \mathbb{Z} \,\}$ with $n_\nu(\varphi) = \nu + \frac{1}{2\pi}\varphi(\pi_P)$. If $\varphi \in J(0)$, the only $n \in \mathcal{C}_P$ for which $n(\varphi) = 0$ may happen is $n = n_0$. We put, for $\varphi_0 \in J(0)$ and $P \in X^\infty$,

$$J_P(\varphi_0) = \begin{cases} (-1,1) & \text{if } \varphi(\pi_P) = 0 \\ (-1,0) & \text{if } -2\pi < \varphi_0(\pi_P) < 0 \\ (0,1) & \text{if } 0 < \varphi_0(\pi_P) < 2\pi \end{cases}$$

$$J_P^0(\varphi_0) = \begin{cases} \{0\} & \text{if } \varphi_0(\pi_P) = 0 \\ J_P(\varphi_0) & \text{otherwise.} \end{cases}$$

Then

$$J(\varphi_0) = \psi^{-1}\left(\prod_{P \in X^\infty} J_P(\varphi_0) \right), \quad J^0(\varphi_0) = \psi^{-1}\left(\prod_{P \in X^\infty} J_P^0(\varphi_0) \right).$$

In particular, $J(0) = \psi^{-1}\left(\prod_{P \in X^\infty} (-1,1) \right)$ and $J^0(0) = \psi^{-1}\{0\}$ (with $0 \in \mathbb{R}^p$).

10.1.6 *Minimal cells.* $J(\varphi_0) = J^0(\varphi_0)$ if and only if $\varphi_0(\pi_P) \neq 0$ for all $P \in X^\infty$. We call such a cell of continuation a *minimal cell.*

10.1.7 *Growth condition.* To a cell of continuation J we assign a growth condition $\mathbf{c}_J$ by

$$\begin{array}{ll} \mathbf{c}_J(P) = \{n_0\} & \text{if } P \in X^\infty,\ \varphi(\pi_P) = 0 \text{ for all } \varphi \in J^0 \\ \mathbf{c}_J(P) = \emptyset & \text{if } P \in X^\infty,\ \varphi(\pi_P) \neq 0 \text{ for all } \varphi \in J^0 \\ \mathbf{c}_J(P) = \emptyset & \text{if } P \in \mathcal{P}_Y. \end{array}$$

This implies that the growth condition $\mathbf{c}_J$ is suitable on $\Omega = \{ (\varphi, s) \in \mathcal{V} \times \mathbb{C} : \operatorname{Re} \varphi \in J \}$. So $\mathcal{A}_l(\mathbf{c})$ is an eigenfunction module on Ω for each $\mathbf{c} \supset \mathbf{c}_J$.

For given $P \in \mathcal{P}$, $n \in \mathcal{C}_P$, we indicate by $\mathbf{c}_{J,n}$ the growth condition $\mathbf{c}_{J,n} \supset \mathbf{c}_J$ given by $\mathbf{c}_{J,n}(Q) = \mathbf{c}_J(Q)$ if $Q \neq P$, and $\mathbf{c}_{J,n}(P) = \mathbf{c}_J(P) \cup \{n\}$.

10.1.8 *Basis.* Let J be a cell of continuation. Put $\Omega = \{ \varphi \in \mathcal{V} : \operatorname{Re} \varphi \in J \} \times \mathbb{C}$. For $P \in \mathcal{P}$, $n \in \mathcal{C}_P(\mathcal{V})$, put

$$\begin{aligned} \eta_l(P,n) &= \eta_l(J,P,n) \\ &= \begin{cases} \tilde{\mu}_l(P,n) & \text{if } P \in X^\infty,\ n = n_0,\ \varphi(\pi_P) = 0 \text{ for } \varphi \in J^0 \\ \omega_l(P,n) & \text{otherwise.} \end{cases} \end{aligned}$$

We shall use $\mu_l(P,n)$, $\eta_l(P,n)$ as $\mathcal{M}$-basis of $(\mathcal{M} \otimes_{\mathcal{O}} \mathcal{W}_l(P,n))(\Omega)$. This is the basis used in Proposition 5.2.1. The tilde has been defined by $\tilde{\mu}_l(P,n;\varphi,s) = \mu_l(P,n;\varphi,-s)$.

10.2 Meromorphic continuation

We are ready to complete the proof of the meromorphic continuation of Poincaré series. We follow the idea sketched in 1.6.5.

10.2.1 Theorem. *Let J be a cell of continuation, let $P \in \mathcal{P}$, $n \in \mathcal{C}_P(\mathcal{V})$. Let $\chi_1 \in \mathcal{X}_0$, l a weight as in 10.1.1. Let $\mathbf{c}_{J,n}$ and $\eta_l(P,n)$ be defined as in 10.1.7 and 10.1.8. Let $w_P = 1$ if $P \in X^\infty$, and $w_P = v_P$ otherwise.*

i) There exists a unique family $Q_l(J,P,n) \in (\mathcal{M} \otimes_{\mathcal{O}} \mathcal{A}_l(\mathbf{c}_{J,n}))(U \times \mathbb{C})$, with U a neighborhood of J in $\mathcal{V}$, such that for all $Q \in \mathcal{P}$, $m \in \mathcal{C}_Q(\mathcal{V})$

$$\tilde{F}_{Q,m} Q_l(J,P,n) \in \delta_{P,Q} \delta_{n,m} w_P \mu_l(P,n) + \mathcal{M}(U \times \mathbb{C}) \cdot \eta_l(Q,m).$$

ii) Let $\varphi_0 \in J^0$. Then the restriction $s \mapsto Q_l(J,P,n;\chi_1,\varphi_0,s)$ exists as a meromorphic family of automorphic forms on the parameter space $(\mathbb{C},\ s \mapsto \chi_1 \cdot \mathbf{exp}(\varphi_0),\ s \mapsto \frac{1}{4} - s^2)$. On $\mathbb{C}_{1/2} = \{ s \in \mathbb{C} : \operatorname{Re} s > \frac{1}{2} \}$ this restriction coincides with the holomorphic family of Poincaré series

$$s \mapsto P_{l(\varphi_0)}(P, n(\varphi_0), \chi_1 \cdot \mathbf{exp}(\varphi_0), s).$$

Remarks. The family $Q_l(J,P,n)$ depends on $\chi_1 \in \mathcal{X}_0$. When confusion threatens, we shall write $Q_l(J,P,n,\chi_1)$.

We have seen in 7.5.3 that Poincaré series define holomorphic families in s for $\mathrm{Re}\, s > \frac{1}{2}$. The theorem implies that these holomorphic families have a meromorphic extension to $\mathbb{C}$. This is well known, see, e.g., Hejhal, [21], Ch. VI, §11, p. 108, for the Eisenstein series, and Neunhöffer, [40], for other Poincaré series with trivial character. As far as I know, it is a new result that this meromorphic family in one variable is the restriction of a meromorphic family of automorphic forms on the open set Ω, which has in general complex dimension larger than one. The family $Q_l(J,P,n)$ has a good claim to the designation 'meromorphic continuation of a Poincaré series in character and eigenvalue jointly'. We shall call it a *Poincaré family.*

$Q_l(J,P,n)$ is uniquely characterized by the coordinates of its Fourier terms corresponding to $\mathbf{c}_{J,n}$, with respect to the basis given in 10.1.8. For cells of continuation $J \subset J_1$, this characterization implies that the families $Q_l(J,P,n)$ and the $Q_l(J_1,P',n')$ satisfy a relation on the intersection of their domains. The uniqueness also implies a functional equation for $s \mapsto -s$, see Section 10.3.

The neighborhood U in part i) is not unique. The uniqueness refers to the germs of $Q_l(J,P,n)$ at the points of $J \times \mathbb{C}$.

The existence of the restriction of $Q_l(J,P,n)$ to vertical lines $\{\varphi_0\} \times \mathbb{C}$ with real φ_0 is not trivial. It takes some work to show that such vertical lines are not contained in the polar set of $Q_l(J,P,n)$. I do not know what happens for non-real φ_0.

The restriction to a vertical line may be holomorphic at points where the full family is singular. This may occur at points of the intersection of the zero set and the polar set. So part ii) does not state that $Q_l(J,P,n)$ is holomorphic at (φ_0, s) with $\varphi_0 \in J^0$ and $\mathrm{Re}\, s > \frac{1}{2}$. Proposition 10.2.12 will give more information concerning the holomorphy at such points.

10.2.2 *Poincaré families on linear subvarieties.* If V_r is a linear subspace of $\mathcal{V}_r$, $V = \mathbb{C} \otimes_{\mathbb{R}} V_r$, and $\varphi_0 \in V_r \cap J^0$, the restriction of the Poincaré family $(\varphi, s) \mapsto Q(J,P,n,\chi_1; \varphi_0 + \varphi, s)$ exists as a meromorphic family of automorphic forms on $U_1 \times \mathbb{C}$ with U_1 a neighborhood of $(\varphi_0 + V_r) \cap J$ in $\varphi_0 + V$. The existence of the restriction to $\{\varphi_0\} \times \mathbb{C}$ implies the existence of the restriction to a larger region. (Here we use the assumption $\varphi_0 \in J^0$.) The uniqueness carries over to the restrictions.

We use the notation $Q_l(J,P,n)$ for this restriction as well. This has the danger that one might forget that the restriction may be holomorphic at some points where the full family is not. On the other hand, the notation is quite heavy already.

10.2.3 *Generalization.* The Poincaré families are characterized by the coordinates of their Fourier terms with respect to a particular basis of the $\mathcal{M} \otimes_{\mathcal{O}} \mathcal{W}_l(P,n)$.

The basis used above is natural in view of Proposition 5.2.1. But it will be useful to have part of the theorem available for other bases.

We first prove a more general result, Proposition 10.2.4. We give the proof of Theorem 10.2.1 in 10.2.9.

In 10.2.3–10.2.8 we use J, P, n, χ_1, and l as in Theorem 10.2.1. We fix $\varphi_0 \in J$ and an $\mathbb{R}$-linear subspace V_r of $\mathcal{V}_r$. We put $V = \mathbb{C} \otimes_{\mathbb{R}} V_r$. We assume that there is a neighborhood U_0 of φ_0 in $\varphi_0 + V$ for which there are $\lambda_{Q,m}, \nu_{Q,m} \in (\mathcal{M} \otimes_{\mathcal{O}} \mathcal{W}_l(Q,m))(U_0 \times \mathbb{C})$ for each $Q \in \mathcal{P}$, $m \in \mathbf{c}_{J,n}(Q)$, such that the restrictions $\dot\lambda_{Q,m} : s \mapsto \lambda_{Q,m}(\varphi_0, s)$ and $\dot\nu_{Q,m} : s \mapsto \nu_{Q,m}(\varphi_0, s)$ exist. We further assume that they are $\mathcal{M}(\mathbb{C})$-linearly independent, and that $\dot\nu_{Q,m}$ is equal to $s \mapsto \eta_l(Q,m)(\varphi_0, s)$, with $\eta_l(Q,m)$ as introduced in 10.1.8.

We shall often denote the restriction to a 'vertical line' $\{\varphi\} \times \mathbb{C}$ by a dot. In the next lemmas we put $N = |\mathbf{c}_{J,n}|$.

10.2.4 Proposition. *Under the assumptions in 10.2.3, there is a unique meromorphic family*

$$R(P,n) \in (\mathcal{M} \otimes_{\mathcal{O}} \mathcal{A}_l(\mathbf{c}_{J,n}))\,(U(\varphi_0) \times \mathbb{C}),$$

with $U(\varphi_0)$ a neighborhood of φ_0 in $\varphi_0 + V$, such that for all $Q \in \mathcal{P}$ and $m \in \mathbf{c}_{J,n}(Q)$:

$$\tilde F_{Q,m} R(P,n) \in \delta_{P,Q}\delta_{m,n} w_P \lambda_{P,n} + \mathcal{M}(U(\varphi_0) \times \mathbb{C}) \cdot \nu_{Q,m}.$$

The restriction $\dot R(P,n;\varphi_0) : s \mapsto R(P,n;\varphi_0,s)$ exists.

Proof. In 10.2.5–10.2.8.

10.2.5 Lemma. *Define $q : \mathcal{M} \otimes_{\mathcal{O}} \mathcal{A}_l(\mathbf{c}_{J,n}) \to \mathcal{M}^N$ on $U_0 \times \mathbb{C}$ by*

$$(qf)_{Q,m} = \frac{\mathrm{Wr}(\tilde F_{Q,m} f, \nu_{Q,m})}{\mathrm{Wr}(\lambda_{Q,m}, \nu_{Q,m})} \quad \text{for } Q \in \mathcal{P},\ m \in \mathbf{c}_{J,n}(Q).$$

This is a morphism of $\mathcal{M}$-modules on $U_0 \times \mathbb{C}$. If the restriction of f to $\{\varphi_0\} \times \mathbb{C}$ exists, then the same holds for the restriction of qf.

Remark. The morphism q computes the coefficients of $\lambda_{Q,m}$ in the Fourier terms of automorphic forms with respect to the basis $\lambda_{Q,m}$, $\nu_{Q,m}$. To prove Proposition 10.2.4 means to invert q.

Proof. Clear from the assumptions on $\lambda_{Q,m}$ and $\nu_{Q,m}$.

10.2.6 Lemma. *Define $F : \mathcal{M} \otimes_{\mathcal{O}} \mathcal{A}_l(\mathbf{c}_{J,n}) \to \mathcal{M}^N$ on $U_0 \times \mathbb{C}$ by*

$$(Ff)_{Q,m} = F_{Q,m} f(a(Q)) \quad \text{for } Q \in \mathcal{P},\ m \in \mathbf{c}_{J,n}(Q).$$

If f is holomorphic, then so is Ff.

There exists a neighborhood $U(\varphi_0)$ of φ_0 in $\varphi_0 + V$, contained in U_0, and an $\mathcal{M}$-morphism $\mathbf{e} : \mathcal{M}^N \longrightarrow \mathcal{M} \otimes_{\mathcal{O}} \mathcal{A}_l(\mathbf{c}_{J,n})$ inverting F on $U(\varphi_0) \times \mathbb{C}$. If the restriction of the section ρ of $\mathcal{M}^N$ to $\{\varphi_0\} \times \mathbb{C}$ exists, then the same holds for the restriction of $\mathbf{e}\rho$.

Remark. Like the morphism q, the morphism F gives Fourier coefficients of Fourier terms of automorphic forms. But this morphism works with respect to the basis $\alpha_{Q,m}$, $\beta_{Q,m}$.

Proof. We have already inverted this morphism in Theorem 9.4.1. Use $\chi_1 \cdot \mathbf{exp}\,(\varphi_0)$ as the $\chi_0 \in \mathcal{X}_u$ in the application of perturbation theory carried out in the previous chapters. Put $U(\varphi_0) = U_0 \cap \big(\varphi_0 + V_0(l(\varphi_0), \chi_1 \cdot \mathbf{exp}\,(\varphi_0), \mathbf{c}_{J,n})\big)$. Then $\mathbf{e} : \rho \mapsto e_\rho$, with e_ρ as in Theorem 9.4.1, determines $\mathbf{e}$ with the desired properties.

10.2.7 Lemma. *Let $\varphi_0 \in J^0$. Then the $\mathcal{M}$-morphism $q\mathbf{e} : \mathcal{M}^N \to \mathcal{M}^N$ on $U(\varphi_0)\times\mathbb{C}$ is invertible. Its inverse can be restricted to $\{\varphi_0\} \times \mathbb{C}$.*

Proof. $q\mathbf{e}$ is given by a $N \times N$-matrix with elements in $\mathcal{M}(U(\varphi_0) \times \mathbb{C})$. We have to show that its determinant does not vanish identically in $\mathcal{M}(U(\varphi_0) \times \mathbb{C})$.

The matrix elements of $q\mathbf{e}$ are not singular along the vertical line $S = \{\varphi_0\} \times \mathbb{C}$. It suffices to show that $\det(q\mathbf{e}) \neq 0$ along S. We denote the restriction along S by a dot. If the determinant would vanish along S, then there would exist a non-empty open set $R \subset \mathbb{C}_{1/2}$ and $f \in \mathcal{O}_{\mathbb{C}}(R)^N$ such that $\dot{q}\dot{\mathbf{e}}f = 0$, $f \neq 0$. The section $\dot{\mathbf{e}}f$ is holomorphic on an open dense subset R_1 of R. This section is a family of automorphic forms. From $\dot{q}\dot{\mathbf{e}}f = 0$ it follows that $(\dot{\mathbf{e}}f)(s) \in H(\chi_1 \cdot \mathbf{exp}\,(\varphi_0), l(\varphi_0))$ for all $s \in R_1$. Here we use $\varphi_0 \in J^0$ and $\mathrm{Re}\, s > 0$. Proposition 6.6.2 and the self-adjointness of $A(\chi_1 \cdot \mathbf{exp}\,(\varphi_0), l(\varphi_0))$ imply that $\dot{\mathbf{e}}f$ vanishes on $\{\, s \in R_1 : s \notin \mathbb{R} \,\}$. Hence $f = 0$.

Remark. Here we need the additional assumption $\varphi_0 \in J^0$ to apply L^2-methods.

10.2.8 *Proof* of Proposition 10.2.4. The $\mathcal{M}$-morphism $\mathbf{e}(q\mathbf{e})^{-1} : \mathcal{M}^N \to \mathcal{M} \otimes_{\mathcal{O}} \mathcal{A}_l(\mathbf{c}_{J,n})$ inverts q over $U(\varphi_0) \times \mathbb{C}$. Take $R(P,n) = \mathbf{e}(q\mathbf{e})^{-1}\rho$ with ρ defined by $\rho_{Q,m} = \delta_{P,Q}\delta_{n,m}w_P$.

10.2.9 *Proof* of Theorem 10.2.1. We apply Proposition 10.2.4 and the Lemmas 10.2.5–10.2.7 with $V_r = \mathcal{V}_r$. If we take $\lambda_{Q,m} = \mu_l(P,n)$, $\nu_{Q,m} = \eta_l(Q,m)$ as in 10.1.8, we can take $U_0 = \{\, \varphi \in \mathcal{V} : \mathrm{Re}\,\varphi \in J \,\}$ for each $\varphi_0 \in J$. These $\lambda_{Q,m}$ and $\nu_{Q,m}$ satisfy the assumptions in 10.2.3. All morphisms q (see Lemma 10.2.5) for the various $\varphi_0 \in J$ are the same, and the morphisms F in Lemma 10.2.6 also coincide for all $\varphi_0 \in J$.

We define U as the connected component of J in $\bigcup_{\varphi_0 \in J} U(\varphi_0)$. As the e_ρ are unique, the various $\mathbf{e}$ may be glued together to give one morphism $\mathbf{e} : \mathcal{M}^N \to \mathcal{M} \otimes_{\mathcal{O}} \mathcal{A}_l(\mathbf{c}_{J,n})$ on $U \times \mathbb{C}$. The morphism $q\mathbf{e} : \mathcal{M}^N \to \mathcal{M}^N$ is given by a matrix with elements meromorphic on $U \times \mathbb{C}$. The determinant of this matrix has a non-zero restriction to $\{\varphi_0\} \times \mathbb{C}$ for $\varphi_0 \in J^0$, hence it is non-zero on $U \times \mathbb{C}$. Thus we see that $q\mathbf{e}$ is invertible on $U \times \mathbb{C}$, and that $\mathbf{e}(q\mathbf{e})^{-1}$ can be used to define $Q_l(J,P,n)$ on $U \times \mathbb{C}$, just as we did in the proof of Proposition 10.2.4. This gives i). If $\varphi_0 \in J^0$, we have $Q_l(J,P,n) = R(P,n)$.

Let $\varphi_0 \in J^0$. We have seen that the restriction of $Q_l(J,P,n)$ to $\{\varphi_0\} \times \mathbb{C}$ exists as a meromorphic family. 7.5.3 implies that

$$s \mapsto Q_l(J,P,n,\chi_1;\varphi_0,s) - P_{l(\varphi_0)}(P, n(\varphi_0), \chi_1 \cdot \mathbf{exp}\,(\varphi_0), s)$$

is a meromorphic family of automorphic forms on $\mathbb{C}_{1/2}$. By Proposition 5.2.1 and part i) of the theorem, it is square integrable at all points where it is holomorphic. The same reasoning as in the proof of Lemma 10.2.7 shows that this family has to vanish. This gives part ii) of the theorem.

10.2.10 *Differentiation.* Part i) of Proposition 7.2.9 implies that the $\mathbf{E}^{\pm}Q_l(J,P,n)$ are meromorphic families of automorphic forms as well. The differentiation formulas in Table 4.1 on p. 63 and the uniqueness imply that

$$\mathbf{E}^{\pm}Q_l(J,P,n;\varphi,s) = \theta(1+2s\pm l)Q_{l\pm 2}(J,P,n;\varphi,s),$$

with $\theta = 1$ if $P \in X^{\infty}$, and $\theta = -1$ otherwise.

10.2.11 *Restricted parameter space.* Statement ii) of Theorem 10.2.1 may be improved if we work on a smaller parameter space.

Let J be a cell of continuation. Define $V_{J,r}$ as the $\mathbb{R}$-linear subspace of $\mathcal{V}_r$ spanned by J^0, and $V_J = \mathbb{C}\otimes_{\mathbb{R}} V_{J,r}$. We shall show that the restriction of $Q_l(J,P,n)$ to $\Omega \cap (V_J \times \mathbb{C})$ behaves nicely at points of $J^0 \times \mathbb{C}_{1/2}$.

We call any subset $U \times \mathbb{C} \subset \mathcal{V} \times \mathbb{C}$ with $U \subset \{\, \varphi \in V_J : \mathrm{Re}\,\varphi \in J \,\}$ a *restricted parameter space.*

$V_J = \mathcal{V}$ if J is a minimal cell, and $V_J(0) = \{0\}$ for the null cell. In general, $V_J \supset V_{J_1}$ if $J \subset J_1$. See 14.5.2 for an example. Non-trivial examples do not occur for the modular group.

10.2.12 Proposition. *Let J, P, n, and U be as in Theorem 10.2.1. Put $\Omega_J = (U \cap V_J) \times \mathbb{C}$; consider it as a parameter space, with χ and s inherited from $(\mathcal{V}\times\mathbb{C}) \supset \Omega_J$. Then the restriction $Q_l(J,P,n)$ of the Poincaré family to Ω_J exists as an element of $\left(\hat{\mathcal{M}} \otimes_{\hat{\mathcal{O}}} \hat{\mathcal{A}}_l(\mathbf{c}_{J,n})\right)(\Omega_J)$. It is holomorphic at each point $(\varphi_0, s_0) \in \Omega_J$ for which $\varphi_0 \in J^0$ and $\mathrm{Re}\, s_0 > \frac{1}{2}$.*

Remarks. The hats denote that we mean sheaves on Ω_J.

If $J = \{0\}$, the proposition amounts to nothing more than part ii) of Theorem 10.2.1. In all other cases it provides us with new information. In particular, if J is a minimal cell, it says that $Q_l(J,P,n)$ itself is holomorphic at all points of $J \times \mathbb{C}_{1/2}$.

Proof. Part ii) of Theorem 10.2.1 implies that the set Ω_J cannot be a subset of the polar set of $Q_l(J,P,n)$. Hence the restriction $\hat{Q}_l(J,P,n)$ exists.

Let $\varphi_0 \in J^0$, $\mathrm{Re}\, s_0 > \frac{1}{2}$. Put $w_0 = (\varphi_0, s_0)$. Suppose that the restricted Poincaré family $Q_l(J,P,n)$ is not holomorphic at w_0. Take $\psi_0 \in \hat{\mathcal{O}}_{w_0}$, minimal with respect to divisibility in $\hat{\mathcal{O}}_{w_0}$, such that $f = \psi_0 \cdot Q_l(J,P,n)$ is holomorphic at w_0. The zero set $N(\psi_0)$ has a discrete intersection with each of the lines $\{\varphi_1\} \times \mathbb{C}$, with $\varphi_1 \in J^0$, φ_1 near φ_0. Consider an irreducible factor ψ of ψ_0 in $\hat{\mathcal{O}}_{w_0}$. For all $w_1 = (\varphi_1, s_1) \in N(\psi) \cap (J^0 \times \mathbb{C}_{1/2})$,

$$f(w_1) = \lim_{s\to s_1} \left(\psi_0(\varphi_1,s)\cdot P_{l(\varphi_1)}(P,n(\varphi_1),\chi_1\cdot \mathbf{exp}\,(\varphi_1),s)\right) = 0.$$

Lemma 9.4.13 implies that f is divisible by ψ. Hence ψ_0 was not minimal after all.

10.2.13 *Holomorphy determined by the Fourier terms.* Let J be a cell of continuation and $w_0 = (\varphi_0, s_0) \in J \times \mathbb{C}$. Let ψ be a meromorphic function at w_0. Then $\psi \cdot Q_l(J,P,n)$ is holomorphic at w_0 if and only if $\psi \cdot \tilde{F}_{Q,m} Q_l(J,P,n)$ is holomorphic at w_0 for all $Q \in \mathcal{P}$, $m \in \mathbf{c}_{J,n}(Q)$. In one direction this follows from part iii) of Proposition 7.2.9. In the other direction apply Theorem 9.4.1 with $\chi_0 = \chi_1 \cdot \mathbf{exp}(\varphi_0)$ and $V = \mathcal{V}$; take $\rho_{Q,m} = \psi \cdot F_{Q,m} Q_l(J,P,n;a(Q))$.

In Theorem 9.4.1 we can take for V any linear subspace of $\mathcal{V}$ defined over $\mathbb{R}$. We obtain:

10.2.14 Proposition. *Let J be a cell of continuation. Let V_r be an $\mathbb{R}$-linear subspace of $\mathcal{V}_r$. Put $V = \mathbb{C} \otimes_{\mathbb{R}} V_r$. Let $w_0 = (\varphi_0, s_0) \in J \times \mathbb{C}$. Let ψ be meromorphic on a neighborhood Ω of w_0 in $(\varphi_0 + V) \times \mathbb{C}$. Then $(\varphi, s) \mapsto \psi(\varphi,s) Q_l(J,P,n;\varphi,s)$ defines an element f of $(\mathcal{M} \otimes_{\mathcal{O}} \mathcal{A}_l(\mathbf{c}_{J,n}))(\Omega_1)$ for some open neighborhood $\Omega_1 \subset (\varphi_0 + V) \times \mathbb{C}$ of (φ_0, s_0). It is holomorphic at w_0 if and only if $\tilde{F}_{Q,m} f$ is holomorphic at w_0 for all $Q \in \mathcal{P}$, $m \in \mathbf{c}_{J,n}(Q)$.*

The same statement holds if one replaces $Q_l(J,P,n)$ by $R(P,n)$, as defined in Proposition 10.2.4.

10.2.15 *Eisenstein families.* Let J be a cell of continuation. We put

$$X^\infty(J) = \left\{ P \in X^\infty : \varphi(\pi_P) = 0 \text{ for } \varphi \in J^0 \right\}.$$

In the modular case $X^\infty(J(0)) = X^\infty$, and $X^\infty(J) = \emptyset$ if J is a minimal cell, i.e., $J = (-12, 0)$ or $(0, 12)$ in the notations of 10.1.4.

For $P \in X^\infty(J)$ we use the notation $E_l(J,P) = Q_l(J,P,n_0)$. We call $E_l(J,P)$ an *Eisenstein family.*

In the modular case, the $E_l(J(0),P)$ are the sole Eisenstein families.

10.2.16 *How large can $\Omega = U \times \mathbb{C}$ be?* The construction of U in 10.2.9 uses the sets $V_0(l_0, \chi, \mathbf{c}_{J,n})$ introduced in Lemma 9.1.6. Here $\mathbf{c}_{J,n}$ is fixed; we cannot enlarge $V_0(\chi, \mathbf{c}_{J,n})$ by changing the growth condition. In general, it is difficult to say how far U extends in imaginary direction in $\mathcal{V}$. See 15.5.2 for a special case.

Question. Can one always take $U = \{ \varphi \in \mathcal{V} : \operatorname{Re} \varphi \in J \}$?

10.3 Relations and functional equations

The uniqueness in Theorem 10.2.1 implies functional equations and relations between Poincaré families on intersecting cells of continuation. In this section we consider some examples. To formulate these relations we need the Fourier coefficients.

10.3.1 *Fourier coefficients.* Let J, P, n and $\mathbf{c}$ be as in Theorem 10.2.1. Proposition 7.2.9 implies that for each pair (Q,m), $Q \in \mathcal{P}$, $m \in \mathcal{C}_Q(\mathcal{V})$, there is a meromorphic function $D_l(Q,m;J,P,n) = D_l(Q,m;J,P,n,\chi_1)$ on Ω such that

$$\tilde{F}_{Q,m}Q_l(J,P,n) = \delta_{P,Q}\delta_{n,m}w_P\mu_l(P,n) + D_l(Q,m;J,P,n)\eta_l(J,Q,m),$$

with η_l as defined in 10.1.8. The restriction

$$\dot{D}_l(Q,m;J,P,n;\varphi_0) : s \mapsto D_l(Q,m;J,P,n,\chi_1;\varphi_0,s)$$

exists as a meromorphic function on $\mathbb{C}$ for each $\varphi_0 \in J^0$. It is holomorphic on $\mathbb{C}_{1/2}$, with the coefficient $c_{l(\varphi)}(Q,m(\varphi_0);P,n(\varphi_0),\chi_1 \cdot \mathbf{exp}\,(\varphi_0),s)$ of Proposition 5.2.1 as its value at s. In the case $Q \in X^\infty$, $m = n_0$, $\varphi(\pi_P) = 0$ for $\varphi \in J^0$, one might fear that the singularities of $\tilde{\mu}_l(Q,m)$ at (φ_0,s) with $s \in \frac{1}{2}\mathbb{N}$, destroy the holomorphy of the restriction. But the restriction of $\tilde{\mu}_l(Q,m)$ to $\{\varphi_0\} \times \mathbb{C}$ has no singularities at half-integral points in $\mathbb{C}$.

For *Eisenstein families* we put $C_l(Q,m;J,P) = D_l(Q,m;J,P,n_0)$ if $P \in X^\infty(J)$, $Q \in \mathcal{P}$ and $m \in \mathcal{C}_Q$. If $Q \in X^\infty(J)$ as well, and $m = n_0$, then we put $C_l(J;Q,P) = C_l(Q,n_0;J,P)$.

10.3.2 *Relations.* Let J and J_1 be two cells of continuation with non-empty intersection. We can express $Q_l(J,P,n)$ in the $Q_l(J_1,Q,m)$ on the intersections of their domains. The following principle works for any $f \in \mathcal{A}_l(\mathbf{c};\Omega_1)$ with $\mathbf{c} \supset \mathbf{c}_J$ and Ω_1 contained in the intersection of the domains of the $Q_l(J,P,n)$ with $P \in \mathcal{P}$, $n \in \mathbf{c}(P)$: *Take the Fourier coefficients $\tilde{F}_{P,n}f$ for those (P,n); let $\psi_{P,n}$ be the first coordinate of $\tilde{F}_{P,n}f$ with respect to the basis $\mu_l(P,n)$, $\eta_l(J,P,n)$. Then $f = \sum_{(P,n)} \psi_{P,n} w_P^{-1} Q_l(J,P,n)$.* Indeed, the Fourier terms of the difference is a multiple of $\eta_l(Q,m)$, so if the difference would be non-zero it would give a contradiction to the uniqueness in Theorem 10.2.1.

The null cell $J(0)$ contains all other cells of continuation. So we restrict our attention to a comparison of Poincaré families for $J(0)$ with those for $J = J(\varphi_1)$. We work on an open set containing $J(\varphi_1) \times \mathbb{C}$ such that all functions under consideration are defined on it.

The difference in the choice of bases in 10.1.8 concerns only the pairs (Q,n_0) with $Q \in X^\infty$ for which $\varphi_1(\pi_Q) \neq 0$, i.e., $Q \in X^\infty \smallsetminus X^\infty(J)$. Let $P \in \mathcal{P}$, $n \in \mathcal{C}_P$. We use the relation $\omega_l(Q,n_0) = \sum_\pm v_l(Q,n_0,\pm s)\mu_l(Q,n_0,\pm s)$ for $Q \in X^\infty$; see 4.2.8. We define $v_l(Q,n_0) : (\varphi,s) \mapsto v_{l(\varphi)}(Q,n_0(\varphi),s)$, and $\tilde{v}_l(Q,n_0;\varphi,s) = v_l(Q,n_0;\varphi,-s)$. Computation of the Fourier terms shows that

$$\begin{aligned} Q_l(J,P,n) &= Q_l(J(0),P,n) \\ &\quad + \sum_{Q \in X^\infty \smallsetminus X^\infty(J)} v_l(Q,n_0) D_l(Q,n_0;J,P,n) E_l(J(0),Q), \\ Q_l(J(0),P,n) &= Q_l(J,P,n) \\ &\quad - \sum_{Q \in X^\infty \smallsetminus X^\infty(J)} \frac{v_l(Q,n_0)}{\tilde{v}_l(Q,n_0)} D_l(Q,n_0;J(0),P,n) Q_l(J,Q,n_0). \end{aligned}$$

In this form the relations are not fully satisfactory. For instance, $Q_l(J, P, n)$ is expressed in terms of Poincaré families on $J(0)$ with its own Fourier coefficients as coefficients in the right hand side. These coefficients might be expressed in Fourier coefficients of the families on $J(0)$. In 10.3.4, we shall carry out these computations for the case $P \in X^\infty$, $n = n_0$. See Equation (13.4) on p. 251 for an example.

10.3.3 *Scattering matrix.* We consider the Eisenstein families for a cell of continuation J. Taking them together for all $P \in X^\infty(J)$ gives a vector $\mathbf{E}_l(J) = (E_l(J,P))_{P \in X^\infty(J)}$; we consider this as a row vector. We combine the corresponding Fourier coefficients into a column vector $\mathbf{F}(J) = \left(\tilde{F}_{Q,n_0}\right)_{Q \in X^\infty(J)}$. Then $\mathbf{F}(J)\mathbf{E}_l(J)$ is the matrix with $\tilde{F}_{Q,n_0} E_l(J,P)$ in the row indexed by Q and the column indexed by P. We obtain

$$\mathbf{F}(J)\mathbf{E}_l(J) = \mu_l(J) + \tilde{\mu}_l(J)\mathbf{C}_l(J).$$

Here $\mu_l(J)$, respectively $\tilde{\mu}_l(J)$, are diagonal matrices with $\mu_l(P, n_0)$, respectively $\tilde{\mu}_l(P, n_0)$, at position P, and Id is the identity matrix. $\mathbf{C}_l(J)$ is called the *scattering matrix.* It has $C_l(J; Q, P)$ at position (Q, P). Its restriction to $\{\varphi_0\} \times \mathbb{C}$ for $\varphi_0 \subset J^0(0)$ is meromorphic in s. That restriction is the usual scattering matrix in the spectral theory of automorphic forms; see, e.g., [29], Ch. XIV, §14.

10.3.4 *Relations, special case.* Let $J \neq J(0)$ be a cell of continuation. We want to express the $Q_l(J, P, n_0)$ with $P \in X^\infty$ in terms of Eisenstein families on $J(0)$.

Consider the row vector $\mathbf{P}_l(J) = (Q_l(J, P, n_0))_{P \in X^\infty}$. As $J \subset J(0)$, there is an open set on which $\mathbf{P}_l(J)$ and $\mathbf{E}_l(J(0))$ are both defined. By the uniqueness in Theorem 10.2.1, there is a unique invertible matrix $\mathbf{X}$ with meromorphic coefficients such that $\mathbf{P}_l(J) = \mathbf{E}_l(J(0))\mathbf{X}$.

We write the cuspidal Fourier terms of order zero in vector notation. This gives $\mathbf{F}(J(0))\mathbf{P}_l(J) = \mu_l(J(0)) + \eta_l(J)\mathbf{D}_l(J)$, with $\mathbf{D}_l(J)$ the matrix of the coefficients $D_l(Q, n_0; J, P, n_0)$, and $\eta_l(J)$ a diagonal matrix with $\tilde{\mu}_l(P, n_0)$ at position $P \in X^\infty(J)$ and $\omega_l(J, P, n_0)$ at position $P \in X^\infty \setminus X^\infty(J)$.

For $q = 0$ or 1, write $\mathbf{v}_l^q(J)$ for the diagonal matrix with q at position $P \in X^\infty(J)$ and $v_l(P, n_0)$ at position $P \in X^\infty \setminus X^\infty(J)$. Put $\tilde{\mathbf{v}}_l^q(J; \varphi, s) = \mathbf{v}_l^q(J; \varphi, -s)$. Then $\eta_l(J) = \mu_l(J(0))\mathbf{v}_l^0(J) + \tilde{\mu}_l(J(0))\tilde{\mathbf{v}}_l^1(J)$. This leads to the following equations:

$$\begin{aligned} \mathbf{X} &= \mathrm{Id} + \mathbf{v}_l^0(J)\mathbf{D}_l(J) \\ \mathbf{C}_l(J(0))\mathbf{X} &= \tilde{\mathbf{v}}_l^1(J)\mathbf{D}_l(J). \end{aligned}$$

Hence $\mathbf{X} = \left(\mathrm{Id} - \mathbf{v}_l^0(J)\tilde{\mathbf{v}}_l^1(J)^{-1}\mathbf{C}_l(J(0))\right)^{-1}$, and

$$\begin{aligned} \mathbf{P}_l(J) &= \mathbf{E}_l(J(0)) \left(\mathrm{Id} - \mathbf{v}_l^0(J)\tilde{\mathbf{v}}_l^1(J)^{-1}\mathbf{C}_l(J(0))\right)^{-1} \\ \mathbf{D}_l(J) &= \tilde{\mathbf{v}}_l^1(J)^{-1}\mathbf{C}_l(J(0)) \left(\mathrm{Id} - \mathbf{v}_l^0(J)\tilde{\mathbf{v}}_l^1(J)^{-1}\mathbf{C}_l(J(0))\right)^{-1}. \end{aligned}$$

10.3.5 *Functional equation for Eisenstein families.* The map $\sigma : (\varphi, s) \mapsto (\varphi, -s)$ is an involutive automorphism of the parameter space $W = \mathcal{V} \times \mathbb{C}$. Composition with σ gives a map $\sigma^\sharp : \mathcal{A}_l(\mathbf{c}; U) \longrightarrow \mathcal{A}_l(\mathbf{c}; \sigma^{-1}U)$ for each open set $U \subset W$. We have already used the convention to denote $\sigma^\sharp f$ by $\tilde{f}$.

Let J be a cell of continuation for which $X^\infty(J) \neq \emptyset$ (if $X^\infty(J)$ is empty there are no Eisenstein families for J). We work on a σ-invariant neighborhood Ω of $J \times \mathbb{C}$ contained in the domain of all $E_l(J, P)$ with $P \in X^\infty(J)$. Then $\tilde{\mathbf{E}}_l(J) = \sigma^\sharp \mathbf{E}_l(J)$ is a row vector of elements of $\mathcal{A}_l(\mathbf{c}_J; \Omega)$ and $\mathbf{F}(J)\tilde{\mathbf{E}}_l(J) = \tilde{\mu}_l(J) + \mu_l(J)\tilde{\mathbf{C}}_l(J)$. On the other hand, $\mathbf{E}_l(J)\tilde{\mathbf{C}}_l(J)$ is a row vector with $\mathbf{F}(J)\left(\mathbf{E}_l(J)\tilde{\mathbf{C}}_l(J)\right) = \mu_l(J)\tilde{\mathbf{C}}_l(J) + \tilde{\mu}_l(J)\mathbf{C}_l(J)\tilde{\mathbf{C}}_l(J)$. The uniqueness in Theorem 10.2.1 implies

$$\tilde{\mathbf{E}}_l(J) = \mathbf{E}_l(J)\tilde{\mathbf{C}}_l(J), \qquad \mathbf{C}_l(J)\tilde{\mathbf{C}}_l(J) = \mathrm{Id}.$$

These are the *functional equations for Eisenstein families.* If we fix our attention on one coordinate, we obtain

$$E_l(J, P; \varphi, -s) = \sum_{Q \in X^\infty(J)} C_l(J; Q, P; \varphi, -s) E_l(J, Q; \varphi, s).$$

For $Q \in \mathcal{P}$, $m \in \mathcal{C}_Q$ such that $Q \notin X^\infty(J)$ or $m \neq n_0$ we define the row vector $\mathbf{C}_l(J, Q, m) = \left(C_l(Q, m; J, P)\right)_{P \in X^\infty(J)}$. The functional equation of these Fourier coefficients takes the form $\tilde{\mathbf{C}}_l(J, Q, m) = \mathbf{C}_l(J, Q, m)\tilde{\mathbf{C}}_l(J)$.

10.3.6 *Functional equation for other families.* Let $P \in \mathcal{P}$, $n \in \mathcal{C}_P$, and either $P \notin X^\infty(J)$ or $n \neq n_0$. As $\tilde{\omega}_l(P, n) = \omega_l(P, n) = v_l(P, n)\mu_l(P, n) + \tilde{v}_l(P, n)\tilde{\mu}_l(P, n)$, we obtain from the uniqueness in Theorem 10.2.1:

$$\begin{aligned}
&v_l(P, n)Q_l(J, P, n) + \tilde{v}_l(P, n)\tilde{Q}_l(J, P, n) \\
&\quad = \tilde{v}_l(P, n) \sum_{Q \in X^\infty(J)} \tilde{D}_l(Q, n_0; J, P, n) E_l(J, Q) \\
&v_l(P, n)D_l(P, n; J, P, n) + \tilde{v}_l(P, n)\tilde{D}_l(P, n; J, P, n) + w_P \\
&\quad = \tilde{v}_l(P, n) \sum_{Q \in X^\infty(J)} \tilde{D}_l(Q, n_0; J, P, n) C_l(P, n; J, Q) \\
&v_l(P, n)D_l(Q, n_0; J, P, n) \\
&\quad = \tilde{v}_l(P, n) \sum_{R \in X^\infty(J)} \tilde{D}_l(R, n_0; J, P, n) C_l(J; Q, R) \quad \text{for } Q \in X^\infty(J).
\end{aligned}$$

To write the second and third line in vector notation, we define a column vector $\mathbf{D}_l(J, P, n) = \left(D_l(Q, n_0; J, P, n)\right)_{Q \in X^\infty(J)}$. We obtain

$$\begin{aligned}
&v_l(P, n)D_l(P, n; J, P, n) + \tilde{v}_l(P, n)\tilde{D}_l(P, n; J, P, n) + w_P \\
&\quad = \tilde{v}_l(P, n)\mathbf{C}_l(J, P, n)\tilde{\mathbf{D}}_l(J, P, n) \\
&v_l(P, n)\mathbf{D}_l(J, P, n) = \tilde{v}_l(P, n)\mathbf{C}_l(J)\tilde{\mathbf{D}}_l(J, P, n),
\end{aligned}$$

and if $R \in \mathcal{P}$, $m \in \mathcal{C}_R$, $(R, m) \neq (P, n)$, such that $R \notin X^\infty(J)$ or $m \neq n_0$:

$$\begin{aligned}
&v_l(P,n)D_l(R,m;J,P,n) + \tilde{v}_l(P,n)\tilde{D}_l(R,m;J,P,n)\\
&\quad = \tilde{v}_l(P,n)\mathbf{C}_l(J,R,m)\tilde{\mathbf{D}}_l(J,P,n).
\end{aligned}$$

10.3.7 *Examples.* See 13.2.5, and Sections 13.4 and 14.5 for examples of the Fourier coefficients in special cases.

Chapter 11
Poincaré families along vertical lines

Theorem 10.2.1 gives the meromorphic continuation of Poincaré series in (φ, s). One may ask where the resulting Poincaré families are holomorphic. Propositions 10.2.12 and 10.2.14 give some results in this direction. The final Chapters 11 and 12 of Part I of this book discuss the singularities of Poincaré families at points (φ, s) with $\varphi \in \mathcal{V}_r$ and $\operatorname{Re} s \geq 0$. At which of these points are the Poincaré families not holomorphic? How bad are the singularities? It turns out that often there is a relation with the presence of eigenfunctions of the selfadjoint extension of the Casimir operator. The restriction to $\varphi \in \mathcal{V}_r$ is essential for these results: in general it is difficult even to define the extension $A(\chi_0 \cdot \mathbf{exp}(\varphi), l_0)$ for other φ. The restriction to points with $\operatorname{Re} s \geq 0$ is less essential; the functional equation can be used to obtain assertions for other s.

In this chapter we consider the singularities of the one-dimensional Poincaré families $s \mapsto Q_l(J, P, n; \varphi_0, s)$. Most of these results are known, but it is quite some work to find them, e.g., in [21]. The study of the singularities of families depending on more than one parameter is based on the one-dimensional results. Hence it is worthwhile to discuss the one-dimensional case in detail as a preparation for the next chapter.

The results for Eisenstein families differ from those for other Poincaré families. In Section 11.1 we give results that are common to both cases. In the other sections we specialize: first to Eisenstein families, and then to other Poincaré families.

11.1 General results

In Theorem 10.2.1 we have characterized the Poincaré families by the coordinates of their Fourier terms with respect to the basis $\mu_l(Q, m)$, $\eta_l(Q, m)$, given in 10.1.8. We shall see that singularities are mainly due to two causes:

i) The presence of $f \in H(\chi_0, l_0) \cap A_{l_0}(\chi_0, \mathbf{c}_{J,n}, s_0)$ with $\tilde{F}_{P,n} f \neq 0$.

ii) The fact that the sections μ_l and η_l form an unsuitable basis at the point (φ_0, s_0).

I consider the first cause as the essential one. Here we avoid the latter one by not working with a Poincaré family, but with a family $R(P, n)$ as introduced in Proposition 10.2.4.

11.1.1 *Conventions.* We keep fixed $\chi_1 \in \mathcal{X}_0$, $\varphi_0 \in J(0)$, $l \in \mathbf{wt}$. We put $J = J(\varphi_0)$, $\chi_0 = \chi_1 \cdot \mathbf{exp}\,(\varphi_0)$, $l_0 = l(\chi_1, \varphi_0)$. We consider $P \in \mathcal{P}$, $n \in \mathcal{C}_P$, and $s_0 \in \mathbb{C}$ with $\mathrm{Re}\, s_0 \geq 0$.

We use a dot to indicate the restriction of functions and sheaves to the vertical line given by $s \mapsto (\varphi_0, s)$. So we shall study the Poincaré family $\dot{Q}_l(J, P, n)$, which is an element of $(\dot{\mathcal{M}} \otimes_{\dot{\mathcal{O}}} \dot{\mathcal{A}}_l(\mathbf{c}_{J,n}))(\mathbb{C})$. In this notation φ_0 is not visible.

Let $Q \in \mathcal{P}$. The elements of $\mathcal{C}_Q$ are constant on the vertical line $\{\varphi_0\} \times \mathbb{C}$. In this chapter we identify elements $m \in \mathcal{C}_Q$ and their value $m(\varphi_0)$.

$Q \in X^\infty$, $m = 0$		
$s_0 \neq 0$	$\lambda_{Q,m} = \dot{\mu}_l(Q, m)$	$i(Q, m) = 0$
$s_0 = 0$	$\lambda_{Q,m} = \dot{\lambda}_l(Q, m)$	$i(Q, m) = 1$
$Q \in X^\infty$, $\varepsilon m > 0$, $\varepsilon \in \{1, -1\}$		
$0 \leq s_0 \leq (\varepsilon l_0 - 1)/2$ and $s_0 \equiv (\varepsilon l_0 - 1)/2 \bmod 1$	$\lambda_{Q,m} = \dot{\omega}_l(Q, m)$	$i(Q, m) = 1$
other s_0	$\lambda_{Q,m} = \dot{\mu}_l(Q, m)$	$i(Q, m) = 0$
$Q \in \mathcal{P}_Y$, $p = \varepsilon(m - l_0)/4 \geq 0$, $q = \zeta(m + l_0)/4 \geq 0$, $\varepsilon, \zeta \in \{1, -1\}$		
$0 \leq s_0 \leq q - p - \frac{1}{2}$ and $s_0 \equiv q - p - \frac{1}{2} \bmod 1$	$\lambda_{Q,m} = \dot{\omega}_l(Q, m, \zeta)$	$i(Q, m) = 1$
other s_0	$\lambda_{Q,m} = \dot{\mu}_l(Q, m)$	$i(Q, m) = 0$

Table 11.1 Choice of $\lambda_{Q,m}$.

11.1.2 *Choice of a basis for the Fourier terms.* We shall apply 10.2.3–10.2.8 with $V_r = V = \{0\}$. For $Q \in \mathcal{P}$, $m \in \mathbf{c}_{J,n}(Q)$ we choose $\lambda_{Q,m}$ and $\nu_{Q,m}$ in the space $\dot{\mathcal{W}}_l(Q, m)(C)$ for a neighborhood C of s_0 in $\mathbb{C}$ such that the conditions in 10.2.3 are satisfied. Here we impose the condition of holomorphy. The meromorphy of families of automorphic forms is the main point of interest of Chapter 10. Holomorphy of the basis elements was not necessary.

These conditions leave no choice for ν: necessarily $\nu_{Q,m} = \dot{\eta}_l(Q, m)$. The section $\dot{\eta}_l(Q, m)$ is holomorphic at s_0, even if $s_0 \in \frac{1}{2}\mathbb{N}$, $Q \in X^\infty$ and $m = 0$.

Table 11.1 shows how we can choose $\lambda_{Q,m}$. See 4.2.8, 4.2.9 and 7.6.13–7.6.15 for the definitions of the families of eigenfunctions that we use. With these choices, both $\lambda_{Q,m}$ and $\nu_{Q,m}$ are holomorphic at s_0, and their Wronskian does not vanish at s_0.

There are holomorphic functions f and g on a neighborhood of s_0 such that $\dot\mu_l(Q,m) = f\lambda_{Q,m} + g\dot\eta_l(Q,m)$. We denote by $i(Q,m) \in \mathbb{Z}$ the order of f at s_0, i.e., $\lim_{s\to s_0}(s-s_0)^{-i(Q,m)}f(s)$ exists and is non-zero.

11.1.3 *The family $R(P,n)$.* Proposition 10.2.4 provides us with the family of automorphic forms $R(P,n)$ in the space $\left(\dot{\mathcal{M}} \otimes_{\dot{\mathcal{O}}} \dot{A}_l(\mathbf{c}_{J,n})\right)(C)$. As we have taken $V_r = \{0\}$, we need not put a dot upon $R(P,n)$ in this chapter. With $\dot\mu_l(P,n) = f\lambda_{P,n} + g\dot\eta_l(P,n)$ as above, the uniqueness implies that $\dot Q_l(J,P,n)$ is equal to $f \cdot R(P,n)$.

We can assume that $R(P,n)$ is holomorphic on $C_0 = C \setminus \{s_0\}$. Let $q \in \mathbb{Z}$ be minimal such that $R_q : s \mapsto (s-s_0)^q R(P,n;s)$ is holomorphic at s_0. Then $R_q \in \dot{A}_l(\mathbf{c}_{J,n})(C)$, and $R_q(s_0) \neq 0$. In this section we investigate what values of q can occur, and what we can say concerning $R_q(s_0)$.

11.1.4 Lemma. $q \geq 0$.

If $q = 0$, then $R_q(s_0) = R(P,n;s_0)$ is not regular at P.

If $q > 0$, then $\tilde F_{Q,m} R_q(s_0) \in \mathbb{C} \cdot \dot\eta_l(Q,m)(s_0)$ for all $Q \in \mathcal{P}$, $m \in \mathcal{C}_Q$.

Remark. In 4.1.9 we have defined regularity as square integrability at cusps, and as smoothness at interior points.

Proof. The holomorphy of the basis $\lambda_{Q,m}$, $\dot\eta_l(Q,m)$ implies that the first coordinate $s \mapsto (s-s_0)^q$ of $\tilde F_{P,n} R_q$ is holomorphic at s_0. Hence $q \geq 0$. If the space $W^0_{l_0}(P,n,s_0)$ is non-zero, then it is spanned by $\dot\eta(s_0)$ (see Proposition 4.2.11). So, if $q = 0$, then $\tilde F_{P,n} R(P,n;s_0)$ is not regular (use 4.1.10). The assertion concerning $q > 0$ is clear.

11.1.5 *Resolvent.* We use spectral theory to obtain more information on $R(P,n)$. Let $\mathrm{R}(\lambda) = (A-\lambda)^{-1}$ be the resolvent of the selfadjoint extension $A = A(\chi_0, l_0)$ of the Casimir operator (see 6.5.2). The proof of the next lemma comes from [12], proof of Théorème 3. We have used the same idea in the proof of Proposition 9.4.2.

11.1.6 Lemma. *For all $s \in \mathbb{C}$, $\mathrm{Re}\, s > -\frac12$, there exists $h(s) \in C^\infty(\tilde G \setminus \mathrm{pr}^{-1}\{P\})$ such that*

i) $s \mapsto (\omega - \frac14 + s^2)h(s)$ is L^2-holomorphic on $\{\, s \in \mathbb{C} : \mathrm{Re}\, s > -\frac12 \,\}$ with values in $K(\chi_0, l_0) \subset H(\chi_0, l_0)$.

ii) $R(P,n)$ is holomorphic at s for each $s \in S = \{\, s \in C_0 : \mathrm{Re}\, s > 0,\ \frac14 - s^2 \notin [-\frac14 l_0^2, \frac14) \,\}$ and its value satisfies

$$R(P,n;s) = h(s) - \mathrm{R}\left(\frac14 - s^2\right)\left((\omega - \frac14 + s^2)h(s)\right).$$

Proof. For $Q \in \mathcal{P}$, $m \in \mathbf{c}_{J,n}(Q)$ we put $\theta_{Q,m}(s) = \delta_{P,Q}\delta_{m,n}w_P \cdot \tau_P \cdot \lambda_{P,n}(s)$, with τ_P as in 9.4.4. Define $h(s)$ just as we defined $h(\varphi, s)$ in 9.4.4. As $\lambda_{Q,m}(s)$ determines an eigenfunction of ω with eigenvalue $\frac{1}{4} - s^2$, the function $(\omega - \frac{1}{4} + s^2)h(s)$ is compactly supported in Y. In fact, it vanishes outside the set $\mathrm{pr}^{-1}(U_P(A_P) \setminus U_P(a(P)))$. Part i) follows.

As $s_0 \notin S$, the family $s \mapsto R(P, n; s)$ is pointwise holomorphic on S.

Put $G(s) = \mathrm{R}(\frac{1}{4} - s^2)(\omega - \frac{1}{4} + s^2)h(s)$. The resolvent $\mathrm{R}(\frac{1}{4} - s^2)$ is holomorphic on S. So G is holomorphic on this region. We want to show that $R(P, n) - h = G$ on S. To do this, it suffices to show that $G(s) + R(P, n; s) - h(s) \in \ker(A - \frac{1}{4} + s^2)$ for all $s \in S$. We have chosen $h(s)$ in such a way, that $R(P, n; s) - h(s)$ is smooth on $\tilde{G}$, even if P is an interior point. As $\mathrm{Re}\, s > 0$ on S, the contribution of the multiple of $\dot{\eta}_{l_0}(Q, m_0)$ at the cusp Q is square integrable. So $R(P, n; s) - h(s) - G(s)$ is square integrable for $s \in S$. We show that $R(P, n) - h - G$ is orthogonal to $(\omega - \frac{1}{4} + \bar{s}^2)k$ for all $k \in K(l_0)$ to complete the proof. The definition of G implies

$$\left(G(s), (\omega - \frac{1}{4} + \bar{s}^2)k\right)_{l_0}$$
$$= \mathfrak{s}[G(s), k] = \left((A - \frac{1}{4} + s^2)G(s), k\right)_{l_0} = \left((\omega - \tfrac{1}{4} + s^2)h(s), k\right)_{l_0}.$$

$(\omega - \frac{1}{4} + s^2)(R(p, n; s) - h(s)) = -(\omega - \frac{1}{4} + s^2)h(s)$ outside $\mathrm{pr}^{-1}\mathcal{P}_Y$, and each point of $\mathrm{pr}^{-1}\mathcal{P}_Y$ has a neighborhood on which this quantity vanishes. Hence

$$\left(R(P, n; s) - h(s), \left(\omega - \frac{1}{4} + \bar{s}^2\right)k\right)_{l_0}$$
$$= \left(\left(\omega - \frac{1}{4} + s^2\right)(R(P, n; s) - h(s)), k\right)_{l_0} = -((\omega - \tfrac{1}{4} + s^2)h(s), k)_{l_0}.$$

This gives the desired equality.

11.1.7. Take $b \geq q$, where q is the order of $R(P, n)$ at s_0. Put $F_b(s) = (s - s_0)^b$ $(R(P, n; s) - h(s))$. This extends as a pointwise holomorphic family on a neighborhood of s_0. It determines an L^2-holomorphic family on S, but at s_0 it may leave $H(\chi_0, l_0)$. Nevertheless, we have the following continuity result:

11.1.8 Lemma. *Let $f \in A^2_{l_0}(\chi_0, s_0)$ and $b \geq q$. Define the family F_b by $F_b(s) = (s - s_0)^b \cdot (R(P, n; s) - h(s))$. Then*

$$\lim_{s \to s_0,\, s \in S} \int_{\tilde{\Gamma} \backslash \mathfrak{H}} F_b(s; p(z)) \overline{f(p(z))}\, d\mu(z) = \int_{\tilde{\Gamma} \backslash \mathfrak{H}} F_b(s_0; p(z)) \overline{f(p(z))}\, d\mu(z).$$

The integral on the right exists if interpreted in the sense of Remark 9.4.16.

Proof. Take truncation data as in Section 8.3, for the growth condition $\mathbf{c}_{J,n}$. As $F \cap \overline{\mathrm{pr}}^{-1}({}^aY)$ is relatively compact in $\mathfrak{H}$, we have

$$\lim_{s \to s_0,\, s \in S} \int_{F \cap \overline{\mathrm{pr}}^{-1}({}^aY)} F_b(s) \overline{f}\, d\mu = \int_{F \cap \overline{\mathrm{pr}}^{-1}({}^aY)} F_b(s_0) \overline{f}\, d\mu.$$

Let $Q \in \mathcal{P}$. As F_b is a section of $\mathcal{A}_l(\mathbf{c}_{J,n})$, the truncated family $F_b[\mathbf{c}_{J,n}, Q]$ corresponds (under transformation as in Chapter 8) to an L^2-holomorphic family on $F \cap \mathrm{pr}^{-1} \dot{U}_Q(a(Q))$. Hence

$$\begin{aligned} \lim_{s \to s_0,\, s \in S} & \int_{F \cap \overline{\mathrm{pr}}^{-1} U_Q(a(Q))} F_b(s)[\mathbf{c}_{J,n}, Q]\, \overline{f[\mathbf{c}_{J,n}, Q]}\, d\mu \\ &= \int_{F \cap \overline{\mathrm{pr}}^{-1} U_Q(a(Q))} F_b(s_0)[\mathbf{c}_{J,n}, Q]\, \overline{f[\mathbf{c}_{J,n}, Q]}\, d\mu. \end{aligned}$$

We are left with finitely many terms corresponding to (Q,m), $Q \in \mathcal{P}$, $m \in \mathbf{c}_{J,n}(Q)$. The presence of $h(s)$ cancels the $\lambda_{Q,m}$-part of $\tilde{F}_{Q,m} R(P,n)$. Hence each term is of the form

$$\int_{{}^a\hat{I}_Q} (s - s_0)^b d_m(s) \dot{\eta}_l(Q,m)(s) \cdot \overline{c_m \dot{\eta}(s_0)}\, d\nu_Q,$$

with $d_m(s)$ the coefficient of $\dot{\eta}_l(Q,m)(s)$ in $\tilde{F}_{Q,m} R(P,n;s)$. The coefficient of $\dot{\eta}_l(Q,m)(s_0)$ in $\tilde{F}_{Q,m} F_b(s_0)$ is equal to $\lim_{s \to s_0} (s-s_0)^b d_m(s)$. If $m \neq 0$ or $Q \notin X^\infty(J)$, then the integrand is holomorphic in s, and the limit $\lim_{s \to s_0,\, s \in S}$ can be taken inside the integral, to give $\int_{{}^a\hat{I}_Q} F_{Q,m} F_b(s_0) \cdot \overline{F_{Q,m} f}\, d\nu_Q$. The same holds for $m = 0$ and $Q \in X^\infty(J)$, provided $\mathrm{Re}\, s_0 > 0$. But if $\mathrm{Re}\, s_0 = 0$, then the square integrability of f implies $c_m = 0$.

11.1.9 Lemma. *Let q_0 be the order of $s \mapsto s^2 - s_0^2$ at $s = s_0$. Then*

i) $q \leq q_0$.

ii) If $\tilde{F}_{P,n} f \neq 0$ for some $f \in A^2_{l_0}(\chi_0, s_0)$, then $q = q_0$.

iii) If $q = q_0$ and $R_q(s_0) \in A^2_{l_0}(\chi_0, s_0)$, then $\tilde{F}_{P,n} R_q(s_0) \neq 0$.

iv) If $R_q(s_0) \in A^2_{l_0}(\chi_0, s_0)$, then $R_q(s_0)$ is orthogonal to $S_{l_0}(\chi_0, \mathbf{c}_{J,n}, s_0)$.

Remarks. $R_q : s \mapsto (s - s_0)^q R(P,n;s)$ has been introduced in 11.1.3.

We have $q_0 = 2$ if $s_0 = 0$, $q_0 = 1$ if $s_0^2 \in \mathbb{R}$ and $s_0 \neq 0$, and $q_0 = 0$ if $s_0^2 \notin \mathbb{R}$.

Proof. The function $s \mapsto \int_{\tilde{\Gamma} \backslash \tilde{G} / \tilde{K}} R_q(s;g) \overline{\psi(g)}\, dg$ is holomorphic on a neighborhood of s_0 for each $\psi \in K(\chi_0, l_0)$. We can choose ψ such that this integral is non-zero at $s = s_0$, and such that ψ vanishes on a neighborhood of $\mathrm{pr}^{-1}\{P\}$ in $\tilde{G}$. As

$q \geq 0$, the function $s \mapsto (s-s_0)^q \int_{\tilde\Gamma\backslash\tilde G/\tilde K} h(s;g)\overline{\psi(g)}\,dg$ is also holomorphic at s_0. Lemma 11.1.6 implies, for $s \in S$ (i.e., $s \neq s_0$, $\operatorname{Re} s > 0$ and $\frac14 - s^2 \notin [-\frac14 l_0^2, \infty)$):

$$\begin{aligned}
&\int_{\tilde\Gamma\backslash\tilde G/\tilde K} \Big(R_q(s;g) - (s-s_0)^q h(s)\Big)\overline{\psi(g)}\,dg \\
&\quad = -(s-s_0)^q \left(\mathrm{R}(\frac14 - s^2)(\omega - \frac14 + s^2)h(s), \psi\right)_{l_0}.
\end{aligned}$$

$\|(\omega - \frac14 + s^2)h(s)\|_{l_0} = \mathcal{O}(1)$ as $s \to s_0$, and $\left\|\mathrm{R}\left(\frac14 - s^2\right)\right\| \leq \frac{1}{\operatorname{Im}(\frac14 - s^2)}$ for $s \in S$, see [25], Ch. V, §3.5, (3.16). This gives

$$\int_{\tilde\Gamma\backslash\tilde G/\tilde K} (R_q(s;g) - (s-s_0)^q h(s;g))\,\overline{\psi(g)}\,dg = \mathcal{O}\left(\frac{(s-s_0)^q}{\operatorname{Im}(\frac14 - s^2)}\right)$$

as $s \to s_0$, $s \in S$. We choose a path in S approaching s_0 on which the values of $(s-s_0)^{q_0}/\operatorname{Im}(\frac14 - s^2)$ stay bounded. If $q > q_0$, then

$$\begin{aligned}
0 &\neq \int_{\tilde\Gamma\backslash\tilde G/\tilde K} R_q(s_0;g)\overline{\psi(g)}\,dg \\
&= \lim_{s \to s_0} \int_{\tilde\Gamma\backslash\tilde G/\tilde K} \Big(R_q(s;g) - (s-s_0)^q h(s;g)\Big)\overline{\psi(g)}\,dg \\
&= \lim_{s \to s_0} \mathcal{O}\left(\frac{(s-s_0)^q}{\operatorname{Im}(\frac14 - s^2)}\right) = 0
\end{aligned}$$

leads to a contradiction. This gives i).

Let $f \in A^2_{l_0}(\chi_0, \mathbf{c}_{J,n}, s_0)$, $f \neq 0$. Hence $s_0 \in i\mathbb{R} \cup (0,\infty)$. For s on the path indicated above:

$$\begin{aligned}
&(-s_0^2 + s^2)\left(R(P,n;s) - h(s), f\right)_{l_0} \\
&= -(-s_0^2 + s^2)\left(\mathrm{R}(\frac14 - s^2)(\omega - \frac14 + s^2)h(s), f\right)_{l_0} \\
&= -(-s_0^2 + s^2)\left((\omega - \frac14 + s^2)h(s), \mathrm{R}(\frac14 - \bar s^2)f\right)_{l_0} \\
&= -(-s_0^2 + s^2)\left((\omega - \frac14 + s^2)h(s), (\bar s^2 - s_0^2)^{-1} f\right)_{l_0} \\
&= -\left((\omega - \frac14 + s^2)h(s), f\right)_{l_0}.
\end{aligned}$$

The square integrability of f implies that $\tilde F_{P,n} f = c\dot\eta_l(P,n)(s_0)$ for some $c \in \mathbb{C}$ (and $c = 0$ if $P \in X^\infty(J)$, $n = 0$ and $\operatorname{Re} s_0 = 0$). We have constructed h in such a way that

$$\begin{aligned}
&\left((\omega - \frac{1}{4} + s^2)h(s), f\right)_{l_0} \\
&= \int_{I_P} (l_{P,n} - \frac{1}{4} + s^2) F_{P,n} h(s) \cdot \overline{c\dot{\eta}_l(P,n)(s_0)}\, d\nu_P \\
&= \begin{cases}
\bar{c} \int_{A_P}^{\infty} \left(-\tau_P''(y)\lambda_{P,n}(s;y) - 2\tau_P'(y)\lambda_{P,n}(s)'(y)\right) \\
\qquad \cdot \overline{\dot{\eta}_l(P,n)(s_0;y)}\, dy & \text{if } P \in X^\infty(J) \\
4\pi\bar{c} \int_0^{A_P} \{-(u^2+u)\tau_P''(u)\lambda_{P,n}(s;u) \\
\qquad - 2(u^2+u)\tau_P'(u)\lambda_{P,n}(s)'(u) \\
\qquad - (2u+1)\tau_P'(u)\lambda_{P,n}(s;u)\}\, \overline{\dot{\eta}_l(P,n)(s_0;u)}\, du \\
 & \text{if } P \in \mathcal{P}_Y.
\end{cases}
\end{aligned}$$

This is holomorphic in s on a neighborhood of s_0. The definitions in Section 4.2 imply $\overline{\dot{\eta}_l(P,n)(s_0)} = \dot{\eta}_l(P,n)(s_0)$ for the present value of s_0. We apply partial integration, and end up with

$$\begin{aligned}
&\left((\omega - \frac{1}{4} + s_0^2)h(s_0), f\right)_{l_0} \\
&= \bar{c} \cdot \mathrm{Wr}(\lambda_{P,n}(s_0), \dot{\eta}_l(P,n)(s_0)) \begin{cases} -\int_{A_P}^{\infty} \tau_P'(y)\, dy & \text{if } P \in X^\infty(J) \\ 4\pi \int_0^{A_P} \tau_P'(u)\, du & \text{if } P \in \mathcal{P}_Y \end{cases} \\
&= -\bar{c} \cdot \mathrm{Wr}(\lambda_{P,n}(s_0), \dot{\eta}_l(P,n)(s_0)) \begin{cases} 1 & \text{if } P \in X^\infty(J) \\ 4\pi & \text{if } P \in \mathcal{P}_Y. \end{cases}
\end{aligned}$$

So the limit $\lim_{s \to s_0} (-s_0^2 + s^2)\, (R(P,n;s) - h(s), f)_{l_0}$ (over the path chosen above) exists. It is non-zero if and only if $\tilde{F}_{P,n} f \neq 0$.

Suppose that $q < q_0$. As $\lim_{s \to s_0} (-s_0^2 + s^2)/(s - s_0)^{-q_0}$ exists and is non-zero, we obtain from Lemma 11.1.8

$$\begin{aligned}
&\lim_{s \to s_0} (-s_0^2 + s^2)\, (R(P,n;s) - h(s), f)_{l_0} \\
&= \lim_{s \to s_0} \frac{-s_0^2 + s^2}{(s - s_0)^{q_0}} \cdot \lim_{s \to s_0} (s - s_0)^{q_0 - q} \lim_{s \to s_0} (R_q(s) - (s - s_0)^q h(s), f)_{l_0} \\
&= 0,
\end{aligned}$$

hence $\tilde{F}_{P,n} f = 0$. This gives ii).

Suppose $R_q(s_0) \in A^2_{l_0}(\chi_0, s_0)$. Then $q > 0$ by Lemma 11.1.4. Application of Lemma 11.1.8 with $b = q$ and $f = R_q(s_0)$ gives

$$\lim_{s \to s_0, s \in S} (s - s_0)^q \int_{\tilde{\Gamma} \backslash \mathfrak{H}} (R(P, n; s) - h(s)) \overline{R_q(s_0)} \, d\mu > 0.$$

Let $q = q_0$. Then

$$\lim_{s \to s_0,\, s \in S} (-s_0^2 + s^2) \int_{\tilde{\Gamma} \backslash \mathfrak{H}} (R(P, n; s) - h(s)) \overline{R_q(s_0)} \, d\mu \neq 0.$$

The computation given above shows that $\tilde{F}_{P,n} R_q(s_0) \neq 0$. This gives iii).

To prove part iv), suppose $R_q(s_0) \in A^2_{l_0}(\chi_0, s_0)$, and take $f \in S_{l_0}(\chi_0, \mathbf{c}_{J,n}, s_0)$. Note that q has to be positive, see Lemma 11.1.4. Lemma 11.1.8 implies

$$(R_q(s_0), f)_{l_0} = \lim_{s \to s_0,\, s \in S} (s - s_0)^q \int_F (R(P, n; s) - h(s)) \overline{f} \, d\mu,$$

with the interpretation of the integral as in Remark 9.4.16. As $R(P, n)$ has been defined as an e_ρ, Proposition 9.4.16 implies that the integral $\int_F R(P, n; s) \overline{f} \, d\mu$ vanishes for all $s \in S$. The integral $\int_F h(s) \overline{f} \, d\mu$ vanishes, as $\tilde{F}_{P,n} f = 0$.

11.1.10 Lemma. *Let F be any family in $\dot{\mathcal{A}}_l(\mathbf{c}_{J,n}; U)$, with U a neighborhood of s_0 in $\mathbb{C}$. If $F(s_0) \in A^2_{l_0}(\chi_0, s_0)$, then $F(s_0)$ is orthogonal to $S_{l_0}(\chi_0, \mathbf{c}_{J,n}, s_0)$.*

Remark. This generalizes part iv) of the previous lemma.
Proof. We follow the ideas in the previous lemmas.

The uniqueness in Theorem 10.2.1 implies that $F = \sum_{Q,m} \beta_{Q,m} R(Q, m)$, with $\beta_{Q,m}$ holomorphic on U, $Q \in \mathcal{P}$ and $m \in \mathbf{c}_{J,n}(Q)$. Define $h(s)$ as in the proof of Lemma 11.1.6, with $\theta_{Q,m}(s) = w_Q \beta_{Q,m}(s) \cdot \tau_Q \cdot \lambda_{Q,m}(s)$. The statements in this lemma hold for the new h, and for F instead of $R(P, n)$. Define $F_b(s) = F(s) - h(s)$, repeat 11.1.7 and the proof of Lemma 11.1.8 for this F_b, with $b = 0$. The last paragraph in the proof of Lemma 11.1.9 gives the desired result.

11.2 Eisenstein families

We apply the results of the previous section in the case $P \in X^\infty$, $n = 0$. This gives well known properties of the meromorphic continuation of the Eisenstein series as a function of the spectral parameter.

We take χ_1, χ_0, φ_0, J, l and l_0 as in 11.1.1. We consider $P \in X^\infty(J)$ and $n = 0$. Hence $\mathbf{c} = \mathbf{c}_J$ in this section.

We investigate the singularities in $\{\, s \in \mathbb{C} : \operatorname{Re} s \geq 0 \,\}$ of the Eisenstein family $\dot{E}_l(J, P) : s \mapsto E_l(J, P, \chi_1; \varphi_0, s)$.

11.2.1 *Maass-Selberg relation.* Take $\mathbf{c} = \mathbf{c}_J$, and apply Theorem 4.6.5 to the automorphic forms

$$f = \dot{E}_l(J, Q, \chi_1; \varphi_0, s), \qquad h = \dot{E}_{-l}(-J, R, \chi_1^{-1}; -\varphi_0, s)$$

with $Q, R \in X^\infty(J)$ and $s \in \mathbb{C}$ such that both functions exist at s. The second dot indicates the restriction to the line $\{-\varphi_0\} \times \mathbb{C}$. We find $\dot{C}_{-l}(-J; Q, R; s) = \dot{C}_l(J; R, Q; s)$, or in matrix notation (see 10.3.3) $\dot{\mathbf{C}}_{-l}(-J) = \dot{\mathbf{C}}_l(J)^t$; the t means the matrix transpose.

11.2.2 *Complex conjugation* interchanges the elements of $\overline{A_l(\chi_1 \cdot \mathbf{exp}\,(\varphi_0), \mathbf{c}_J, s)}$ and $A_{-l}(\chi_1^{-1} \cdot \mathbf{exp}\,(-\varphi_0), \mathbf{c}_{-J}, \bar{s})$. This implies that $h : s \mapsto \overline{\dot{E}_l(J, Q, \chi_1; \varphi_0, \bar{s})}$ is a section of $\mathcal{M} \otimes_{\dot{\mathcal{O}}} \dot{\mathcal{A}}_{-l}(\mathbf{c}_{-J}; \{-\varphi_0\} \times \mathbb{C})$. We inspect the Fourier terms, and apply the uniqueness in Theorem 10.2.1 to conclude that $h(s) = \dot{E}_{-l}(-J, Q, \chi_1^{-1}; -\varphi_0, s)$. Thus we obtain $\dot{\mathbf{C}}_l(J; s)^* = \dot{\mathbf{C}}_l(J; \bar{s})$, where $*$ denotes conjugate transpose. In particular, $\dot{\mathbf{C}}_l(J; s)$ is a *unitary* matrix for all $s \in i\mathbb{R}$ at which it is holomorphic. (Use the functional equation in 10.3.5.)

11.2.3 Proposition. *$\dot{E}_l(J, P)$ is holomorphic at all $s_0 \in i\mathbb{R}$.*

Proof. By Proposition 10.2.14 it suffices to prove the holomorphy of

$$s \mapsto \mathbf{F}(J)\dot{E}_l(J, P; s) = \dot{\mu}_l(J, s)\mathrm{Id} + \dot{\mu}_l(J, -s)\dot{\mathbf{C}}_l(J, s).$$

The restrictions $\dot{\mu}_l(Q, n_0)$ are holomorphic on $\mathbb{C}$, as they are given by $s \mapsto y^{s+1/2}$. (It does not matter here that the unrestricted families may have singularities at (φ_0, s) with $s \in -\frac{1}{2}\mathbb{N}$.) The scattering matrix $\dot{\mathbf{C}}_l(J, s)$ is unitary for all $s \in i\mathbb{R}$ at which it is holomorphic. Hence its matrix elements are bounded on $i\mathbb{R}$. So poles at imaginary points are impossible. (Here it is important to have only one complex parameter.)

11.2.4 Proposition. Eisenstein families. *Let $P \in X^\infty(J)$.*

i) The meromorphic extension $\dot{E}_l(J, P)$ of $s \mapsto P_{l_0}(P, 0, \chi_0, s)$ is holomorphic at all points $s_0 \in \mathbb{C}$ that satisfy $\mathrm{Re}\, s_0 \geq 0$ *and $s_0 \notin \Sigma^e(\chi_0, l_0)$.*

Let s_0 be such a point.

- *a) If $s_0 \neq 0$, then the value $\dot{E}_l(J, P; s_0)$ is not an element of $H(\chi_0, l_0)$.*
- *b) At $s_0 = 0$ the Eisenstein family may have a zero of at most order one. If $\dot{E}_l(J, P; 0) \neq 0$, then it is not in $H(\chi_0, l_0)$; in the other case $\lim_{s \to 0} \frac{1}{s}\dot{E}_l(J, P; s) \notin H(\chi_0, l_0)$.*

ii) There are two possibilities at points $s_0 \in \Sigma^e_c(\chi_0, l_0)$:

- *a) $\dot{E}_l(J, P)$ has a first order pole at s_0. Then $\lim_{s \to s_0}(s - s_0)\dot{E}_l(J, P; s)$ is a smooth square integrable automorphic form, orthogonal to the cusp forms; its Fourier coefficient at P of order zero does not vanish.*

b) $\dot{E}_l(J,P)$ *is holomorphic at* s_0. *In this case* $\dot{E}_l(J,P;s_0) \notin H(\chi_0, l_0)$ *and* $\tilde{F}_{P,0}f = 0$ *for all* $f \in A^2_{l_0}(\chi_0, s_0)$.

Remarks. $\Sigma^e(\chi_0, l_0)$ is the set of s with $\mathrm{Re}\, s \geq 0$, for which there are smooth square integral automorphic forms that are not cusp forms.

In the previous section we distinguished the spaces $H(\chi_0, l_0) \cap A_{l_0}(\chi_0, \mathbf{c}_J, s_0)$ and $A^2_{l_0}(\chi_0, s_0)$. When P is an interior point that distinction is necessary; square integrable automorphic forms are not necessarily smooth. Here both spaces are equal.

Proof. First consider $s_0 = 0$. In this case $\dot{E}_l(J,P;s) = 2sR(P,0;s)$, with $R(P,0)$ as in Proposition 10.2.4 with the basis chosen in the previous section. Take $a \in \mathbb{Z}$ such that $F_1 = \lim_{s\to 0} s^a \dot{E}_l(J,P;s) \neq 0$. The q and R_q in 11.1.2 satisfy $q = a+1$, and $R_q(0) = \frac{1}{2}F_1$. Lemmas 11.1.4 and 11.1.9 imply $-1 \leq a \leq 1$. Proposition 11.2.3 implies $a \leq 0$. If $a = 0$, then $\tilde{F}_{Q,0}\dot{E}_l(J,P;0) \neq 0$ for some $Q \in X^\infty(J)$ (apply Proposition 10.2.14 with $V_r = \{0\}$). Hence $\dot{E}_l(J,P;0)$ cannot be square integrable. If $a = -1$ we conclude $F_1 \notin H(\chi_0, l_0)$ from Lemma 11.1.4.

Consider $\mathrm{Re}\, s_0 \geq 0$, $s_0 \neq 0$. The choice $\lambda_{Q,0} = \mu_l(Q,0)$ for all $Q \in X^\infty(J)$ leads to $\dot{E}_l(J,P) = R(P,0)$. We apply Lemma 11.1.4 and part i) of Lemma 11.1.9. This gives $q = 0$ or 1. If $q = 0$, then $\dot{E}_l(J,P)$ is holomorphic at s_0, and not in $H(\chi_0, l_0)$. If $q = 1$, then $\dot{E}_l(J,P)$ has a singularity at s_0, and F_1, defined as $\lim_{s\to s_0}(s-s_0)\dot{E}_l(J,P;s)$, satisfies $\tilde{F}_{Q,0}F_1 \in \mathbb{C}\cdot\dot{\mu}_{l_0}(Q,0,-s_0)$ for all $Q \in X^\infty$. Proposition 11.2.3 shows that $q = 1$ cannot occur for $\mathrm{Re}\, s_0 = 0$. Hence, if $q = 1$, then $\mathrm{Re}\, s_0 > 0$, and $F_1 \in H(\chi_0, l_0)$. This implies $s_0 > 0$. Let $s_0 > 0$, and suppose $\dot{E}_l(J,P)$ is singular at s_0. Then $q = 1$. Parts iii) and iv) of Lemma 11.1.9 imply that $\tilde{F}_{P,0}F_1 \neq 0$ and that F_1 is orthogonal to $S_{l_0}(\chi_0, \mathbf{c}_J, s_0)$. As $F_1 \neq 0$, this shows that $s_0 \in \Sigma^e(\chi_0, l_0)$.

The statements i) and ii)a) have been proved. Assertion ii)b) follows from part ii) of Lemma 11.1.9.

11.2.5 *Finiteness of* Σ^e. If $s_0 \in \Sigma^e$, then some $\dot{E}_{l_0}(J,P)$ is singular at s_0. As each meromorphic family of automorphic forms on $\mathbb{C}$ has a singular set that is discrete in $\mathbb{C}$, we conclude that Σ^e is a finite set.

11.2.6 Proposition. *Let* $s_0 \in \Sigma^e(\chi_0, l_0)$. *The residues* $\lim_{s\to s_0}(s-s_0)\dot{E}_l(J,P;s)$, *with* $P \in X^\infty(J)$, *span the orthogonal complement of the space of cusp forms* $S_{l_0}(\chi_0, \mathbf{c}_J, s_0)$ *in the finite dimensional space* $A^2_{l_0}(\chi_0, s_0)$.

Proof. Put $m = |\mathbf{c}_J|$. In 9.5.5 we have seen that there are $f_1, \dots f_m \in \dot{\mathcal{A}}_l(\mathbf{c}_J, \Omega_0)$ on a neighborhood Ω_0 of s_0 such that

$$A_{l_0}(\chi_0, \mathbf{c}_J, s_0) = S_{l_0}(\chi_0, \mathbf{c}_J, s_0) \oplus \bigoplus_{j=1}^{m} \mathbb{C}\cdot f_j(s_0).$$

Put $D = \bigoplus_{j=1}^{m} \mathbb{C} \cdot f_j(s_0)$ and $D_2 = D \cap A^2_{l_0}(\chi_0, s_0)$. A suitable $\mathbb{C}$-linear transformation of the f_j gives $D_2 = \bigoplus_{j=1}^{k} \mathbb{C} \cdot f_j(s_0)$ with $1 \leq k \leq m$. In the sense of Remark 9.4.16 the $f_j(s_0)$ are orthogonal to $S_{l_0}(\chi_0, \mathbf{c}_J, s_0)$; for $j \leq k$ this coincides with L^2-orthogonality. So we have to express the $f_j(s_0)$ with $j \leq k$ as linear combinations of the residues at s_0 of some $\dot{E}_l(J, P)$ with $P \in X^\infty(J)$.

For each $P \in X^\infty(J)$ we have $\tilde{F}_{P,0} f_j = a_{P,j} \dot{\mu}_l(Q, n_0) + b_{P,j} \dot{\tilde{\mu}}_l(P, n_0)$. Note that $s_0 > 0$, hence we are working with a basis of $\dot{\mathcal{W}}_l(P, n_0)$. The $a_{P,j}$ and $b_{P,j}$ are holomorphic on Ω. We obtain $f_j = \sum_{P \in X^\infty(J)} a_{P,j} \dot{E}_l(J, P)$ from the Fourier expansions. Let $j \leq k$. For $Q \in X^\infty(J)$,

$$\begin{aligned} &\tilde{F}_{Q,0} f_j(s_0) \\ &\quad = \quad a_{Q,j}(s_0) \dot{\mu}_l(P, n_0, s_0) + \sum_{P \in X^\infty(J)} a_{P,j}(s_0) \dot{C}_l(J; Q, P; s_0) \dot{\mu}_l(Q, n_0, -s_0). \end{aligned}$$

The square integrability implies $a_{Q,j}(s_0) = 0$. So only the $\dot{E}_l(J, P)$ that are singular at s_0 can contribute to $f_j(s_0)$.

11.2.7 *Residue at $\frac{1}{2}$.* For $\chi_0 = 1$, $l_0 = 0$:

$$\lim_{s \to 1/2} (s - \frac{1}{2}) \dot{E}_0(J, P, s) = \operatorname{vol}(\tilde{\Gamma} \backslash \mathfrak{H})^{-1} \cdot \mathbf{1}.$$

Indeed, we know that this limit is a smooth square integrable automorphic form; it has the form $\alpha_P \cdot \mathbf{1}$ with $\alpha_P \in \mathbb{C}$ (see the proof of Proposition 6.7.8). In the proof of Lemma 11.1.9 take $f = \mathbf{1}$, $R(P, 0) = \dot{E}_0(J, P)$, and obtain

$$\begin{aligned} &\lim_{s \to 1/2} (\omega - \frac{1}{4} + s^2) \Big(\dot{E}_0(J, P, s) - h(s), \mathbf{1} \Big)_0 \\ &\quad = \quad \lim_{s \to 1/2} -((\omega - \frac{1}{4} + s^2) h(s), \mathbf{1})_0 \\ &\quad = \quad \bar{c} \operatorname{Wr}(\dot{\mu}_0(P, 0; \frac{1}{2}), \dot{\mu}_0(P, 0; -\frac{1}{2})) \;=\; \bar{c}, \end{aligned}$$

with $\tilde{F}_{P,0} \mathbf{1} = c \dot{\mu}_0(P, 0; -\frac{1}{2})$, hence $c = 1$. This gives $(\alpha_P \cdot \mathbf{1}, \mathbf{1})_0 = 1$, and $\alpha_P = 1 / \operatorname{vol}(\tilde{\Gamma} \backslash \mathfrak{H})$.

If $|X^\infty(J)| > 1$, the difference of the Eisenstein families for two cusps is holomorphic at $\frac{1}{2}$. See 14.4.7 for an example.

11.2.8 *Definition.* For $s \in \mathbb{C}$, and J a cell of continuation,

$$\begin{aligned} &A^*_{l_0}(\chi_0, J, s) \\ &\quad = \quad \Big\{ f \in A_{l_0}(\chi_0, \mathbf{c}_J, s) : \tilde{F}_{Q,0} f \in \mathbb{C} \cdot \dot{\mu}_l(P, n_0; -s) \text{ for all } Q \in X^\infty(J) \Big\}. \end{aligned}$$

Clearly $S_{l_0}(\chi_0, \mathbf{c}_J, s) \subset A^*_{l_0}(\chi_0, J, s) \subset A_{l_0}(\chi_0, \mathbf{c}_J, s)$. The space $A^*_{l_0}(\chi_0, J, s)$ contains the residues of Eisenstein families.

11.2.9 Lemma. *If* $\mathrm{Re}\, s \geq 0$, $s \neq 0$, *then* $A^*_{l_0}(\chi_0, J, s) = A^2_{l_0}(\chi_0, s)$.

The space $A^*_{l_0}(\chi_0, J, 0)$ *is spanned by* $S_{l_0}(\chi_0, \mathbf{c}_J, 0)$ *and the* $\dot{E}_l(J, P; 0)$ *with* $P \in X^\infty(J)$.

Remark. This implies that $A^*_{l_0}(\chi_0, J, -s) = A^*_{l_0}(\chi_0, J, s)$ for $\mathrm{Re}\, s = 0$. Moreover, for $s \neq 0$, $\mathrm{Re}\, s = 0$, this space is equal to the space of cusp forms $S_{l_0}(\chi_0, \mathbf{c}_J, s)$.

The $\dot{E}_l(J, P; 0)$ may vanish. This is the case for the modular group, see 13.2.6.

Proof. Let $\mathrm{Re}\, s > 0$. Then $A^*(\chi_0, J, s) \subset A^2_{l_0}(\chi_0, s)$, as we see from the Fourier expansion. If $A^2_{l_0}(\chi_0, s) \neq S_{l_0}(\chi_0, \mathbf{c}_J, s)$, then $s \in \Sigma^e(\chi_0, l_0)$, and Proposition 11.2.6 implies the converse conclusion.

For $\mathrm{Re}\, s = 0$, $s \neq 0$, the $\dot{E}_l(P, J)$, $P \in X^\infty(J)$, are holomorphic at s, and the values at s induce a basis of $A_{l_0}(\chi_0, \mathbf{c}_J, s)$ mod $S_{l_0}(\chi_0, \mathbf{c}_J, s)$. From the Fourier expansions we see that they induce a basis of $A_{l_0}(\chi_0, \mathbf{c}_J, s)$ mod $A^*_{l_0}(\chi_0, J, s)$ as well.

Let $s = 0$. We start as in the proof of Proposition 11.2.6. Let $m = |\mathbf{c}_J|$. There are $f_1, \ldots, f_m \in \dot{\mathcal{A}}_l(\mathbf{c}_J, \Omega_0)$ on a neighborhood Ω of 0 in the s-plane, such that

$$A_{l_0}(\chi_0, \mathbf{c}_J, 0) = S_{l_0}(\chi_0, \mathbf{c}_J, 0) \oplus \bigoplus_{j=1}^{m} f_j(0)$$

$$A^*_{l_0}(\chi_0, J, 0) = S_{l_0}(\chi_0, \mathbf{c}_J, 0) \oplus \bigoplus_{j=1}^{k} f_j(0)$$

with $0 \leq k \leq m$. For $P \in X^\infty(J)$ we use the basis $\lambda_{P,0}, \dot{\tilde{\mu}}_l(P, n_0)$ of $\dot{\mathcal{W}}_l(P, n_0)$ discussed in 11.1.2. There are holomorphic functions $a_{P,j}, b_{P,j}$ on Ω such that $\tilde{F}_{P,0} f_j = a_{P,j} \lambda_{P,0} + b_{P,j} \dot{\tilde{\mu}}_l(P, n_0)$ for $P \in X^\infty(J)$, $1 \leq j \leq m$. Hence $f_j(s) = \sum_{P \in X^\infty(J)} \alpha_{P,j}(s) R(P, 0; s) = \sum_{P \in X^\infty(J)} \frac{\alpha_{P,j}(s)}{2s} \dot{E}_l(P, 0; s)$. Suppose $1 \leq j \leq k$. Then $\alpha_{P,j}(0) = 0$ for each $P \in X^\infty(J)$. So $f_j(0) = \sum_{P \in X^\infty(J)} \frac{1}{2} \alpha'_{P,j}(0) E_l(P, 0; 0)$.

11.2.10 Lemma. *If* $\mathrm{Re}\, s_0 < 0$, *the space* $A^*_{l_0}(\chi_0, J, s_0)$ *is spanned by the cusp forms in* $S_{l_0}(\chi_0, \mathbf{c}_J, s_0)$ *and the automorphic forms*

$$\lim_{s \to -s_0} \sum_{Q \in X^\infty(J)} \beta_Q(s) \dot{E}_l(J, Q; s),$$

where $(\beta_Q)_{Q \in X^\infty(J)}$ *runs through the vectors of holomorphic functions on a neighborhood of* $-s_0$ *for which the maps*

$$\tau_R : s \mapsto \sum_{Q \in X^\infty(J)} \beta_Q(s) \dot{C}_l(J; Q, R; s) \qquad \text{for } R \in X^\infty(J)$$

are holomorphic at $-s_0$ *with value* 0.

If, for $\mathrm{Re}\, s_0 < 0$, *the space* $A^*_{l_0}(\chi_0, J, s_0)$ mod $S_{l_0}(\chi_0, \mathbf{c}_J, s_0)$ *is non-zero, then the scattering matrix* $\dot{\mathbf{C}}_{l_0}(J)$ *is not holomorphic at* s_0.

Remark. This is the only result we state concerning the behavior in the left half plane $\mathrm{Re}\, s < 0$. We need it in the next chapter.

Proof. The condition on the τ_R implies the existence of the limit, see Proposition 10.2.14. As $\dot{\eta}_l(Q, 0; s_0) = \dot{\mu}_l(Q, 0, -s_0)$, the limit is an element of $A^*_{l_0}(\chi_0, J, s_0)$.

We take $m = |\mathbf{c}_J|$ and families $f_1, \ldots, f_m$ on a neighborhood Ω of $-s_0$ such that $f_1(-s_0), \ldots, f_m(-s_0)$ induce a basis of $A_{l_0}(\chi_0, \mathbf{c}_J, -s_0) \bmod S_{l_0}(\chi_0, \mathbf{c}_J, -s_0)$. There are holomorphic functions $\alpha_{Q,j}$ on Ω such that $f_j = \sum_{Q \in X^\infty(J)} \alpha_{Q,j} \dot{E}_l(J, Q)$ as an identity of meromorphic families on Ω. Up to a cusp form, each element of $A^*_{l_0}(\chi_0, J, s_0)$ is the value at $-s_0$ of a $\mathbb{C}$-linear combination $f = \sum_{j=1}^m c_j f_j$. The family f is holomorphic at $-s_0$. We conclude from Proposition 10.2.14 that

$$\tau_R : s \mapsto \sum_{j=1}^m c_j \sum_{Q \in X^\infty(J)} \alpha_{Q,j}(s) \dot{C}_l(J; R, Q; s)$$

is holomorphic at $-s_0$ for each $R \in X^\infty(J)$. From $\tilde{F}_{R,0} f(-s_0) \in \mathbb{C} \cdot \dot{\mu}_l(R, 0, -s_0)$ we obtain $\tau_R(-s_0) = 0$. Take $\beta_Q = \sum_{j=1}^m c_j \alpha_{Q,j}$.

For each $P \in X^\infty(J)$ the functional equation, see 10.3.5, gives

$$\begin{aligned} \beta_P(-s) &= \sum_{R \in X^\infty(J)} \dot{C}_l(J, P, R; s) \sum_{Q \in X^\infty(J)} \beta_Q(-s) \dot{C}_l(J, R, Q; -s) \\ &= \sum_{R \in X^\infty(J)} \dot{C}_l(J, P, R; s) \tau_R(-s). \end{aligned}$$

If $\dot{\mathbf{C}}_l(J)$ is holomorphic at s_0, then $\beta_P(-s_0) = 0$ for each $P \in X^\infty(J)$. This implies that the automorphic form $f(-s_0) = \lim_{s \to -s_0} \beta_Q(s) \dot{E}_l(J, Q; s)$ is an element of $A^*_{l_0}(\chi_0, J, -s_0)$. As this space intersects $A^*_{l_0}(\chi_0, J, s_0)$ in $S_{l_0}(\chi_0, \mathbf{c}_J, s_0)$, we see that $f(-s_0)$ does not contribute to $A^*_{l_0}(\chi_0, J, s_0) \bmod S_{l_0}(\chi_0, \mathbf{c}_J, s_0)$.

11.3 Other Poincaré families

The main result in this section is Proposition 11.3.9. It describes the singularities of those Poincaré families that are not Eisenstein families. Here we have to separate more cases than we did for the Eisenstein families in Proposition 11.2.4. The main difference stems from the fact that general Poincaré families may have singularities at points of $i\mathbb{R}$ as well as at points in $(0, \frac{1}{2}]$. This can be expected from Lemma 11.1.9. That result suggests that singularities at s_0 (with $\mathrm{Re}\, s_0 \geq 0$) are due to the presence of elements f of $A^2_{l_0}(\chi_0, s_0)$ with non-vanishing Fourier term $\tilde{F}_{P,n} f$. In the case of an Eisenstein family, $P \in X^\infty(J)$, $n = 0$, this cannot happen for $\mathrm{Re}\, s_0 = 0$.

11.3.1 *Notations.* Take χ_1, φ_0, l, J, χ_0 and l_0 as in 11.1.1. In this section we consider $\dot{Q}_l(J, P, n) : s \mapsto Q_l(J, P, n, \chi_1; \varphi_0, s)$ for $P \in \mathcal{P}$, $n \in \mathcal{C}_P$, with either

$P \notin X^\infty$, or $n(\chi_0) \neq 0$. This is the meromorphic continuation in s of the Poincaré series $P_{l_0}(P, n, \chi_0, s)$.

The growth conditions $\mathbf{c}_{J,n}$ and $\mathbf{c}_J$ are different in this section (see 10.1.7).

11.3.2 *Poincaré element.* The points where $\dot{Q}_l(J, P, n)$ may have a singularity will turn out to be elements of $\Sigma_c(\chi_0, l_0) \cup \{0, |\tau(\chi_0)|\}$. The main criterion determining the order of $\dot{Q}_l(J, P, n)$ at s_0 will be

$$\tilde{F}_{P,n} f \neq 0 \text{ for some } f \in A^2_{l_0}(\chi_0, s_0).$$

Let us define the *Poincaré element* $\mathbf{p}_{l_0}(P, n; s_0)$ of order (P, n) in $A^2_{l_0}(\chi_0, s_0)$. It is the element of $A^2_{l_0}(\chi_0, s_0)$ such that $\tilde{F}_{P,n} f = \big(f, \mathbf{p}_{l_0}(P, n; s_0)\big)_{l_0} \omega_{l_0}(P, n, s_0)$ for all $f \in A^2_{l_0}(\chi_0, s_0)$. As the linear form $f \mapsto \tilde{F}_{P,n} f / \omega_{l_0}(P, n, s_0)$ is continuous on this finite dimensional Hilbert space, such an element exists and is unique. The criterion stated above can be formulated as $\mathbf{p}_{l_0}(P, n; s_0) \neq 0$.

At $s_0 = 0$ we also need to consider the following element of $A_{l_0}(\chi_0, \mathbf{c}_J, s_0)$:

$$E_{P,n} = \sum_{Q \in X^\infty(J)} \overline{\dot{C}_l(P, n; J, Q; 0)}\, \dot{E}_l(J, Q; 0).$$

Suppose that $E_{P,n} \neq 0$. As $E_{P,n}$ is the value at $s = 0$ of a family $\dot{e}_\rho$, Proposition 9.4.16 shows that $\tilde{F}_{Q,0} E_{P,n} \neq 0$ for some $Q \in X^\infty(J)$. The space $W_{l_0}(Q, 0, 0)$ contains no regular elements other than 0. So we conclude that $E_{P,n} \notin H(\chi_0, l_0)$ if it is non-zero. A computation shows that $E_{P,n} \neq 0$ implies $\tilde{F}_{P,n} E_{P,n} \neq 0$.

The role played by the Poincaré element $\mathbf{p}_{l_0}(P, n; s_0)$ in the case $s_0 \neq 0$ is partly taken over by $E_{P,n}$ when $s_0 = 0$.

11.3.3 *Consequences of Section 11.1.* Let $s_0 \in \mathbb{C}$, $\operatorname{Re} s_0 \geq 0$.

We consider the family $R(P, n)$ of Section 11.1. We suppose that $\lambda_{P,n}$ and $\lambda_{Q,0}$ for $Q \in X^\infty(J)$ have been chosen as indicated in 11.1.2. We take q, q_0 and R_q as indicated in 11.1.3 and Lemma 11.1.9. From the Lemmas 11.1.4 and 11.1.9 we obtain:

i) $0 \leq q \leq q_0$.

ii) If $\mathbf{p}_{l_0}(P, n; s_0) \neq 0$, then $q = q_0$.

iii) If $R_q(s_0) \in A^2_{l_0}(\chi_0, s_0)$, then $\mathbf{p}_{l_0}(P, n; s_0) \neq 0$ and $R_q(s_0)$ is a multiple of $\mathbf{p}_{l_0}(P, n; s_0)$.

11.3.4 *Spaces of automorphic forms.* In this section we use many spaces of automorphic forms, see Table 11.2. The growth conditions $\mathbf{c}_{J,n}$ and $\mathbf{c}_J$ satisfy $\mathbf{c}_{J,n} \supset \mathbf{c}_J$, and $|\mathbf{c}_{J,n}| = |\mathbf{c}_J| + 1$. This explains the top and bottom inclusions in the table. The inclusions in between are determined by the Fourier terms of order $(Q, 0)$, with $Q \in X^\infty(J)$.

$s_0 = 0$	$\mathrm{Re}\, s_0 = 0,\ s_0 \neq 0$	$\mathrm{Re}\, s_0 > 0$
$A_{l_0}(\chi_0, \mathbf{c}_{J,n}, 0)$	$A_{l_0}(\chi_0, \mathbf{c}_{J,n}, s_0)$	$A_{l_0}(\chi_0, \mathbf{c}_{J,n}, s_0)$
$\cup$	$\cup$	$\cup$
$A_{l_0}(\chi_0, \mathbf{c}_J, 0)$	$A_{l_0}(\chi_0, \mathbf{c}_J, s_0)$	$A_{l_0}(\chi_0, \mathbf{c}_J, s_0)$
$\cup$	$\cup$	$\cup$
$A^*_{l_0}(\chi_0, J, 0)$	$A^*_{l_0}(\chi_0, J, s_0)$	$A^*_{l_0}(\chi_0, J, s_0)$
$\cup$	$\|$	$\|$
$A^2_{l_0}(\chi_0, 0)$	$A^2_{l_0}(\chi_0, s_0)$	$A^2_{l_0}(\chi_0, s_0)$
$\|$	$\|$	$\cup$
$S_{l_0}(\chi_0, \mathbf{c}_J, 0)$	$S_{l_0}(\chi_0, \mathbf{c}_J, s_0)$	$S_{l_0}(\chi_0, \mathbf{c}_J, s_0)$
$\cup$	$\cup$	$\cup$
$S_{l_0}(\chi_0, \mathbf{c}_{J,n}, 0)$	$S_{l_0}(\chi_0, \mathbf{c}_{J,n}, s_0)$	$S_{l_0}(\chi_0, \mathbf{c}_{J,n}, 0)$

Table 11.2 Inclusion and equality for some spaces of automorphic forms.

The Poincaré element $\mathbf{p}_{l_0}(P, n; s_0)$ is an element of $A^2_{l_0}(\chi_0, s_0)$. It is orthogonal to $S_{l_0}(\chi_0, \mathbf{c}_{J,n}, s_0)$. The automorphic form $R_q(s_0)$ is always an element of the largest space $A_{l_0}(\chi_0, \mathbf{c}_{J,n}, s_0)$. We investigate under what conditions it is an element of the smaller spaces in the table. $\tilde{F}_{P,n} R_q(s_0)$ is regular if and only if $R_q(s_0)$ is an element of the second space, $A_{l_0}(\chi_0, \mathbf{c}_J, s_0)$.

11.3.5 Lemma.

i) Let $R_q(s_0) \in A_{l_0}(\chi_0, \mathbf{c}_J, s_0)$, but $R_q(s_0) \notin A^2_{l_0}(\chi_0, s_0)$. Then $s_0 = 0$, $q = 1$, and $R_q(0)$ is a multiple of $E_{P,n}$.

ii) Let $s_0 = 0$. Then $\tilde{F}_{Q,0} R_q(s) = c_Q(s) \dot{\mu}_l(Q, 0, -s)$ for all $Q \in X^\infty(J)$, with

$$c_Q(s) = \frac{1}{4} u_P w_P s^{q-1} \overline{\dot{C}_l(P, n; J, Q, \bar{s})} \mathrm{Wr}(\lambda_{P,n}(s), \dot{\omega}_l(P, n, s)).$$

Proof. Let $R \in X^\infty(J)$. There is a holomorphic function c_R on a neighborhood of s_0 such that $\tilde{F}_{R,0} R_q(s) = c_R(s) \dot{\mu}_l(R, 0; -s)$. We apply the Maass-Selberg relation (Theorem 4.6.5) to $R_q(s)$ and $\dot{E}_{-l}(R, -J, s)$. The non-zero contributions come from the following Fourier terms:

$$\tilde{F}_{P,n} R_q(s) \quad = \quad (s - s_0)^q w_P \lambda_{P,n}(s) + \text{multiple of } \dot{\omega}_l(P, n, s)$$

$$\begin{aligned}\tilde{F}_{P,-n}\dot{E}_{-l}(R,-J;s) &= \dot{C}_{-l}(P,-n;-J,R;s)\dot{\omega}_{-l}(P,-n,s)\\ \tilde{F}_{R,0}R_q(s) &= c_R(s)\dot{\mu}_l(R,0;-s)\\ \tilde{F}_{R,0}\dot{E}_{-l}(R,-J;s) &= \dot{\mu}_{-l}(R,0;s)+\text{multiple of }\dot{\mu}_{-l}(R,0,-s).\end{aligned}$$

The Maass-Selberg relation and the definition of the map ι in 4.2.12 give

$$\begin{aligned}&u_P(s-s_0)^q w_P\dot{C}_{-l}(P,-n;-J,R;s)\mathrm{Wr}(\lambda_{P,n}(s),\dot{\omega}_l(P,n,s))\\ &= 2c_R(s)\mathrm{Wr}(\dot{\mu}_l(R,0,s),\dot{\mu}_l(R,0,-s)).\end{aligned}$$

In 11.2.2 we see that $\dot{C}_{-l}(P,-n;-J,R;s)$ can be replaced by $\overline{\dot{C}_l(P,n;J,R;\bar{s})}$. For $s_0=0$ we obtain the expression in part ii).

We use the notations of 10.3.3. We write $\mathbf{F}(J)R_q(s)=\dot{\mu}_l(J;-s)\mathbf{a}(s)$, where $\mathbf{a}$ is a column vector of length $|\mathbf{c}_J|=|X^\infty(J)|$, holomorphic in s. The Q-th coordinate of $\mathbf{a}$ is c_Q. Use $\mathbf{C}_l(J,Q,n)$, as defined in 10.3.5, to get

$$\begin{aligned}\mathbf{a}(s) &= s^{-1}(s-s_0)^q\alpha(s)\dot{\mathbf{C}}_l(J,P,n;\bar{s})^*\\ \alpha(s) &= \frac{1}{4}u_Pw_P\mathrm{Wr}(\lambda_{P,n}(s),\dot{\omega}_l(P,n,s)).\end{aligned}$$

The holomorphic function α is non-zero at s_0. The * means conjugate transpose.

Suppose that the assumptions in i) are satisfied: $R_q(s_0)\in A_{l_0}(\chi_0,\mathbf{c}_J,s_0)$, but $R_q(s_0)\notin A^2_{l_0}(\chi_0,s_0)$. Lemma 11.1.4 implies $q>0$. As $R_q(s_0)$ is smooth on $\tilde{G}$ by the assumption $R_q(s_0)\in A_{l_0}(\chi_0,\mathbf{c}_J,s_0)$, we see that not all coordinates of $\mathbf{F}(J)R_q(s_0)$ can be regular; otherwise we would have $R_q(s_0)\in A^2_{l_0}(\chi_0,s_0)$. As $\dot{\mu}_l(W,0,-s;g_Qp(z))=y^{1/2-s}$, we conclude $\mathrm{Re}\,s=0$. Remark 9.5.7 implies that $R_q(s_0)$ may be expressed as the sum of a cusp form and the value at s_0 of a family $f\in\dot{\mathcal{A}}_l(\mathbf{c}_J)_{s_0}$. The uniqueness in Theorem 10.2.1 implies that f is an $\dot{\mathcal{M}}_{s_0}$-linear combination of the $\dot{E}_l(J,Q)$ with $Q\in X^\infty(J)$. We write $f=\dot{\mathbf{E}}_l(J)\beta$, with β a column vector, meromorphic in s.

Let $s_0\neq 0$ (and $\mathrm{Re}\,s_0=0$). The $\dot{E}_l(J,Q)$ are holomorphic at s_0 (see Proposition 11.2.4). As f is holomorphic as well, this implies that β is holomorphic at s_0. We obtain the relation $\mu_l(J;-s_0)\mathbf{c}(s_0)=\mathbf{F}(J)f(s_0)=\mathbf{F}(J)\dot{\mathbf{E}}(J;s_0)\beta(s_0)=\dot{\mu}_l(J,s_0)\beta(s_0)+\dot{\mu}_l(J,-s_0)\mathbf{C}_l(J;s_0)\beta(s_0)$. This gives $\beta(s_0)=0$, and $f(s_0)=0$, in contradiction to the assumption $R_q(s_0)\notin A^2_{l_0}(\chi_0,s_0)$.

We are left with $s_0=0$ and $q>0$. In this situation

$$\mathbf{a}(s)=s^{q-1}\alpha(s)\dot{\mathbf{C}}_l(J,P,n;\bar{s})^*.$$

Proposition 11.2.3 implies that $s\mapsto\dot{\mathbf{C}}_l(J,P,n;\bar{s})^*$ is holomorphic at 0. Hence $q=2$ would imply $\mathbf{a}(0)=0$, again in contradiction to the assumption that $R_q(0)$ is not a square integrable function. Thus we obtain $q=1$, and $\mathbf{c}(0)=\alpha(0)\dot{\mathbf{C}}_l(J,P,n;0)^*\neq$

0. There exists $f \in \dot{\mathcal{A}}_l(\mathbf{c}_J)_0$ such that $R_q(0) - f(0) \in S_{l_0}(\chi_0, \mathbf{c}_J, 0)$. Again take β such that $f = \mathbf{E}_l(J)\beta$. Now

$$\begin{aligned} \mathbf{F}(J)f(s) &= \left(\dot{\mu}_l(J;s) + \dot{\mu}_l(J;-s)\dot{\mathbf{C}}_l(J;s)\right)\beta(s) \\ &= 2s\dot{\lambda}_l(J;s)\beta(s) + \dot{\mu}_l(J;-s)(\mathrm{Id} + \dot{\mathbf{C}}_l(s))\beta(s), \end{aligned}$$

where $\lambda_l(J;s) = (2s)^{-1}(\mu_l(J,s) - \mu_l(J;-s))$ is the diagonal matrix with $\lambda_l(Q,0,s)$ at position $Q \in X^\infty(J)$. The term with $\lambda_l(J)$ has to vanish at $s = 0$, and we obtain the holomorphy of β at $s = 0$. Hence

$$\mathbf{F}(J)f(0) = \alpha(0)\dot{\mu}_l(J;0)\dot{\mathbf{C}}_l(J,P,n;0)^* = \dot{\mu}_l(J;0)(\mathrm{Id} + \dot{\mathbf{C}}_l(J;0))\beta(0).$$

The functional equations in 10.3.5, and the relation $\dot{\mathbf{C}}_l(J;0)^* = \dot{\mathbf{C}}_l(J;0)$ in 11.2.2, give $(\mathrm{Id} - \dot{\mathbf{C}}_l(J;0))\dot{\mathbf{C}}_l(J,P,n;0)^* = 0$. Put $\mathbf{b}_1 = \frac{1}{2}\alpha(0)\dot{\mathbf{C}}_l(J,P,n;0)^*$. Then

$$\mathbf{F}(J)f(0) - \dot{\mu}_l(J;0)(\mathrm{Id} + \dot{\mathbf{C}}_l(J;0))\mathbf{b}_1 = \frac{1}{2}\alpha(0)(\mathrm{Id} - \dot{\mathbf{C}}_l(J;0))\dot{\mathbf{C}}_l(J,p,n;0)^* = 0,$$

and $\mathbf{F}(J)R_q(0) = \mathbf{F}(J)f(0) = \mathbf{F}(J)\dot{\mathbf{E}}_l(J;0)\mathbf{b}_1$. This means that there is a cusp form $g_1 \in S_{l_0}(\chi_0, \mathbf{c}_J, 0)$ such that

$$R_q(0) - g_1 = \dot{\mathbf{E}}(J;0)\mathbf{b}_1 = \frac{1}{2}\alpha(0)\dot{\mathbf{E}}_l(J;0)\dot{\mathbf{C}}_l(J,P,n;0)^* = \frac{1}{2}\alpha(0)E_{P,n}.$$

As $q \neq 2$, we have $\mathbf{p}_{l_0}(P,n;0) = 0$ (part ii) of Lemma 11.1.9). This means that the cusp form $q_1 = R_q(0) - \frac{1}{2}\alpha E_{P,n}$ is an element of $S_{l_0}(\chi_0, \mathbf{c}_{J,n}, 0)$. We apply Proposition 9.4.16 twice. First we take $\mathbf{c} = \mathbf{c}_J$ to see that g_1 is orthogonal to $E_{P,n}$. Next we use $\mathbf{c} = \mathbf{c}_{J,n}$ to conclude that g_1 is orthogonal to $R_q(0)$ as well. Both times orthogonality is understood in the sense of Remark 9.4.16. Hence g_1 is orthogonal to itself, even in the sense of Lebesgue integration, and vanishes; $R_q(0)$ is a multiple of $E_{P,n}$. As $R_q(0) \notin A^2_{l_0}(\chi_0, 0)$, we obtain $E_{P,n} \neq 0$

11.3.6 Lemma.

i) If $\mathbf{p}_{l_0}(P,n;s_0) \neq 0$, *then* $s_0 \in \Sigma(\chi_0, l_0)$, $q = q_0$, *and* $R_q(s_0) \in \mathbb{C}\cdot\mathbf{p}_{l_0}(P,n;s_0)$.

ii) If $s_0 = 0$, $\mathbf{p}_{l_0}(P,n;0) = 0$ *and* $E_{P,n} \neq 0$, *then* $q = 1$ *and* $R_q(s_0) \in \mathbb{C}\cdot E_{P,n}$.

iii) In all other cases $q = 0$ *and* $R_q(s_0) \notin A_{l_0}(\chi_0, \mathbf{c}_J, s_0)$.

Proof. Let $\mathbf{p}_{l_0}(P,n;s_0) \neq 0$. Then $q = q_0$ (see 11.3.3), and $A^2_{l_0}(\chi_0, s_0) \neq \{0\}$, hence $s_0 \in \Sigma(\chi_0, l_0)$. Lemma 11.1.4 implies that $R_q(s_0)$ is an element of $A_{l_0}(\chi_0, \mathbf{c}_J, s_0)$. If $R_q(s_0)$ were not in $A^2_{l_0}(\chi_0, s_0)$, then $s_0 = 0$, and $q = 1$, see part i) of Lemma 11.3.5. This would contradict $q = q_0$. Statement iii) in 11.3.3 shows that $R_q(s_0)$ is a multiple of $\mathbf{p}_{l_0}(P,n,s_0)$.

Let $s_0 = 0$, $\mathbf{p}_{l_0}(P, n, s_0) = 0$. From 11.3.3 we obtain $R_q(0) \notin A^2_{l_0}(\chi_0, s_0)$. Suppose $q = 0$. We use the formula for the holomorphic function c_R in part ii) of the previous lemma. This formula shows that $\dot{C}_l(P, n; J, R; 0)$ has to be zero for all $R \in X^\infty(J)$ to counteract the factor s^{-1}. Hence $E_{P,n}$ vanishes. If $q = 2$, then $\tilde{F}_{Q,0}R_q(0) = 0$ for all $Q \in X^\infty(J)$ by the same formula. Lemma 11.1.4 would imply $R_q(0) \in A^2_{l_0}(\chi_0, 0)$. We end up with $q = 1$. Lemma 11.3.5 states that $R_q(0)$ is a multiple of $E_{P,n}$.

Let $\mathbf{p}_{l_0}(P, n, s_0) = 0$, and $E_{P,n} = 0$ if $s_0 = 0$. Assertion iii) in 11.3.3 excludes $R_q(s_0) \in A^2_{l_0}(\chi_0, s_0)$, and part i) of Lemma 11.3.5 shows that $R_q(s_0) \in A_{l_0}(\chi_0, \mathbf{c}_J, s_0)$ is impossible as well.

11.3.7 *Reformulation.* Now we turn to the singularities of the family $\dot{Q}_l(J, P, n)$ itself.

We denote by r the order of $\dot{Q}_l(J, P, n)$ at s_0; i.e., r is the integer for which $F_1 = \lim_{s \to s_0} (s - s_0)^{-r} \dot{Q}_l(J, P, n; s)$ exists and is non-zero. In 11.1.2 and 11.1.3 we see that $r = i(P, n) - q$.

<table>
<tr><td rowspan="2"></td><td colspan="2">$i(P,n) = 0$</td><td colspan="2">$i(P,n) = 1$</td></tr>
<tr><td>$s_0 \neq 0$</td><td>$s_0 = 0$</td><td>$s_0 \neq 0$</td><td>$s_0 = 0$</td></tr>
<tr><td rowspan="2">$\mathbf{p}_{l_0}(P, n; s_0) \neq 0$</td><td>$r = -1$</td><td>$r = -2$</td><td>$r = 0$</td><td>$r = -1$</td></tr>
<tr><td colspan="4">$F_1 \in \mathbb{C} \cdot \mathbf{p}_{l_0}(P, n; s_0)$</td></tr>
<tr><td>$\mathbf{p}_{l_0}(P, n; 0) = 0$
and $E_{P,n} \neq 0$</td><td></td><td>$r = -1$
$F_1 \in \mathbb{C} \cdot E_{P,n}$</td><td></td><td>$r = 0$
$F_1 \in \mathbb{C} \cdot E_{P,n}$</td></tr>
<tr><td rowspan="2">$\mathbf{p}_{l_0}(P, n; s_0) = 0$,
and if $s_0 = 0$
then $E_{P,n} = 0$</td><td colspan="2">$r = 0$</td><td colspan="2">$r = 1$</td></tr>
<tr><td colspan="4">$F_1 \notin A_{l_0}(\chi_0, \mathbf{c}_J, s_0)$</td></tr>
</table>

Table 11.3 The order of Poincaré families along vertical lines.

Lemma 11.3.6 implies the results in Table 11.3. A singularity corresponds to $r < 0$. We see that singularities can occur only for $s_0 \in \{0, |\tau(\chi_0)|\} \cup \Sigma_c$. Indeed, $\mathrm{Re}\, s_0 > \frac{1}{2}$ is impossible, as the Poincaré series converge in that region. Further $\mathbf{p}_{l_0}(P, n; s_0) \neq 0$ implies $s_0 \in \Sigma$. The sole element of Σ_d in $[0, \frac{1}{2}]$ is $|\tau(\chi_0)|$.

The next lemma gives more information on the case $s_0 = |\tau(\chi_0)|$. Remember that $\tau = \tau(\chi_0)$ characterizes the weight modulo 2: it is determined by $-\frac{1}{2} < \tau \leq \frac{1}{2}$ and $\tau \equiv \frac{1}{2}(l_0 - 1) \bmod 1$.

11.3.8 Lemma. *If $\tau(\chi_0) \neq 0$ and $\dot{Q}_l(J, P, n)$ is singular at $|\tau(\chi_0)|$, then either $P \in \mathcal{P}_Y$, $n = l_0 = 0$, and $\chi_0 = 1$, or $l_0 \notin \mathbb{Z}$ and $\tau(\chi_0) l_0 < 0$.*

Proof. Let $\tau = \tau(\chi_0) \neq 0$, hence $l_0 \notin 1+2\mathbb{Z}$; put $s_0 = |\tau|$. Suppose that $\dot{Q}_l(J,P,n)$ is singular at $|\tau|$. Table 11.3 shows that $\mathbf{p}_{l_0}(P,n;|\tau|) \neq 0$, and $i(P,n) = 0$. The idea of the proof is to combine the conditions for $i(P,n) = 0$ in Table 11.1 on p. 192, with the results in Propositions 6.7.3 and 6.7.6.

	$P \in X^\infty$, $n \neq 0$	$P \in \mathcal{P}_Y$
A	$l_0 \operatorname{sign} n < 2\|\tau\| + 1$ or $l_0 \operatorname{sign} n \not\equiv 2\|\tau\| + 1 \bmod 2$	$\frac{1}{2}\|n+l_0\| - \frac{1}{2}\|n-l_0\| < 2\|\tau\| + 1$ or $\frac{1}{2}\|n+l_0\| - \frac{1}{2}\|n-l_0\| \not\equiv 2\|\tau\| + 1 \bmod 2$
B	$\|l_0\| \geq b > 0$	$b = 2$ if $\tau = \frac{1}{2}$; $b = 2\tau \operatorname{sign} l_0$ otherwise
C	$n \operatorname{sign} l_0 > 0$	$n \operatorname{sign} l_0 \geq b$
D	$\|l_0\| < 2\|\tau\| + 1$ or $\|l_0\| \not\equiv 2\|\tau\| + 1 \bmod 2$	$\min(\|l_0\|, \|n\|) < 2\|\tau\| + 1$ or $\min(\|l_0\|, \|n\|) \not\equiv 2\|\tau\| + 1 \bmod 2$
E	$\|l_0\| < b$	$\min(\|l_0\|, \|n\|) < b$

Table 11.4 Conditions in the proof of Lemma 11.3.8.

Let us first draw conclusions from Table 11.1. In the case $P \in X^\infty$ we have $n \neq 0$. This means that $\varepsilon = \operatorname{sign} n$. The condition $0 \leq s_0 = |\tau|$ is satisfied automatically. Hence we have $l_0 \operatorname{sign} n < 2|\tau| + 1$ or $l_0 \operatorname{sign} n \not\equiv 2|\tau| + 1 \bmod 2$. In the other case, $P \in \mathcal{P}_Y$, there are the quantities $p = \frac{1}{4}|n-l_0|$ and $q = \frac{1}{4}|n+l_0|$. From $i(P,n) = 0$ we conclude that $\frac{1}{2}|n+l_0| - \frac{1}{2}|n-l_0| < 2|\tau|+1$ or $\frac{1}{2}|n+l_0| - \frac{1}{2}|n-l_0| \not\equiv 2|\tau| + 1 \bmod 2$. Thus we have obtained line A in Table 11.4.

$\mathbf{p}_{l_0}(P,n;|\tau|) \neq 0$ implies $\ker(A(\chi_0,l_0) - \frac{1}{4} + \tau^2) \neq \{0\}$. Proposition 6.7.3 shows that this is possible in four cases. The first case is immediately seen not to apply. In the last case $l_0 = 0$, $\tau = \frac{1}{2}$, we use that $A^2_{l_0}(\chi_0, \frac{1}{2})$ is contained in the space of constant functions. This implies $n = 0$, and hence $P \in \mathcal{P}_Y$. Moreover, $\mathbf{p}_{l_0}(P,n;|\tau|)$ has to vanish if the character χ_0 is not trivial. We have obtained the first possibility in the lemma. It satisfies the top conditions in part A of Table 11.4.

In the remaining two cases in Proposition 6.7.3 there is a positive number b such that $\left|\frac{b-1}{2}\right| = |\tau|$, $b \equiv \pm l_0 \bmod 2$, and $b \leq \pm l_0$. This gives $l_0 \neq 0$ as new information. In the case $\tau = \frac{1}{2}$ we conclude $b = 2$. Let $0 < |\tau| < \frac{1}{2}$. Then $0 < b \leq l_0|$, $b \equiv |l_0| \bmod 2$, and $b = 1+2\zeta|\tau|$, with $\zeta \in \{1,-1\}$. So $\operatorname{sign} l_0\,(2\tau + 1) \equiv b \equiv 1 + 2\zeta|\tau| \bmod 2$. Hence $b = 2\tau \operatorname{sign} l_0 + 1$. We have put this information in line B of the table.

The condition $\mathbf{p}_{l_0}(P,n,|\tau|) \neq 0$ implies $\tilde{F}_{P,n}\mathbf{p}_{l_0}(P,n,|\tau|) \neq 0$. According to Proposition 6.7.6, this can happen only if the conditions in part C of Table 11.4 are satisfied, for b as defined above. (We have used that $n \neq 0$ if $P \in X^\infty$.)

From B and C we conclude that $\operatorname{sign} l_0 = \operatorname{sign} n$. Hence $\frac{1}{2}|n+l_0| - \frac{1}{2}|n-l_0|$ in part A is equal to $\min(|l_0|,|n|)$. We replace A by D. We have to show that B, C and D together imply $0 < |\tau| < \frac{1}{2}$ and $\tau l_0 < 0$

Let us assume that $\tau = \frac{1}{2}$. This corresponds to $l_0 \in 2\mathbb{Z}$. Now the bottom assertions in D in Table 11.4 are false; use $n \equiv l_0 \bmod 2$ in the case $P \in \mathcal{P}_Y$. If $P \in X^\infty$, the top assertion in D becomes $|l_0| < 2$, in contradiction to $l_0 \neq 0$. If $P \in \mathcal{P}_Y$ we obtain $|n| \geq 2$ from C, and obtain a contradiction to D as well.

We are left with $0 < |\tau| < \frac{1}{2}$. The definition of τ implies $2|\tau| + 1 \equiv l_0 \operatorname{sign} \tau$. This means that the bottom line in D is equivalent to $\operatorname{sign} l_0 \neq \operatorname{sign} \tau$. (In the case $P \in \mathcal{P}_Y$ we use that C implies $|n| \equiv |l_0| \bmod 2$.) This gives the second alternative in the lemma. To complete the proof assume that $\operatorname{sign} l_0 = \operatorname{sign} \tau$. The top line in D gives the statements in E. These are in contradiction with $|l_0| \geq b$, and in the case $P \in \mathcal{P}_Y$, $|n| \geq b$; see B and C.

11.3.9 Proposition. Poincaré families. *Take χ_0, l, l_0 and J as in 11.1.1. Take $P \in \mathcal{P}$, $n \in \mathcal{C}_P$; suppose $n(\chi_0) \neq 0$ if $P \in X^\infty(J)$. The meromorphic continuation $\dot{Q}_l(J,P,n)$ of the family of Poincaré series $s \mapsto P_{l_0}(P,n,\chi_0,s)$ has no singularities in the closed half plane* $\operatorname{Re} s \geq 0$, *except for:*

i) a first order pole at $s_0 \in \Sigma_c(\chi_0) \smallsetminus \{0\}$ if $\tilde{F}_{P,n} f \neq 0$ for some $f \in A^2_{l_0}(\chi_0, s_0)$;

ii) a singularity at $s_0 = 0$ in the following cases:

a) if $\tilde{F}_{P,n} f \neq 0$ for some $f \in A^2_{l_0}(\chi_0, 0)$. The order of the singularity is 1 *if $l_0 \in 1 + 2\mathbb{Z}$ and $l_0 n > 0$; otherwise the order is* 2.

b) if $\tilde{F}_{P,n} f = 0$ for all $f \in A^2_{l_0}(\chi_0, 0)$, but $E_{P,n} \neq 0$, and either $l_0 \notin 1 + 2\mathbb{Z}$ or $l_0 n \leq 0$. The order of the singularity is 1.

iii) at $s_0 = |\tau(\chi_0)|$, with $\tau(\chi_0) \neq 0$, a pole of order 1 *in the following cases:*

a) $P \in \mathcal{P}_Y$, $n = l_0 = 0$ and $\chi_0 = 1$;

b) $l_0 \notin \mathbb{Z}$, $\tau(\chi_0) l_0 < 0$ and $\tilde{F}_{P,n} f \neq 0$ for some automorphic form $f \in A^2_{l_0}(\chi_0, \tau(\chi_0))$.

Remarks. $E_{P,n} = \sum_{Q \in X^\infty(J)} \overline{\dot{C}_l(P,n;J,Q;0)} \dot{E}_l(J,Q;0)$, see 11.3.2.

More detailed information, concerning not only the singularities but the zeros as well, may be found in 11.3.7, and in Lemma 11.3.8.

Examples are discussed in 13.4.5–13.4.6.

Proof. Let $\dot{Q}_l(J,P,n)$ be singular at s_0 with $\operatorname{Re} s_0 \geq 0$. This means that $r < 0$ in Table 11.3. We have $\operatorname{Re} s_0 \leq \frac{1}{2}$ from the convergence of the Poincaré series for $\operatorname{Re} s > \frac{1}{2}$.

First consider $s_0 \neq 0$, $s_0 \neq |\tau(\chi_0)|$. This implies $i(P,n) = 0$; see Table 11.1. Table 11.3 shows that in this case $r < 0$ if and only if $i(P,n) = 0$ and $\mathbf{p}_{l_0}(P,n,s_0) \neq 0$. If this holds, then $r = -1$. Part i) of the proposition follows from part i) of Lemma 11.3.6 together with the definition of the Poincaré element $\mathbf{p}_{l_0}(P,n,s_0)$.

Next consider $s_0 = |\tau(\chi_0)| > 0$. Lemma 11.3.8 leaves open two possibilities for a singularity to occur. The first possibility is $n = l_0 = 0$, $\chi_0 = 1$. Then the constant function $\mathbf{1}$ satisfies $\tilde{F}_{P,0}\mathbf{1} \neq 0$, and $i(P,0) = 0$. Hence a first order singularity really occurs. This is part iii)a). The other possibility is $l_0 \notin \mathbb{Z}$, $\tau l_0 < 0$. A reasoning similar to the proof of Lemma 11.3.8 gives $i(P,n) = 0$. Table 11.3 gives part iii)b).

We consider the case $s_0 = 0$. The condition $\mathbf{p}_{l_0}(P,n,0) \neq 0$ leads to the condition in ii)a). In Table 11.1 we see for $P \in X^\infty$ that $i(P,n) = 1$ if and only if $l_0 \in 1+2\mathbb{Z}$ and $\operatorname{sign} n = \operatorname{sign} l_0$. The case $P \in \mathcal{P}_Y$ is a bit more complicated. We use the fact that $n \equiv l_0 \bmod 2$ to obtain $q-p-\frac{1}{2} \equiv 0 \bmod 1 \iff \frac{1}{2}|n+l_0|-\frac{1}{2}|n-l_0| \in 1+2\mathbb{Z} \iff \min(|n|,|l_0|)\cdot\operatorname{sign}(nl_0) \in 1+2\mathbb{Z} \iff l_0 \in 1+2\mathbb{Z}$. The condition $q-p-\frac{1}{2} \geq 0$ amounts to $\min(|n|,|l_0|)\cdot\operatorname{sign}(nl_0) \geq 1$. This is equivalent to $nl_0 \geq 1$, if we already know that $n, l_0 \in 1+2\mathbb{Z}$. This gives ii)a).

Finally, consider $s_0 = 0$, $\mathbf{p}_{l_0}(P,n,0) = 0$, but $E_{P,n} \neq 0$. Now $r = -1$ if and only if $i(P,n) = 0$. This is the negation of the condition that we have considered above.

11.3.10 *Resolvent kernel.* Case iii)a) of Proposition 11.3.9 states that the resolvent kernel in weight zero (see 5.1.8) has a singularity at $s = \frac{1}{2}$ if and only if the character χ_0 is trivial. Suppose $\chi_0 = 1$. The same reasoning as in Remark 11.2.7 shows, for $P \in \mathcal{P}_Y$:

$$\lim_{s\to 1/2}(s-\frac{1}{2})\dot{Q}_0(J(0),P,0;s) = 4\pi\operatorname{vol}(\tilde{\Gamma}\backslash\mathfrak{H})^{-1}\cdot\mathbf{1}.$$

This implies that $s \mapsto \dot{Q}_0(J(0),P,0;s) - 4\pi\dot{E}_0(J(0),Q;s)$ is holomorphic at $s = \frac{1}{2}$ for all $Q \in X^\infty$.

Chapter 12
Singularities of Poincaré families

In this last chapter of Part I we consider singularities of Poincaré families depending on the character and the spectral parameter jointly. The results are by no means complete. We consider singularities at points (φ_0, s_0) with $\varphi_0 \in \mathcal{V}_r$ and $\operatorname{Re} s_0 \geq 0$. Our point of view is local.

Singularities of families of eigenfunctions on a parameter space of complex dimension more than one may be quite complicated. The first two sections serve to get some hold on these complications. The main idea is to approach a given point w_0 of the parameter space along a one-dimensional path. The family of automorphic forms to be studied is restricted to this path. The resulting one-dimensional families are not as nice as the 'vertical' ones of the previous chapter. In particular, they are no longer determined by their non-regular Fourier terms. But a suitable multiple of the family still has a non-zero limit at w_0. This limit is an automorphic form with weight, character and eigenvalue determined by w_0. If we vary the path through w_0, the value of the limit may change. The resulting automorphic forms may span a space of dimension larger than one. If this dimension is high, the family of automorphic forms behaves in a complicated way at the point w_0.

As in the previous chapter, the Poincaré families for regular and singular (P, n) cannot be handled in exactly the same way. We shall give a unified treatment as far as is feasible. We treat restricted and more general parameter spaces separately (see 10.2.11 for the definition of restricted parameter spaces). In the non-restricted case there are singularities that are not at all related to spectral features, but to the fact that at a cusp the condition of regularity does not always determine a submodule of $\mathcal{W}_l(Q, m)$ with $\mathcal{O}$-rank one.

The main results on the singularities of Poincaré families on general parameter spaces we state in Propositions 12.4.2 and 12.4.3, and those on restricted parameter spaces in Propositions 12.5.8 and 12.5.9.

12.1 Local curves

In 7.4.15 we have seen that for a one-dimensional parameter space a section of an eigenfunction module has a well determined order at each point of its domain. This does not hold in higher dimensions. For a local study of families of automorphic forms on a higher dimensional parameter space it is interesting to probe them along a path of complex dimension one. This means that we compose these families with a morphism of parameter spaces. That gives families on an easier parameter space. In this section we consider such morphisms.

12.1.1 *Minimal holomorphic maps.* W denotes a parameter space. We fix $w_0 \in W$.

We consider holomorphic maps $j : U \to W$, with U a neighborhood of 0 in $\mathbb{C}$, such that $j(0) = w_0$. If one such map is given, we can form other ones by taking the composition $j \circ h : U_1 \to W$, with $h : U_1 \to U$ be holomorphic, U_1 a neighborhood of 0 in $\mathbb{C}$. If $h'(0) \neq 0$, then h has an inverse near 0, so we can get back j from $j \circ h$ by the same method.

We call j *minimal* if it cannot be written in the form $j = j_1 \circ h$ with $j_1 : U_1 \to W$ and $h : U \to U_1$ holomorphic, $h(0) = h'(0) = 0$. If j is minimal, then it cannot be constant.

12.1.2 *Definition.* A *local curve* at w_0 is a holomorphic map $j : U \to W$, with U a neighborhood of 0 in $\mathbb{C}$, $j(0) = w_0$, that is minimal in the sense stated above.

Example. In the previous chapter we studied Poincaré families along 'vertical' local curves $u \mapsto (\varphi_0, s_0 + u)$.

Actually, the essential objects are germs of local curves; local curves j_1 and j_2 through w_0 determine the same germ if they coincide on a neighborhood of 0 in $\mathbb{C}$. But we shall work with representatives of such germs.

If h is holomorphic at 0, $h(0) = 0$ and $h'(0) \neq 0$, one might also call j and $j \circ h$ equivalent. We shall not identify them, as j and $j \circ h$ may lead to different limits of families of automorphic forms.

12.1.3 *Parameter space.* Let $j : U \to W$ be a local curve at w_0 in the parameter space (W, χ, s). Define $s_{(j)} = s \circ j$, $\chi_{(j)}(\gamma) = \chi(\gamma) \circ j$ for all $\gamma \in \tilde{\Gamma}$. This makes U into a parameter space, and j into a morphism of parameter spaces. For $Q \in \mathcal{P}$ we denote the set $\{ n \circ j : n \in \mathbf{c}(Q) \}$ by $\mathbf{c}_{(j)}(Q)$. So $\mathbf{c}_{(j)}$ is a growth condition on U.

12.1.4 *Definition.* Let ψ be a non-zero holomorphic function on a neighborhood of w_0 with $\psi(w_0) = 0$. We call a local curve $j : U \to W$ at w_0 a *local curve* at w_0 *along* $N(\psi)$ if $\psi(j(u)) = 0$ for all $u \in U$.

The Theorem of Puiseux, see Lemma 9.4.12, provides us with such local curves:

12.1.5 Lemma. *If $\tau(z,t) = t^k + \sum_{j=0}^{k-1} a_j(z) t^j$, as in Lemma 9.4.12, represents an irreducible element of the stalk $\mathcal{O}_0$, then there exists a holomorphic function h on a neighborhood U of 0 in $\mathbb{C}$ such that $u \mapsto (u^k, h(u))$ is a local curve at 0 along $N(\tau)$. There is a neighborhood Ω of 0 in $\mathbb{C}^2$ such that $N(\tau) \cap \Omega \subset j(U)$.*

Remarks. This gives the existence of a local curve at 0 along $N(\tau)$. This holds even if τ is not irreducible: apply the lemma to an irreducible factor.

It is essential that the number of variables is two.

Proof. Take d and h as given by Lemma 9.4.12, and $j : u \mapsto (u^d, h(u))$. In the proof in [14], Chap. III, §1.4–6, we find at the end of §1.6, on p. 135, that

$$\tau_1(z,t) = \prod_{m=1}^{k} \left(t - h(e^{2\pi i m/d} z^{1/d}) \right)$$

determines a holomorphic function on a neighborhood Ω of 0 in $\mathbb{C}^2$ dividing τ in the stalk $\mathcal{O}_0$. For an irreducible τ this gives $d = k$, and $N(\tau) = N(\tau_1)$. If Ω is small enough, then for each $w_1 = (z_1, t_1) \in \Omega \cap N(\tau)$, we have $t_1 = h(e^{2\pi i m/d} u_0)$ for some m and some u_0 with $u_0^d = z_1$. Hence $w_1 = j(e^{-2\pi i m/d} u_0)$.

Moreover, if h is of the form $h(u) = h_1(u^l)$ with h holomorphic at 0 in $\mathbb{C}$ and l dividing d, then τ_1 is the product of l holomorphic factors. Hence $l \geq 2$ is impossible.

Suppose $j : u \mapsto (u^d, h(u))$ can be written as $j = j_1 \circ c$, with $j_1 : v \mapsto (a(v), b(v))$, a, b and c holomorphic at 0 in $\mathbb{C}$ with value 0. We have to show that $c'(0) \neq 0$.

Write $c(u) = u^\gamma c_1(u)$ and $a(v) = v^\alpha a_1(v)$, with $\alpha, \gamma \geq 1$, a_1 and c_1 holomorphic at 0 with a non-zero value. Then $d = \alpha\gamma$. As $a_1(0) \neq 0$, there is a_2, holomorphic at 0, such that $a_1 = a_2^\alpha$. Put $\varphi : v \mapsto v a_2(v)$, then $a = \varphi^\alpha$, and φ has a holomorphic inverse $\varphi^\leftarrow$ on a neighborhood of 0. Put $j_2 = j_1 \circ \varphi^\leftarrow$. It has the form $j_2 : t \mapsto (t^\alpha, b(\varphi^\leftarrow(t)))$. For each t near 0, there is a number u such that $t = \varphi(c(u))$. From $j = j_2 \circ \varphi \circ c$ we conclude that $\tau(j_2(t)) = 0$ for all t in a neighborhood of 0. We have seen that irreducibility of τ implies $\alpha = k = d$, hence $\gamma = 1$. This means that j is minimal.

12.1.6 *Existence of local curves in higher dimensions.* Consider $\dim(W) = n > 2$. Suppose that $\psi \in \mathcal{O}_{w_0}$ is not a unit, and $\psi \neq 0$. We pick coordinates $z_1, \ldots, z_n$ with $z_j(w_0) = 0$ on a neighborhood of w_0 in W, in such a way that ψ does not vanish identically along the line given by $z_1 = z_2 = \ldots = z_{n-1} = 0$. This is always possible. The zero set $N(\psi)$ does not change if we multiply ψ by a unit at w_0. So we may assume that ψ has Weierstrass form in z_n, see, e.g., [23], Corollary 6.1.2. We apply Lemma 12.1.5 to an irreducible factor of $\tau(u,t) = \psi(up_1, \ldots, up_{n-1}, t)$ for each non-zero $(p_1, \ldots, p_{n-1}) \in \mathbb{C}^{n-1}$ sufficiently close to 0. Thus we obtain a local curve at w_0 along $N(\psi)$.

12.1.7 Proposition. *Suppose* $\dim(W) \geq 2$. *Let* ψ *and* η *be non-zero holomorphic functions on a neighborhood of* w_0 *such that* $\psi(w_0) = 0$, *and* η *is relatively prime to* ψ *in the ring* $\mathcal{O}_{w_0}$. *Then there is a local curve* $j : U \to W$ *along* $N(\psi)$ *such that* $\eta \circ j$ *is not the zero function.*

Proof. Take coordinates as above. Replacing η by $\eta + \psi$ (if necessary) we arrange that both ψ and η do not vanish along the line $z_j = 0$, $j = 1, \ldots, n-1$. Again, we can assume that ψ and η have Weierstrass form in z_n. As in the proof of Lemma 9.4.13, the fact that ψ and η are relatively prime leads to the existence of holomorphic functions l, f_1 and f_2 on a neighborhood of w_0 with $l = f_1\psi + f_2\eta$, such that l is not identically zero, and does not depend on z_n. Take a point w_1 near w_0 such that $u \mapsto l(uz_1(w_1), \ldots, uz_{n-1}(w_1))$ is not the zero function. Apply Lemma 12.1.5 to an irreducible factor of $\tilde{\psi}(u,t) = \psi(uz_1(w_1), \ldots, uz_{n-1}(w_0), t)$. This provides us with a local curve $j : v \mapsto (v^k, h(v))$ along $N(\tilde{\psi})$; it can be considered as a local curve along $N(\psi)$ as well. As $l(j(v)) = l(v^k z_1(w_1), \ldots, v^k z_{n-1}(w_1))$, we conclude that $\eta \circ j$ cannot vanish identically.

12.1.8 *Parameter spaces for Poincaré families.* We specialize W to have the form $(\tilde{\varphi} + V) \times \mathbb{C}$, with $V = \mathbb{C} \otimes_{\mathbb{R}} V_r$ for an $\mathbb{R}$-linear subspace V_r of $\mathcal{V}_r$, and $\tilde{\varphi} \in \mathcal{V}_r$.

Let $w_0 = (\varphi_0, s_0) \in (\tilde{\varphi} + V) \times \mathbb{C}$, and let $\varphi_1 \in V$. We define a local curve at w_0 with *direction* φ_1 to be a local curve of the form $u \mapsto (\varphi_0 + u^d\varphi_1, s_0 + h(u))$. If $d = 1$ we call this local curve *linear*, if $d = 2$ *quadratic*.

12.1.9 *Real type.* Let $w_0 = (\varphi_0, s_0) \in (\tilde{\varphi} + V) \times \mathbb{C}$, with $\varphi_0 \in \mathcal{V}_r$. We call $\psi \in \mathcal{O}_{w_0}$ a germ of *real type* if $\psi(\varphi, s) = 0$ and $\varphi \in V_r$ imply $\frac{1}{4} - s^2 \in \mathbb{R}$.

12.1.10 Lemma. *Suppose* $\dim_{\mathbb{C}} V = 1$, $\varphi_0 \in \mathcal{V}_r$, *and suppose that* $\psi \in \mathcal{O}_{w_0}$ *is a germ of real type that does not vanish identically along* $\{\varphi_0\} \times \mathbb{C}$. *Any local curve along* $N(\psi)$ *given by Lemma 12.1.5 (with* z *corresponding to a coordinate on* V, *and* t *to a coordinate on* $\mathbb{C}$*) is linear if* $s_0 \neq 0$, *and linear or quadratic if* $s_0 = 0$.

Proof. We can assume that ψ has Weierstrass form in the s-direction. As $\frac{1}{4} - s_0^2 \in \mathbb{R}$, we have $s_0 \in \mathbb{R} \cup i\mathbb{R}$. Let $\varphi_1 \in V_r$, $\varphi_1 \neq 0$, and let $\tau(z,t) = \psi(\varphi_0 + z\varphi_1, s_0 + ta)$, with $a = 1$ if $s_0 \in \mathbb{R}$, and $a = -i$ otherwise. Consider the local curve $j : u \mapsto (u^k, h(u))$ along $N(\tau)$ given by Lemma 12.1.5. Put $h_0(u) = h(u)^2$ if $s_0 = 0$, and $h_0 = h$ otherwise. Then $h_0(u) \in \mathbb{R}$ for all u near 0 such that $u^k \in \mathbb{R}$. Take $u \in \mathbb{R}$ to see that $\overline{h_0} = h_0$. We have $h_0(re^{\pi i m/k}) = \overline{h_0(re^{\pi i m/k})} = \overline{h_0}(re^{-\pi i m/k}) = h_0(re^{-\pi i m/k})$ for all real r near 0, for each $m \in \mathbb{Z}$. This implies that h_0 is of the form $h_0(u) = h_1(u^k)$ with h_1 holomorphic at 0. Note that $h_1(0) = 0$.

In the case $s_0 \neq 0$ this gives $\tau(u^k, h_1(u^k)) = 0$ for all u near 0 in $\mathbb{C}$. The minimality in the definition of local curve implies $k = 1$. Hence $u \mapsto (\varphi_0 + u\varphi_1, s_0 + ah(u))$ is a linear local curve along $N(\psi)$ with direction φ_1.

Let $s_0 = 0$. If k is odd, the fact that $u \mapsto h_1(u^k)$ is the square of the holomorphic function h implies that $h(u) = h_2(u^k)$, with h_2 holomorphic at 0. In that case $k = 1$ by minimality. Suppose k to be even. Write $h_1(u) = u^m h_3(u)$, with $m \geq 1$ and h_3 holomorphic at 0 with non-zero value. Then h_3 is the square of a holomorphic function h_4. This gives $h(u) = \pm u^{km/2} h_4(u^k)$, and $k = 2$ by minimality.

12.1.11 Proposition. *Let* $\dim V \geq 1$, *and let* $w_0 = (\varphi_0, s_0) \in (\tilde{\varphi} + V) \times \mathbb{C}$. *Suppose that* ψ *and* η *are holomorphic on a neighborhood of* w_0, *such that* $\psi(w_0) = 0$, *but* ψ *is not identically zero along the line* $\{\varphi_0\} \times \mathbb{C}$, *and* η *is relatively prime to* ψ *in the stalk* $\mathcal{O}_{w_0}$. *Then there are* $\varphi_1 \in \tilde{\varphi} + V_r$ *and* $s_1 \in \mathbb{C}$ *such that* $(\varphi_1, s_1) \in N(\psi)$ *and*

$\eta(\varphi_1, s_1) \neq 0$. *For each* (φ_1, s_1) *of this type, there is a local curve* j *along* $N(\psi)$ *with direction* $\varphi_1 - \varphi_0$ *such that* $\eta \circ j$ *is not the zero function.*

Suppose in addition that ψ *has real type. If* $s_0 = 0$*, then* j *is linear or quadratic, otherwise it is linear.*

Proof. We can assume that ψ is irreducible in the stalk $\mathcal{O}_{w_0}$. By replacing η by $\eta + \psi$ (if necessary), we arrange that η does not vanish along $\{\varphi_0\} \times \mathbb{C}$. Use the same proof as for Proposition 12.1.7, but take coordinates $z_1, \ldots, z_{n-1}$ that are real on V_r, and $z_n = s - s_0$. The point $(z_1(w_1), \ldots, z_{n-1}(w_1))$ may be taken to be real, c.f. Lemma 9.4.13. Apply Lemma 12.1.10 to the local curve obtained in this way.

12.1.12 *Remark.* If $\dim V > 1$, one might want to construct holomorphic maps $\varphi \mapsto (\varphi, h(\varphi))$, with h holomorphic on an open set in V, with image contained in some given $N(\psi)$. Even if ψ is of real type this need not be possible. Kato, [25], Ch. II, Remark 5.12, gives the following example: Take $\dim V = 2$, and coordinates (v_1, v_2) on a piece of V; $\psi(v_1, v_2, s) = (s - s_0)^2 - v_1^2 - v_2^2$. Along lines $u \mapsto (u\tilde{v}_1, u\tilde{v}_2)$ we can take a holomorphic square root of $v_1^2 + v_2^2$, but not on a two-dimensional neighborhood of $(0, 0)$.

12.2 Value sets

Holomorphic families of automorphic forms are determined by infinitely many holomorphic functions: a family f is determined by the pointwise evaluations $f_g : w \mapsto f(w; g)$, with g running through $\tilde{G}_{\mathcal{P}}$. If the parameter space W has complex dimension larger than one, then different f_g may be relatively prime in a stalk $\mathcal{O}_w$, without being units at w. So, if f has a zero at w, it might happen that f cannot be divided by any non-unit in $\mathcal{O}_w$. But if we approach w along a path with complex dimension 1, we may divide f by a power of a local coordinate at w, and end up with a non-zero automorphic form as its value.

In this section we use local curves to probe germs of families of automorphic forms. Actually, we proceed a bit more generally and consider families of eigenfunctions, as discussed in Section 7.3.

12.2.1. Let w_0 be a point of the parameter space W, $\mathbf{c}$ a growth condition on W, and l a weight. In this section we do not restrict our attention to the germs in $\mathcal{A}_l(\mathbf{c})_{w_0}$, but consider also the germs in $\mathcal{A}_{l \circ j}(\mathbf{c}_{(j)})_0$, and the values $f(w_0), g(0) \in A_{l(w_0)}(\chi(w_0), \mathbf{c}(w_0), s(w_0))$ of $f \in \mathcal{A}_l(\mathbf{c})_{w_0}$ and $g \in \mathcal{A}_{l \circ j}(\mathbf{c}_{(j)})_0$ for all local curves $j : W_1 \to W$ at w_0. In 7.3.14 and 7.5.4 we have indicated how to relate the sheaves $\mathcal{A}_l(\mathbf{c})$ and $\mathcal{A}_{l \circ j}(\mathbf{c}_{(j)})$. We need to work with this relation, not only for modules of families of automorphic forms, but also for Fourier term modules. Perhaps the aesthetically most pleasing way to handle this would be to work with the functor assigning an $\mathcal{A}_l(\mathbf{c})$ to each parameter space, and similar functors. Here we are content to proceed under the following assumptions:

12.2.2 *Assumptions.* In this section we consider an eigenfunction module $\mathcal{H}$ on W. This eigenfunction module is a submodule of some $\mathcal{F}_l(U)$ with $l \in \mathbf{wt}$, and U an open subset of $\tilde{G}$ satisfying $U\tilde{K} = U$. We assume that for all local curves $j : W_1 \to W$ at w_0 there is an eigenfunction module $\mathcal{H}_{(j)} \subset \mathcal{F}_{l\circ j}(U)$ on W_1 such that $j^\sharp \mathcal{H} \subset j_* \mathcal{H}_{(j)}$; see 7.3.14. We take for each $w \in W$ a subspace $H(w)$ of $C^\infty(U)$ that contains all values $f(w)$ with $f \in \mathcal{H}_w$, as well as all $g(0)$ where g runs through $\left(\mathcal{H}_{(j)}\right)_0$ and j runs through the local curves at w.
Examples. $\mathcal{H} = \mathcal{A}_l(\mathbf{c})$, $\mathcal{H}_{(j)} = \mathcal{A}_{l\circ j}(\mathbf{c}_{(j)})$, $H(w) = A_{l(w)}(\chi(w), \mathbf{c}(w), s(w))$. Or $\mathcal{H} = \mathcal{W}_l(P, n)$, $\mathcal{H}_{(j)} = \mathcal{W}_{l\circ j}(P, n \circ j)$, $H(w) = W_{l(w)}(P, n(w), s(w))$.

12.2.3 *Primitive germs.* We call a germ $f \in \mathcal{H}_{w_0}$ *primitive* if $f \neq 0$ and $\mathcal{H}_{w_0} \cap (\mathcal{M}_{w_0} \cdot f) = \mathcal{O}_{w_0} \cdot f$.

Let $f \in (\mathcal{M} \otimes_{\mathcal{O}} \mathcal{H})_{w_0}$. If $\psi \in \mathcal{M}_{w_0}$ has been chosen such that ψf is primitive, then we call ψf a *primitive multiple* of f.

On a parameter space W of dimension one, the value at w_0 of a primitive germ is not zero. On parameter spaces of higher dimension this may happen if there are elements g_1 and g_2 in the set U (on which $\mathcal{H}$ lives), such that $w \mapsto f(w; g_1)$ and $w \mapsto f(w; g_2)$ are relatively prime in $\mathcal{O}_{w_0}$.

12.2.4 Lemma. *Each non-zero element of $(\mathcal{M} \otimes_{\mathcal{O}} \mathcal{H})_{w_0}$ has a primitive multiple. This multiple is unique up to multiplication by an element of $\mathcal{O}^*_{w_0}$.*

Proof. We show that for each $\mathcal{M}_{w_0}$-linear subspace X of $(\mathcal{M} \otimes_{\mathcal{O}} \mathcal{H})_{w_0}$ with $\mathcal{M}_{w_0}$-dimension 1, the $\mathcal{O}_{w_0}$-module $X_0 = X \cap \mathcal{H}_{w_0}$ is generated by one element.

Take $f \in X_0$, $f \neq 0$. If f vanishes on $N(\psi)$ for some irreducible $\psi \in \mathcal{O}_{w_0}$, then Lemma 7.3.9 and Definition 7.3.11 show that $\frac{1}{\psi} f \in X_0$. Replace f by $\frac{1}{\psi} f$. After a finite number of steps f does not vanish on any $N(\psi)$ with $\psi \in \mathcal{O}_{w_0}$, $\psi(w_0) = 0$, $\psi \neq 0$. Suppose $g \in X_0$. Then $g = \frac{\nu}{\delta} f$, with ν and δ relatively prime in $\mathcal{O}_{w_0}$. Now $\delta g = \nu f$ implies that f vanishes on $N(\delta)$. Hence δ is a unit in $\mathcal{O}_{w_0}$, and $g \in \mathcal{O}_{w_0} \cdot f$.

12.2.5 Lemma. *Let $f \in \mathcal{H}(\Omega_0)$, with Ω_0 a neighborhood of w_0. If the germ of f at w_0 is primitive, then there exists a neighborhood $\Omega \subset \Omega_0$ of w_0 such that the germ of f at w is primitive for all $w \in \Omega$.*

Proof. If $f(w_0) \neq 0$, then clearly $f(w) \neq 0$ for all w in some neighborhood $\Omega \subset \Omega_0$ of w_0. This gives primitivity at all points of Ω.

Let $f(w_0) = 0$. For $g \in U$ we define $f_g : w \mapsto f(w; g)$. Take $g_1 \in U$ such that $f_{g_1} \neq 0$. As f is primitive at w_0, there is some $g_2 \in U$ such that f_{g_1} and f_{g_2} have no common factor in $\mathcal{O}_{w_0}$. Then f_{g_1} and f_{g_2} are relatively prime on a neighborhood $\Omega \subset \Omega_0$ of w_0, see [23], Theorem 6.2.3. Consider $w_1 \in \Omega$, and $\alpha, \beta \in \mathcal{O}_{w_1}$, relatively prime, such that $\frac{\alpha}{\beta} f \in \mathcal{H}_{w_1}$. Then β divides f_{g_1} and f_{g_2} in $\mathcal{O}_{w_1}$, hence β is a unit, and $\frac{\alpha}{\beta} f \in \mathcal{O}_{w_1} \cdot f$. This shows that f is primitive at w_1.

12.2.6 *Remark.* Let $f \in (\mathcal{M} \otimes_{\mathcal{O}} \mathcal{H})_{w_0}$, $f \neq 0$, and let ψf be a primitive multiple. The reasoning in the proof of Lemma 7.4.14 shows that $N(\psi)$ and $\mathrm{Pol}(f)$ coincide as germs at w_0.

12.2.7 *Definition.* Let f be a *primitive germ* in $\mathcal{H}_{w_0}$. For a local curve $j : V \to W$ at w_0, the germ $f \circ j \in \left(\mathcal{H}_{(j)}\right)_0$ need not be primitive; it may even vanish if $\dim W > 2$. If $f \circ j = 0$, we define $f[j] = 0 \in H(w_0)$. Otherwise we take $m \geq 0$ such that $v \mapsto v^{-m}(f \circ j)(v)$ is primitive, and define $f[j] \in H(w_0)$ as the non-zero value at $0 \in \mathbb{C}$ of this primitive germ. Thus we have defined $f[j] \in H(w_0)$ for each local curve j at w_0.

We define the *value set* of f at w_0 as

$$f\{w_0\} = \{\, f[j] : j \text{ local curve at } w_0 \,\}.$$

12.2.8 Lemma. *Let $f \in \mathcal{H}_{w_0}$ be primitive. Then there are the following possibilities for $f\{w_0\}$:*

i) $f\{w_0\}$ consists of one non-zero element of $H(w_0)$. This occurs if and only if $f(w_0) \neq 0$. If this holds, then $f\{w_0\} = \{f(w_0)\}$.

ii) $f\{w_0\}$ is a linear subspace of $H(w_0)$ minus the origin.

iii) $f\{w_0\}$ is a linear subspace of $H(w_0)$.

In the cases ii) and iii) the dimension of $f\{w_0\} \cup \{0\}$ is at least 1.

Proof. If $f(w_0) \neq 0$, then $f[j] = f(w_0)$ for each local curve j at w_0. This gives the forward implication in i).

Suppose $f(w_0) = 0$ in the remainder of the proof. As $f \neq 0$, there is a local curve j such that $f \circ j \neq 0$. From $f \circ j(0) = f(w_0) = 0$ it follows that $0 \neq f[j] = \lim_{u \to 0} u^{-m}(f \circ j)(u)$ for some positive m. Take $j_1(u) = j(tu)$ with $t \in \mathbb{C}^*$; then $f[j_1] = t^m f[j]$. This shows that $\mathbb{C}^* \cdot f\{w_0\} = f\{w_0\}$, and gives the reverse implication in i).

Suppose $h_i = \lim_{v \to 0} v^{-m_i}(f \circ j_i)(v)$, $m_i > 0$, with j_i a local curve at w_0 for $i = 1, 2$, and let $h_1 \neq 0$, $h_2 \neq 0$ and $h_1 \neq -h_2$. Take $a_1 = m_2/(m_1, m_2)$ and $a_2 = m_1/(m_1, m_2)$. Choose coordinates at w_0 such that w_0 corresponds to $0 \in \mathbb{C}^d$. Consider the expansion in $x, y \in \mathbb{C}^d$ on a neighborhood of 0. We have $f(x+y; g) = f(x; g) + f(y; g) +$ higher order terms in x and y, for each $g \in U$. Put $j_0(v) = j_1(v^{a_1}) + j_2(v^{a_2})$. We obtain $f(j_0(v); g) = v^{a_1 m_1} h_1(g) + \mathcal{O}(v^{a_1(m_1+1)}) + v^{a_2 m_2} h_2(g) + \mathcal{O}(v^{a_2(m_2+1)}) = v^m \left(h_1(g) + h_2(g)\right) + \mathcal{O}\left(v^{m+1}\right)$, with $m = m_1 m_2/(m_1, m_2)$. Hence $\lim_{v \to 0} v^{-m} f(j_0(v); g) = h_1(g) + h_2(g)$. Apply Lemma 7.3.8 and the fact that we are working with an eigenfunction module to see that $h_1 + h_2$ is in $H(w_0)$. If j_0 is not minimal in the sense of 12.1.2, write $j_0 = j \circ h$. We obtain $h_1 + h_2$ as the value at 0 of $v \mapsto v^{m/t} f \circ j(v)$, where t is the order of h at 0.

Thus we see that if for two non-zero elements h_1 and h_2 of $f\{w_0\}$ the sum h_1+h_2 does not vanish, then $h_1+h_2 \in f\{w_0\}$. Together with $\mathbb{C}^* \cdot f\{w_0\} = f\{w_0\}$ this leaves ii) and iii) as the possibilities.

12.2.9 *Remark.* We have defined $f\{w_0\}$ for a primitive germ. Replacing f by ψf, with ψ a unit in $\mathcal{O}_{w_0}$, has the effect that all elements of $f\{w_0\}$ are multiplied by $\psi(w_0)$. In the cases ii) and iii) of the lemma this does not change $f\{w_0\}$.

12.2.10 *Definition.* Let $f \in (\mathcal{M} \otimes_{\mathcal{O}} \mathcal{H})_{w_0}$, $f \neq 0$. Let f_1 be a primitive multiple of f. If $f_1(w_0) \neq 0$ we call the germ f of *determinate type*, otherwise f is of *indeterminate type.*

If f is of indeterminate type the set $f_1\{w_0\}$ does not depend on the choice of the primitive multiple f_1. In this case it is sensible to define $f\{w_0\} = f_1\{w_0\}$. We do not define $f\{w_0\}$ for a non-primitive germ f of determinate type.

12.2.11 *Remarks.* All non-zero germs in $\mathcal{M}_{w_0} \cdot f$ are of the same type. In particular, a holomorphic germ may have indeterminate type.

If $\dim W \leq 1$, then each non-zero element of $(\mathcal{M} \otimes_{\mathcal{O}} \mathcal{H})_{w_0}$ has determinate type. This should not be confused with w_0 being a point of indeterminacy of a meromorphic function h, where the zero set and the polar set intersect each other. In that situation h takes any value in any neighborhood of w_0; c.f. [23], Theorem 6.2.3.

12.2.12 Proposition. *Suppose* $\dim W \geq 2$. *Let* $f \in (\mathcal{M} \otimes_{\mathcal{O}} \mathcal{H})_{w_0}$ *be of indeterminate type at* w_0. *Then* $f\{w_0\}$ *spans a linear subspace of* $H(w_0)$ *of dimension at least 2.*

Proof. We assume f to be primitive. Take a local curve j_1 at w_0 such that $f[j_1] \neq 0$. There is a $g_1 \in U$ such that $f[j_1](g_1) \neq 0$. Proposition 12.1.7 gives a local curve j_2 at w_0 along $N(f_{g_1})$ such that $f \circ j_2$ is not identically zero. But $u \mapsto f(j_2(u), g_1) = 0$ for all u near 0, hence $f[j_2](g_1) = 0$, whereas $f[j_2] \neq 0$. As $f[j_1](g_1) \neq 0$, we have obtained two linearly independent elements of $f\{w_0\}$.

Example. In 15.6.21 we give an explicit example of a two-dimensional value set.

12.2.13 Proposition. *Let* $f \in (\mathcal{M} \otimes_{\mathcal{O}} \mathcal{H})(\Omega_0)$, $f \neq 0$, Ω_0 *a neighborhood of* w_0. *Define* $\mathrm{Indet}(f) = \{w \in \Omega_0 :$ *the germ of* f *at* w *is of indeterminate type*$\}$. *This is an analytic subset of* Ω_0, *and* $\dim_{w_0} \mathrm{Indet}(f) \leq \dim(W) - 2$.

Proof. We can assume that $f \in \mathcal{H}(\Omega_0)$, and that it is primitive at all points of Ω_0, see Lemma 12.2.5. Again we use f_g as defined in the proof of that lemma. Definition 12.2.10 amounts to

$$w \in \mathrm{Indet}(f) \iff \forall g \in U : f_g(w) = 0.$$

Hence $\mathrm{Indet}(f)$ is the intersection of the analytic sets $N(f_g)$ for $g \in U$. In [17], Ch. 5, §6.1 we see that any intersection of analytic sets is analytic.

Take $g_1, g_2 \in U$ such that f_{g_1} and f_{g_2} are relatively prime at w_0. In [17], Ch. 5, §2.4 we see that $\dim_{w_0} N(f_{g_i}) = \dim(W) - 1$ for $i = 1, 2$. If $N(f_{g_1}) \cap N(f_{g_2})$ would have the same dimension at w_0 as $N(f_{g_1})$, then f_{g_2} would have an irreducible factor in common with f_{g_1} in $\mathcal{O}_{w_0}$, see [17], Ch. 5, §3.2. Hence $\dim_{w_0} N(f) \leq \dim_{w_0} (N(f_{g_1}) \cap N(f_{g_2})) < \dim(W) - 1$.

12.3 General results on singularities

We turn to the study of singularities of Poincaré families depending both on character and spectral parameter. The previous sections have given us some idea of the complications we may expect when dealing with families of automorphic forms on a parameter space with dimension more than 1.

Like we did in the previous chapter, we first consider the families $R(P, n)$ obtained in Proposition 10.2.4. These families behave well with respect to a holomorphic basis of the Fourier term modules. We use these results in the later sections to obtain information concerning singularities of the actual Poincaré families.

12.3.1 *Notations.* We fix an $\mathbb{R}$-linear subspace V_r of $\mathcal{V}_r$ with $\dim_{\mathbb{R}} V_r > 0$. We put $V = \mathbb{C} \otimes_{\mathbb{R}} V_r$.

We take J, P, n, χ_1, l and $\mathbf{c}_{P,n}$ as in Theorem 10.2.1. We fix $\varphi_0 \in J^0$; this is a stronger condition than the one in 10.2.3. We put $\chi_0 = \chi_1 \cdot \mathbf{exp}(\varphi_0)$, $l_0 = l(\chi_1, \varphi_0)$, $n_0 = n(\varphi_0)$, and take $s_0 \in \mathbb{C}$, $\mathrm{Re}\, s_0 \geq 0$, and put $w_0 = (\varphi_0, s_0)$. We take a neighborhood U_0 of φ_0 in $\varphi_0 + V$ and a neighborhood Ω_0 of w_0 contained in $U_0 \times \mathbb{C}$.

12.3.2 *Basis.* We choose $\lambda_{Q,m}$ and $\nu_{Q,m}$ in $\mathcal{M} \otimes_{\mathcal{O}} \mathcal{W}_l(U_0 \times \mathbb{C})$ for all $Q \in \mathcal{P}$, $m \in \mathbf{c}_{P,n}(Q)$, such that

i) $\lambda_{Q,m}, \nu_{Q,m} \in \mathcal{W}_l(\Omega_0)$.

ii) $\mathrm{Wr}(\lambda_{Q,m}, \nu_{Q,m})(w) \neq 0$ for each $w \in \Omega_0$.

iii) $\nu_{Q,m}(\varphi_0, s) = \dot{\eta}_l(Q, m; \varphi_0; s)$ for all $s \in \mathbb{C}$ such that $(\varphi_0, s) \in \Omega_0$. The section $\eta_l(Q, m)$ has been defined in 10.1.8; $\dot{\eta}_l(Q, m; \varphi_0)$ denotes its restriction to the vertical line $\{\varphi_0\} \times \mathbb{C}$.

Outside $\{\varphi_0\} \times \mathbb{C}$ the sections $\nu_{Q,m}$ and $\eta_l(Q, m)$ need not have the same restrictions to vertical lines. The sections $\lambda_{Q,m}$ and $\nu_{Q,m}$ satisfy the conditions in 10.2.3. We have imposed the additional condition of holomorphy on Ω_0.

12.3.3 *Transformation.* The standard sections $\mu_l(Q, m)$ and $\eta_l(Q, m)$ can be expressed in this new basis by

$$\begin{pmatrix} \mu_l(Q, m) \\ \eta_l(Q, m) \end{pmatrix} = \begin{pmatrix} \alpha_{R,m} & \beta_{R,m} \\ \gamma_{R,m} & \delta_{R,m} \end{pmatrix} \begin{pmatrix} \lambda_{Q,m} \\ \nu_{Q,m} \end{pmatrix}$$

with $\alpha_{R,m},\ldots,\delta_{R,m}$ meromorphic on Ω_0 for all (R,m) with $R\in\mathcal{P}$ and $m\in\mathbf{c}_{P,n}(R)$. For these (R,m) we have

$$\begin{aligned}&\tilde F_{R,m}Q_l(J,P,n)\\ &\quad= \delta_{P,R}\delta_{n,m}w_P\mu_l(P,n)+D_l(R,m;J,P,n)\eta_l(R,m)\\ &\quad\in (\delta_{P,R}\delta_{n,m}w_P\alpha_{P,n}+D_l(R,m;J,P,n)\gamma_{R,m})\,\lambda_{R,m}+\mathcal{M}(\Omega_0)\cdot\nu_{R,m};\end{aligned}$$

hence

$$Q_l(J,P,n)=\alpha_{P,n}R(P,n)+\sum_{(R,m)}\frac{\gamma_{R,m}}{w_R}D_l(R,m;J,P,n)R(R,m),$$

where $R(R,m)$ denotes the families of automorphic forms that we obtained in Proposition 10.2.4. The D_l occurring in the right hand side are Fourier coefficients of some Q_l. So this relation does not explicitly express the Poincaré families in terms of the families $R(Q,m)$.

We define the Fourier coefficients $B(Q_2,m_2;Q_1,m_1)$ of $R(Q_1,m_1)$ by

$$\tilde F_{Q_2,m_2}R(Q_1,m_1)=w_{Q_1}\delta_{Q_2,Q_1}\delta_{m_2,m_1}\lambda_{Q_1,m_1}+B(Q_2,m_2;Q_1,m_1)\nu_{Q_2,m_2}$$

for $Q_1,Q_2\in\mathcal{P}$, $m_1\in\mathbf{c}_{J,n}(Q_1)$, $m_2\in\mathbf{c}_{J,n}(Q_2)$. The $B(Q_2,m_2;Q_1,m_1)$ are meromorphic functions on Ω_0. The weight l is suppressed in this notation. When we shall work with more than one weight at the same time, we shall write $R_l(P,n)$ and $B_l(Q_2,m_2;Q_1,m_1)$.

12.3.4 Lemma. *If U_0 is taken small enough there is a unique polynomial*

$$\psi(\varphi,s)=\sum_{l=0}^{k-1}a_l(\varphi)(s-s_0)^l+(s-s_0)^k$$

with coefficients a_l holomorphic on U_0, $a_l(\varphi_0)=0$, such that the germ of $\psi\cdot R(P,n)$ at w_0 is primitive.

Take q minimal in $\mathbb{Z}$ such that the family $s\mapsto(s-s_0)^q\dot R(P,n;\varphi_0)$ is holomorphic at s_0. Then $q\le k$.

- *i) $k=q$ if and only if $R(P,n)$ has determinate type at w_0.*
- *ii) Let $k>q$. Then $s_0\in\{0\}\cup\Sigma(\chi_0,l_0)$; the value set $R(P,n)\{w_0\}$ contains $\lim_{s\to s_0}(s-s_0)^q\dot R(P,n;\varphi_0)(s)$, and non-zero elements of $A^*_{l_0}(\chi_0,J,s_0)$.*

Remarks. $\dot R(P,n;\varphi_0)$ denotes the restriction $s\mapsto R(P,n)(\varphi_0,s)$. In 11.1.3 we called this function $R(P,n)$. The number q is the same as the q in Chapter 11.

See 11.2.8 for the definition of $A^*_{l_0}(\chi_0,J,s_0)$.

Proof. Take $\psi\in\mathcal{M}_{w_0}$ such that $\psi\cdot R(P,n)$ is a primitive multiple of $R(P,n)$. This meromorphic function ψ is determined up to a unit in $\mathcal{O}_{w_0}$. We conclude from $\tilde{F}_{P,n}\,\psi R(P,n) = \psi\lambda_{P,n} + \psi B(P,n;P,n)\nu_{P,n}$ that ψ and $\psi B(P,n;P,n)$ are in $\mathcal{O}_{w_0}$. Proposition 10.2.4 implies that the restriction $\dot{R}(P,n;\varphi_0)$ of $R(P,n)$ to $(\{\varphi_0\}\times\mathbb{C})\cap\Omega_0$ exists as a meromorphic section; hence ψ does not vanish identically along this line. We take ψ in Weierstrass form with respect to $s-s_0$, see [23], Corollary 6.1.2. This determines ψ uniquely. The holomorphic restriction $s\mapsto(\psi R(P,n))(\varphi_0,s)$ is equal to $s\mapsto(s-s_0)^k\dot{R}(P,n;\varphi_0)$. Hence $k\geq q$, and $k=q$ if and only if the value of $\psi R(P,n)$ at w_0 is non-zero.

Let $k>q$. Take a local curve j at w_0 along $N(\psi)$. Inspection of the Fourier terms shows that $(\psi R(P,n))[j]\in A^*_{l_0}(\chi_0,J,s_0)$, and that $s_0\in\{0\}\cup\Sigma(\chi_0,l_0)$ (use Lemma 11.2.9). On the other hand, the vertical local curve $j_0: u\mapsto(\varphi_0,s_0+u)$ gives $(\psi R(P,n))\,[j_0]=\lim_{s\to s_0}(s-s_0)^q\dot{R}(P,n)$.

12.3.5 *Choice of a basis for the Fourier terms.* Up till now we have not used the precise choice of bases for the Fourier terms. We find in the Lemmas 7.6.13 and 7.6.15 a choice of a basis satisfying the conditions in 12.3.2 for all points w_0 under consideration. The results are summarized in Table 12.1.

In cases **4** and **5** in the table the expression for $\alpha_{R,m}$ has been omitted. In both cases it has the form $\alpha_{R,m}=\tilde{v}/(\tilde{v}\hat{v}-v\tilde{\hat{v}})$, with v and $\hat{v}$ as in 4.2.8, respectively 4.2.9. A computation gives for case **4**

$$\alpha_{R,m}(\varphi,s) = \frac{(4\pi\varepsilon m(\varphi))^{-s-1/2}\Gamma(2s+1)e^{\pi i\varepsilon l(\varphi)/2}}{\Gamma(\frac12+s-\frac12\varepsilon l(\varphi))},$$

and for case **5**

$$\alpha_{R,m}(\varphi,s)=i\frac{\Gamma(2s+1)}{\Gamma(1+2q)}e^{\pi i(q(\varphi)-p)}\frac{\Gamma(\frac12-s+p+q(\varphi))}{\Gamma(\frac12+s+p-q(\varphi))},$$

with $q(\varphi)=\frac14\zeta(n(\varphi)+l(\varphi))$.

The choices in Table 12.1 are not canonical, but we use them as our standard choice in the next sections. Except for case **3**, these $\lambda_{R,m}$ and $\nu_{R,m}$ satisfy the conditions in 12.3.2, not only for w_0, but for all $w_1\in(\varphi_0+V_r)\times\mathbb{C}$ in some neighborhood of w_0.

12.3.6 *Large* s_0. In the remainder of this section we assume $\operatorname{Re}s_0>|\tau(\chi_0)|$. This leads to more explicit statements concerning ψ and k in Lemma 12.3.4.

Proposition 11.2.4 and Lemma 11.3.6 imply that $q>0$ is possible only if $s_0\in\Sigma_d(\chi_0,s_0)$. Lemma 12.3.4 shows that $k>0$ also implies $s_0\in\Sigma_d(\chi_0,l_0)$. So we are led to the discrete series case, for which we find conditions in Propositions 6.7.3 and 6.7.6. In particular, $s_0=\left|\frac{b-1}{2}\right|$ with $0<b\leq|l_0|$, $b\equiv l_0 \bmod 2$. As $2\tau(\chi_0)+1\equiv l_0 \bmod 2$, we get $\tau(\chi_0)\cdot\operatorname{sign}(l_0)\equiv\frac{b-1}{2} \bmod 1$. The assumption $\operatorname{Re}s_0>|\tau(\chi_0)|$ implies $b>2$.

Let $l\in\mathbf{wt}(\Omega_0)$. We denote by $\psi_l(Q,m)$ the function ψ associated to $R_l(Q,m)$ in Lemma 12.3.4, and by $k_l(Q,m)$ and $q_l(Q,m)$ the corresponding numbers k and q.

(R,m) singular, $\operatorname{Re} s_0 \geq 0$, $s_0 \notin \frac{1}{2}\mathbb{Z}$						**1**
$\lambda_{R,m}$	$=$	$\mu_l(R,m)$	$\nu_{R,m}$	$=$	$\eta_l(R,m) = \tilde{\mu}_l(R,m)$	
$\alpha_{R,m}(\varphi,s)$	$=$	1	$\gamma_{R,m}(\varphi,s)$	$=$	0	
(R,m) singular, $s_0 = 0$						**2**
$\lambda_{R,m}$	$=$	$\lambda_l(R,m)$	$\nu_{R,m}$	$=$	$\eta_l(R,m) = \tilde{\mu}_l(R,m)$	
$\alpha_{R,m}(\varphi,s)$	$=$	$2s$	$\gamma_{R,m}(\varphi,s)$	$=$	0	
(R,m) singular, $s_0 = \frac{b}{2}$, $b \in \mathbb{N}$						**3**
$\lambda_{R,m}$	$=$	$\mu_l(R,m)$	$\nu_{R,m}$	$=$	$\nu_l^b(R,m)$	
$\alpha_{R,m}(\varphi,s)$	$=$	1	$\gamma_{R,m}(\varphi,s)$	$=$	$-w_l^b(m(\varphi),s)$	
				$=$	$-(4\pi m(\varphi))^b \frac{\Gamma(-2s)}{\Gamma(2s)} \cdot \left(\frac{1}{2} - s - \frac{l(\varphi)}{2}\right)_b$	
$R \in X^\infty(J)$, $\varepsilon \in \{\pm 1\}$, $\varepsilon m(\varphi_0) > 0$, $0 \leq s_0 \leq \frac{\varepsilon l_0 - 1}{2}$, $s_0 \equiv \frac{\varepsilon l_0 - 1}{2} \bmod 1$						**4**
$\lambda_{R,m}$	$=$	$\hat{\omega}_l(R,m)$	$\nu_{R,m}$	$=$	$\eta_l(R,m) = \omega_l(R,m)$	
$\alpha_{R,m}$ complicated			$\gamma_{R,m}(\varphi,s)$	$=$	0	
$R \in \mathcal{P}_Y$, $\varepsilon, \zeta \in \{\pm 1\}$, $p = \frac{\varepsilon}{4}(m(\varphi_0) - l_0) \geq 0$, $q_0 = \frac{\zeta}{4}(m(\varphi_0) + l_0) \geq 0$, $0 \leq s_0 \leq q_0 - p - \frac{1}{2}$, $s_0 \equiv q_0 - p - \frac{1}{2} \bmod 1$						**5**
$\lambda_{R,m}$	$=$	$\hat{\omega}_l(P,m,\zeta)$	$\nu_{R,m}$	$=$	$\eta_l(R,m) = \omega_l(R,m)$	
$\alpha_{R,m}$ complicated			$\gamma_{R,m}(\varphi,s)$	$=$	0	
(R,m) regular, $\operatorname{Re} s_0 \geq 0$, not case **4** or **5**						**6**
$\lambda_{R,m}$	$=$	$\mu_l(R,m)$	$\nu_{R,m}$	$=$	$\eta_l(R,m) = \omega_l(R,m)$	
$\alpha_{R,m}(\varphi,s)$	$=$	1	$\gamma_{R,m}(\varphi,s)$	$=$	0	

Table 12.1 Holomorphic bases for the Fourier terms. (We call (R,m) *singular* if $R \in X^\infty(J)$ and $m(\varphi_0) = 0$; we call it *regular* in all other cases.)

12.3.7 Lemma. *Let* $\mathrm{Re}\, s_0 > |\tau(\chi_0)|$, *hence* $s_0 > \frac{1}{2}$. *Then* $\psi_l(P,n) = 1$, $k_l(P,n) = 0$, *except possibly if the following conditions hold:*

$$2s_0 + 1 \equiv |l_0| \bmod 2 \text{ and } 2 < 2s_0 + 1 \leq |l_0|.$$

Proof. From $k > 0$ we obtain $s_0 \in \Sigma_d(\chi_0, l_0)$, and $s_0 = \frac{b-1}{2}$, with $2 < b \leq |l_0|$, $b \equiv |l_0| \bmod 2$. This gives the conditions in the lemma.

12.3.8 *Other weights.* If the condition in Lemma 12.3.7 is satisfied for $l_0 = l(\chi_1, \varphi_0)$ for some $l \in \mathbf{wt}(\Omega_0)$, then this holds for infinitely many $l \in \mathbf{wt}(\Omega)$. Let $\varepsilon \in \{1, -1\}$, and $b > 2$. In the next discussion we use

$$\mathbf{wt}_b^\varepsilon(\Omega_0) = \{\, l \in \mathbf{wt}(\Omega_0) : \varepsilon l_0 \geq b,\, \varepsilon l_0 \equiv b \bmod 2 \,\}.$$

If $\mathbf{wt}_b^\varepsilon(\Omega_0) \neq \emptyset$, then it is of the form $\{\, l_b^\varepsilon + 2\kappa : \kappa \in \mathbb{Z},\, \kappa \geq 0 \,\}$, with $l_b^\varepsilon \in \mathbf{wt}(\Omega_0)$ determined by $l_b^\varepsilon(\varphi_0) = \varepsilon b$.

12.3.9 Lemma. *Let* $s_0 > |\tau(\chi_0)|$, $\varepsilon \in \{1, -1\}$. *Put* $b = 2s_0 + 1$. *Suppose that* $\mathbf{wt}_b^\varepsilon(\Omega_0) \neq \emptyset$. *Then exactly one of the following statements holds:*

A) $\psi_l(P,n) = 1$, $k_l(P,n) = 0$, *for all* $l \in \mathbf{wt}_b^\varepsilon(\Omega_0)$*;*

B) $\psi_l(P,n;\varphi,s) = s + \frac{1}{2} - \frac{1}{2}\varepsilon l(\varphi)$, $k_l(P,n) = 1$, *for all* $l \in \mathbf{wt}_b^\varepsilon(\Omega_0)$. *In this case* $\varepsilon n(\varphi) > 0$ *if* $P \in X^\infty$, *and* $\varepsilon n(\varphi_0) \geq b$ *if* $P \in \mathcal{P}_Y$.

Remarks. This implies that statement A) holds in the singular case $P \in X^\infty(J)$, $n_0 = n(\varphi_0) = 0$.

In the situation of statement B) it may happen that $\dot{Q}_l(J,P,n;s_0) = 0$. Then $R_l(P,n)$ is of indeterminate type at w_0. See 13.4.6 for an example.

Proof. In 12.3.10–12.3.14.

12.3.10 *Differentiation results.* If the neighborhood Ω_0 of w_0 is small enough, there are holomorphic functions $a_l^\pm(Q,m)$ and $b_l^\pm(Q,m)$ such that

$$\begin{aligned} \mathbf{E}^\pm \lambda_{Q,m} &\in a_l^\pm(Q,m)\lambda_{Q,m}^\pm + \mathcal{O}(\Omega_0)\nu_{Q,m}^\pm \\ \mathbf{E}^\pm \nu_{Q,m} &\in b_l^\pm(Q,m)\lambda_{Q,m}^\pm + \mathcal{O}(\Omega_0)\nu_{Q,m}^\pm, \end{aligned}$$

where $\lambda_{Q,m}^\pm$ and $\nu_{Q,m}^\pm$ denote the corresponding sections in weight $l \pm 2$. The differentiation results in Table 4.1 on p. 63 imply that $b_l^\pm(Q,m) = 0$, except in the case that (Q,m) is singular and $s_0 \in \frac{1}{2}\mathbb{N}$; in that case the value $b_l^\pm(Q,m;w_0)$ is nevertheless zero. If l_0 or $l_0 \pm 2$ do not come under case **4** or **5** in Table 12.1, we have $a_l^\pm(Q,m;\varphi,s) = \zeta(1 + 2s \pm l(\varphi))$, with $\zeta = 1$ or -1, according to $P \in X^\infty$ or $P \in \mathcal{P}_Y$.

12.3.11 *Singular case.* Take $Q \in X^\infty(J)$. Let m_Q denote the element of $\mathcal{C}_Q$ that vanishes on J^0. The uniqueness in Proposition 10.2.4 implies

$$\begin{aligned} \mathbf{E}^\pm R_l(Q,m_Q) &= a_l^\pm(Q,m_Q)R_{l\pm2}(Q,m_Q) \\ &+ \sum_{R\in X^\infty(J)} b_l^\pm(R,m_R)B_l(R,m_R;Q,m_Q)R_{l\pm2}(R,m_R). \end{aligned} \tag{12.1}$$

Hence the vector of derivatives $(\mathbf{E}^\pm R_l(Q,m_Q))_{Q\in X^\infty(J)}$ depends linearly on the vector $(R_{l\pm2}(Q,m_Q))_{Q\in X^\infty(J)}$; the relation is described by a holomorphic matrix that has the value $(1+2s_0\pm l_0)\cdot \mathrm{Id}$ at $w=w_0$.

We proceed by induction. We start at $l=l_b^\varepsilon-2\varepsilon$. As $|l_0|=|\varepsilon b-2\varepsilon|=b-2<2s_0+1$, Lemma 12.3.7 implies the holomorphy of $R_l(R,m_R)$ for all $R\in X^\infty(J)$. We apply $\mathbf{E}^\pm$ repeatedly, with $\pm$ chosen such that $\pm\varepsilon=1$. At each step the matrix found above has a holomorphic inverse on a neighborhood of w_0, and expresses $R_{l\pm2}(R,m_R)$ holomorphically in the holomorphic $\mathbf{E}^\pm R_l(Q,m_Q)$. This gives the holomorphy of $R_l(R,m_R)$, $R\in X^\infty(J)$ for all $l\in \mathbf{wt}_b^\varepsilon(\Omega_0)$, and hence $\psi_l(R,m_R)=1$, and $k_l(R,m_R)=0$.

12.3.12 *Regular case.* Let (P,n) be regular. We have seen that the $\psi_l R_l(Q,m_Q)$ are holomorphic at w_0. As $\mathbf{E}^\pm\psi_l(P,n)R_l(P,n)$, and all $\psi_l(P,n)B_l(Q,m_Q;P,n)$ with $Q\in X^\infty(J)$, are holomorphic at w_0, we obtain from (12.1) on p. 226 the holomorphy of $a_l^\pm(P,n)\psi_l(P,n)R_{l\pm2}(P,n)$. Thus we see that $\psi_{l\pm2}(P,n)$ divides $a_l^\pm(P,n)\psi_l(P,n)$ in $\mathcal{O}_{w_0}$.

For $l=l_b^\varepsilon-2\varepsilon$ we know that $R_l(P,n)$ is holomorphic at w_0, and $\psi_l(P,n)=1$. We choose again $\pm$ such that $\pm\varepsilon=1$, and apply $\mathbf{E}^\pm$. We have $(1\,\mathrm{or}\,-1)\cdot a_l^\pm(P,n;w_0)=1+2s_0\pm l_0>0$, hence $\psi_{l\pm2}(P,n)=1$, as long as l_0 and $l_0\pm2$ do not come under case **4** or **5** in Table 12.1. We investigate the cases $P\in X^\infty(J)$ and $P\in\mathcal{P}_Y$ separately.

12.3.13 *Cuspidal case.* Let $P\in X^\infty(J)$. If $\operatorname{sign} n_0=-\varepsilon$, we never come under case **4** in Table 12.1. Indeed, $\operatorname{sign}(n_0)\cdot l_0=-\varepsilon l_0<0$ for all $l\in\mathbf{wt}_b^\varepsilon(\Omega_0)$. Suppose $\varepsilon n_0>0$. For $l=l_b^\varepsilon-2\varepsilon$

$$\begin{aligned} \mathbf{E}^\pm\lambda_{P,n}(\varphi,s) &= \mathbf{E}^\pm\mu_l(P,n;\varphi,s)=(1+2s\pm l(\varphi))\mu_{l\pm2}(P,n;\varphi,s) \\ &= (1+2s\pm l(\varphi))\alpha_{P,n}(\varphi,s)\hat{\omega}_{l\pm2}(P,n;\varphi,s)+\text{ a multiple of }\nu_{P,n}^\pm, \end{aligned}$$

with $\alpha_{P,n}$ as in 12.3.5, for the weight $l\pm2$. Thus we obtain

$$a_l^\pm(P,n;\varphi,s)=\frac{-(1+2s+\varepsilon l(\varphi))(4\pi\varepsilon n(\varphi))^{-s-1/2}\Gamma(2s+1)e^{\pi i\varepsilon l(\varphi)/2}}{\Gamma(-\frac12+s-\frac12\varepsilon l(\varphi))}.$$

This is not a unit in $\mathcal{O}_{w_0}^*$. From the factor in the denominator we conclude that $\psi_{l+2\varepsilon}(P,n)$ is equal to $s-\frac12-\frac12\varepsilon l=s+\frac12-\frac{\varepsilon}{2}(l\pm2)$ or to 1.

Once $l\in\mathbf{wt}_b^\varepsilon(\Omega_0)$, both l_0 and $l_0\pm2$ come under case **4**. Table 4.1 on p. 63 implies that $a_l^\pm(P,n;\varphi,s)=\frac12(\varepsilon l+1)^2-2s^2$, which is a unit in $\mathcal{O}_{w_0}$. Hence $\psi_{l\pm2}$

divides ψ_l. Going from $l \pm 2$ to l by $\mathbf{E}^{\mp}$ shows that $\psi_l(P,n)$ and $\psi_{l\pm 2}(P,n)$ are equal (again use Table 4.1).

12.3.14 *Interior case.* Let $P \in \mathcal{P}_Y$. For a given $s_0 = \frac{b-1}{2} > \frac{1}{2}$, the condition for case **5** in Table 12.1 amounts to

$$\frac{1}{2}|n_0 + l_0| - \frac{1}{2}|n_0 - l_0| \geq b, \quad \frac{1}{2}|n_0 + l_0| - \frac{1}{2}|n_0 - l_0| \equiv b \bmod 2,$$

or, equivalently, (use $n_0 \equiv l_0 \bmod 2$)

$$\operatorname{sign}(l_0) = \operatorname{sign}(n_0), \quad |n_0| \equiv |l_0| \equiv b \bmod 2, \quad \min(|n_0|, |l_0|) \geq b.$$

We have to apply this to the $l = l_b^\varepsilon + 2\kappa\varepsilon$, with $\kappa \geq -1$, $\kappa \in \mathbb{Z}$, for a fixed n, satisfying $n_0 \equiv \varepsilon b \bmod 2$.

If $|n_0| < b$, case **5** does not occur, and $\psi_l(P,n) = 1$ for all l that we consider. The same holds for $\varepsilon n_0 \leq -b$. (Use $b > 2$ to see that $\varepsilon l_0 > 0$ even if $\kappa = -1$.) Finally, we consider $\varepsilon n_0 \geq b$. We get into case **5** as soon as $\kappa \geq 0$, and stay there. For the starting weight $l = l_b^\varepsilon - 2\varepsilon$ we obtain

$$a_l^{\pm}(P,n;\varphi,s) = i(1 + 2s + \varepsilon l(\varphi)) \frac{\Gamma(2s+1) e^{\pi i \varepsilon l(\varphi)/2} \Gamma(\frac{1}{2} - s + \frac{1}{2}\varepsilon n(\varphi))}{\Gamma(2 + \frac{1}{2}\varepsilon n(\varphi) + \frac{1}{2}\varepsilon l(\varphi)) \Gamma(-\frac{1}{2} + s - \frac{1}{2}\varepsilon l(\varphi))}.$$

As in the cuspidal case, we conclude that $\psi_{l_b^\varepsilon}(P,n;\varphi,s)$ may be equal to 1 or to $s - \frac{1}{2} - \frac{1}{2}\varepsilon l$. Again apply Table 4.1 to see that ψ_l does not change when we go to $k > 0$.

12.4 General parameter spaces

Singularities of Poincaré families at points (φ_0, s_0) with $\varphi_0 \in \mathcal{V}_r$ and $\operatorname{Re} s_0 \geq 0$, $s_0 \neq 0$, may be due to two reasons: to spectral features, and to μ_l and η_l not being a suitable basis. The second reason we have not seen in the previous section. In fact, we have introduced the $R(P,n)$ in order to avoid this complication.

In this section we suppose the linear space V to be as 'bad' as possible, i.e., V is not contained in V_J. This gives singularities of both types. In particular, the Poincaré families may be singular at points (φ_0, s_0) with $\operatorname{Re} s_0 > \frac{1}{2}$. The presence of singularities of this type gives rise to distribution results of the type discussed in Section 13.6. Perhaps this is a reason not to call these V 'bad' after all.

12.4.1 *Notations.* Let J be a cell of continuation. We take P, n, χ_1, l and $\mathbf{c}_{J,n}$ as in Theorem 10.2.1, and assume that $X^\infty(J) \neq \emptyset$. We take an $\mathbb{R}$-linear subspace $V_r \subset \mathcal{V}_r$ with $\dim_{\mathbb{R}} V_r > 0$. *We assume that V_r is not contained in $V_{J,r}$.* Put $V = \mathbb{C} \otimes_{\mathbb{R}} V_r$.

We take $\varphi_0 \in J^0$, $s_0 \in \mathbb{C}$ with $\operatorname{Re} s_0 \geq 0$, and take χ_0, l_0, n_0, λ_0, w_0, U_0 and Ω_0 as in 12.3.1. For $w \in \Omega_0$ we write $w = (\varphi, s)$.

Let $Q \in X^\infty(J)$. We denote by m_Q the unique element of $\mathcal{C}_Q$ that satisfies $m_Q(\varphi_0) = 0$. The assumption $V_r \not\subset V_{r,J}$ implies that at least one m_Q is not identically zero on U_0.

12.4.2 Proposition. Singularities of Eisenstein families. *We use the assumptions and notations in 12.4.1. Let $P \in X^\infty(J)$.*

i) If the Eisenstein family $E_l(J,P)$ on Ω_0 has a singularity at w_0, then $s_0 \in \Sigma_c(\chi_0) \cup \{0, |\tau(\chi_0)|\} \cup \frac{1}{2}\mathbb{N}$.

ii) a) If $E_l(J,P)$ is of indeterminate type at w_0, and $s_0 \in \frac{1}{2}\mathbb{N}$, $s_0 \geq 1$, then $|X^\infty(J)| > 1$.

b) Suppose $E_l(J,P)$ is singular and of determinate type at w_0. Then $s_0 \in \Sigma^e(\chi_0, l_0) \cup \{0\} \cup \frac{1}{2}\mathbb{N}$; in particular, s_0 is real.

iii) Suppose $s_0 = \frac{b}{2} \geq 1$, $b \in \mathbb{N}$. Let $\mathbf{w}_l^b(\varphi, s)$ be the diagonal matrix of size $|\mathbf{c}_J| \times |\mathbf{c}_J|$ with

$$w_l^b(m_Q(\varphi), s) = \left(4\pi m_Q(\varphi)\right)^b \left(\frac{1}{2} - s - \frac{1}{2}l(\varphi)\right)_b \frac{\Gamma(-2s)}{\Gamma(2s)}$$

at position $Q \in X^\infty(J)$. Then $\det(\mathrm{Id} - \mathbf{w}_l^b \mathbf{C}_l(J))$ is a non-zero meromorphic function on Ω_0. The elements of the row vector $\mathbf{E}_l(J)\left(\mathrm{Id} - \mathbf{w}_l^b \mathbf{C}_l(J)\right)^{-1}$ are meromorphic families of automorphic forms on Ω_0 and are holomorphic at w_0. The value at w_0 is given by Eisenstein series.

Remarks. The row vector $\mathbf{E}_l(J)$ of Eisenstein families, and the scattering matrix $\mathbf{C}_l(J)$ have been introduced in 10.3.3. See Definition 12.2.10 for the concept '(in)determinate type'.

Proof. Consider $P \in X^\infty(J)$ and $n = m_P$.

Suppose first that $s_0 \notin \frac{1}{2}\mathbb{Z}$. In Table 12.1 we see that the bases λ_{Q,m_Q}, ν_{Q,m_Q} and $\mu_l(Q,m_Q)$, $\tilde{\mu}_l(Q,m_Q)$ coincide for all (Q,m_Q) with $Q \in X^\infty(J)$. Hence $R(P,n) = E_l(J,P)$ in this case. Lemma 12.3.4 shows that if $E_l(J,P)$ has determinate type at w_0, then it is singular at w_0 if and only if $\dot{E}_l(J,P;\varphi_0) : s \mapsto E_l(J,P;\varphi_0,s)$ is singular at s_0. This can only happen if $s_0 \in \Sigma^e(\chi_0, l_0)$; see Proposition 11.2.4. Let $E_l(J,P)$ be of indeterminate type at w_0. Lemma 12.3.4 shows that $k > q \geq 0$, and that $A^*_{l_0}(\chi_0, J, s_0) \neq \{0\}$. Hence $s_0 \in \Sigma(\chi_0, l_0)$, see Lemma 11.2.9. Lemma 12.3.9 shows that $s_0 > |\tau(\chi_0)|$ is impossible (use $n_P(\varphi_0) = 0$).

The proposition leaves all possibilities open in the case $s_0 = 0$. Hence we are left with $s_0 = \frac{1}{2}b$, $b \in \mathbb{N}$, $b \geq 2$.

Consider $Q \in X^\infty(J)$. Lemma 12.3.9 implies that $R(Q, m_Q)$ is holomorphic and has determinate type at w_0. In this case there is interaction between the cusps; so we use the vector notation discussed in 10.3.3. Suppose that the meromorphic function $\det(\mathrm{Id} - \mathbf{w}_l^b \mathbf{C}_l(J))$ on Ω_0 vanishes. Then there is a non-zero vector $\mathbf{b} \in$

$\mathcal{O}(\Omega_0)$ such that $\mathbf{b} = \mathbf{w}_l^b \mathbf{C}_l(J)\mathbf{b}$. (It may be necessary to replace Ω_0 by a smaller neighborhood of w_0.) The non-zero vector-valued family of automorphic forms $f = \mathbf{E}_l(J)\mathbf{b}$ has the columns of

$$\begin{aligned} \mathbf{F}(J)f &= \mathbf{F}(J)\mathbf{E}_l(J)\mathbf{b} = \mu_l(J)\mathbf{b} + \tilde{\mu}_l(J)\mathbf{C}_l(J)\mathbf{b} \\ &= \left(\mu_l(J)\mathbf{w}_l^b + \tilde{\mu}_l(J)\right)\mathbf{C}_l(J)\mathbf{b} \end{aligned}$$

as its Fourier terms of order zero at the cusps in $X^\infty(J)$. So all its Fourier terms $\tilde{F}_{Q,m_Q} f$ with $Q \in X^\infty(J)$ are multiples of ν_{Q,m_Q}, in contradiction to the uniqueness of the $R(P, m_P)$, c.f. Proposition 10.2.4. So $\det(\mathrm{Id} - \mathbf{w}_l^b \mathbf{C}_l(J))$ is not the zero function.

The relation in 12.3.3 amounts to

$$\mathbf{E}_l(J) = \mathbf{R}(J)\left(1 - \mathbf{w}_l^b \mathbf{C}_l(J)\right),$$

with $\mathbf{R}(J)$ the row vector with $R(Q, m_Q)$ at position $Q \in X^\infty(J)$. The choice in Table 12.1 on page 224 implies $\dot{R}(Q, m_Q; \varphi_0) = \dot{E}_l(J, P; \varphi_0)$. This gives part iii) of the proposition.

To obtain part ii)a) suppose that $|X^\infty(J)| = 1$. Then $E_l(J, P)$ is a meromorphic multiple of $R(P, n)$, and hence of determinate type.

Question. Can an Eisenstein family really be of indeterminate type at $s_0 \in \frac{1}{2}\mathbb{N}$, $s_0 > |\tau(\chi_0)|$?

12.4.3 Proposition. Singularities of other Poincaré families. *We use the assumptions and notations in 12.4.1. Suppose that $P \notin X^\infty$ or that $n(\varphi_0) \neq 0$.*

i) If the Poincaré family $Q_l(J, P, n)$ on Ω_0 has a singularity at w_0, then $s_0 \in \Sigma_c(\chi_0, l_0) \cup \{0, |\tau(\chi_0)|\} \cup \frac{1}{2}\mathbb{N}$.

ii) If $Q_l(J, P, n)$ is singular and of determinate type at w_0, then $s_0 \in \Sigma_c(\chi_0, l_0) \cup \{0, |\tau(\chi_0)|\}$, and the restriction $\dot{Q}_l(J, P, n; \varphi_0)$ has a singularity at s_0.

iii) Suppose $s_0 = \frac{b}{2} \geq 1$, $b \in \mathbb{N}$. Let $\mathbf{D}_l(J, P, n)$ denote the column vector with $D_l(Q, m_Q; J, P, n)/w_Q$ in position $Q \in X^\infty(J)$. Take $\mathbf{w}_l^b$ as in Proposition 12.4.2. Then the family of automorphic forms

$$Q_l(J, P, n) + \mathbf{E}_l(J)\left(\mathrm{Id} - \mathbf{w}_l^b \mathbf{C}_l(J)\right)^{-1} \mathbf{w}_l^b \mathbf{D}_l(J, P, n)$$

is holomorphic at w_0, with value $P_{l_0}(P, n(\varphi_0), \chi_0)$.

Remark. See Proposition 11.3.9 for singularities of the restriction $\dot{Q}_l(J, P, n; \varphi_0)$ to the vertical line $\{\varphi_0\} \times \mathbb{C}$.

Proof. Consider first $s_0 \notin \frac{1}{2}\mathbb{N}$. Then $Q_l(J, P, n) = \alpha_{P,n} R(P, n)$, with $\alpha_{P,n}$ holomorphic at w_0; see 12.3.5 and Table 12.1 on p. 224. This means that $Q_l(J, P, n)$ can have a singularity at w_0 only if $R(P,n)$ has one as well. Furthermore, $Q_l(J,P,n)$ and $R(P,n)$ have the same type of determinacy at w_0.

Part ii) of Lemma 12.3.4 implies that $s_0 \in \Sigma(\chi_0, l_0) \cup \{0\}$ in the indeterminate case. If $s_0 > |\tau(\chi_0)|$, then Proposition 6.7.3 shows that we are in the discrete series case treated in Lemma 12.3.9. If $\psi \neq 1$, it divides $\alpha_{P,n}$. Hence $Q_l(J,P,n) = \alpha_{P,n} R(P,n)$ is holomorphic at w_0. This gives part i) in the indeterminate case.

If $Q_l(J,P,n)$ has determinate type at w_0, then $R(P,n)$ has a singularity at w_0 if and only if $\dot{R}(P,n,\varphi_0)$ is singular at s_0. We consult 12.3.5 and the Tables 12.1 (page 224) and 11.1 (page 192) to see that $i(P,n)$ is equal to the order of $\dot{\alpha}_{P,n}(\varphi_0)$ at s_0. Hence a singularity of $Q_l(J,P,n)$ at w_0 corresponds to a singularity of $\dot{Q}_l(J,P,n,\varphi_0)$ at s_0; use Proposition 11.3.9 to see that $s_0 > |\tau(\chi_0)|$ is impossible. This gives part ii), even if $s_0 \in \frac{1}{2}\mathbb{N}$.

Consider $s_0 = \frac{1}{2}b \in \frac{1}{2}\mathbb{N}$, $s_0 \geq 1$. We have to prove the assertions in part iii). From the relation in 12.3.3 and the fact that $\mathbf{E}_l(J)\left(\mathrm{Id} - \mathbf{w}_l^b \mathbf{C}_l(J)\right)^{-1}$ is the row vector with $R(Q, m_Q)$ at position Q, we derive that the family in part iii) is equal to $\alpha_{P,n} R(P,n)$, with $\alpha_{P,n}$ as in Table 12.1. In particular, $\alpha_{P,n}$ is holomorphic at $w_0 = (\varphi_0, \frac{1}{2}b)$. In Table 12.1, 12.3.5 and Lemma 12.3.9 we see that $\psi_l(P,n)$ divides $\alpha_{P,n}$, hence $\alpha_{P,n} R(P,n)$ is holomorphic at w_0. Its value follows from Theorem 10.2.1, and the fact that $\mathbf{w}_l^b$ vanishes along $\varphi = \varphi_0$.

12.5 Restricted parameter spaces

In Proposition 10.2.12 we have seen that the restriction of $Q_l(J,P,n)$ to $\Omega_J = (U \cap V_J) \times \mathbb{C}$ is holomorphic at points (φ_0, s_0) with $\varphi_0 \in J^0$ and $\mathrm{Re}\, s_0 > \frac{1}{2}$. This shows that the singularities at (φ_0, s_0) with $s_0 \in \frac{1}{2}\mathbb{N}$, $s_0 > \frac{1}{2}$ (that turned up in the previous section) are not present on a restricted parameter space, as defined in 10.2.11. This can be understood easily: on a restricted parameter space each m_Q with $Q \in X^\infty(J)$ is identically zero. Hence in case **3** in Table 12.1 we have $\nu_{Q,m_Q} = \eta_l(Q, m_Q) = \tilde{\mu}_l(Q, m_Q)$. So the standard basis μ_l, $\tilde{\mu}_l$ is suitable at (φ_0, s_0), and does not cause singularities.

For restricted parameter spaces we look more closely at the connection between singularities of Poincaré families and the existence of smooth square integrable automorphic forms with non-vanishing Fourier term of order (P,n).

12.5.1 *Notations.* Let J be a cell of continuation. We take P, n, χ_1, l and $\mathbf{c}_{J,n}$ as in Theorem 10.2.1. We take an $\mathbb{R}$-linear subspace $V_r \subset V_{J,r}$, see 10.2.11; we assume that $\dim V_r \geq 1$. Let $V = \mathbb{C} \otimes_{\mathbb{R}} V_r$, and fix $\varphi_0 \in J^0$, $s_0 \in \mathbb{C}$, with $\mathrm{Re}\, s_0 \geq 0$, and take χ_0, l_0, n_0, w_0, U_0 and Ω_0 as in 12.3.1. For $w \in \Omega_0$ we write $w = (\varphi, s)$.

12.5.2 *Basis.* We take bases for the Fourier term modules as indicated in Table 12.1 on page 224. In case **3** we have $\alpha_{Q,m_Q} = 1$ and $\gamma_{Q,m_Q} = 0$ for all $Q \in X^\infty(J)$.

12.5.3 *Principle aim.* The leading idea in this section is the relation between singularities of the Poincaré family $Q_l(J,P,n)$ and smooth square integrable automorphic forms f with $\tilde{F}_{P,n} f \neq 0$. It will not always be possible to obtain such a form in $A_{l_0}(\chi_0, \mathbf{c}_J, s_0)$, but we try to get a family f of automorphic forms for which $\tilde{F}_{P,n}f$ does not vanish identically. (The point (φ_0, s_0) itself may fall in the zero set of $\tilde{F}_{P,n}f$ for all such families.) More formally, we aim at the following condition:

There is a morphism of parameter spaces $j : U \to \Omega_0$ with U a neighborhood of 0 in $\mathbb{C}^b$, $b \geq 0$, and there exists on U a family of automorphic forms $f \in \mathcal{A}_{l \circ j}(\mathbf{c}_J \circ j; U)$ such that: (12.2)
$j(0) = w_0$, and $j(U \cap \mathbb{R}^b) \subset J^0 \times \mathbb{C}$,
$\tilde{F}_{P,n \circ j} f$ is not identically zero on U,
$f(u) \in A^2_{l(w)}(\chi(w), s(w))$ for all $w = j(u)$, $u \in j(U \cap \mathbb{R}^b)$.

The morphism j is not necessarily a local curve at w_0. The dimension of U may be larger than 1; it is also useful to include the case that U consists of one point. In the latter case the existence of a family of cusp forms is just the existence of one cusp form.

Note that the condition $j(U \cap \mathbb{R}^b) \subset J^0 \times \mathbb{C}$ implies that $j(u) \in V_r \times (\mathbb{R} \cup i\mathbb{R})$ for all $u \in U \cap \mathbb{R}^b$.

If j happens to be a local curve ($b = 1$), the square integrability implies that $\frac{1}{4} - s(j(u))^2 \in \mathbb{R}$ for all $u \in U \cap \mathbb{R}$. Lemma 12.1.10 shows that in this case j is linear (or possibly quadratic if $s_0 = 0$).

12.5.4 Lemma. *Suppose Condition (12.2) holds. Then $R(P, n)$ is singular at w_0, and $j(U) \subset \mathrm{Pol}(R(P, n))$.*

Proof. Write $j(u) = (\varphi(u), s(u))$. The Lemmas 11.1.9 and 12.3.4 imply that $\{\, j(u) : u \in \mathbb{R}^b,\ \tilde{F}_{P,n}f(u) \neq 0 \,\}$ is contained in $\mathrm{Pol}(R(P, n))$. As $\tilde{F}_{P,n}f$ does not vanish on a neighborhood of 0 in $\mathbb{R}^b \cap U$, we conclude that $j(U)$ is contained in the analytic set $\mathrm{Pol}(R(P, n))$.

12.5.5 Lemma. *Let $w_1 = (\varphi_1, s_1) \in \Omega_0$ with $\varphi_1 \in J^0$. Suppose that $w_1 \in \mathrm{Pol}(R(P, n))$, but $w_1 \notin \mathrm{Indet}(R(P, n))$. Take ψ as in Lemma 12.3.4. Then the automorphic form $(\psi R(P, n))(w_1)$ is an element of $A^*_{l(\varphi_1)}(\chi(\varphi_1), J, s_1)$, and, if w_1 is sufficiently near to w_0, then at least one of the following statements holds:*

a) $(\psi R(P, n))(w_1) \in A^2_{l(\varphi_1)}(\chi(\varphi_1), s_1)$, *and* $\tilde{F}_{P,n(\varphi_1)}(\psi R(P, n))(w_1) \neq 0$.

b) $s_1 = 0$, $X^\infty(J) \neq \emptyset$, *$\psi(\varphi, s)$ has degree 1 in the variable s, and at least one of the Eisenstein series $\dot{E}_l(J, Q; \varphi_1; 0)$ with $Q \in X^\infty(J)$ is non-zero.*

c) $\mathrm{Re}\, s_1 < 0$, $X^\infty(J) \neq \emptyset$, *and the scattering matrix $\mathbf{C}_l(J)$ has a singularity at w_1.*

Proof. Lemma 12.2.5 shows that $\psi R(P,n)$ is primitive at w_1 if w_1 is sufficiently near w_0. Let us assume this. We denote the automorphic form $(\psi R(P,n))(w_1)$ by r. It is non-zero, as $R(P,n)$ has determinate type at w_1. We conclude $w_1 \in N(\psi)$ from $w_1 \in \mathrm{Pol}(R(P,n))$. Hence $r \in A^*_{l(\varphi_1)}(\chi(\varphi_1), s_1)$.

The family $R(P,n)$ has been constructed corresponding to the basis given in Table 12.1 for the point w_0. As we work on a restricted parameter space, there is a neighborhood of w_0 in $V_r \times \mathbb{C}$ for which this basis satisfies the conditions in 12.3.2. We assume that w_1 is in this neighborhood.

Let $\mathrm{Re}\, s_1 \geq 0$. We apply Lemma 12.3.4, and see that r is a non-zero multiple of the value at s_1 of the restriction $s \mapsto (s-s_1)^q \dot{R}(P,n;\varphi_1;s)$ (this is $R_q(s)$ in the notation of 11.1.3). Part iii) of Lemma 11.1.9 and Lemma 11.2.9 show that statement a) holds, except possibly if $s_1 = 0$ and $q = 1$. If in the latter case all Eisenstein series $\dot{E}_l(J,Q;\varphi_1;0)$ would vanish, or if $X^\infty(J)$ would be empty, then Lemma 11.2.9 shows that r is a cusp form. But part iv) of Lemma 11.1.9 implies that if $r \in S_{l(\varphi_1)}(\chi(\varphi_1), \mathbf{c}_J, 0)$, then $\tilde{F}_{P,n(\varphi_1)} r \neq 0$, and condition a) is satisfied after all.

Let $\mathrm{Re}\, s_1 < 0$. Lemma 11.2.10 shows that if r is not a cusp form, then $X^\infty(J) \neq \emptyset$ and $\dot{\mathbf{C}}_l(J;\varphi_1)$ is not holomorphic at s_1, and hence $\mathbf{C}_l(J)$ cannot be holomorphic at w_1. We are left with the case that r is a cusp form with $\tilde{F}_{P,n(\varphi_1)} r = 0$. Lemma 11.1.10 would imply $r = 0$.

12.5.6 Lemma. *Let $w_0 \in \mathrm{Pol}(R(P,n))$. We take ψ as in Lemma 12.3.4. Then at least one of the following statements holds:*

i) There are points $w_1 = (\varphi_1, s_1)$ in each neighborhood $\Omega_1 \subset \Omega_0$ of w_0, such that $\varphi_1 \in J^0$, $\varphi_1 \neq \varphi_0$, $w_1 \in \mathrm{Pol}(R(P,n))$, and statement a) or c) of the previous lemma holds for w_1.

ii) $\psi(\varphi, s) = s$, $X^\infty(J) \neq \emptyset$, and the restriction $\varphi \mapsto E_l(J,Q;\varphi,0)$ is non-zero for at least one $Q \in X^\infty(J)$.

Proof. Define $c = 0$ if ψ is of the form $(\varphi, s) \mapsto s^k$, and $c = 1$ otherwise. Let ψ_1 be an irreducible factor of ψ in the stalk $\mathcal{O}_{w_0}$, not dividing $(\varphi, s) \mapsto s^c$. Then $\dim_{w_0} N(\psi_1) = \dim_{\mathbb{C}} V > \dim_{w_0} \mathrm{Indet}(R(P,n))$, see Proposition 12.2.13. Hence there exists a holomorphic function $\eta_1 \in \mathcal{O}_{w_0}$ such that $\mathrm{Indet}(R(P,n)) \cap N(\psi_1) \subset N(\eta_1) \cap N(\psi_1)$, and η_1 and ψ_1 are relatively prime at w_0. If $X^\infty(J) \neq \emptyset$, the analytic sets $\mathrm{Indet}(E_l(Q, m_Q))$ with $Q \in X^\infty(J)$ also have dimension strictly smaller than $\dim_{\mathbb{C}} V$. If we take the union of these sets with $\mathrm{Indet}(R(P,n))$, the dimension at w_0 stays smaller than $\dim_{\mathbb{C}} V$. So we may arrange $\mathrm{Indet}(E_l(J,Q)) \cap N(\psi_1) \subset N(\eta_1) \cap N(\psi_1)$ as well.

We apply Lemma 9.4.13 to ψ_1 and $h(\varphi, s) = s^c \eta_1(\varphi, s)$ to see that w_0 may be approached by points $w_1 = (\varphi_1, s_1) \in (\varphi_0 + V_r) \times \mathbb{C}$, with $\psi_1(w_1) = 0$, and $s_1^c \eta_1(\varphi_1, s_1) \neq 0$. For these points we have $w_1 \in \mathrm{Pol}(R(P,n))$, $w_1 \notin \mathrm{Indet}(R(P,n))$, and also $w_1 \notin \mathrm{Indet}(E_l(J,Q))$ for $Q \in X^\infty(J)$. Proposition 10.2.4 implies that the vertical line $\{\varphi_0\} \times \mathbb{C}$ is not contained in $\mathrm{Pol}(R(P,n))$; hence infinitely many of the w_1 satisfy $\varphi_1 \neq \varphi_0$.

If $c = 1$ the points $w_1 = (\varphi_1, s_1)$ satisfy all conditions of the previous lemma, and $s_1 \neq 0$. This gives statement i).

Consider $c = 0$. At each $w_1 = (\varphi_1, 0)$ constructed above we are in one of the cases of the previous lemma. If statement i) does not hold, then a) and c) of Lemma 12.5.5 occur only a finite number of times. This leaves case b) for the remaining points, which converge to $w_0 = (\varphi_0, 0)$. We obtain $k = 1$, $X^\infty(J) \neq \emptyset$, and some Eisenstein series $E_l(J, Q; \varphi_1; 0)$ is non-zero. This gives ii). (We have omitted the dot, as we have taken care to give all Eisenstein series determinate type at the w_1.)

12.5.7 Lemma. *Let $w_0 \in \mathrm{Pol}(R(P,n))$. Then $s_0 \in \Sigma(\chi_0, l_0) \cup \{0\}$, and at least one of the following statements is satisfied, according to the rules in Table 12.2.*

(α) *Condition (12.2), on p. 231, holds, with j a linear local curve.*

(β) *Condition (12.2) holds, with j a quadratic local curve.*

(γ) *The scattering matrix $\mathbf{C}_l(J)$ has a singularity at w_0.*

(δ) *There exists $Q \in X^\infty(J)$ for which $\varphi \mapsto E_l(J, Q; \varphi, 0)$ is a non-zero family on a neighborhood of φ_0.*

	$X^\infty(J) = \emptyset$	$X^\infty(J) \neq \emptyset$
$s_0 = 0$	(α) or (β)	(α), (β), (γ) or (δ)
$\mathrm{Re}\, s_0 = 0$, $s_0 \neq 0$	(α)	(α) or (γ)
$s_0 > 0$	(α)	(α)

Table 12.2 Possibilities for a singularity of $R(P,n)$ at w_0.

Proof. Suppose $w_0 = (\varphi_0, s_0) \in \mathrm{Pol}(R(P,n))$; take ψ as in Lemma 12.3.4. That lemma implies $s_0 \in \Sigma(\chi_0, l_0) \cup \{0\}$ if $w_0 \in \mathrm{Indet}(R(P,n))$. If $R(P,n)$ has determinate type at w_0, part i) of Lemma 12.3.4 implies that $\dot{R}(P,n;\varphi_0)$ is singular at s_0. In the case that (P,n) is singular, we have either $s_0 = 0$, or $\dot{R}(P,n;\varphi_0) = \dot{E}_l(J,P;\varphi_0)$; hence $s_0 \in \Sigma^e(\chi_0, l_0) \cup \{0\}$. If (P,n) is regular and $s_0 \neq 0$, then Lemma 11.3.6 implies that $s_0 \in \Sigma(\chi_0, l_0)$.

If w_0 satisfies statement ii) in the previous lemma, we get possibility (δ); of course this can happen only if $s_0 = 0$ and $X^\infty(J) \neq \emptyset$.

If statement i) holds, consider a sequence of points w_1 as indicated, approaching w_0. If for infinitely many of these points case c) in Lemma 12.5.5

holds, then the limit w_0 is an element of the polar set of $\mathbf{C}_l(J)$ as well. This can happen if $X^\infty(J) \neq \emptyset$ and $\operatorname{Re} s_0 = 0$ (note that $\operatorname{Re} s_1 < 0$ for these w_1). Suppose case a) of Lemma 12.5.5 holds for all w_1 sufficiently near to w_0. We have seen in the proof of the previous lemma that we can arrange $w_1 \in N(\psi_1)$ for all these w_1, for some irreducible factor ψ_1 of ψ. Define the holomorphic function η by $\tilde{F}_{P,n}(\psi R(P,n)) = \psi \lambda_{P,n} + \eta \nu_{P,n}$. Then $\eta(w_1) \neq 0$. Hence ψ_1 and η are relatively prime at w_0. Apply Proposition 12.1.11 to obtain a local curve $j : U \to \Omega_0$ at w_0 along $N(\psi_1)$ with direction $\varphi_1 - \varphi_0$ such that $\eta \circ j$ is not the zero function. We define $f \in \mathcal{A}_{l\circ j}(\mathbf{c}_J; U)$ by $f = (\psi R(P,n)) \circ j$. Then $\tilde{F}_{P,n\circ j} f$ is not the zero function. Let $u \in U \cap \mathbb{R}$ with $\tilde{F}_{P,n} f(u) \neq 0$. If $j(u) \notin \operatorname{Indet}(R(P,n))$, then $j(u)$ has the properties of the w_1, and hence $f(u) \in A^2_{l(j(u))}(\chi(j(u)), s(j(u)))$. As $f(u) \neq 0$, this implies in particular that $\frac{1}{4} - s(j(u))^2 \in \mathbb{R}$. Such u are dense in $U \cap \mathbb{R}$, hence j is linear, or possibly quadratic if $s_0 = 0$, see Proposition 12.1.11. Finally, consider $u \in U \cap \mathbb{R}$ such that $j(u) \in \operatorname{Indet}(R(P,n))$. Suppose that statements (γ) and (δ) are not satisfied for w_0. As $f(u) \in A^*_{l(j(u))}(\chi(j(u)), J, s(j(u)))$, we can apply the Lemmas 11.2.9 and 11.2.10 to see that $f(u)$ has to be square integrable. Hence j satisfies Condition (12.2) and is linear, or possibly quadratic in the case that $s_0 = 0$.

12.5.8 Proposition. Eisenstein family on a restricted parameter space. *Under the assumptions and notations in 12.5.1, let $P \in X^\infty(J)$, and $n = 0$.*

i) If $E_l(J,P)$ is singular at w_0, then $s_0 \in \{0, |\tau(\chi_0)|\} \cup \Sigma_c(\chi_0)$.

ii) Let $s_0 > 0$. Then $E_l(J,P)$ is singular at w_0 if and only if the following condition is satisfied:

> *There is a morphism of parameter spaces $j : U \to V \times \mathbb{C}$ with U a neighborhood of 0 in $\mathbb{C}^b$ for some $b \geq 0$, satisfying $j(0) = w_0$ and $j(U \cap \mathbb{R}^b) \subset V_r \times \mathbb{C}$, and there exists $f \in \mathcal{A}_{l\circ j}(\mathbf{c}_J \circ j; U)$ with $\tilde{F}_{P,0} f$ not the zero function on U, and $f(u) \in A^2_{l(w)}(\chi(w), s(w))$ for all $w \in j(U \cap \mathbb{R}^b)$.*

iii) Let $\operatorname{Re} s_0 = 0$, $s_0 \neq 0$. Then $E_l(J,P)$ is singular at w_0 if and only if it is of indeterminate type at w_0. If $E_l(J,R)$ is singular at w_0, then the scattering matrix $\mathbf{C}_l(J)$ is not holomorphic at w_0.

iv) The restriction $\varphi \mapsto E_l(J,P;\varphi,0)$ exists as a meromorphic family of automorphic forms.

Remarks. If $\operatorname{Re} s_0 = 0$, $s_0 \neq 0$, the restriction $\dot{\mathbf{C}}_l(J;\varphi_0)$ is always holomorphic at s_0. This does not imply that the unrestricted scattering matrix is holomorphic at w_0.

In the situation of iii) we have $A^2_{l(\varphi)}(\chi(\varphi), s) = S_{l(\varphi)}(\chi(\varphi), \mathbf{c}_J(\varphi), s)$ for all $(\varphi, s) \in (\varphi_0 + V_r) \times \mathbb{C}$ near w_0. Hence the assertion $\tilde{F}_{P,0} f \neq 0$ in Condition (12.2) cannot hold at w_0 for $(P,0)$, although Lemma 12.5.7 allowed (α) and (γ).

If $w_0 = (\varphi_0, 0) \in \operatorname{Pol}(E_l(J,P))$, then at least one of the four conditions in Lemma 12.5.7 holds. But only (α) and (β) imply $(\varphi_0, 0) \in \operatorname{Pol}(E_l(J,P))$.

Proof. For $s_0 \neq 0$ the choice of $\lambda_{P,0}$ in Table 12.1 implies $R(P,0) = E_l(J,P)$, see 12.3.3. Lemma 12.5.7 implies that $s_0 \in \{0\} \cup \Sigma(\chi_0, l_0)$. Use Proposition 10.2.12 to obtain $s_0 \leq \frac{1}{2}$, and hence $s_0 \in \Sigma_c(\chi_0, l_0) \cup \{0, |\tau(\chi_0)|\}$.

Part ii) follows from the Lemmas 12.5.4 and 12.5.7.

Let $\operatorname{Re} s_0 = 0$, $s_0 \neq 0$. In Proposition 11.2.3 we see that the restriction $\dot{E}_l(J,P;\varphi_0)$ is holomorphic at s_0. From the Fourier expansion at P we see that $\dot{E}_l(J,P;\varphi_0;s_0) \neq 0$. Take $\psi \in \mathcal{O}_{w_0}$ in Weierstrass form with respect to $s - s_0$ of degree k, such that $\psi E_l(J,P)$ is primitive at w_0. If $w_0 \in \operatorname{Pol}(E_l(J,P))$, then $k \geq 1$, and

$$(\psi E_l(J,P))(w_0) = \lim_{s \to s_0} (s - s_0)^k \dot{E}_l(J,P;\varphi_0;s) = 0,$$

hence $E_l(J,P)$ is of indeterminate type at w_0. If $E_l(J,P)$ is holomorphic at w_0, then $k = 0$, and $E_l(J,P;w_0) = \dot{E}_l(J,P;\varphi_0;s_0) \neq 0$, hence it has determinate type at w_0. This gives the equivalence in part iii). Use Proposition 10.2.14 to complete the proof of this part.

Suppose that iv) does not hold. Then we can take $w_0 = (\varphi_0, 0)$, such that $E_l(J,P)$ is singular at all $(\varphi_1, 0)$ for φ_1 in a neighborhood of φ_0 in $\varphi_0 + V_r$. The function ψ in Lemma 12.3.4 contains the factor $(\varphi, s) \mapsto s$ at least twice; once as $E_l(J,P)$ is singular along $(\varphi_0 + V) \times \{0\}$, and again as $R(P,0) = \frac{1}{2s} E_l(J,P)$. Proposition 11.2.4 shows that $\dot{E}_l(J,P;\varphi_1)$ is holomorphic at $s = 0$. So $\dot{R}(P,0)$ has at most a first order pole. The second part of Lemma 12.3.4 shows that $E_l(J,P)$ is of indeterminate type at $(\varphi_1, 0)$ for all $\varphi_1 \in \varphi_0 + V_r$ near φ_0. This contradicts Proposition 12.2.13.

12.5.9 Proposition. Poincaré family on a restricted parameter space. *Under the assumptions in 12.5.1, let $P \in \mathcal{P}_Y$ or $n \neq 0$.*

i) If $Q_l(J,P,n)$ is singular at w_0, then $s_0 \in \Sigma_c(\chi_0) \cup \{0, |\tau(\chi_0)|\}$.

ii) If $s_0 \neq |\tau(\chi_0)|$, then the following condition a) implies $w_0 \in \operatorname{Pol}(Q_l(J,P,n)$.

a) There is a morphism of parameter spaces $j : U \to V \times \mathbb{C}$ with U a neighborhood of 0 in some $\mathbb{C}^b$ with $b \geq 0$, and $j(U \cap \mathbb{R}^b) \subset J^0 \times \mathbb{C}$, and there exists a family of automorphic forms $f \in \mathcal{A}_{l \circ j}(\mathbf{c}_J \circ j; U)$ such that $\tilde{F}_{P,n \circ j} f$ is not identically zero on U, and such that $f(u) \in A^2_{l(w)}(\chi(w), s(w))$ for all $w \in j(U \cap \mathbb{R}^b)$.

iii) If $s_0 > 0$ and $w_0 \in \operatorname{Pol}(Q_l(J,P,n))$, then condition a) holds.

iv) If $\operatorname{Re} s_0 = 0$, $s_0 \neq 0$, then $w_0 \in \operatorname{Pol}(Q_l(J,P,n))$ implies condition a) or the following condition b).

b) The scattering matrix $\mathbf{C}_l(J)$ has a singularity at w_0.

v) If $s_0 = 0$ and $w_0 \in \mathrm{Pol}(Q_l(J,P,n))$, then at least one of the conditions a), b) and c) hold.

c) There exists $Q \in X^\infty(J)$ for which $\varphi \mapsto E_l(J,Q;\varphi,0)$ is not the zero family.

Proof. Use Lemma 12.5.7 and Proposition 10.2.12 to get i). That lemma implies also iii)–v). For ii) we use Lemma 12.5.4, and the relation $Q_l(J,P,n) = R(P,n)$, valid for $s \in \Sigma_c(\chi_0)$, but not necessarily for $s_0 = |\tau(\chi_0)|$.

12.5.10 Corollary. Poincaré family on a minimal cell. *Under the assumptions in 12.5.1 let $P \in \mathcal{P}_Y$ or $n \neq 0$. Suppose that $X^\infty(J) = \emptyset$.*

i) If $Q_l(J,P,n)$ is singular at w_0, then $s_0 \in \Sigma_c(\chi_0) \cup \{|\tau(\chi_0)|\}$.

ii) Let $s_0 \neq |\tau(\chi_0)|$. Then $w_0 \in \mathrm{Pol}(Q_l(J,P,n))$ if and only if there are $j : U \to (\varphi_0 + V) \times \mathbb{C}$ and $f \in \mathcal{S}_{l_0 j}(\mathbf{n} \circ j; U)$, with U a neighborhood of 0 in some $\mathbb{C}^b$, $b \geq 0$, such that $\tilde{F}_{P,\mathbf{n} \circ j} f$ is not identically zero on U.

Proof. Clear from the previous proposition, and from the fact that $A^*_{l_0}(\chi_0, \mathbf{n}, 0) = S_{l_0}(\chi_0, \mathbf{n}, 0)$ for the empty growth condition $\mathbf{n}$.

Part II

Examples

Chapter 13
Automorphic forms for the modular group

Part II gives examples for the general theory in Part I. This first chapter of Part II complements the discussion of the modular group in the introductory Chapter 1.

Sections 13.1–13.4 give examples for some of the concepts in Part I.

Sections 13.5 and 13.6 give an application of Proposition 12.4.2. We turn to the Eisenstein family $E : (r,s) \mapsto E(r,s)$, introduced in 1.5.8. Theorem 10.2.1 shows that it is meromorphic, jointly in the weight r and the spectral parameter s. We consider the derivatives $\partial_r^m E$, and restrict these to the line $\{0\} \times \mathbb{C}$. In this way we obtain meromorphic families $s \mapsto \partial_r^m E(0,s)$ of functions that are not modular forms, but still have some automorphic transformation behavior. Proposition 12.4.2 has the remarkable consequence that for $m \geq 2$ these families have singularities at some points $s = \frac{1}{2}l$, $l \in \mathbb{Z}$, $l \geq 2$, although the Eisenstein series converges for $\operatorname{Re} s > \frac{1}{2}$. These singularities lead to distribution results, for instance those considered in Section 13.6.

13.1 The covering group and its characters

In Chapter 2 we have considered automorphic forms as functions on the universal covering group $\tilde{G}$ of $\mathrm{SL}_2(\mathbb{R})$. In this section we define the full original $\tilde{\Gamma}_{\mathrm{mod}}$ in $\tilde{G}$ of the modular group $\Gamma_{\mathrm{mod}} = \mathrm{SL}_2(\mathbb{Z})$. We give generators and relations for $\tilde{\Gamma}_{\mathrm{mod}}$, and describe the character group. It turns out that all characters can be expressed in terms of Dedekind sums.

The generators we give are easy to use, but they are not canonical in the sense of Section 3.3. In 13.1.9 we shall give canonical generators for $\tilde{\Gamma}_{\mathrm{mod}}$ as well, and specialize the discussion in Chapter 3 to the modular case.

13.1.1 *Definition.* $\tilde{\Gamma}_{\mathrm{mod}}$ is the full original in $\tilde{G}$ of the modular group $\Gamma_{\mathrm{mod}} = \mathrm{SL}_2(\mathbb{Z})$. It consists of all $k(\pi m)\widetilde{\begin{pmatrix} a & b \\ c & d \end{pmatrix}}$ with $m, a, b, c, d \in \mathbb{Z}$, $ad - bc = 1$. The projection $\tilde{\Gamma}_{\mathrm{mod}} \to \Gamma_{\mathrm{mod}}$ maps $k(\pi m)\widetilde{\begin{pmatrix} a & b \\ c & d \end{pmatrix}}$ onto $(-1)^m \begin{pmatrix} a & b \\ c & d \end{pmatrix}$.

13.1.2 *Generators.* In 1.1.4 we have remarked that Γ_{mod} is generated by $U = \begin{pmatrix} 1 & 1 \\ 1 & 0 \end{pmatrix}$ and $W = \begin{pmatrix} 0 & 1 \\ -1 & 0 \end{pmatrix}$. Consider in $\tilde{\Gamma}_{\mathrm{mod}}$ the elements $n = n(1)$ and $w = k(\pi/2)$, that project to U and W. These two elements generate $\tilde{\Gamma}_{\mathrm{mod}}$; to see this note first that n, w and $\zeta = k(\pi)$ generate $\tilde{\Gamma}_{\mathrm{mod}}$, and next that $\zeta = w^2$.

The relations in Γ_{mod} are $(UW)^3 = W^4 = \mathrm{Id}$ and centrality of W^2. Clearly w^2 is central in $\tilde{\Gamma}_{\mathrm{mod}}$, and $w^4 = \zeta^2$. The computation of $(nw)^3$ can be done with

Relation (2.1) on p. 28, and the fact that $k(\varphi) = \widetilde{\begin{pmatrix} \cos\varphi & \sin\varphi \\ -\sin\varphi & \cos\varphi \end{pmatrix}}$ for $-\pi \leq \varphi < \pi$:

$$
\begin{aligned}
(n(1)w)^3 &= n(1)\widetilde{\begin{pmatrix} 0 & 1 \\ -1 & 0 \end{pmatrix}} p(1+i)wn(1)w \\
&= n(1)p((i-1)/2)k(-\arg(-1-i)+\pi/2)n(1)w \\
&= n(1)n(-1/2)a(1/2)k(5\pi/4)n(1)w \\
&= n(1/2)a(1/2)\zeta\widetilde{\begin{pmatrix} 1/\sqrt{2} & 1/\sqrt{2} \\ -1/\sqrt{2} & 1/\sqrt{2} \end{pmatrix}} p(1+i)k(\pi/2) \\
&= n(1/2)a(1/2)p(-1+2i)k(-\arg(-i)+3\pi/2) \\
&= p(i)k(2\pi) \;=\; w^4.
\end{aligned}
$$

We conclude that the generators n and w of $\tilde{\Gamma}_{\mathrm{mod}}$ satisfy the relations

$$
w^2 \text{ is central} \quad \text{and} \quad (n(1)w)^3 = w^4,
$$

and that all relations can be derived from these.

Another way of obtaining $(n(1)w)^3$ is to start from the fact that $\widetilde{n(1)w} = \begin{pmatrix} -1 & 1 \\ -1 & 0 \end{pmatrix}$ leaves invariant the point $z_2 = e^{\pi i/3} \in \mathfrak{H}$. Hence $p(z_2)^{-1}n(1)wp(z_2)$ fixes i, and should be of the form $k(\varphi)$. The relation above predicts $\varphi = \frac{2}{3}\pi$. Indeed

$$
\begin{aligned}
&p(z_2)^{-1}n(1)wp(z_2) \\
&= \left(n\left(\frac{1}{2}\right)a\left(\frac{1}{2}\sqrt{3}\right)\right)^{-1} n(1)\widetilde{\begin{pmatrix} 0 & 1 \\ -1 & 0 \end{pmatrix}} p(z_2) \\
&= a(2/\sqrt{3})n(1/2)p\left(-e^{-\pi i/3}\right)k\left(-\arg\left(-e^{\pi i/3}\right)\right) \\
&= a(2/\sqrt{3})n(1/2)n(-1/2)a(\sqrt{3}/2)k(-(-\pi+\pi/3)) \;=\; k\left(\frac{2}{3}\pi\right).
\end{aligned}
$$

13.1.3 *Characters.* All characters $\chi \in \mathcal{X}(\tilde{\Gamma}_{\mathrm{mod}})$ are determined by their values on the generators n and w. The relations imply the condition $\chi(w) = \chi(n)^3$. So $r \mapsto \chi_r$, determined by

$$
\chi_r(n(1)) = e^{\pi i r/6}, \qquad \chi_r(w) = e^{\pi i r/2},
$$

gives an isomorphism $\mathbb{C} \bmod 12\mathbb{Z} \to \mathcal{X}(\tilde{\Gamma}_{\mathrm{mod}})$. Clearly, χ_r is an element of the subgroup $\mathcal{X}_u$ of unitary characters if and only if $r \in \mathbb{R} \bmod 12\mathbb{Z}$. The character χ_r belongs to the weights $l \equiv r \bmod 2$.

13.1.4 *The multiplier system v_r.* In 2.3.6 we have seen that $v : g \mapsto \chi_r(\tilde{g})$ is a multiplier system for Γ_{mod}. It turns out to be the multiplier system v_r discussed in 1.5.5. We shall check this by considering the generators U and W.

13.1.5 *The logarithm of the eta function of Dedekind.* In, e.g., [30], Ch. IX we find the function

$$\log \eta(z) = \frac{\pi i}{12} + \sum_{m=1}^{\infty} \log(1 - e^{2\pi imz}).$$

It satisfies $\log \eta(z+1) = \log \eta(z) + \pi i/12$ and $\log \eta(-1/z) = \log \eta(z) + \frac{1}{2}\log(z/i)$. We take $\log(z/i)$ with the argument in $(-\frac{\pi}{2}, \frac{\pi}{2}]$.

This implies for η_r, as defined in 1.5.5:

$$\begin{aligned} \eta_r(U \cdot z) &= e^{\pi ir/6}\eta_r(z) \\ \eta_r(W \cdot z) &= e^{\pi ir/2}e^{ir\arg(-z)}\eta_r(z). \end{aligned}$$

This is sufficient to identify v_r as the multiplier system obtained from χ_r.

13.1.6 *Relation with Dedekind sums.* There is an explicit description of v_r in terms of Dedekind sums. In [30], Chapter IX, one finds for $c \geq 1$

$$\log \eta\left(\frac{az+b}{cd+d}\right) = \log \eta(z) + \frac{1}{2}\log\left(\frac{cz+d}{i}\right) + \pi i\left(\frac{a+d}{12c} - S(d,c)\right),$$

with $S(d,c)$ the *Dedekind sum:*

$$\begin{aligned} S(d,c) &= \sum_{x \bmod c} \left(\!\left(\frac{x}{c}\right)\!\right)\left(\!\left(\frac{xd}{c}\right)\!\right) \\ ((t)) &= \begin{cases} 0 & \text{if } t \in \mathbb{Z} \\ t - \text{entier}(t) - \frac{1}{2} & \text{otherwise,} \end{cases} \end{aligned}$$

see, e.g., [30], Chap. IX, Theorem 2.1 on p. 147.

We obtain

$$\chi_r\left(\widetilde{\begin{pmatrix} a & b \\ c & d \end{pmatrix}}\right) = v_r\begin{pmatrix} a & b \\ c & d \end{pmatrix} = e^{\pi ir(a+d)/6c - \pi ir/2 - 2\pi irS(d,c)}.$$

13.1.7 *Powers of the eta function of Dedekind, as functions on $\tilde{G}$.* The automorphic form η_r, discussed in 1.5.5, gives rise to a function on $\tilde{G}$, which we denote also by η_r:

$$\eta_r(p(z)k(\theta)) = y^{r/2}e^{2r\log\eta(z)}e^{ir\theta}.$$

The logarithm of the eta function corresponds to a function on $\tilde{G}$ that has almost all the properties of an automorphic form. Put $L(p(z)k(\theta)) = \frac{1}{2}\log y + 2\log \eta(z) + i\theta$. We consider left translation over the generators n and w of $\tilde{\Gamma}_{\text{mod}}$:

$$\begin{aligned} &L(np(z)k(\theta)) - L(p(z)k(\theta)) \\ &\quad = \; 2\log \eta(z+1) - 2\log \eta(z) \; = \; \frac{\pi i}{6} \end{aligned}$$

$$\begin{aligned} &L(wp(z)k(\theta)) - L(p(z)k(\theta)) \\ &\quad= L(p(-1/z)k(-\arg(-z)+\theta)) - L(p(z)k(\theta)) \\ &\quad= -\log|-z| + 2\log\eta(-1/z) - i\arg(-z) - 2\log\eta(z) \\ &\quad= -\log(-z) + \log(z/i) = \frac{\pi i}{2}. \end{aligned}$$

L satisfies the left-$\tilde\Gamma_{\mathrm{mod}}$-behavior $L(\gamma g) = L(g) + i\alpha(\gamma)$ for all $\gamma \in \tilde\Gamma_{\mathrm{mod}}$ and $g \in \tilde G$, for the homomorphism $\alpha : \tilde\Gamma_{\mathrm{mod}} \to \frac{\pi}{6}\mathbb{Z}$ determined by $\alpha(n(1)) = \frac{\pi}{6}$, $\alpha(k(\pi/2) = \frac{\pi}{2}$. On the right we have the additive $\tilde K$-behavior $L(gk(\theta)) = L(g) + i\theta$. The action of the Casimir operator gives $\omega L = \frac{1}{2}$.

The homomorphism α satisfies $\chi_r = e^{ir\alpha}$. For $\gamma = \begin{pmatrix} a & b \\ c & d \end{pmatrix} \in \Gamma_{\mathrm{mod}}$, $c \geq 1$:

$$\alpha(\tilde\gamma) = 2\pi\left(\frac{a+d}{12c} - \frac{1}{4} - S(d,c)\right). \tag{13.1}$$

13.1.8 *Canonical fundamental domain.* The standard fundamental domain for the modular group in Figure 1.1 on p. 2 is not canonical. $\tilde\Gamma_{\mathrm{mod}}$ has genus $g = 0$, number of cusps $p = 1$, and number of elliptic orbits $q = 2$. To obtain the canonical fundamental domain in Figure 13.1 we take:

l_1 the geodesic from $i\infty$ to $e^{7\pi i/12}$, $\pi_1 = n(1)$
h_1 the geodesic from $z_1 = i$ to $e^{5\pi i/12}$, $\varepsilon_1 = w = k(\pi/2)$
h_2 the geodesic from $z_2 = e^{\pi i/3}$ to $1 + e^{7\pi i/12}$,
$\varepsilon_2 = p(z_2)k(\pi/3)p(z_2)^{-1}$.

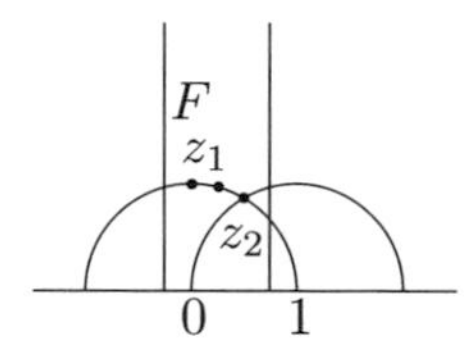

Figure 13.1 A canonical fundamental domain for $\tilde\Gamma_{\mathrm{mod}}$.

13.1.9 *Canonical generators.* We have already seen that π_1 and ε_1 generate $\tilde\Gamma_{\mathrm{mod}}$. We compute:

$$\begin{aligned} \pi_1\varepsilon_1\varepsilon_2 &= n(1)\widetilde{\begin{pmatrix} 0 & 1 \\ -1 & 0 \end{pmatrix}} p(z_2)k(\pi/3)a(2/\sqrt{3})n(-1/2) \\ &= n(1)p(-2/(1+i\sqrt{3}))k(-\arg(-1-i\sqrt{3})+\pi/3) \\ &\qquad a(2/\sqrt{3})n(-1/2) \\ &= n(-1/2)a(\sqrt{3}/2)k(\pi)a(\sqrt{3}/2)n(-1/2) = \zeta. \end{aligned}$$

This immediately implies that $\varepsilon_2 \in \tilde{\Gamma}_{\text{mod}}$; further we obtain

$$\begin{aligned} (n(1)w)^3 &= (\pi_1\varepsilon_1\varepsilon_2\varepsilon_2^{-1})^3 = (\zeta\varepsilon_2^{-1})^3 \\ &= \zeta^3 k(-\pi) = \zeta^2 = w^4. \end{aligned}$$

This is the relation we have seen in 13.1.2. We conclude that ζ, π_1, ε_1 and ε_2 generate $\tilde{\Gamma}_{\text{mod}}$, with the relations:

$$\zeta \text{ is central}, \qquad \varepsilon_j^{v_j} = \zeta \text{ for } j = 1, 2, \qquad \pi_1\varepsilon_1\varepsilon_2 = \zeta.$$

Thus we have found canonical generators and relations as discussed in 3.3.3.

13.1.10 *Cuspidal and elliptic orbits.* The cusps for $\tilde{\Gamma}_{\text{mod}}$ form one orbit. There are two elliptic orbits: $\tilde{\Gamma}_{\text{mod}} \cdot i$ and $\tilde{\Gamma}_{\text{mod}} \cdot e^{\pi i/3}$. The elements $\gamma \in \Gamma_{\text{mod}}$ that leave fixed an element of $\tilde{\Gamma}_{\text{mod}} \cdot i$, all satisfy $\text{trace}(\hat{\gamma}) = 0$. The other elliptic orbit is characterized by the condition $|\text{trace}(\hat{\gamma})| = 1$ for all γ that leave fixed some element of the orbit.

13.1.11 *Exceptional points.* All sets $\mathcal{P}$ of exceptional points for $\tilde{\Gamma}_{\text{mod}}$, see Section 3.5, have to contain $X^\infty = \{\tilde{\Gamma}_{\text{mod}} \cdot i\infty\}$. In this chapter we use the notation $P = \tilde{\Gamma}_{\text{mod}} \cdot i\infty$.

We shall mostly work with the minimal choice $\mathcal{P} = \{P\}$. For the continuation of the resolvent kernel, one needs $\mathcal{P} = \{P, Q_w\}$, with $Q_w = \tilde{\Gamma}_{\text{mod}} \cdot w$, $w \in \mathfrak{H}$.

13.1.12 *Cutting up X_{mod} as in Section 3.5.* If $\mathcal{P} = \{P\}$, then $A_P = 2$ is the minimal choice, see Figure 13.2.

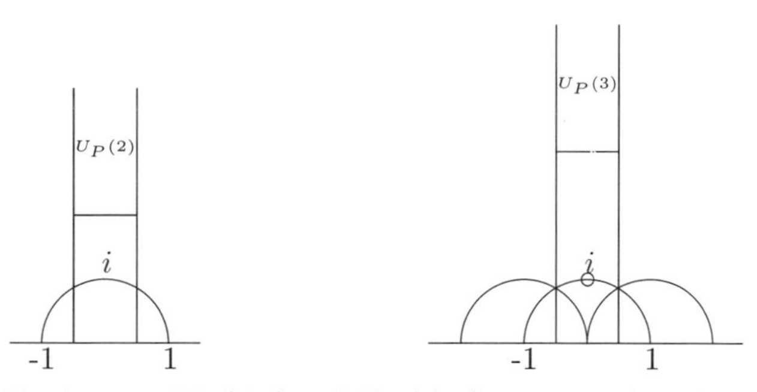

Figure 13.2 Possible choices of $U_P(A_P)$ and $U_{Q_i}(A_{Q_i})$ represented in the standard fundamental domain.

This choice is not right for larger $\mathcal{P}$. Consider for instance $\mathcal{P} = \{P, Q_i\}$. Put $u = \tilde{A}_P = A_P - 1$ and $v = \sqrt{\tilde{A}_{Q_i}/(\tilde{A}_{Q_i}+1)}$. The condition $U_P(\tilde{A}_P) \cap U_{Q_i}(\tilde{A}_{Q_i}) = \emptyset$ is equivalent to the condition that the disk $\left|\frac{z-i}{z+i}\right| < v$ should not contain a point with $\text{Im}\, z > u$. This is equivalent to

$$u \geq \frac{1+v}{1-v}.$$

As $v < 1$ this implies $u \geq 1$, and hence $Q_P(u)$ is small enough. We still the condition that the disk $\left|\frac{z-i}{z+i}\right| < v$ should be contained in $F_{\text{mod}} \cup wF_{\text{mod}}$. This means $v \leq \sqrt{5} - 2$. All this is satisfied if we take, e.g., $A_P = 3$ and $A_{Q_i} = \frac{1}{50}$. (In Figure 13.2 the disk corresponding to $U_{Q_i}(A_{Q_i})$ is drawn too large.)

13.2 Fourier expansions of modular forms

In Section 1.4 we have discussed the Fourier expansions of some modular forms. Here we reconsider this in the terminology of Chapter 4.

We use $P = \tilde{\Gamma}_{\text{mod}} \cdot i\infty$ in this chapter. The minimal growth condition for the trivial character is $\mathbf{c}_0$, given by $\mathbf{c}_0(P) = \{0\}$, and $\mathbf{c}_0(Q) = \emptyset$ for any other element Q of $\mathcal{P}$.

13.2.1 *The discriminant* Δ has the expansion $\Delta(z) = \sum_{n\geq 1} \tau(n) e^{2\pi i n z}$, stated in 1.4.2.

For the trivial character the terms in Fourier expansions are indexed by $\mathcal{C}_P(1) = \mathbb{Z}$, see 4.1.2. From the definitions in Section 4.2 we derive

$$\begin{array}{rcll} \omega_l(P, n, \frac{l-1}{2}; a(y)) & = & (4\pi n y)^{l/2} e^{-2\pi n y} & \text{for } \operatorname{Re} n > 0 \\ \hat{\omega}_l(P, n, \frac{l-1}{2}; a(y)) & = & e^{\pi i l/2} (-4\pi n y)^{l/2} e^{-2\pi n y} & \text{for } \operatorname{Re} n < 0. \end{array}$$

(These facts concerning Whittaker functions can also be derived from the differential equations $\mathbf{E}^- \omega_l(P, n, \frac{l-1}{2}) = 0$, $\mathbf{E}^- \hat{\omega}_l(P, n, \frac{l-1}{2}) = 0$, and the asymptotic behavior discussed in 4.2.8.)

The function $\tilde{\Delta} : p(z)k(\theta) \mapsto y^6 \Delta(z) e^{12i\theta}$, corresponding to Δ, is given by

$$\sum_{n=1}^{\infty} e^{2\pi i n x} \frac{\tau(n)}{(4\pi n)^6} \omega_l(P, n, \tfrac{l-1}{2}; a(y)) e^{12i\theta}.$$

Thus $\tilde{F}_{P,n} \tilde{\Delta} = \tau(n)(4\pi n)^{-6} \omega_l(P, n, \frac{l-1}{2})$ if $n \geq 1$, and it vanishes for $n \leq 0$. We see that Δ corresponds to an element of the space of cusp forms $S_{12}(1, \mathbf{c}_0, \frac{11}{2})$.

13.2.2 *Powers of the eta function.* The function $\tilde{\Delta}$ on $\tilde{G}$ is equal to η_{12}, see 13.1.7.

For the character χ_r the possible Fourier term orders at P are given by $\mathcal{C}_P(\chi_r) = \frac{r}{12} + \mathbb{Z}$. Write $n = \nu + \frac{r}{12}$. The expansion in 1.5.5 leads to

$$\tilde{F}_{P,n} \eta_r = \begin{cases} p_\nu(r)(4\pi n)^{-r/2} \omega_r(P, n, (r-1)/2) & \text{if } 12\nu > -\operatorname{Re} r,\ \nu \geq 0 \\ p_\nu(r) \mu_r(P, 0, (r-1)/2) & \text{if } n = 0,\ r \in 12\mathbb{N}_{\geq 0} \\ p_\nu(r) e^{\pi i r/2} (-4\pi n)^{-r/2} \hat{\omega}_r(P, n, (r-1)/2) & \text{if } 0 \leq 12\nu < -\operatorname{Re} r \\ 0 & \text{if } \nu < 0. \end{cases}$$

So η_r is a cusp form in $S_r(\chi_r, \mathbf{n}, \frac{r-1}{2})$ for $0 < \mathrm{Re}\, r < 12$, with the *empty growth condition* $\mathbf{n}$ given by $\mathbf{n}(Q) = \emptyset$ for all $Q \in \mathcal{P}$. Note that $\mathbf{n}$ is the minimal growth condition for each non-trivial character of $\tilde{\Gamma}_{\mathrm{mod}}$. For larger $\mathrm{Re}\, r$ we have a cusp form for even larger growth conditions. On the other hand, if $\mathrm{Re}\, r \leq 0$ we have $\eta_r \in A_r(\chi_r, \mathbf{c}, \frac{r-1}{2})$ with a growth condition $\mathbf{c}$ that has to satisfy $\mathbf{c}(P) \supset \{\nu + \frac{r}{12} \in \mathcal{C}_P(\chi_r) : 0 \leq 12\nu < \mathrm{Re}\, r\}$. So η_r is not a cusp form for all values of r.

13.2.3 *The modular invariant.* In Shimura, [53], Theorem 2.9, we see that the modular invariant $J = (2\pi)^{-12} 12^3 (60 G_4)^3/\Delta(z)$ is a $\tilde{\Gamma}_{\mathrm{mod}}$-invariant holomorphic function on $\mathfrak{H}$ with an expansion $J(z) = \sum_{n=-1}^{\infty} c_n e^{2\pi i n z}$, where $c_n \in \mathbb{Z}$, $c_{-1} = 1$. Notice that $\Delta_{\mathrm{Shimura}} = (2\pi)^{12} \Delta_{\mathrm{here}}$. The Fourier expansion at P of the corresponding function on $\tilde{G}$ is

$$
\begin{aligned}
p(z)k(\theta) \quad \mapsto \quad & \hat{\omega}_0(P, -1, \tfrac{1}{2}; a(y)) e^{-2\pi i x} + c_0 \mu_0(P, 0, -\tfrac{1}{2}; a(y)) \\
& + \sum_{n=1}^{\infty} c_n \omega_0(P, n, -\tfrac{1}{2}; a(y)) e^{2\pi i n x}.
\end{aligned}
$$

J determines an element of $A_0(1, \mathbf{c}, \frac{1}{2})$ for each growth condition $\mathbf{c}$ for which $\mathbf{c}(P) \supset \{-1, 0\}$.

13.2.4 *Singularity at an interior point.* We may consider also $f = \frac{1}{J - J(i)}$. It has a second order pole at i and is invariant under $z \mapsto \frac{-1}{z}$. Hence it has a Laurent series expansion

$$
f(z) = \sum_{m=-1}^{\infty} b_m \left(\frac{z - i}{z + i} \right)^{2m}
$$

for $z \neq i$ in a neighborhood of i. This neighborhood is much smaller than $\mathfrak{H}$, as f has a pole at each point of $\tilde{\Gamma}_{\mathrm{mod}} \cdot i$. Consider $Q_i = \tilde{\Gamma}_{\mathrm{mod}} \cdot i$. This is an elliptic point with order $v_{Q_i} = 2$. So $\mathcal{C}_{Q_i}(1) = 4\mathbb{Z}$, and

$$
\begin{aligned}
k(\eta) a(t_u) k(\psi) & \mapsto f(k(\eta) \cdot i t_u) \\
&= \sum_{m=-1}^{\infty} b_m \left(e^{2i\eta} \sqrt{\frac{u}{u+1}} \right)^{2m} = \sum_{m=-1}^{\infty} b_m e^{4im\eta} \left(\frac{u}{u+1} \right)^m \\
&= i b_{-1} e^{-4i\eta} \hat{\omega}_0(Q_i, -4, -1, -\tfrac{1}{2}; a(t_u)) \\
&\quad + \sum_{n=0}^{\infty} b_m e^{4im\eta} \omega_0(Q_i, 4m, -\tfrac{1}{2}; a(t_u)).
\end{aligned}
$$

The singularity at Q_i involves only the Fourier term of order -4. For large y we have an expansion of the form $f(z) = \sum_{n=1}^{\infty} a_n e^{2\pi i n z}$. Hence f is regular at P. Take the growth condition $\mathbf{c}$ on $\{P, Q_i\}$ given by $\mathbf{c}(P) = \{0\}$, $\mathbf{c}(Q_i) = \{-4\}$. Then $f \in A_0(1, \mathbf{c}, \frac{1}{2})$.

13.2.5 *Eisenstein series in the domain of absolute convergence.* In even weight the series in Proposition 5.2.9 can be expressed in terms of well known functions. We consider $E_\kappa(s) = P_\kappa(P, 0, 1, s)$, with $\kappa \in 2\mathbb{Z}$, and $\mathrm{Re}\, s > \frac{1}{2}$.

The cusp is $P = \tilde{\Gamma}_{\mathrm{mod}} \cdot i\infty$. So we have $\Delta_P = \{\, n(l)k(\pi m) : l, m \in \mathbb{Z}\,\}$, and $\Delta_P\backslash\tilde{\Gamma}_{\mathrm{mod}}$ has the system of representatives

$$\{1\} \cup \left\{ \widetilde{\begin{pmatrix} a & b \\ c & d \end{pmatrix}} : c \geq 1,\ d \in \mathbb{Z},\ (c,d) = 1 \right\},$$

where for each (c,d) we have chosen a and b such that $\begin{pmatrix} a & b \\ c & d \end{pmatrix} \in \Gamma_{\mathrm{mod}} = \mathrm{SL}_2(\mathbb{Z})$. As

$$\begin{aligned} &\widetilde{\begin{pmatrix} a & b \\ c & d \end{pmatrix}} p(z)k(\theta) \\ &= n\left(\frac{(ax+b)(cx+d)+acy^2}{|cz+d|^2} \right) a\left(\frac{y}{|cz+d|^2} \right) k(-\arg(cz+d)+\theta), \end{aligned}$$

we obtain

$$\begin{aligned} E_\kappa(s; p(z)k(\theta)) &= \frac{1}{2} \sum_{c,d\in\mathbb{Z},\,(d,c)=1} \left(\frac{y}{|cz+d|^2} \right)^{s+1/2} \left(\frac{c\bar{z}+d}{cz+d} \right)^{\kappa/2} e^{i\kappa\theta} \qquad (13.2) \\ &= \frac{y^{s+1/2}}{2\zeta(2s+1)} \sum_{n,m\in\mathbb{Z}}{}' |mz+n|^{-2s-1} \left(\frac{m\bar{z}+n}{mz+n} \right)^{\kappa/2} e^{i\kappa\theta}. \end{aligned}$$

This shows that the Eisenstein series in weight 0, see Equation (1.1) on p. 4, is given by $e(s;z) = E_0(s; p(z))$, and also that for even $\kappa \geq 4$ the holomorphic Eisenstein series, see 1.3.2, satisfies

$$G_\kappa(z) = 2\zeta(\kappa) y^{-\kappa/2} E_\kappa\left(\tfrac{\kappa-1}{2}; p(z)\right).$$

13.2.6 *Kloosterman sums.* Consider $n, m \in \mathbb{Z}$. The Kloosterman sum in Equation (5.1) on p. 80 is given by

$$\mathrm{S}_1(P, m; P, n; c) = \sum_{d \bmod c}{}^{*} e^{2\pi i(na+md)/c},$$

where for each $d \bmod c$ we have chosen a such that $ad \equiv 1 \bmod c$. For $mn \neq 0$ this is the classical Kloosterman sum. In the case of the Eisenstein series we have $n = 0$, and obtain

$$\sum_{c\geq 1} \mathrm{S}_1(P, m; P, 0; c) c^{-1-2s} = \sum_{c=1}^{\infty} c^{-1-2s} \sum_{d \bmod c}{}^{*} e^{2\pi i m d/c}$$

$$= \zeta(2s+1)^{-1} \sum_{N=1}^{\infty} \sum_{c|N} \sum_{d \bmod c}^{*} e^{2\pi i m d/c}$$

$$= \zeta(2s+1)^{-1} \sum_{N=1}^{\infty} \sum_{x|N} e^{2\pi i m x/N}$$

$$= \begin{cases} \zeta(2s)/\zeta(2s+1) & \text{if } m = 0 \\ \sigma_{-2s}(|m|)/\zeta(2s+1) & \text{if } m \neq 0, \end{cases}$$

where $\sigma_a(k) = \sum_{d|k} d^a$. This implies

$$c_\kappa(P, m; P, 0; 1, s)$$
$$= \begin{cases} \dfrac{(1/2-s)_{|\kappa|/2}}{(1/2+s)_{|\kappa|/2}} \dfrac{\Lambda(2s)}{\Lambda(2s+1)} & \text{if } m = 0 \\ \dfrac{(-1)^{\kappa/2}\Gamma(s+1/2)}{\Gamma(1/2+s+\kappa\,\mathrm{sign}(m)/2)} \dfrac{|m|^{-s-1/2}\sigma_{2s}(|m|)}{\Lambda(2s+1)}, & \text{if } m \neq 0, \end{cases}$$

with $\Lambda(u) = \pi^{-u/2}\Gamma(u/2)\zeta(u)$.

Take $\kappa = 0$ to obtain the Fourier expansion of $e(s)$ in 1.4.4, and $\kappa = k$, $s = \frac{1}{2}(k-1)$ to obtain the result in 1.4.2.

13.3 The modular spectrum

In 6.7.7 we have introduced the spectral set $\Sigma(\chi, l)$, and various subsets, for unitary characters χ and corresponding real weights l. The subset $\Sigma_d(\chi, l)$ describes the part of the spectrum due to automorphic forms of discrete series type; it is completely known if we know the holomorphic and antiholomorphic automorphic forms. Its complement in $\Sigma(\chi, l)$ is $\Sigma_c(\chi, l) = \Sigma_c(\chi)$. This does not depend on the weight.

We discuss $\Sigma_c(\chi_r)$ and $\Sigma_d(\chi_r, l)$ for the modular case. Here $r \in \mathbb{R} \bmod 12\mathbb{Z}$, and $l \equiv r \bmod 2$.

13.3.1 *Symmetry.* In 4.4.4 we have already mentioned the automorphism j of $\tilde{G}$ that sends $p(z)k(\theta)$ to $p(-\bar{z})k(-\theta)$. This automorphism satisfies $j^2 = \mathrm{Id}$. The corresponding involution j on the functions on $\tilde{G}$ is given by $(jf)(g) = f(jg)$ and satisfies

$$\mathbf{W}jf = -j\mathbf{W}f, \quad \mathbf{E}^{\pm}jf = j\mathbf{E}^{\mp}f, \quad \omega jf = j\omega f.$$

This can be used to relate automorphic forms of opposite weight. The drawback is that it changes the group $\tilde{\Gamma}$ into $j\tilde{\Gamma}$. But if $\tilde{\Gamma}$ is invariant under j, then this involution gives additional information on the spectral sets.

13.3.2 *The symmetry in the modular case.* As $jn(1) = n(-1)$ and $jk(\pi/2) = k(-\pi/2)$, the group $\tilde{\Gamma}_{\mathrm{mod}}$ is invariant under j, and $\chi_r \circ j = \chi_{-r}$. Moreover,

$jg_P = g_P$, and in Iwasawa coordinates on $U_P(\tilde{A}_P)$ we have $j : (x,y,\theta) \mapsto (-x,y,-\theta)$. Hence $\tilde{F}_{P,-n} jf = j\tilde{F}_{P,n} f$. This gives isomorphisms $j : A_l(\chi_r, \mathbf{c}, s) \to A_{-l}(\chi_{-r}, -\mathbf{c}, s)$ and $j : S_l(\chi_r, \mathbf{c}, s) \to S_{-l}(\chi_{-r}, -\mathbf{c}, s)$, provided $\mathcal{P} = \{P\}$. By $-\mathbf{c}$ we denote the growth condition $(-\mathbf{c})(P) = \{\, -n : n \in \mathbf{c}(P)\,\}$. (If $\mathcal{P}$ contains points of Y the corresponding results are a bit more complicated.)

We find $jP_l(P,n,\chi_r,s) = P_{-l}(P,-n,\chi_{-r},s)$ for the Poincaré series at the cusp; use $j\Delta_P = \Delta_P$, and $j\mu_l(P,n,s) = \mu_{-l}(P,-n,s)$.

We also get an isometry $j : K(\chi_r, l) \to K(\chi_{-r}, -l)$ and, after completion, an isometry $j : H(\chi_r, l) \to H(\chi_{-r}, -l)$. This implies that j maps $A_l^2(\chi_r, s)$ into $A_{-l}^2(\chi_{-r}, s)$. The action of j on the spectral sets gives $\Sigma(\chi_r, l) = \Sigma(\chi_{-r}, -l)$, and similarly for the subsets $\Sigma_{\cdot}$ and $\Sigma^{\cdot}$.

13.3.3 *The modular spectrum of continuous series type.* $\Sigma_c(\chi_r) = \Sigma_c(\chi_{r+12}) = \Sigma_c(\chi_{-r})$ for all $r \in \mathbb{R}$. For $\Sigma^e(\chi_r, l)$ to be non-empty, it is necessary that P is singular for χ_r, i.e., $r \equiv 0 \bmod 12$. In 13.4.2 we shall see that $\Sigma^e(1,0) \subset \{\frac{1}{2}\}$, hence $\Sigma_c(\chi_r) = \Sigma_c^0(\chi_r)$ in all cases.

13.3.4 *Cuspidal spectrum of continuous series type.* In the case $r = 0$ we have the well known eigenvalue estimate $\frac{1}{4} - s^2 \geq \frac{3}{2}\pi^2$ for all $s \in \Sigma_c^0(1)$, see, e.g., Proposition 2.1 on p. 511 of [21]. The same idea is used in [7], Proposition 10.1 to obtain

$$\frac{1}{4} - s^2 \geq \max_{l \equiv r \bmod 2,\, 2 \leq r-l \leq 10} \left(\frac{1}{8} \left(\pi\sqrt{3} - \left| \pi\sqrt{3}\left(1 - \frac{1}{6}r\right) + l \right| \right)^2 - \frac{1}{4}l^2 \right), \quad (13.3)$$

for all $s \in \Sigma_c^0(\chi_r)$, $0 < r < 12$.

This estimate shows that $\Sigma_c^0(\chi_r) \subset i(\mathbb{R} \setminus \{0\})$ for r in a large part of $(0,12)$. Proposition 10.2 in [7] gives an estimate that implies this inclusion for all $r \in (0,12)$. So in the modular spectrum there are no 'exceptional eigenvalues' of continuous series type. Moreover, in [5], Proposition 2.8, one finds that for $\chi_r \neq 1$ the elements of $\Sigma(\chi_r, r)$ are given by real analytic functions of the weight on the interval $(0,12)$.

13.3.5 *Spectrum of discrete series type.* This part of the spectrum is completely known, as it corresponds to holomorphic and antiholomorphic modular forms. We restrict our attention to $l \geq 0$, and hence to holomorphic modular forms. From $\Sigma(\chi_r, l) = \Sigma(\chi_{-r}, -l)$ we get the corresponding results for negative weight.

Let $l = 0$. Proposition 6.7.8 gives

$$\Sigma_d(1,0) = \left\{ \frac{1}{2} \right\}, \qquad \Sigma_d(\chi_r, 0) = \emptyset \quad \text{for } r \equiv 0 \bmod 2,\ r \not\equiv 0 \bmod 12.$$

Let $l > 0$. We may take $0 < r \leq 12$. In Proposition 6.7.3 we have seen that

$$\Sigma_d(\chi_r, l) \subset \left\{ \left| \frac{b-1}{2} \right| : 0 < b \leq l,\ b \equiv l \bmod 2 \right\}.$$

Use of the operator $\mathbf{E}^-$ shows us that for $b \in (0, l]$, $b \equiv l \bmod 2$, we have $\left|\frac{b-1}{2}\right| \in \Sigma_d(\chi_r, l)$ if and only if there exists a non-zero $f \in A_b^2(\chi_r, \frac{b-1}{2})$ with $\mathbf{E}^- f = 0$. Such an f corresponds to a square integrable holomorphic modular form $\tilde{f}$ of weight b for the multiplier system $v_r : \left(\begin{smallmatrix} a & b \\ c & d \end{smallmatrix}\right) \mapsto \chi_r\left(\widetilde{\left(\begin{smallmatrix} a & b \\ c & d \end{smallmatrix}\right)}\right)$. The square integrability implies that the Fourier series of $\tilde{f}$ starts with $c_q e^{2\pi i q z}$ with $q \equiv \frac{r}{12} \bmod 1$, $q \geq r$. We divide $\tilde{f}$ by η^{2r}, and obtain a holomorphic modular form of weight $b-r$ for the trivial multiplier system. Such forms can only be non-zero if $b - r \geq 4$ or $b - r = 0$ (see, e.g., [53], Theorem 2.23). Hence

$$\Sigma_d(\chi_r, l) = \left\{ \left|\frac{b-1}{2}\right| : b \equiv l \bmod 2,\ r \leq b \leq l,\ b \neq r + 2 \right\}.$$

We obtain

$$\begin{aligned}
\Sigma_d(1, 0) &= \left\{\frac{1}{2}\right\} \\
\Sigma_d(\chi_r, 0) &= \emptyset \qquad \text{if } r \subset 2\mathbb{Z},\ 0 < r < 12 \\
\Sigma_d(1, l) &= \left\{ \frac{b-1}{2} : b \in 2\mathbb{Z},\ 12 \leq b \leq |l|,\ b \neq 14 \right\} \qquad \text{if } l \in 2\mathbb{Z},\ l \neq 0 \\
\Sigma_d(\chi_r, l) &= \left\{ \left|\frac{b-1}{2}\right| : b \equiv l \bmod 2,\ r \leq b \leq l,\ b \neq r + 2 \right\} \\
&\qquad \text{if } 0 < r < 12,\ l > 0 \\
\Sigma_d(\chi_r, l) &= \left\{ \left|\frac{b-1}{2}\right| : b \equiv -l \bmod 2,\ 12 - r \leq b \leq -l,\ b \neq 14 - r \right\} \\
&\qquad \text{if } 0 < r < 12,\ l < 0.
\end{aligned}$$

For $0 < r < 12$ the set $\Sigma_d(\chi_r, l)$ is empty if $r - 12 < l < r$. These weights are exactly those that occur in the estimate (13.3). The reasoning given above leads to the dimension of the corresponding eigenspace as well. For instance, the constants are the only holomorphic modular forms in weight 0, hence $\dim A_r^2(\chi_r, \frac{1-r}{2}) = 1$ for $0 < r < 12$.

13.4 Families of modular forms

This section gives a few examples of families of modular forms. First, we consider the Eisenstein family as a meromorphic function of the weight and the spectral parameter. Its domain is determined by the null cell. We carry out the computation indicated in 10.3.2, which relates this Eisenstein family to a Poincaré family for a minimal cell. In Section 13.5 we shall look at the singularities of the Eisenstein family at the points $(0, \frac{b}{2})$, $b \in \mathbb{N}$.

Proposition 11.3.9 discusses the singularities of Poincaré families along vertical lines. In 13.4.5 we show that in the modular case we really have singularities of the type described in parts i), ii)a), and iii)b) of that proposition. In 13.4.3 we shall see that case ii)b) does not occur for the modular group. In 11.2.7 we have seen that case iii)a) occurs for all groups.

Proposition 12.5.9 gives several conditions under which Poincaré families on a restricted parameter space have a singularity. Condition a) is the presence of a family of cusp forms for which the Fourier term corresponding to the Poincaré family does not vanish identically. The dimension of the parameter space of that family is allowed to be any non-negative integral number. One might think that a family of dimension zero would suffice. This would mean the existence of a single cusp form. In 13.4.6 we give an example where we need a one-dimensional family. I have no examples in which one needs a family of higher dimension.

13.4.1 *Continuation of Eisenstein series.* The null cell $J(0)$ corresponds to the interval $(-12, 12)$ on the r-line, under the parametrization $r \mapsto r\alpha$ of $\mathcal{V}$. Take $\kappa \in 2\mathbb{Z}$, and consider the weight $l_\kappa : r \mapsto \kappa + r$. Theorem 10.2.1, applied with $P = \tilde{\Gamma}_{\text{mod}} \cdot i\infty$, $n = n_0 : r \mapsto \frac{r}{12}$, and $J = J(0)$, implies that there exists a family of modular forms

$$E_\kappa = E_{l_\kappa}(J(0), P) \in (\mathcal{M} \otimes_{\mathcal{O}} \mathcal{A}_l(\mathbf{c}_0))(U \times \mathbb{C}),$$

with U a neighborhood of $(-12, 12)$ in $\mathbb{C}$, that extends the Eisenstein series discussed in 13.2.5: E_κ as defined there is equal to the restriction $\dot{E}_\kappa(0) : s \mapsto E_\kappa(0, s)$. (We avoid the notation E_{l_κ} for the family of two variables; we pay for it by a notational ambiguity.) The growth condition $\mathbf{c}_0$ on $\mathcal{P} = \{P\}$ is given by $\mathbf{c}_0(P) = \{n_0\}$.

The family E_0 is equal to the family E discussed in Section 1.6.

13.4.2 *Scattering 'matrix'.* Put $C_\kappa^\nu = C_{l_\kappa}(P, n_\nu; J(0), P)$. In 13.2.6 we have shown that for $\operatorname{Re} s > \frac{1}{2}$ the function $\dot{C}_\kappa^0(0) : s \mapsto C_\kappa(0, s)$ is given by

$$\dot{C}_\kappa^0(0; s) = \frac{(1/2 - s)_{|\kappa|/2}}{(1/2 + s)_{|\kappa|/2}} \frac{\Lambda(2s)}{\Lambda(1 + 2s)},$$

with $\Lambda(u) = \Lambda(1 - u) = \pi^{-u/2}\Gamma(u/2)\zeta(u)$. This equality extends to $s \in \mathbb{C}$, as both both sides are meromorphic functions.

From $\dot{C}_\kappa^0(0) = -1$ we see that $\tilde{F}_{P,0}\dot{E}_\kappa(0; 0) = 0$. There are no cusp forms with eigenvalue $\frac{1}{4}$ (see 13.3.4), hence $\dot{E}(0) = 0$. Case ii)b) in Proposition 11.3.9 does not occur for the modular group, see the definition of $E_{P,n}$ in 11.3.2.

13.4.3 *Non-cuspidal spectrum.* Proposition 11.2.4 shows that Σ^e is determined by the singularities of the Eisenstein family. Hence $\Sigma^e(1, \kappa) = \{\frac{1}{2}\}$ if $\kappa = 0$, and $\Sigma^e(1, \kappa)$ is empty for other $\kappa \in 2\mathbb{Z}$.

13.4.4 *Poincaré families on a minimal cell.* Let $\nu \in \mathbb{Z}$, $\kappa \in 2\mathbb{Z}$, and $l_\kappa : r \mapsto r + \kappa$. The interval $(0, 12)$ corresponds to a minimal cell for the modular group. For this cell J^+, and $P = \tilde{\Gamma}_{\mathrm{mod}} \cdot i\infty$, $n = n_\nu : r \mapsto \nu + \frac{r}{12}$, Theorem 10.2.1 gives the existence of a meromorphic family $Q^\nu_\kappa = Q_{l_\kappa}(J^+, P, n_0) \in (\mathcal{M} \otimes_{\mathcal{O}} \mathcal{A}_{l_\kappa}(\mathbf{c}_\nu))(U^+ \times \mathbb{C})$, with U^+ a neighborhood of $(0, 12)$ in $\mathbb{C}$, that satisfies $Q^\nu_\kappa(r,s) = P_{\kappa+r}(P, \nu + \frac{r}{12}, \chi_r, s)$ for real $r \in (0,12)$ and $\mathrm{Re}\, s > \frac{1}{2}$. This family extends the Poincaré series with weight in $(0, 12)$.

Let $U \supset (-12, 12)$ be an open set in $\mathbb{C}$ such that the Eisenstein family lives on $U \times \mathbb{C}$. We may assume $U^+ \subset U$. In 10.3.2 we see how to relate Q^ν_κ to families on the null cell. In the case $\nu = 0$, with the notation $Q_\kappa = Q^0_\kappa$, we obtain

$$Q_\kappa(r,s) = \Big(1 + v_{\kappa+r}(P, r/12, s) D_\kappa(r,s)\Big) E_\kappa(r,s),$$

on $U^+ \times \mathbb{C}$, with $D_\kappa = D_{l_\kappa}(P, n_0; J^+, P, n_0)$. See 4.2.8 for $v_{\kappa+r}$. A computation of the Fourier term for (P, n_0) leads to

$$Q_\kappa(r,s) = \frac{v_{\kappa+r}(P, r/12, -s)}{v_{\kappa+r}(P, r/12, -s) - v_{\kappa+r}(P, r/12, s) C^0_\kappa(r,s)} E_\kappa(r,s). \tag{13.4}$$

13.4.5 *Singularities.* Consider $0 < r < 12$. The space $H(\chi_r, r)$ is equal to the space ${}^aH$ built in Section 8.3 for $\chi_0 = \chi_r$, $l_0 = r$, $\mathbf{c} = \mathbf{n}$, and, e.g., $a > 5$. (As $\mathbf{c}$ is empty, truncation has no effect.) The extension of the Casimir operator $A(\chi_r, r)$ and the pseudo Casimir operator ${}^aA(0)$ for r coincide. Proposition 9.2.2 implies that $H(\chi_r, r)$ has an orthonormal basis of eigenfunctions of $A(\chi_r, r)$; these eigenfunctions correspond to elements of $A_r(\chi_r, \mathbf{n}, s)$, see Proposition 9.2.6. The number s runs through a discrete countable subset of $i(0,\infty)$ satisfying $\frac{1}{4} - s^2 \geq \frac{r}{2}(1 - \frac{r}{2})$. Each of these automorphic forms has non-zero Fourier terms. Except for the first one, they come under part i) of Proposition 11.3.9. Hence there are many Poincaré families Q^ν_κ with a singularity at the corresponding s, see part i) of Proposition 11.3.9.

A similar result occurs in weight zero for the trivial character. Hecke theory even implies that for each $s \in \Sigma_c(1)$ there is an element $f \in S_0(1, \mathbf{c}_0, s)$ with $\tilde{F}_{P,1} f \neq 0$; see, e.g., [35], p. 258–259.

For $0 < r < 12$ the smallest eigenvalue is $\frac{1}{2}r - \frac{1}{4}r^2$. The corresponding eigenspace is spanned by η_r, see 13.1.7. For $0 < r \leq 2$ we have $\tau(\chi_r) = \frac{r-1}{2}$. Consider Q^0_κ, and apply Proposition 11.3.9 with $n = n_0$. For $0 < r < 1$ the conditions in part iii)b) are satisfied, and for $r = 1$ the first case in part ii)a) occurs.

13.4.6 *A Poincaré family of indeterminate type.* The concept *(in)determinate type* has been introduced in Definition 12.2.10. The Fourier coefficients of η_r have an explicit description, see 13.2.2 and 1.5.5. This enables us to give an example of a Poincaré family that has indeterminate type at some point.

In the Fourier expansion of η_r we used polynomials p_n, see 1.5.5. A computation shows that $p_3(r) = -\frac{4}{3}r^3 + 6r^2 - \frac{8}{3}r$. Hence $\tilde{F}_{P,n_3}\eta_4 = 0$, whereas $\tilde{F}_{P,n_3}\eta_r \neq 0$ for $r \neq 4$, r in a neighborhood of 4.

We work on the parameter space $W_{\mathrm{mod}} = \mathbb{C}^2$, $\chi(r,s) = \chi_r$, $s(r,s) = s$, with the weight $r : (r,s) \mapsto r$, and $n_\nu \in \mathcal{C}_P$ given by $n_\nu(r,s) = \nu + \frac{r}{12}$. We take $w_0 = (4, \frac{3}{2}) \in W_{\mathrm{mod}}$. We apply Proposition 10.2.4 with the basis for the Fourier terms given in Table 12.1 on p. 224. This gives us a family $R_r(P, n_3)$. Table 12.1 gives the same basis of $\mathcal{W}_r(P, n_3)$ at $(r, \frac{r-1}{2})$ for all $r \in (3,5)$. Hence all $R_r(P, n_3)$ for these r coincide. Apply Lemma 12.3.4 to $R_r(P, n_3)$ at $(r, \frac{r-1}{2})$ for $3 < r < 5$. At such a point $q_r(P, n_3) = 1$ if and only if there is an element f in $A_r^2(\chi_r, \frac{1}{2}(r-1)) = S_r(\chi_r, \mathbf{n}, \frac{r-1}{2})$ for which $\tilde{F}_{P,n_3} f \neq 0$, see Lemma 11.1.9. In Proposition 6.7.3 we see that such an f_r satisfies $\mathbf{E}^- f_r = 0$, hence it corresponds to a holomorphic modular form of weight r and multiplier system v_r. This means that f is a multiple of η_r. We conclude that $q_r(P, n_3) = 1$ for $r \neq 4$. The points $(r, \frac{r-1}{2})$ with $3 < r < 5$, $r \neq 4$ are in the analytic set $\mathrm{Pol}(R_r(P, n_3))$. Hence $(4, \frac{3}{2}) \in \mathrm{Pol}(R_r(P, n_3))$ as well, and at this point $k_r(P, n_3) \geq 1 > q_r(P, n_3) = 0$.

13.5 Derivatives of Eisenstein families

Part iii) of Proposition 12.4.2 shows that we may expect singularities of the Eisenstein family E_κ at points $(r,s) = (0, \frac{b}{2})$, with $b \in \mathbb{N}$. This is not in contradiction to the fact that for $\mathrm{Re}\, s > \frac{1}{2}$ we are in the domain of absolute convergence of the Eisenstein family. The singularity need not be visible after restriction to a vertical line. In this section we shall see that such singularities really occur. We use the derivatives $s \mapsto \partial_r^m E_\kappa(0,s)$, $m \in \mathbb{N}$. These derivatives are meromorphic functions in one variable. If one of them has a singularity at $s = \frac{b}{2}$, then E_κ surely has a singularity at $(0, \frac{b}{2})$. The transformation behavior of these derivatives under $\tilde{\Gamma}_{\mathrm{mod}}$ is almost that of a modular form. In fact, these functions invite us to extend our concept of modular form, see 13.5.5 and 13.5.6.

The main result of this section is Proposition 13.5.10. There we see that the $\partial_r^m E_\kappa(0,s)$ have a series representation for large $\mathrm{Re}\, s$. This means that each $\partial_r^m E_\kappa(0,s)$ is singular at $s = \frac{b}{2}$ for only a finite number of $b \geq 2$. Moreover, we get an explicit hold on the right-most singularity.

These results give a rather explicit description of the behavior of E_κ at the points $(0, \frac{b}{2})$, and form the basis for the distribution results in the next section.

13.5.1 *Singularities at half-integral points.* We keep $\kappa \in 2\mathbb{Z}$. Proposition 12.4.2 shows that E_κ is holomorphic at points $(0,s)$ with $\mathrm{Re}\, s > \frac{1}{2}$, except possibly for the half-integral values of s. Those singularities are not connected to the presence of square integrable modular forms, but are due to the fact that $\mu_{\kappa+r}$, $\tilde{\mu}_{\kappa+r}$ is unsuitable as a basis for the space of Fourier terms of order $(P, \frac{r}{12})$ near $(r,s) = (0, \frac{b}{2})$, $b \in \mathbb{N}$.

For $b \in \mathbb{N}$, $b \geq 2$, Proposition 12.4.2 shows that

$$E_\kappa^{(b)} : (r,s) \mapsto \frac{1}{1 - w_{r+\kappa}^b(\frac{r}{12}, s) C_\kappa^0(r,s)} E_\kappa(r,s)$$

is holomorphic at $(0, \frac{b}{2})$, with value given by the absolutely converging Eisenstein series $P^0_\kappa(0,s)$, equal to $E_\kappa(s)$ in Equation (13.2) on p. 246. The proof of Proposition 12.4.2 shows that $E^{(b)}_\kappa$ is the family $R(P,n_0)$ from Proposition 10.2.4. Hence $E^{(b)}_\kappa$ is characterized by its Fourier term of order (P,n_0):

$$\tilde F_{P,n_0} E^{(b)}_\kappa(r,s) = \mu_{\kappa+r}(P, \frac{r}{12}, s) + B^{(b)}_\kappa(r,s)\nu^b_{\kappa+r}(P, \frac{r}{12}, s),$$

with $B^{(b)}_\kappa$ a meromorphic function.

At $(0,\frac12)$ we have more or less the same situation, provided $\kappa \neq 0$. Indeed, $\dim A_\kappa(1, \mathbf{c}_0, \frac12) = 1$, but if $\kappa \neq 0$ its non-zero elements are not square integrable. Proposition 12.2.12 implies that E_κ has determinate type at $(0,\frac12)$, and that the same holds for R_{P,n_0}. As $\dot R_{P,n_0}(0) = \dot E_\kappa(0)$, we conclude from Proposition 11.2.4 and Lemma 12.3.4 that $\tilde E^{(1)}_\kappa : (r,s) \mapsto \left(1 - w^1_{r+\kappa}(\frac{r}{12}, s) C^0_\kappa(r,s)\right)^{-1} E_\kappa(r,s)$ is holomorphic at $(0,\frac12)$. Its value at $(0,\frac12)$ is not given by an absolutely converging Eisenstein series.

One may consult [7], Proposition 6.5, for the singularity of E_0 at $(0,\frac12)$.

13.5.2 *Derivatives.* For $m \geq 0$ and $g \in \tilde G$ consider

$$e^m_\kappa(s;g) = \frac{1}{m!}\partial^m_r E_\kappa(r,s;g)\Big|_{r=0}.$$

This defines meromorphic families e^m_κ of functions on $\tilde G$ depending on the parameter s. Note that $e^0_\kappa(s) = \dot E_\kappa(0;s)$, but that $e^m_\kappa(s)$ does not satisfy our definition of a modular form on $\tilde G$ if $m > 0$. These families have no singularities in the region $\operatorname{Re} s > 0$, except possibly at points of $\frac12\mathbb{N}$.

13.5.3 *Singularities of e^m_κ.* Let $b \in \mathbb{N}$, and suppose $b \geq 2$ or $\kappa \neq 0$. We have defined $B^{(b)}_\kappa(r,s)$ as the coefficient of $\nu^b_{r+\kappa}(P, \frac{r}{12}, s)$ in $\tilde F_{P,n_0} E^{(b)}_\kappa(r,s)$. Hence $B^{(b)}_\kappa(r,s) = C^0_\kappa(r,s)/(1 - w^b_{r+\kappa}(\frac{r}{12}, s) C^0_\kappa(r,s))$. It is holomorphic at $(0,\frac{b}{2})$, and $B^{(b)}_\kappa(0,s) = \dot C^0_\kappa(0;s)$ for s near $\frac{b}{2}$. On the other hand $C^0_\kappa = B^{(b)}_\kappa/(1 + w^b_{\kappa+r} B^{(b)}_\kappa)$, and

$$E_\kappa(r,s) = \left(1 + w^b_{r+\kappa}(P, \frac{r}{12}, s) B^{(b)}_\kappa(r,s)\right)^{-1} E^{(b)}_\kappa(r,s).$$

The expansion in r of $w^b_{r+\kappa}(\frac{r}{12}, s)$ starts with

$$\begin{aligned} & r^b \left(\frac{\pi}{3}\right)^b \left(\frac12 - s - \frac{\kappa}{2}\right)_b \frac{\Gamma(-2s)}{\Gamma(2s)} \\ &+ \frac12 r^{b+1} \left(\frac{\pi}{3}\right)^b \partial_s \left(\left(\frac12 - s - \frac{\kappa}{2}\right)_b\right) \frac{\Gamma(-2s)}{\Gamma(2s)} + \dots \end{aligned}$$

The factor $(\frac12 - s - \frac{\kappa}{2})_b \Gamma(-2s)/\Gamma(2s)$ has a first order pole at $s = \frac{b}{2}$ if b is even. If b is odd this factor may be holomorphic at $\frac{b}{2}$; in that case, the next term has a first order pole at $\frac{b}{2}$. From

$$\sum_{m=0}^{\infty} r^m e_\kappa^m(s) = \left(1 - r^b \left(\frac{\pi}{3}\right)^b \left(\frac{1}{2} - s - \frac{\kappa}{2}\right)_b \frac{\Gamma(-2s)}{\Gamma(2s)} B_\kappa^{(b)}(0,s) + \dots\right)$$
$$\cdot \left(\text{the expansion in } r \text{ of } E_\kappa^{(b)}(r,s)\right)$$

we conclude that e_κ^m is holomorphic at $s = \frac{b}{2}$ for $0 \leq m \leq b$ if b is odd and $b \geq 1 + |\kappa|$, and for $0 \leq m \leq b-1$ otherwise. In 13.2.6 we see that

$$B_\kappa^{(b)}(0;s) = \dot{C}_\kappa^0(0;s) = \frac{\left(\frac{1}{2}-s\right)_{|\kappa|/2}}{\left(\frac{1}{2}+s\right)_{|\kappa|/2}} \frac{\Lambda(2s)}{\Lambda(2s+1)}.$$

The factor $\left(\frac{1}{2}-s\right)_{|\kappa|/2}$ ensures that e_κ^m is holomorphic at $\frac{b}{2}$ for odd b, even if $m = b$ and $b \leq |\kappa| - 1$.

For fixed m this means that e_κ^m can have a pole at $\frac{b}{2}$ only if $\frac{b}{2} \leq \left[\frac{m}{2}\right]$.

13.5.4 Proposition. *If $\kappa \in 2\mathbb{Z}$, and $\kappa \neq 0$, then e_κ^1 is holomorphic on* $\operatorname{Re} s > 0$.

Let $\kappa \in 2\mathbb{Z}$, and $m \geq 2$. Poles of e_κ^m in $\operatorname{Re} s > 0$ *may occur only at $\frac{b}{2}$ with $b \in \mathbb{Z}$, $\frac{1}{2} \leq \frac{b}{2} \leq \left[\frac{m}{2}\right]$. The right-most singularity occurs at $\frac{b}{2} = \left[\frac{m}{2}\right]$, and has order* 1. *Its residue is*

$$\frac{(-1)^{b/2}}{b!} \left(\frac{\pi}{6}\right)^b \frac{\zeta(b)}{\zeta(b+1)} e_\kappa^0\left(\frac{b}{2}\right).$$

Proof. We still need to compute, for $m \geq 2$,

$$\lim_{s \to b/2} -\left(s - \frac{b}{2}\right)\left(\frac{\pi}{3}\right)^b \frac{\left(\frac{1}{2}-s-\frac{\kappa}{2}\right)_b \Gamma(-2s)\left(\frac{1}{2}-s\right)_{|\kappa|/2} \Lambda(2s)}{\Gamma(2s)\left(\frac{1}{2}+s\right)_{|\kappa|/2} \Lambda(2s+1)},$$

with $b = 2\left[\frac{m}{2}\right]$. Use

$$\left(\frac{1}{2} - s - \frac{\kappa}{2}\right)_b \frac{\left(\frac{1}{2}-s\right)_{|\kappa|/2} \pi^{1/2}\Gamma(s)}{\Gamma(2s)\left(\frac{1}{2}+s\right)_{|\kappa|/2}\Gamma(s+\frac{1}{2})} = \frac{\pi(-1)^{\kappa/2} 2^{1-2s}}{\Gamma(\frac{1}{2}+s-\frac{\kappa}{2})\Gamma(\frac{1}{2}+s+\frac{\kappa}{2}-b)}.$$

13.5.5 *Reformulation of the definition of modular form.* If we differentiate the relation $E_\kappa(r,s;\gamma g k(\theta)) = e^{ir\alpha(\gamma)} E_\kappa(r,s;g) e^{i(r+\kappa)\theta}$ a number of times with respect to r at $r = 0$, we get a complicated transformation behavior for the e_κ^m. This becomes more transparent if we impose only a part of the condition of χ_r-l-equivariance, stated in 4.1.1. We carry this out for the modular case.

If $\gamma \in \tilde{\Gamma}_{\text{mod}}$, then $\gamma^{-1} g k(\alpha(\gamma))$ does not depend of the choice of γ in its class $\gamma\tilde{Z}$. Hence

$$(g,x) \mapsto g * x = \gamma^{-1} g k(\alpha(\gamma)) \qquad \text{with } \gamma \in x \in \tilde{\Gamma}_{\text{mod}}/\tilde{Z} = \bar{\Gamma}_{\text{mod}}$$

defines a right action of $\bar{\Gamma}_{\text{mod}} = \tilde{\Gamma}/\tilde{Z} \cong \text{PSL}_2(\mathbb{Z})$ on $\tilde{G}$. If $r \in \mathbb{C}$ and $\kappa \in 2\mathbb{Z}$, the condition of χ_r-equivariance on a function f of weight $r+\kappa$ is equivalent to

$$f(g * x) = \chi_\kappa(x) f(g) \qquad \text{for all } x \in \bar{\Gamma}_{\text{mod}}. \tag{13.5}$$

The character χ_κ of $\tilde{\Gamma}_{\text{mod}}$ is trivial on the center $\tilde{Z}$, hence it can be viewed as a character of $\bar{\Gamma}_{\text{mod}}$.

We define the representation $x : f \mapsto x \cdot f$ of $\bar{\Gamma}_{\text{mod}}$ by $(x \cdot f)(g) = f(g * x)$. This representation commutes with right translation by elements of $\tilde{K}$, and with the action of the differential operators ω and $\mathbf{W}$.

13.5.6 *The e_κ^m as a generalized type of modular form.* The weight r is not involved in the relation $E_\kappa(r, s; g * x) = \chi_\kappa(x) E_\kappa(r, s; g)$; hence $\partial_r^m E_\kappa(r,s)$ satisfies (13.5), and $(\omega - \frac{1}{4} + s^2)\partial_r^m E_\kappa(r,s) = 0$, but it is not a weight function. This holds in particular for $e_\kappa^m(s) = \partial_r^m E_\kappa(0,s)$. So it might be worthwhile to extend the definition of automorphic form as an eigenfunction of ω on $\tilde{G}$ (or $\tilde{G}_{\mathcal{P}}$) satisfying (13.5) and some growth conditions, near the exceptional points, and also as $|\theta| \to \infty$. The $e_\kappa^m(s)$ are examples. The function $p(z)k(\theta) \mapsto 2\log\eta(z) + \log y + i\theta$ has this transformation behavior, but it is not an eigenfunction of the Casimir operator.

13.5.7 *Series representation for e_κ^m.* Equation (13.4) on p. 251 implies for $r \in U^+$:

$$E_\kappa(r,s) = \left(1 - \left(\frac{\pi r}{3}\right)^{2s} \frac{\Gamma(-2s)\Gamma(\frac{1}{2}+s-\frac{r+\kappa}{2})}{\Gamma(2s)\Gamma(\frac{1}{2}-s-\frac{r+\kappa}{2})} C_\kappa^0(r,s)\right) Q_\kappa(r,s). \tag{13.6}$$

Fix $g \in \tilde{G}$ and $s \in \mathbb{C}$ with $\text{Re}\, s = \sigma \in (k, k+\frac{1}{2})$, $k \in \mathbb{N}$, $k \geq 2$. Then

$$E_\kappa(r,s;g) = Q_\kappa(r,s;g) + \mathcal{O}(r^{2\sigma}) \quad \text{for } r \downarrow 0.$$

Hence the $e_\kappa^m(s;g)$ with $0 \leq m < 2\sigma$ can be read off from the expansion in r of $Q_\kappa(r,s;g)$ for $r \downarrow 0$.

Let $0 < r < 12$. The absolutely converging Poincaré series

$$P_{\kappa+r}(P, \tfrac{r}{12}, \chi_r, s; g) = \sum_{\gamma \in \Delta_P \backslash \tilde{\Gamma}_{\text{mod}}} \chi_r(\gamma)^{-1} \mu_{r+\kappa}(P, \tfrac{r}{12}, s; \gamma g)$$

is equal to $Q_\kappa(r,s;g)$ for $\text{Re}\, s > \frac{1}{2}$. We may write it as

$$P_{\kappa+r}(P, \tfrac{r}{12}, \chi_r, s; g) = \sum_x \chi_\kappa(x)^{-1} \mu_{r+\kappa}(P, \tfrac{r}{12}, s; g * x),$$

with x running through representatives of $\bar{\Gamma}_{\text{mod}}/\bar{\Gamma}_{\text{mod}}^\infty$; by $\bar{\Gamma}_{\text{mod}}^\infty$ we denote the image of $\tilde{N} \cap \tilde{\Gamma}_{\text{mod}}$ in $\bar{\Gamma}_{\text{mod}} = \text{PSL}_2(\mathbb{Z})$.

To identify the series for e_κ^m we define the functions β, η and τ on $\tilde{G}$ by

$$\beta(p(z)k(\theta)) = z + \frac{6}{\pi}\theta, \quad \eta(p(z)k(\theta)) = y, \quad \tau(p(z)k(\theta)) = e^{i\theta}.$$

Then

$$\mu_{r+\kappa}(P, \tfrac{r}{12}, s; g) = \tau(g)^\kappa e^{\pi i r \beta(g)/6} \eta(g)^{s+1/2} {}_1F_1 \left[\begin{array}{c} \frac12 + s - \frac{r+\kappa}{2} \\ 1+2s \end{array} \middle| \frac{\pi r}{3}\eta(g) \right].$$

In the sum we take the representatives $x = \pm \left(\begin{smallmatrix} 1 & 0 \\ 0 & 1 \end{smallmatrix}\right)$ and

$$x^{-1} \in \left\{ \pm \begin{pmatrix} \cdot & \cdot \\ c & d \end{pmatrix} : c \geq 1,\, d \in \mathbb{Z},\, (d,c) = 1 \right\}.$$

As $\eta\left(p(z)k(\theta) * \left(\begin{smallmatrix} \cdot & \cdot \\ c & d \end{smallmatrix}\right)^{-1}\right) = \dfrac{y^{s+1/2}}{|cz+d|^{2s+1}}$, the sum $\sum_x \chi_\kappa(x)^{-1} h(g*x)$ converges absolutely, even if we omit the ${}_1F_1$-factor. For $0 < r < 12$ and $\mathrm{Re}\, s > \frac12$ we obtain the absolutely converging double series

$$\begin{aligned} &P_{\kappa+r}(P, \frac{r}{12}, \chi_r, s; g) \qquad (13.7) \\ &= \sum_{n=0}^\infty \frac{\left(\frac12 + s - \frac{r+\kappa}{2}\right)_n}{(1+2s)_n} \left(\frac{\pi}{3}\right)^n \frac{r^n}{n!} \sum_x \chi_\kappa(x)^{-1} \left(\tau^\kappa \cdot e^{\pi i r \beta/6} \cdot \eta^{s+n+1/2}\right)(g*x). \end{aligned}$$

13.5.8 *Another type of series.* To insert the power series for the exponential function $e^{\pi i r\beta/6}$, and interchange the order of summation, we shall need the convergence of sums like

$$S_{q,\sigma} = \sum_x \chi_\kappa(x)^{-1} \beta(g*x)^q \eta(g*x)^{\sigma+1/2} \tau(g*x)^\kappa.$$

Take $g = p(z)k(\theta)$ fixed, and $\left(\begin{smallmatrix} a & b \\ c & d \end{smallmatrix}\right) \in \Gamma_{\mathrm{mod}}$, with $c \geq 1$.

$$\begin{aligned} &\beta\left(p(z)k(\theta) * \begin{pmatrix} d & -b \\ -c & a \end{pmatrix}\right) \\ &= \frac{az+b}{cz+d} + \frac{6}{\pi}\left(-\arg(cz+d) + \theta - \frac{\pi}{6}\frac{a+d}{c} + \frac{\pi}{2} + 2\pi S(d,c)\right) \\ &= -\frac{d}{c} - \frac{1}{c(cz+d)} + 3 + 12 S(d,c) + \frac{6}{\pi}\theta - \frac{6}{\pi}\arg(cz+d) \\ &= \mathcal{O}(|cz+d|) = \mathcal{O}\left(\eta\left(p(z)k(\theta) * \begin{pmatrix} d & -b \\ -c & a \end{pmatrix}\right)^{-1/2}\right). \end{aligned}$$

As $|\tau| = 1$, this gives the absolute convergence of $S_{q,\sigma}$ for $q = 0, 1, \ldots, 2k-1$. We have obtained the next result:

13.5.9 Lemma. *Let $q \geq 0$, $q \in \mathbb{Z}$. The following series converges absolutely, and defines a family of functions ε^q_κ on $\tilde{G}$, pointwise holomorphic in s, for* $\mathrm{Re}\, s >$

$\frac{1}{2}(q+1)$.

$$\begin{aligned} \varepsilon_\kappa^q(s;g) &= \sum_x \chi_\kappa(x) \left(\tau^\kappa \cdot \beta^q \cdot \eta^{s+1/2}\right)(g*x) \\ &= \sum_{\gamma\in\Delta_P\backslash\tilde{\Gamma}_{\mathrm{mod}}} \chi_\kappa(\gamma)\tau(\gamma g k(-\alpha(\gamma)))^\kappa \beta(\gamma g k(-\alpha(\gamma)))^q \eta(\gamma g)^{s+1/2}. \end{aligned}$$

These functions satisfy the modular transformation behavior (13.5) on p. 255.

Remark. The ε_κ^q are not weight functions, and no eigenfunctions of the Casimir operator.

13.5.10 Proposition. *Define* $\pi_{n,j}(a)$ *by* $(a-x)_n = \sum_{j=0}^n \pi_{n,j}(a)x^j$. *For* $m \geq 0$, $\kappa \in 2\mathbb{Z}$, *and* $\mathrm{Re}\, s$ *large,*

$$e_\kappa^m(s) = \sum_{q=0}^m \sum_{0\leq j\leq (m-q)/2} \frac{\pi_{m-q-j,j}\left(\frac{1}{2}+s-\frac{\kappa}{2}\right)\left(\frac{\pi}{3}\right)^{m-j}\left(\frac{i}{2}\right)^q}{2^j(2s+1)_{m-j-q}(m-j-q)!q!}\varepsilon_\kappa^q(s+m-q-j).$$

Proof. For $\varepsilon > 0$ and $a > \frac{1}{2}$ (and again g fixed)

$$\sum_{x,\,\eta(x*g)<\varepsilon} |\eta(g*x)|^{a+1/2} = \mathcal{O}\left(\varepsilon^{a-1/2}\right) \quad (\varepsilon \downarrow 0).$$

Let $\sigma = \mathrm{Re}(s)$, $2 \leq k < \sigma < k+\frac{1}{2}$, and let $n \geq 0$. We take $\varepsilon = r^2$ for $r \downarrow 0$. The following quantity occurs in Equation (13.7):

$$\begin{aligned} &\sum_x \chi_\kappa(x)^{-1}\left(\tau^\kappa \cdot e^{\pi i r\beta/6} \cdot \eta^{s+n+1/2}\right)(g*x) \\ &\quad = \sum_{x,\,\eta(g*x)\geq\varepsilon} \chi_\kappa(x)^{-1}\left(\tau^\kappa \cdot e^{\pi i r\beta/6} \cdot \eta^{s+n+1/2}\right)(g*x) + \mathcal{O}\left(\varepsilon^{\sigma+n-1/2}\right). \end{aligned}$$

In the sum with $\eta(g*x) \geq \varepsilon$ we have $\beta = \mathcal{O}(\eta^{-1/2}) = \mathcal{O}(1/r)$ and $r\beta = \mathcal{O}(1)$. Insert $e^{\pi i r\beta/6} = \sum_{q=0}^{2k-2} \frac{1}{q!}(\pi i r\beta/6)^q + \mathcal{O}\left(r^{2k-1}\eta^{-k+1/2}\right)$.

$$\begin{aligned} &\sum_x \chi_\kappa(x)^{-1}\left(\tau^\kappa \cdot e^{\pi i r\beta/6} \cdot \eta^{s+n+1/2}\right)(g*x) \\ &\quad = \sum_{q=0}^{2k-2}\left(\frac{\pi i r}{6}\right)^q \frac{1}{q!} \sum_{x,\eta(g*x)\geq\varepsilon} \chi_\kappa(x)^{-1}\left(\tau^\kappa \cdot \beta^q \cdot \eta^{s+n+1/2}\right)(g*x) \\ &\qquad + \mathcal{O}\left(r^{2k-1}\right) + \mathcal{O}\left(r^{2\sigma+2n-1}\right) \\ &\quad = \sum_{q=0}^{2k-2}\left(\frac{\pi i}{6}\right)^q \frac{r^q}{q!}\varepsilon_\kappa^q(s+n;g) + \mathcal{O}\left(r^{2\sigma+2n-1}\right) + \mathcal{O}\left(r^{2k-1}\right) \end{aligned}$$

$$= \sum_{q=0}^{2k-2} \left(\frac{\pi i}{6}\right)^q \frac{r^q}{q!} \varepsilon_\kappa^q(s+n;g) + \mathcal{O}\left(r^{2k-1}\right).$$

In view of (13.6) and (13.7) we obtain $e_\kappa^m(s;g)$ with $0 \leq m \leq 2k-2$ as the coefficient of r^m in

$$\sum_{n=0}^{\infty} \frac{\left(\frac{1}{2}+s-\frac{r+\kappa}{2}\right)_n}{(1+2s)_n} \left(\frac{\pi}{3}\right)^n \frac{r^n}{n!} \sum_{q=0}^{2k-2} \left(\frac{\pi i}{6}\right)^q \frac{r^q}{q!} \varepsilon_\kappa^q(s+n;g).$$

13.5.11 Corollary. *Let $\kappa \in 2\mathbb{Z}$, and $m \geq 0$. The family ε_κ^m has a meromorphic extension to $\mathbb{C}$. In the region $\operatorname{Re} s > 0$ it can have singularities only at half-integral points.*

The families ε_κ^0 and ε_κ^1 are holomorphic on $\operatorname{Re} s > \frac{1}{2}$.

Let $m \geq 2$. The right-most singularity of ε_κ^m occurs at $\left[\frac{m}{2}\right]$; it is a first order pole, with residue

$$\begin{cases} \dfrac{\zeta(m)}{\zeta(m+1)} \varepsilon_\kappa^0\left(\dfrac{m}{2}\right) & \text{if } m \text{ is even} \\ \dfrac{6m}{\pi i} \dfrac{\zeta(m-1)}{\zeta(m)} \varepsilon_\kappa^0\left(\dfrac{m-1}{2}\right) & \text{if } m \text{ is odd.} \end{cases}$$

Proof. We already know the statement concerning $\varepsilon_\kappa^0 = e_\kappa^0 = \dot{E}_\kappa(0)$. Proposition 13.5.10 gives $e_\kappa^1(s) = \frac{\pi}{3} \frac{\frac{1}{2}+s-\frac{\kappa}{2}}{2s+1} \varepsilon_\kappa^0(s+1) + \frac{\pi i}{6} \varepsilon_\kappa^1(s)$ for $m = 1$. Proposition 13.5.4 implies that ε_κ^1 is holomorphic on $\operatorname{Re} s > \frac{1}{2}$.

For $m \geq 2$ we use

$$e_\kappa^m(s) = \left(\frac{\pi i}{6}\right)^m \frac{1}{m!} \varepsilon_\kappa^m(s)$$
$$+ \sum_{q=0}^{m-1} \sum_{0 \leq j \leq (m-q)/2} (\text{holomorphic on } \operatorname{Re} s > 0) \cdot \varepsilon_\kappa^q(s+m-q-j).$$

For all terms in the sum $m-q-j \geq \frac{1}{2}$. The terms with $q = 0, 1$ are holomorphic on $\operatorname{Re} s > \frac{1}{2}$. Proceeding inductively, we assume that the terms with $2 \leq q \leq m-1$ can have singularities only at half-integral points $s \leq \left[\frac{q}{2}\right] - (m-q-j) \leq \frac{q}{2} - m + q + \frac{m-q}{2} \leq \left[\frac{m}{2}\right] - \frac{1}{2}$. Hence these terms do not matter for the statement of the corollary. Apply Proposition 13.5.4 to complete the proof.

13.6 Distribution results

In [8] I carried out the computations of the previous section, not for the family E_κ, but for its Fourier coefficients C_κ^ν. This gave the meromorphic continuation of

Dirichlet series with powers of Dedekind sums in its coefficients. The information on the right-most singularities led to distribution results for $S(d,c)/c$, where $S(d,c)$ is the Dedekind sum. In this section the same method is applied to the Eisenstein family itself; this produces similar distribution results for a more complicated quantity.

13.6.1 *Notations.* We fix $g = p(z)k(\theta) \in \tilde{G}$. Define $T = T_g = \{\, cz+d : c, d \in \mathbb{Z}, (d,c) = 1 \,\}$. For $cz+d \in T$ with $c > 0$ we put

$$\sigma_1(cz+d) = 12S(d,c) - \frac{d}{c} + 3 - \frac{6}{\pi}\arg(cz+d) + \frac{6}{\pi}\theta - \mathrm{Re}\left(\frac{1}{c(cz+d)}\right).$$

We extend this definition by $\sigma_1(1) = 0$ and $\sigma_1(-t) = \sigma_1(t)$ to get a function $\sigma_1 : T \to \mathbb{R}$.

We study the distribution of the values of the map $\sigma = \sigma_g : T \to \mathbb{C}$ given by

$$\sigma(cz+d) = \frac{\sigma_1(cz+d)}{cz+d} e^{i\theta}.$$

It turns out that we can obtain distribution results for $\sigma(cz+d)$ from the analytic properties of the $\varepsilon_\kappa^m(s;g)$. In comparison with $S(d,c)/c$, considered in [8], the quantity $\sigma(cz+d)$ is a rather strange one to to study. But the following distribution results can be obtained, and we shall show how to do that.

13.6.2 Proposition. *Fix $g = p(z)k(\theta) \in \tilde{G}$, and define σ_g and T as above. Let $f : \mathbb{C} \to \mathbb{C}$ be a continuous function.*

i) First distribution result.

$$\lim_{X\to\infty} \frac{1}{X} \sum_{cz+d\in T,\, |cz+d|<X} \frac{f(\sigma_g(cz+d))}{|cz+d|} = \frac{12}{\pi y} f(0).$$

ii) Second distribution result. *Let $\varphi(l)$ denote the number of x* mod l *relatively prime to l, and define $\varphi : \mathbb{Z}\cdot z + \mathbb{Z} \to \mathbb{N}$ by $\varphi(\omega) = \varphi(l)$, where l is the largest natural number for which $\frac{1}{l}\omega \in T$. Suppose that $u \mapsto f(u)/|u|^2$ has a continuous extension to $u = 0$. Then*

$$\lim_{X\to\infty} \frac{1}{X} \sum_{cz+d\in T,\, |cz+d|<X} f(\sigma_g(cz+d)) = 2 \sum_{\omega\in\mathbb{Z}\cdot z+\mathbb{Z}}{}' \frac{\varphi(\omega)}{|\omega|} f\left(e^{i\theta}/\omega\right).$$

Remarks. The prime in $\sum'$ indicates that the term with $\omega = 0$ is omitted.

The first distribution result states that the majority of the points $\sigma_g(cz+d)$ is concentrated near 0. The second result gives information on the minority staying away from 0. To get an illustration we have picked $z = i \in \mathfrak{H}$, $\theta = 0$, and plotted

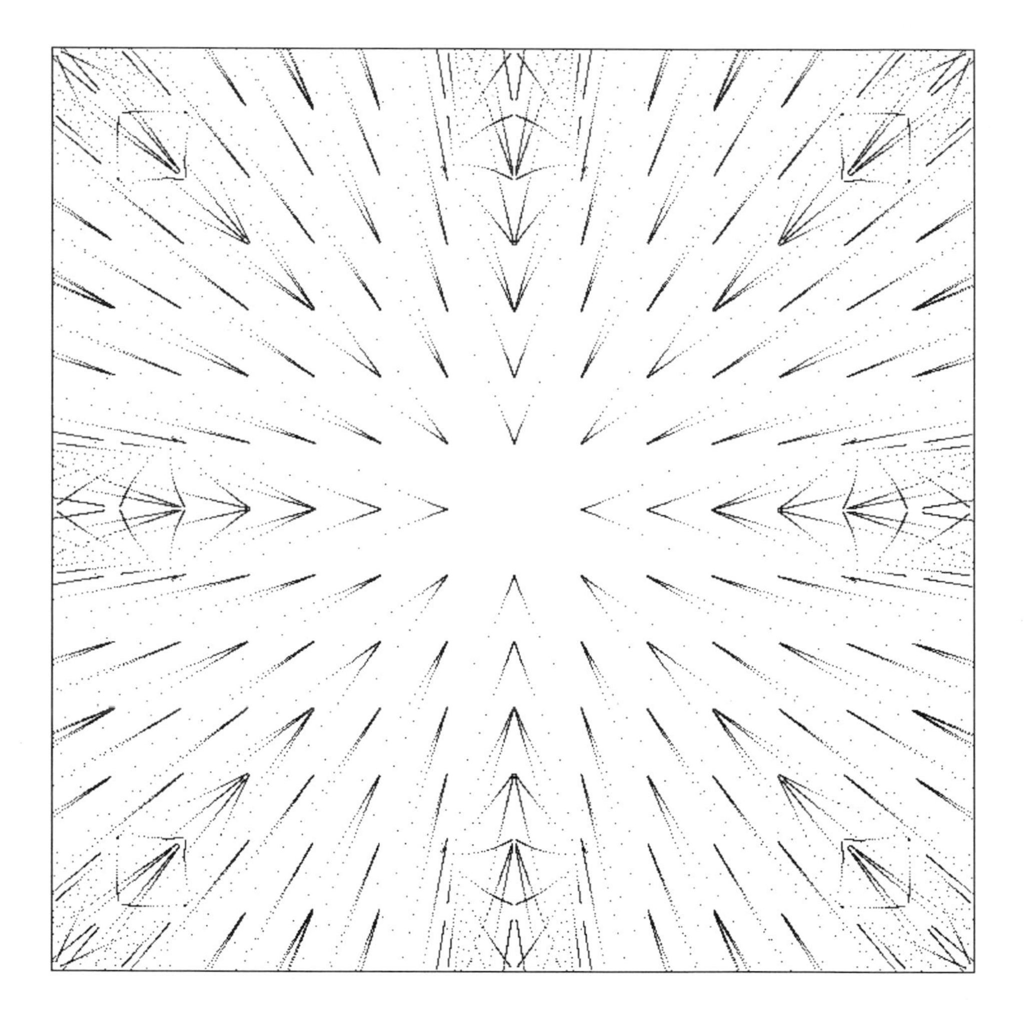

Figure 13.3 The points $1/\sigma(ci+d)$ with $(d,c)=1$, $|ci+d| \leq 1000$, that fall in the region $|\operatorname{Re} w| < 7$, $|\operatorname{Im} w| < 7$ of the complex plane.

some $\frac{1}{\sigma_g(cz+d)}$ in Figure 13.3. In this way we have blown up the majority near 0. The second distribution result explains some of the structure of the resulting picture.
Question. Can one explain more of the structure, for instance by methods like those in [9] and [10]?
Proof. In the remainder of this section we give a proof. We view the distribution results as statements on limits of functionals on a space of continuous functions. It suffices to prove these statements for a set of test functions that is dense in the whole space for a suitable topology. This is done by applying a Tauberian theorem for Dirichlet series.

Of course, the way the distribution results were found is opposite to the direction of this proof.

13.6.3 *Reformulation.* Let $X > 0$. We define functionals μ_X, ν_X, μ, and ν on the continuous functions $f : \mathbb{C} \to \mathbb{C}$ by

$$\begin{aligned} \mu_X(f) &= \frac{1}{X} \sum_{t\in T,\, |t|<X} \frac{f(\sigma(t))}{|t|} \\ \nu_X(f) &= \frac{1}{X} \sum_{t\in T,\, |t|<X} |\sigma(t)|^2 f(\sigma(t)). \\ \mu(f) &= \frac{12}{\pi y} f(0) \\ \nu(f) &= 2 \sum_{\omega\in\mathbb{Z}\cdot z+\mathbb{Z}}{}' \frac{\varphi(\omega)}{|\omega|^3} f\left(\frac{e^{i\theta}}{\omega}\right). \end{aligned}$$

Proving the distribution results amounts to showing that for all continuous functions $f : \mathbb{C} \to \mathbb{C}$:

$$\lim_{X\to\infty} \mu_X(f) = \mu(f), \qquad \lim_{X\to\infty} \nu_X(f) = \nu(f).$$

13.6.4 *Reduction to a smaller class of functions.* First we show that $\sigma_1(t) = \mathcal{O}(|t|)$ for t large, for fixed g. To see this we consider $t = cz + d$ with $c \geq 1$. We note that $12S(d, c) + 3 - \frac{6}{\pi}\arg(cz + d) + \frac{6}{\pi}\theta = \mathcal{O}(c + 1) = \mathcal{O}(|t|)$, and $\left|\mathrm{Re}\, \frac{1}{c(cz+d)}\right| = \frac{|cx+d|}{c|t|^2} \leq \frac{|t|}{1\cdot|t|^2} = \mathcal{O}(1)$. Finally, $\frac{d}{c} = \frac{cx+d}{c} - x = \mathcal{O}\left(\frac{|t|}{1} + 1\right) = \mathcal{O}(|t|)$.

So the values of σ are contained in a closed disk around 0 in $\mathbb{C}$, of radius $C(g) > 0$. We work with the space of all continuous functions on this disk, provided with the supremum norm $\|\,.\,\|_\infty$.

From $|\mu_X(f)| \leq \|f\|_\infty \mu_X(1)$ we conclude that the μ_X are bounded functionals on this space. The functional μ is bounded as well. If the limit formula $\lim_{X\to\infty} \mu_X(f) = \mu(f)$ holds for all f in a dense subspace containing the constant function 1, then the limit formula holds for all continuous functions. Indeed, the formula $\lim_{X\to\infty} \mu_X(1) = \mu(1)$ implies that the norms $\|\mu_X\|$ are bounded uniformly in X. We use the dense subspace spanned by the functions $f_{n,k} : u \mapsto u^n \bar{u}^k$, with $n, k \geq 0$. (To see that it is dense, use the Stone-Weierstrass theorem; see, e.g., [26], Ch. III, §1, Theorem 1.1.)

We have $|\nu_X(f)| \leq \|f\|_\infty \nu_X(1)$ for ν_X as well. Take $C(g)$ larger than the maximal value of $1/|\omega|$ for $\omega \in \mathbb{Z}\cdot z+\mathbb{Z}$, $\omega \neq 0$, to see that ν is bounded. To obtain the limit formula for ν_X and ν, it suffices to consider functions in the same dense subspace. We have to show for $n, k \geq 0$:

$$\lim_{X\to\infty} \mu_X(f_{n,k}) = \begin{cases} 12/\pi y & \text{if } n = k = 0 \\ 0 & \text{otherwise} \end{cases}$$

$$\lim_{X\to\infty} \nu_X(f_{n,k}) = \begin{cases} 4\dfrac{\zeta(n+k+2)}{\zeta(n+k+3)} y^{-(n+k+3)/2} \varepsilon^0_{n-k}\left(\dfrac{n+k}{2}+1; g\right) & \\ & \text{if } n-k \text{ is even} \\ 0 & \text{if } n-k \text{ is odd.} \end{cases}$$

To obtain the assertion for ν, note that

$$2\sum_\omega{}' \frac{\varphi(\omega)}{|\omega|^3}\frac{e^{in\theta-ik\theta}}{\omega^n\bar{\omega}^k} = 2\sum_{l=1}^\infty \frac{\varphi(l)}{l^{3+n+k}} \sum_{cz+d\in T} \frac{e^{-i(n-k)\arg(cz+d)}}{|cz+d|^{3+n+k}} e^{i(n-k)\theta}.$$

If $n-k$ is odd, this vanishes. For even $n-k$, use $e^0_\kappa = \dot{E}(0)$, and (13.2) on p. 246.

13.6.5 *Restriction to even $n-k$.* Let $n-k$ be odd. From the definitions of μ_X and ν_X we see immediately that $\mu_X(f_{n,k}) = \nu_X(f_{n,k}) = 0$. Hence the limit statements hold for $n \not\equiv k \bmod 2$.

In the sequel we restrict ourselves to even $n-k$.

13.6.6 *Reformulation for the even case.* For $X > 0$, $p, q \geq 0$, and $\kappa \in 2\mathbb{Z}$, we define

$$A_X(p,q,\kappa) = \frac{1}{X} \sum_{t\in T,\, |t|<X} \frac{\sigma_1(t)^p e^{i\kappa(\theta-\arg(t))}}{|t|^q}.$$

Then

$$\begin{aligned} \mu_X(f_{n,k}) &= A_X(n+k, n+k+1, n-k), \\ \nu_X(f_{n,k}) &= A_X(n+k+2, n+k+2, n-k). \end{aligned}$$

Write $p = n+k$ and $\kappa = n-k$. Our aim is to show, for even $p \geq 2$ and even κ:

$$\begin{aligned} \lim_{X\to\infty} A_X(0,1,0) &= 12/\pi y \\ \lim_{X\to\infty} A_X(p,p+1,\kappa) &= 0 \\ \lim_{X\to\infty} A_X(p,p,\kappa) &= 4\frac{\zeta(p)}{\zeta(p+1)} y^{-(p+1)/2} \varepsilon^0_\kappa(p/2; g). \end{aligned}$$

13.6.7 *Dirichlet series.* To carry this out, we consider the Dirichlet series

$$F(p,\kappa;u)=\sum_{t\in T}\frac{\sigma_1(t)^p e^{i\kappa(\theta-\arg(t))}}{|t|^u}$$

for even p and κ, $p\geq 0$. The estimate $\sigma_1(t)=\mathcal{O}(|t|)$ for large $|t|$, and the convergence of the Eisenstein series, imply that $F(p,\kappa;u)$ converges for at least $\operatorname{Re} u>p+2$, and gives a holomorphic function on this region.

Define $\sigma_2 : T\to\mathbb{C}$ by

$$\sigma_2(cz+d)=12S(d,c)-\frac{d}{c}+3-\frac{6}{\pi}\arg(cz+d)+\frac{6}{\pi}\theta-\frac{1}{c(cz+d)}$$

if $c>0$, and $\sigma_2(-t)=\sigma_2(t)$, $\sigma_2(1)=0$. As $\sigma_1(t)-\sigma_2(t)=\mathcal{O}(|t|^{-2})$, the series for $F(p,\kappa;u)$ differs from

$$\tilde F(p,\kappa;u)=\sum_{t\in T}\frac{\sigma_2(t)^p e^{i\kappa(\theta-\arg(t))}}{|t|^u}$$

by a series representing a holomorphic function on $\operatorname{Re} u>p-1$.

We apply Lemma 13.5.9 with $s=\frac{u-1}{2}$:

$$\begin{aligned}\varepsilon^p_\kappa(s;g)&= y^{s+1/2}e^{i\kappa\theta}\left(z+\frac{6}{\pi}\theta\right)^p\\&\quad+\sum_\gamma\frac{e^{i\kappa(\theta-\arg(cz+d))}\left(\gamma\cdot z+\frac{6}{\pi}(\theta-\arg(cz+d)-\alpha(\gamma))\right)^p}{y^{-s-1/2}|cz+d|^{2s+1}}\\&= y^{s+1/2}\left(e^{i\kappa\theta}\left(z+\frac{6}{\pi}\theta\right)^p+\frac{1}{2}\sum_t\frac{\sigma_2(t)^p e^{i\kappa(\theta-\arg(t))}}{|t|^{2s+1}}\right)\\&= y^{s+1/2}\left(e^{i\kappa\theta}\left(z+\frac{6}{\pi}\theta\right)^p+\frac{1}{2}\tilde F(p,\kappa;u)\right).\end{aligned}$$

This gives the meromorphic continuation of $\tilde F(p,\kappa;u)$ to $u\in\mathbb{C}$, and of $F(p,\kappa;u)$ to $\operatorname{Re} u>p-1$. Moreover, Corollary 13.5.11 gives information on the rightmost singularity in the case $p\geq 2$. This information is given in Table 13.1, with the notation $u_{p,\kappa}$ for the position of the right-most singularity, and $\rho_{p,\kappa}$ for the residue. The singularity at $u_{p,\kappa}$ is a first order pole, and it is the sole singularity in the region $\operatorname{Re} u>u_{p,\kappa}-1$. The table shows also where to find the results that cover the case $p=0$.

13.6.8 *Tauberian theorem.* We apply the Tauberian theorem for Dirichlet series; see, e.g., [27], Ch. XV, §2–3. This theorem implies that if the functions $g(u)=\sum_{t\in T}B(t)|t|^{-u}$ and $f(u)=\sum_{t\in T}A(t)|t|^{-u}$ satisfy

	$u_{p,\kappa}$	$\rho_{p,\kappa}$	see
$p=\kappa=0$	2	$\frac{12}{\pi y}$	11.2.7
$p\geq 2$	$p+1$	$4\frac{\zeta(p)}{\zeta(p+1)}\frac{\varepsilon_\kappa^0(p/2)}{y^{(p+1)/2}}$	Corollary 13.5.11
$p=0,\ \kappa\neq 0$	$F(p,\kappa;u)$ holomorphic on $\mathrm{Re}\,u>1$		13.2.6

Table 13.1 The rightmost singularity of $F(p,\kappa;u)$ for even p and κ.

i) $|B(t)|\leq CA(t)$ for all $t\in T$, for some $C>0$,

ii) f converges absolutely on $\mathrm{Re}\,u>1$ (hence the same holds for g),

iii) f and g have a meromorphic continuation to a slightly larger half plane $\mathrm{Re}\,u>1-\varepsilon$, without singularities outside $u=1$,

iv) f has a first order pole, and g at most a first order pole at $u=1$;

then

$$\lim_{X\to\infty}\frac{1}{X}\sum_{t\in T,\,|t|<X}B(t)=\operatorname*{Res}_{u\to 1}g(u).$$

If g is holomorphic at $u=1$ its residue is understood to be zero.

13.6.9 *First distribution result.* We take $p\geq 0$, p even. Apply the Tauberian theorem with $B(t)=\sigma_1(t)^p e^{i\kappa(\theta-\arg(t))}|t|^{-p-1}$; note that $B(t)=\mathcal{O}(|t|^{-1})$. Take $A(t)=|t|^{-1}$; then $f(u)=F(0,0;u+1)$ has the right properties. As $g(u)=F(p,\kappa;u+p+1)$, it satisfies the conditions of the Tauberian theorem, see Table 13.1. We obtain $\lim_{X\to\infty}A_X(p,p+1;\kappa)=\mathrm{Res}_{u\to p+2}F(p,\kappa;u)$, which gives the desired result, see 13.6.6.

13.6.10 *Second distribution result.* Take p even, $p\geq 2$. For $\kappa=0$ we take $B(t)=\sigma_1(t)^p|t|^{-p}$. Then $g(u)=F(p,0;u+p)$ is holomorphic on $\mathrm{Re}\,u>1$. As $B(t)\geq 0$ for all $t\in T$, we take $f(u)=g(u)$. The Tauberian theorem gives $\lim_{X\to\infty}A(p,p,0)=\rho_{p,0}$, which is the value stated in 13.6.6.

For $\kappa\neq 0$ we take $B(t)=\sigma_1(t)^p|t|^{-p}e^{i\kappa(\theta-\arg(t))}$, and $A(t)=\sigma_1(t)^p|t|^{-p}$. We have seen that this gives a suitable $f(u)$. As $g(u)=F(p,\kappa;u+p)$, we obtain the right value for $\lim_{X\to\infty}A(p,p,\kappa)$ in this case as well.

Chapter 14
Automorphic forms for the theta group

In this chapter we consider automorphic forms on a well known subgroup of the modular group, the theta group. We give examples to illustrate the ideas in Part I. The main difference with the modular group is the presence of two cusps.

14.1 Theta function and theta group

The theta group is the group of transformations in $\mathfrak{H}$ for which the theta function is an automorphic form. In this section we consider the theta group as a subgroup of $G = \mathrm{SL}_2(\mathbb{R})$. We shall lift it to $\tilde{G}$ in the next section.

14.1.1 *Theta function.* The classical *theta function* on the upper half plane is defined by $\theta(z) = \theta_3(z) = \sum_{m=-\infty}^{\infty} e^{\pi i z m^2}$. It clearly satisfies $\theta(z+2) = \theta(z)$. Application of the formula of Poisson shows that $\theta(-1/z) = (-iz)^{1/2}\theta(z)$, c.f. [49], Theorem 7.1.1.

14.1.2 *Theta group.* These transformation properties explain why the subgroup of Γ_{mod} generated by $\left(\begin{smallmatrix}1&2\\0&1\end{smallmatrix}\right)$ and $\left(\begin{smallmatrix}0&1\\-1&0\end{smallmatrix}\right)$ is called the *theta group*. We denote this group by Γ_θ. One can show that it consists of those $\left(\begin{smallmatrix}a&b\\c&d\end{smallmatrix}\right) \in \mathrm{SL}_2(\mathbb{Z})$ that satisfy either $\left(\begin{smallmatrix}a&b\\c&d\end{smallmatrix}\right) \equiv \left(\begin{smallmatrix}1&0\\0&1\end{smallmatrix}\right) \bmod 2$, or $\left(\begin{smallmatrix}a&b\\c&d\end{smallmatrix}\right) \equiv \left(\begin{smallmatrix}0&1\\1&0\end{smallmatrix}\right) \bmod 2$; see [32], p. 353 and p. 365.

Conjugation by $\left(\begin{smallmatrix}0&1\\-1&-1\end{smallmatrix}\right)$ sends Γ_θ to the Hecke congruence subgroup $\Gamma_0(2)$ of the modular group.

14.1.3 *Theta multiplier system.* The transformation behavior of the theta function is

$$\theta\left(\frac{az+b}{cz+d}\right) = v_\theta\begin{pmatrix}a&b\\c&d\end{pmatrix}(cz+d)^{1/2}\theta(z)$$

for all $\left(\begin{smallmatrix}a&b\\c&d\end{smallmatrix}\right) \in \Gamma_\theta$, with v_θ a multiplier system for weight $\frac{1}{2}$.

All automorphic forms for Γ_{mod} are, of course, automorphic forms for Γ_θ as well. The theta function corresponds to the automorphic form $z \mapsto y^{1/4}\theta(z)$ on $\mathfrak{H}$ (in the sense of Section 2.1) for the theta group Γ_θ, of weight $\frac{1}{2}$, with eigenvalue $\frac{3}{16}$, and multiplier system v_θ.

14.1.4 *Fundamental domain.* A fundamental domain for a subgroup Δ of finite index in Γ_{mod} can be obtained as a union $\bigcup_{\gamma\in R} \gamma \cdot F_{\mathrm{mod}}$, where R is a system of representatives of $\Delta\backslash\Gamma_{\text{mod}}$; see, e.g., [52], Ch. IV, §4.1. For Γ_θ this leads to

$$F_\theta = F_{\text{mod}} \cup \begin{pmatrix} 1 & -1 \\ 0 & 1 \end{pmatrix} F_{\text{mod}} \cup \begin{pmatrix} 1 & 1 \\ -1 & 0 \end{pmatrix} F_{\text{mod}}.$$

One can change one fundamental domain into another by moving around pieces

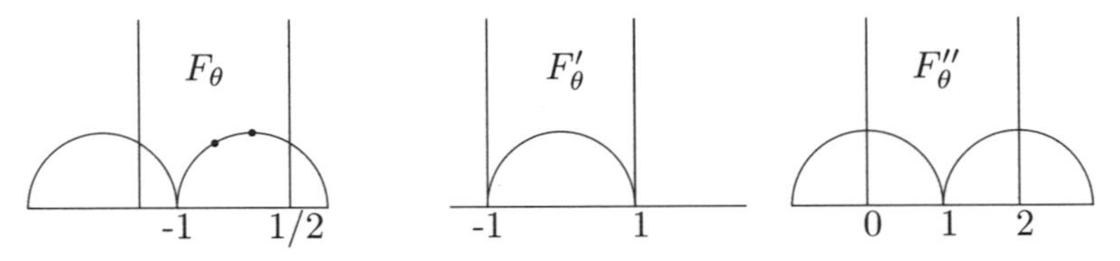

Figure 14.1 Fundamental domains for the theta group.

of a fundamental domain by elements of Γ_θ.

14.1.5 *Cusps.* The set of cusps $\mathbb{Q}\cup\{i\infty\}$ consists of two orbits for the theta group: $P = \Gamma_\theta \cdot i\infty$ and $Q = \Gamma_\theta \cdot (-1)$. Use the fundamental domain F'_θ to conclude that $Y_\theta = \Gamma_\theta\backslash\mathfrak{H}$ is homeomorphic to a plane with one point deleted, and that $X_\theta = \Gamma_\theta\backslash\mathfrak{H}^* = Y_\theta \cup \{P, Q\}$ is homeomorphic to a sphere. See [49], §7.2, for an automorphic function f_2 on $\mathfrak{H}$ that induces an isomorphism of Riemann surfaces $X_\theta \cong \mathbf{P}^1_{\mathbb{C}}$.

14.2 The covering group

We lift the theta group to a subgroup of $\tilde{G}$, and consider the concepts discussed in Chapter 3.

14.2.1 *Definition.* $\tilde{\Gamma}_\theta$ is the full original in $\tilde{G}$ of the theta group Γ_θ.

The fundamental domain F'_θ in Figure 14.1 shows that the group $\bar{\Gamma}_\theta = \Gamma_\theta/\{\pm\text{Id}\}$ of transformations in $\mathfrak{H}$ is generated by $\pm\begin{pmatrix} 1 & 2 \\ 0 & 1 \end{pmatrix}$ and $\pm\begin{pmatrix} 0 & 1 \\ -1 & 0 \end{pmatrix} = W$, with $(\pm W)^2 = \pm\text{Id}$ as the only relation. (Consult [32], Ch. VII, §2, p. 234, to obtain generators and relations from a fundamental domain.) Hence $\tilde{\Gamma}_\theta$ is generated by $n(2)$ and $w = k(\pi/2)$, with the centrality of w^2 as the only relation.

14.2.2 *Characters.* The group of characters $\mathcal{X} = \mathcal{X}(\tilde{\Gamma}_\theta)$ is isomorphic to $(\mathbb{C}^*)^2$: we may pick $\chi(n(2))$ and $\chi(w)$ arbitrarily in $\mathbb{C}^*$ to determine the character χ. We define $\chi_{r_1,r_2} \in \mathcal{X}$ by $\chi_{r_1,r_2}(n(2)) = e^{\pi i r_1/2}$ and $\chi_{r_1,r_2}(w) = e^{\pi i(r_1+r_2)/2}$, for $r_1, r_2 \in \mathbb{C}$. The advantage of this parametrization will become clear in 14.2.5 and 14.4.1. The character χ_{r_1,r_2} is unitary if and only if $r_1, r_2 \in \mathbb{R}$. The map $(r_1, r_2) \mapsto \chi_{r_1,r_2}$ induces an isomorphism $\mathcal{X} \cong \mathbb{C}^2 \bmod 4\pi\mathbb{Z}^2$. The character χ_{r_1,r_2} belongs to the weights in $r_1 + r_2 + 2\mathbb{Z}$.

14.2.3 *Modular forms on $\tilde{G}$.* All examples in the Chapters 1 and 13 give examples of automorphic forms for $\tilde{\Gamma}_\theta$ as well. Note that the character χ_r of $\tilde{\Gamma}_{\text{mod}}$ restricted to $\tilde{\Gamma}_\theta$ equals $\chi_{2r/3,r/3}$.

The automorphic form

$$\tilde{\theta}(p(z)k(\eta)) = y^{1/4}\theta(z)e^{i\eta/2}$$

of weight $\frac{1}{2}$ for the character $\chi_{0,1/2}$ and eigenvalue $\frac{3}{16}$ corresponds to the theta function. Take powers of $\tilde{\theta}$ to construct a one-parameter family similar to η_r.

14.2.4 *Canonical generators.* The fundamental domain F_θ in Figure 14.1 is canonical. For this choice:

> l_1 is the geodesic from $i\infty$ to $-\frac{3}{2}+\frac{i}{2}\sqrt{3} = -1+e^{2\pi i/3}$, $\pi_1 = n(2)$, $g_1 = n(-1)a(2)$.
>
> l_2 is the geodesic from -1 to $-\frac{1}{2}+\frac{i}{2}\sqrt{3} = e^{2\pi i/3}$, $\pi_2 = g_2 n(1) g_2^{-1}$, $g_2 = n(-1)w$.
>
> h_1 is the geodesic from $z_1 = i$ to $\frac{1}{2}+\frac{i}{2}\sqrt{3} = e^{\pi i/3}$, $\varepsilon_1 = w = k(\pi/2)$.

The last of the relations in 3.3.3 takes the form $\pi_1\pi_2\varepsilon_1 = \zeta^{0-2+2+1} = \zeta$. Indeed, with 13.1.2 we obtain

$$\pi_1\pi_2\varepsilon_1 \;=\; n(2)n(-1)wn(1)w^{-1}n(1)w \;=\; (n(1)w)^3 k(-\pi) \;=\; k(2\pi-\pi) \;=\; \zeta.$$

We have chosen $g_1 = n(-1)a(2)$, and not $a(2)$. This choice has the advantage that $g_\theta := g_1 g_2^{-1}$ interchanges the cusps. As $g_\theta^2 = k(\pi)$, conjugation by g_θ gives an involution of $\tilde{\Gamma}_\theta$, $\Gamma_\theta\backslash\mathfrak{H}$, and also of $\mathcal{X}$ and $\mathcal{V}$. To see this, note that π_1 and π_2 generate $\tilde{\Gamma}_\theta$, and that $g_\theta\pi_1 g_\theta^{-1} = \pi_2$, and $g_\theta\pi_2 g_\theta^{-1} = \pi_1$.

14.2.5 *Homomorphisms.* Any homomorphism $\tilde{\Gamma}_\theta \to \mathbb{C}$ is determined by its values on π_1 and π_2. If $\varphi \in \mathcal{V}$ is given by $\varphi(\pi_j) = r_j\pi/2$, then $\mathbf{exp}(\varphi)(\pi_1) = e^{\pi i r_1/2}$ and $\mathbf{exp}(\varphi)(w) = \mathbf{exp}(\varphi)(\pi_1\pi_2) = e^{\pi i(r_1+r_2)/2}$, hence $\mathbf{exp}(\varphi) = \chi_{r_1,r_2}$. The coordinates (r_1, r_2) are suitable if one is interested in the behavior of φ on the cuspidal generators π_1 and π_2. Note that $\chi_{r_1,r_2}(g_\theta\gamma g_\theta^{-1}) = \chi_{r_2,r_1}(\gamma)$ for all $\gamma \in \tilde{\Gamma}_\theta$.

From the standpoint of number theory, it is interesting to take the basis α, β, with $\alpha \in \hom(\tilde{\Gamma}_{\mathrm{mod}}, \mathbb{C})$ defined in 13.1.7, and $\beta : \gamma \mapsto \alpha(g_\theta\gamma g_\theta^{-1})$. This leads to

$$\begin{array}{llll} \alpha(\pi_1) = \frac{\pi}{3} & \alpha(\pi_2) = \frac{\pi}{6} & \alpha(w) = \frac{\pi}{2} & \alpha(\zeta) = \pi \\ \beta(\pi_1) = \frac{\pi}{6} & \beta(\pi_2) = \frac{\pi}{3} & \beta(w) = \frac{\pi}{2} & \beta(\zeta) = \pi. \end{array}$$

With Equation (13.1) the values of $\beta\widetilde{\left(\begin{smallmatrix} a & b \\ c & d \end{smallmatrix}\right)}$ for $\left(\begin{smallmatrix} a & b \\ c & d \end{smallmatrix}\right) \in \Gamma_\theta$ can be expressed in terms of Dedekind sums. Note that $\mathbf{exp}(u\alpha + v\beta) = \chi_{(2u+v)/3,(u+2v)/3}$.

14.2.6 *Exceptional points.* The cuspidal orbits for $\tilde{\Gamma}_\theta$ we have denoted by $P = \tilde{\Gamma}_\theta.\cdot i\infty$ and $Q = \tilde{\Gamma}_\theta\cdot(-1)$. They should be contained in every choice of $\mathcal{P}$.

14.2.7 *Decomposition of X_θ.* For $\mathcal{P} = \{P, Q\}$ we consider the conditions on $\tilde{A}_P$ and $\tilde{A}_Q$. The condition that $U_P(\tilde{A}_P)$ should be represented in F_θ by the image under g_P of an rectangle, leads to $\tilde{A}_P \geq \frac{1}{2}$ and, similarly, $\tilde{A}_Q \geq 1$. As the image of $i\tilde{A}_P + \mathbb{R}$ under $g_P : \mathfrak{H} \to \mathfrak{H}$ is $2i\tilde{A}_P + \mathbb{R}$, and as Im has maximum value $1/\tilde{A}_Q$ on $g_Q(i\tilde{A}_Q + \mathbb{R})$, the disjointness imposes the condition $\tilde{A}_P\tilde{A}_Q \geq \frac{1}{2}$, but this is already implied by the other conditions. The minimal choice for $\mathcal{P}$ is $\tilde{A}_P = \frac{1}{2}$ and $\tilde{A}_Q = 1$. This means $A_P = \frac{3}{2}$ and $A_Q = 2$.

14.3 Fourier expansions

Each automorphic form for $\tilde{\Gamma}_\theta$ has a Fourier expansion at both cusps. We write them down for $\tilde{\theta}$. We need the elements g_P and g_Q to describe the Fourier expansions at cusps. In the modular case of the previous chapter the automorphic forms have period 1 in the Iwasawa coordinate x. For the theta group we use g_P and g_Q to bring them into this form. This insistence on period 1 is a normalization that is especially suitable if one does not restrict oneself to $\tilde{\Gamma}_{\text{mod}}$ and its subgroups. For $\tilde{\Gamma}_\theta$, and other subgroups of $\tilde{\Gamma}_{\text{mod}}$, it leads to a complication. For the modular group we have chosen the element $g_{\tilde{\Gamma}_{\text{mod}} \cdot i\infty} \in \tilde{G}$ to be the neutral element of $\tilde{G}$. But here we have to take $g_P \in Na(2)$. If we deal with a modular form for $\tilde{\Gamma}_{\text{mod}}$, we have to distinguish its Fourier expansion at $\tilde{\Gamma}_{\text{mod}} \cdot i\infty$ from its Fourier expansion at $P = \tilde{\Gamma}_\theta \cdot i\infty$. At Q we need g_Q anyhow.

14.3.1 *Theta function.* $\chi_{0,1/2}(\pi_P) = 1$, hence $\mathcal{C}_P(\chi_{0,1/2}) = \mathbb{Z}$. From $\theta(z) = \sum_{-\infty}^{\infty} e^{\pi i m^2 z}$ we obtain immediately the Fourier expansion at $P = \tilde{\Gamma}_\theta \cdot i\infty$:

$$\begin{aligned} g_P p(z)k(\varphi) &= p(2z-1)k(\varphi) \mapsto (2y)^{1/4}\theta(2z-1)e^{i\varphi/2} \\ &= 2^{1/4} \sum_{m=-\infty}^{\infty} (-1)^m e^{2\pi i m^2 x} y^{1/4} e^{-2\pi m^2 y} e^{i\varphi/2} \\ &= a_0 \mu_{1/2}(P, 0, -\tfrac{1}{4}; g_P p(z)k(\varphi)) + \sum_{n=1}^{\infty} a_n \omega_{1/2}(P, n, \tfrac{1}{4}; g_P p(z)k(\varphi)), \end{aligned}$$

with $a_0 = 2^{1/4}$ and $a_n = 2(-1)^n(2\pi n)^{-1/4}$ if $n \geq 1$, n a square, and $a_n = 0$ otherwise.

For the expansion at $Q = \tilde{\Gamma}_\theta \cdot (-1)$ we use

$$\theta\left(\frac{z+1}{-z}\right) = \sqrt{-iz} \sum_{m=-\infty}^{\infty} e^{\pi i z(m+1/2)^2},$$

see [49], §7.1, Theorem 7.1.2. As $\chi_{0,1/2}(\pi_Q) = e^{\pi i/4}$ we obtain $\mathcal{C}_Q(\chi_{0,1/2}) = \frac{1}{8} + \mathbb{Z}$.

The expansion is:

$$\begin{aligned} g_Q p(z)k(\varphi) &= p\left(\frac{z+1}{-z}\right) k(-\arg(-z)+\varphi) \mapsto \\ & \left(\frac{y}{|z|^2}\right)^{1/4} \theta\left(\frac{z+1}{-z}\right) e^{-i\arg(-z)/2+i\varphi/2} \\ &= y^{1/4} \sum_{m=-\infty}^{\infty} \sqrt{-iz}\, e^{\pi i z(m+1/2)^2} (-z)^{-1/2} e^{i\varphi/2} \\ &= \sum_{m=-\infty}^{\infty} e^{\pi i/4} e^{\pi i(m+1/2)^2 x} y^{1/4} e^{-\pi(m+1/2)^2 y} e^{i\varphi/2} \\ &= \sum_{n\equiv 1/8 \bmod 1,\, n>0} b_n \omega_{1/2}(Q, n, \tfrac14; g_Q p(z)k(\varphi)), \end{aligned}$$

with $b_n = 2(4\pi n)^{-1/4} e^{\pi i/4}$ if $n = \frac12 (m+\frac12)^2$ for some $m \geq 0$, $m \in \mathbb{Z}$, and $b_n = 0$ otherwise.

The theta function θ corresponds to an element of $A_{1/2}(\chi, \mathbf{c}^{(\chi)}, \frac14)$, with $\chi = \chi_{0,1/2}$. It is a smooth square integrable automorphic form, but not a cusp form.

14.3.2 *Changing the group.* If f is a modular form for $\tilde{\Gamma}_{\text{mod}}$, it takes some care to obtain the Fourier expansions at P and Q from the expansion at $\tilde{\Gamma}_{\text{mod}} \cdot i\infty$.

As an example, we consider a modular form f for $\tilde{\Gamma}_{\text{mod}}$ with even weight, trivial character and expansion

$$\begin{aligned} & f(p(z)k(\varphi)) \\ &= d_0 \mu_l(\tilde{\Gamma}_{\text{mod}} \cdot i\infty, 0, s; p(z)k(\varphi)) + c_0 \mu_l(\tilde{\Gamma}_{\text{mod}} \cdot i\infty, 0, -s; p(z)k(\varphi)) \\ &\quad + \sum_{n\neq 0} c_n \omega_l(\tilde{\Gamma}_{\text{mod}} \cdot i\infty, n, s; p(z)k(\varphi)) \\ &= \left(d_0 y^{s+1/2} + c_0 y^{-s+1/2} + \sum_{n\neq 0} c_n e^{2\pi i n x} W_{l\,\text{sign}(n)/2, s}(4\pi |n| y) \right) e^{il\varphi}. \end{aligned}$$

At P we have $\mu_l(P, 0, \pm s; g_P p(z)k(\varphi)) = y^{1/2\pm s} e^{il\varphi}$, and $\omega_l(P, m, s; g_P p(z)k(\varphi)) = e^{2\pi i m x} W_{l\,\text{sign}(m)/2,s}(4\pi|m|y) e^{il\varphi}$. We obtain

$$\begin{aligned} f(g_P p(z)k(\varphi)) &= f(p(2z)k(\varphi)) \\ &= 2^{s+1/2} d_0 \mu_l(P, 0, s; g_P p(z)k(\varphi)) + 2^{-s+1/2} c_0 \mu_l(P, 0, -s; g_P p(z)k(\varphi)) \\ &\quad + \sum_{m\in 2\mathbb{Z},\, m\neq 0} (-1)^{m/2} c_{m/2} \omega_l(P, m, s; g_P p(z)k(\varphi)). \end{aligned}$$

At Q we use the fact that $g_Q = n(-1)w \in \tilde{\Gamma}_{\text{mod}}$, and obtain

$$\begin{aligned} f(g_Q p(z)k(\varphi)) &= f(p(z)k(\varphi)) \\ &= d_0\mu_l(Q,0,s;g_Q p(z)k(\varphi)) + c_0\mu_l(Q,0,-s;g_Q p(z)k(\varphi)) \\ &\quad + \sum_{n\neq 0} c_n\omega_l(Q,n,s;g_Q p(z)k(\varphi)). \end{aligned}$$

14.3.3 *Two harmonic automorphic forms.* Automorphic forms with weight and eigenvalue both equal to 0 we call *harmonic automorphic forms*, as they correspond to harmonic functions on $\mathfrak{H}$.

In 4.6.7 we have concluded from the Maass-Selberg relation that the dimension of the space $A_0(1,\mathbf{c},s)$ mod $S_0(1,\mathbf{c},s)$ is at most $|\mathbf{c}|$ for all s and $\mathbf{c}$. Here we show, by producing explicit examples, that the dimension of $A_0(1,\mathbf{c}_0,\frac{1}{2})$ mod $S_0(1,\mathbf{c}_0,\frac{1}{2})$ is at least equal to $|\mathbf{c}_0| = 2$ in the case of $\tilde{\Gamma}_\theta$. As bounded harmonic functions on $\mathfrak{H}$ are constant, we have $S_0(1,\mathbf{c}_0,\frac{1}{2}) = \{0\}$.

One element is easily given: the constant function $\mathbf{1} : g \mapsto 1$. To obtain a non-constant harmonic automorphic form we start with the logarithm of the eta function: the function F defined by $F(p(z)k(\varphi)) = \operatorname{Re}\log\eta(z) + \frac{1}{4}\log y$, has weight 0, is $\tilde{\Gamma}_{\text{mod}}$-invariant on the left, but satisfies $\omega F = \frac{1}{4}\mathbf{1}$. We take $F_1 : g \mapsto F(g_\theta g)$, with $g_\theta = n(-1)a(2)w^{-1}n(1)$ as above. The fact that $g_\theta\tilde{\Gamma}_\theta g_\theta^{-1} = \tilde{\Gamma}_\theta$ implies that F_1 is $\tilde{\Gamma}_\theta$-invariant on the left. It has weight 0 and satisfies $\omega F_1 = \frac{1}{4}\mathbf{1}$, as ω commutes with left (and right) translations. This means that $F_\theta = F - F_1$ is a harmonic automorphic form for $\tilde{\Gamma}_\theta$.

To see that $F_\theta \in A_0(1,\mathbf{c}_0,\frac{1}{2})$ we look at its Fourier expansions at P and Q. We work modulo regular Fourier terms of non-zero order.

$$F(p(z)k(\varphi)) = -\frac{\pi}{12}y - \sum_{n\neq 0}\frac{1}{2}\sigma_{-1}(|n|)e^{2\pi i n x}e^{-2\pi|n|y} + \frac{1}{4}\log y;$$

$$\begin{aligned} F_\theta(g_P p(z)k(\varphi)) &= F(n(-1)a(2)p(z)k(\varphi)) - F(n(-1)w^{-1}p(z)k(\varphi)) \\ &= F(p(2z-1)) - F(p(z)) \\ &= -\frac{\pi}{12}2y - \frac{1}{2}\sum_{n\neq 0}\sigma_{-1}(|n|)e^{4\pi i n x - 4\pi|n|y} + \frac{1}{4}\log(2y) \\ &\quad + \frac{\pi}{12}y + \frac{1}{2}\sum_{n\neq 0}\sigma_{-1}(|n|)e^{2\pi i n x - 2\pi|n|y} - \frac{1}{4}\log y \\ &\equiv -\frac{\pi}{12}\mu_0(P,0,\tfrac{1}{2};g_P p(z)k(\varphi)) + \frac{1}{4}\log 2\,\mu_0(P,0,-\tfrac{1}{2};g_P p(z)k(\varphi)); \\ F_\theta(g_Q p(z)k(\varphi)) &= F(n(-1)wp(z)) - F(n(-1)a(2)p(z)) \\ &= F(p(z)) - F(p(2z)) \\ &\equiv \frac{\pi}{12}\mu_0(Q,0,\tfrac{1}{2};g_Q p(z)k(\varphi)) - \frac{1}{4}\log 2\,\mu_0(Q,0,-\tfrac{1}{2};g_Q p(z)k(\varphi)). \end{aligned}$$

Clearly $\mathbf{1}$ and F_θ are linearly independent.

We check the Maass-Selberg relation (Theorem 4.6.5). As $l = n = 0$ we leave out the ι.

$$\begin{aligned}
&2\mathrm{Wr}(\tilde F_{P,0}\mathbf{1}, \tilde F_{P,0}F_\theta) + 2\mathrm{Wr}(\tilde F_{Q,0}\mathbf{1}, \tilde F_{Q,0}F_\theta)\\
&= 2\mathrm{Wr}\left(\mu_0(P,0,-\tfrac12), -\frac{\pi}{12}\mu_0(P,0,\tfrac12) + \frac14\log 2\,\mu_0(P,0,-\tfrac12)\right)\\
&\quad + 2\mathrm{Wr}\left(\mu_0(Q,0,-\tfrac12), \frac{\pi}{12}\mu_0(Q,0,\tfrac12) - \frac14\log 2\,\mu_0(Q,0,-\tfrac12)\right)\\
&= 2\left(-\frac{\pi}{12}\right)2\left(-\frac12\right) + 2\frac{\pi}{12}2\left(-\frac12\right) = 0.
\end{aligned}$$

14.4 Eisenstein series

For $\tilde\Gamma_\theta$ one may work out the formulas in Proposition 5.2.9. We do not carry this out completely, but mention some results.

14.4.1 *Eisenstein series* occur if the cusp is singular for the character under consideration. We summarize the possibilities for the theta group in the following scheme:

character χ_{r_1,r_2}	$r_2 \not\equiv 0 \bmod 4$	$r_2 \equiv 0 \bmod 4$
$r_1 \not\equiv 0 \bmod 4$	no Eis. series	Eisenstein series for Q
$r_1 \equiv 0 \bmod 4$	Eis. series for P	Eis. series for P and for Q

14.4.2 *Involution.* We consider the Eisenstein series in weight 0 for the trivial character. For $m \in \mathbb{Z}$ let $\mathbf{c}(m,s) = \begin{pmatrix} c_0(P,m;J(0),P;1,s) & c_0(P,m;J(0),Q;1,s) \\ c_0(Q,m;J(0),P;1,s) & c_0(Q,m;J(0),Q);1,s \end{pmatrix}$. As both cusps are singular, the scattering matrix $\mathbf{c}(0,s)$ is really a matrix.

Before we apply Proposition 5.2.9 we use g_θ, which interchanges the cusps. Let Θ be defined by $\Theta f(g) = f(g_\theta g)$. On functions of weight zero $\Theta^2 = \mathrm{Id}$. Note that $\Theta\mu_0(P,0,s) = \mu_0(Q,0,s)$, and $\Theta\omega_0(P,m,s) = \omega_0(Q,m,s)$. Hence $\Theta P_0(P,0,1,s) = P_0(Q,0,1,s)$, and $\begin{pmatrix} 0&1\\1&0\end{pmatrix}\mathbf{c}(m,s)\begin{pmatrix} 0&1\\1&0\end{pmatrix} = \mathbf{c}(m,s)$. So we only need to consider the Fourier coefficients of $P_0(P,0,1,s)$.

14.4.3 *Case PP.* The $\widetilde{\begin{pmatrix} A&B\\C&D\end{pmatrix}} \in g_P^{-1}\tilde\Gamma_\theta g_P$ occurring in Proposition 5.2.9 are obtained as $\begin{pmatrix} A&B\\C&D\end{pmatrix} = \begin{pmatrix} a+c & (b+d-a-c)/2\\ 2c & d-c\end{pmatrix}$ with $\begin{pmatrix} a&b\\c&d\end{pmatrix} \in \Gamma_\theta$. (Note that $\widetilde{\begin{pmatrix} A&B\\C&D\end{pmatrix}}$ is the group element that we called $\widetilde{\begin{pmatrix} a&b\\c&d\end{pmatrix}}$ in Proposition 5.2.9.) The number $C = 2c$ runs through $2\mathbb{N}$, and $D = d-c$ is obtained from $d \bmod 2c$, $(d,c) = 1$, and $d \not\equiv c \bmod 2$. This gives $\mathrm{S}_1(P,m;P,0;2c) = \sum_d (-1)^m e^{\pi i m d/c}$.

Consider $\operatorname{Re} u > 1$, and let $\zeta_2(u) = \sum_{n\geq 1,\, n\equiv 1 \bmod 2} n^{-u} = (1-2^{-u})\zeta(u)$. Then

$$\begin{aligned}
&\zeta_2(u) \sum_{c=1}^{\infty} \mathrm{S}_1(P,m;P,0;2c)(2c)^{-u} \\
&= 2^{-u}(-1)^m \zeta_2(u) \sum_{c\equiv 1 \bmod 2} c^{-u} \sum_{\delta \bmod c,\, (\delta,c)=1} e^{2\pi i m\delta/c} \\
&\quad + 2^{-2u}(-1)^m \zeta_2(u) \sum_{\gamma\geq 1} \gamma^{-u} \sum_{d \bmod 4\gamma,\, (d,4\gamma)=1} e^{2\pi i m d/4\gamma} \\
&= 2^{-u}(-1)^m \sum_{N\equiv 1 \bmod 2} N^{-u} \sum_{x \bmod N} e^{2\pi i m x/N} \\
&\quad + 2^{-2u}(-1)^m (1-2^{-u}) \sum_{N\geq 1} N^{-u} \sum_{x \bmod 4N,\, (x,4N)|N} e^{2\pi i m x/4N} \\
&= 2^{-u}(-1)^m \sigma^1_{1-u}(m) \\
&\quad + 2^{-2u}(-1)^m(1-2^{-u}) \left(4\sigma_{1-u}(m/4) - \frac{2}{1-2^{-u}} \sigma^1_{1-u}(m/2) \right) \\
&= 2^{2-2u}(1-2^{-u})\sigma_{1-u}(\tfrac{m}{4}) + 2^{-u}(-1)^m \sigma^1_{1-u}(m) - 2^{1-2u}\sigma^1_{1-u}(\tfrac{m}{2}),
\end{aligned}$$

where $\sigma^1_a(m) = \sum_{d|m,\, d\equiv 1 \bmod 2} d^a$, $\sigma^1_a(0) = \zeta_2(-a)$, $\sigma_a(0) = \zeta(-a)$, $\sigma_a(m) = \sum_{d|m} d^a$ if $m \in \mathbb{Z}$; if $m \notin \mathbb{Z}$, then $\sigma_a(m) = \sigma^1_a(m) = 0$.

Proposition 5.2.9 gives

$$c_0(P,0;P,0;1,s) = \Lambda(2s)\Lambda(2s+1)^{-1}(2^{1+2s}-1)^{-1},$$

and for $m \neq 0$:

$$c_0(P,m;P,0;1,s) = \Lambda(2s+1)^{-1}|m|^{s-1/2}\left(\sigma_{-2s}(m) - \frac{\sigma^1_{-2s}(m)}{1-2^{-1-2s}}\right),$$

with $\Lambda(u) = \pi^{-u/2}\Gamma(u/2)\zeta(u)$.

14.4.4 *Case QP.* The $\widetilde{\begin{pmatrix} A & B \\ C & D \end{pmatrix}} \in g_P^{-1}\tilde{\Gamma}_\theta g_Q$ are obtained in the form $\begin{pmatrix} A & B \\ C & D \end{pmatrix} = \begin{pmatrix} (a-b-\gamma)/\sqrt{2} & (a-\delta)/\sqrt{2} \\ \gamma\sqrt{2} & \delta\sqrt{2} \end{pmatrix}$, with $\begin{pmatrix} a & b \\ \delta & \delta-\gamma \end{pmatrix} \in \Gamma_\theta$; hence $\gamma \equiv 1 \bmod 2$, and $(\gamma,\delta) = 1$. This gives $\mathrm{S}_1(Q,m;P,0;\gamma\sqrt{2}) = \sum_{\delta \bmod \gamma,\, (\delta,\gamma)=1} e^{2\pi i m\delta/\gamma}$.

$$\begin{aligned}
&\zeta_2(u) \sum_{\gamma\geq 1,\, \gamma\equiv 1 \bmod 2} \mathrm{S}_1(Q,m;P,0;\gamma\sqrt{2})(\gamma\sqrt{2})^{-u} \\
&= 2^{-u/2} \sum_{N\geq 1,\, N\equiv 1 \bmod 2} N^{-u} \sum_{x \bmod N} e^{2\pi i m x/N} \\
&= 2^{-u/2}\sigma^1_{1-u}(m).
\end{aligned}$$

Hence

$$\begin{aligned} c_0(Q,0;P,0;1,s) &= \Lambda(2s)\Lambda(2s+1)^{-1}(2^{2+1/2}-2^{-s+1/2})(2^{2s+1}-1)^{-1}, \\ &\text{and for } m\neq 0: \\ c_0(Q,m;P,0;1,s) &= \Lambda(2s+1)^{-1}|m|^{s-1/2}\sigma^1_{-2s}(m)2^{s+1/2}(2^{1+2s}-1)^{-1}. \end{aligned}$$

14.4.5 *Scattering matrix.* The scattering matrix for the trivial character turns out to be

$$\mathbf{c}(0,s) = \frac{1}{2^{1+2s}-1}\cdot\frac{\Lambda(2s)}{\Lambda(1+2s)}\begin{pmatrix} 1 & 2^{1/2+s}-2^{1/2-s} \\ 2^{1/2+s}-2^{1/2-s} & 1 \end{pmatrix}.$$

One can check this in [21], Ch. 11, (3.1) on p. 527. (Remember the shift in the spectral parameter s.)

The scattering matrix for other characters has size one or does not exist at all. For the theta-character $\chi_{0,1/2}$ there are more explicit results on the Fourier coefficients of the Eisenstein series in [55], and in §4 of [37].

14.4.6 *Other Fourier terms.* For $m\neq 0$ we find

$$\begin{aligned} &\mathbf{c}(m,s) \\ &= \frac{|m|^{s+1/2}}{\Lambda(2s+1)}\left(\sigma_{-2s}(m)\begin{pmatrix} 1\,0 \\ 0\,1 \end{pmatrix} + \frac{2^{s+1/2}}{2^{1+2s}-1}\sigma^1_{-2s}(m)\begin{pmatrix} -2^{s+1/2} & 1 \\ 1 & -2^{s+1/2} \end{pmatrix}\right). \end{aligned}$$

A reader who likes computations may want to check the functional equation for $s\mapsto -s$.

14.4.7 *Difference of Eisenstein families.* In weight zero there are two Eisenstein families, $\dot E_0(J(0),P)$ and $\dot E_0(J(0),Q)$, with the same residue in $s=\frac12$. Hence their difference is holomorphic at $s=\frac12$. The form of the Fourier coefficients of order zero is

$$\tilde F_{R,0}\Big(\dot E_0(J(0),P;s)-\dot E_0(J(0),Q;s)\Big) = t_R\mu_0(R,0,s)+c_R(s)\mu_0(R,0,-s),$$

with $t_P=1$, $t_Q=-1$ and

$$c_R(s) = \frac{1}{2^{1+2s}-1}\frac{\Lambda(2s)}{\Lambda(1+2s)}\begin{cases} 1-\big(2^{1/2+s}-2^{1/2-s}\big) & \text{if } R=P \\ \big(2^{1/2+s}-2^{1/2-s}\big)-1 & \text{if } R=Q. \end{cases}$$

This gives $c_P(1/2)=-3/\pi$ and $c_Q(1/2)=3/\pi$. Comparison with 14.3.3 shows that

$$\lim_{s\to 1/2}\Big(\dot E_0(J(0),P;s)-\dot E_0(J(0),Q;s)\Big) = -\frac{12}{\pi}F_\theta.$$

14.4.8 *Poincaré series.* For all unitary characters there are Poincaré series at both cusps. Their Fourier coefficients at the cusps can be expressed in terms of generalized Kloosterman sums.

14.5 More than one parameter

This section contains examples for two topics mentioned in Chapter 10.

14.5.1 *Cells of continuation.* As $|X^\infty| = 2$ for the theta group, there are several types of cells of continuation. Under the identification $\mathcal{V} \cong \mathbb{C}^2$ by $r(2\alpha - \beta) + q(2\beta - \alpha) \mapsto (r, q)$ we obtain, e.g.,

$$\begin{aligned} &J(0,0) = (-4,4)^2, \quad J^0(0,0) = \{(0,0)\}, \\ &J(0,q) = (-4,4) \times (0,4), \quad J^0(0,q) = \{0\} \times (0,4) \quad \text{if } 0 < q < 4, \\ &J(r,q) = J^0(r,q) = (0,4)^2 \quad \text{if } r,\, q \in (0,4), \end{aligned}$$

see Figure 14.2. The minimal cells are $J(\varepsilon, \zeta)$, with $\varepsilon, \zeta \in \{1, -1\}$.

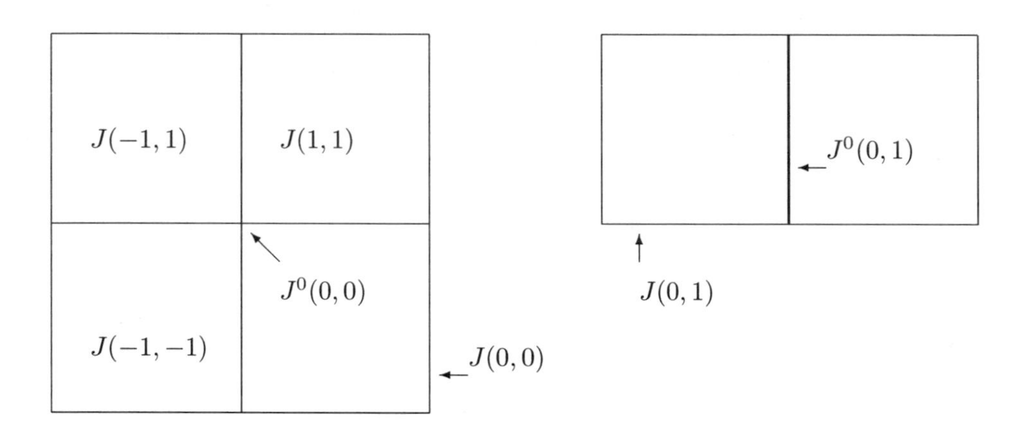

Figure 14.2 Cells of continuation for the theta group.

14.5.2 *Restricted parameter space.* Take $J = J(0,1)$. Then $W = \{\, (q,s) \in \mathbb{C}^2 : \operatorname{Re} q \in (0,4) \,\}$, $\chi : (q,s) \mapsto \chi_{0,q}$, $s : (q,s) \mapsto s$, is an example of a restricted parameter space that is not related to a minimal cell.

The character depends on only one complex parameter. One might hope that part of the results in [7] on the analyticity of the eigenvalues can be carried over to this situation. I think, however, that problems arise from possible singularities of the Eisenstein family.

Chapter 15
Automorphic forms for the commutator subgroup

The commutator subgroup of the full modular group is the last case to which we apply the theory of Part I. It is chosen as an example of a group of positive genus. The fact that the completion X of $\Gamma_{\mathrm{com}}\backslash\mathfrak{H}$ is an elliptic curve gives additional structure, not present for the groups Γ_{mod} and Γ_θ of genus 0. Apart from giving examples for Part I, this chapter discusses some new aspects as well.

In weight 0 the commutator subgroup has an Eisenstein family depending on three parameters: the spectral parameter s, and two parameters describing the characters for weight zero. These characters form a two-dimensional torus. Like all Eisenstein families, this family is given by explicit Eisenstein series in the domain of absolute convergence. There is another region on which there is an explicit description: consider the restriction to the plane $s = \frac{1}{2}$. This gives a meromorphic family in two variables of harmonic automorphic forms. This family can be described explicitly with the help of Jacobi theta functions on the elliptic curve X, see Section 15.6.

In Section 13.6 we have used derivatives of the Eisenstein family for the modular group to obtain distribution results for quantities related to Dedekind sums. We shall see in Section 15.5 that this method does not work here. It breaks down for the Eisenstein family in weight zero.

An interesting question is whether cusp forms occur in families, see, e.g., [46] and [13]. In Section 15.7 we give one result that has a relation to this question.

I thank F. Beukers, B. van Geemen, and D. Zagier for help and information concerning the subjects considered in this chapter.

15.1 Commutator subgroup

15.1.1 *Definition.* We define $\bar\Gamma_{\mathrm{com}} \subset \bar G = \mathrm{PSL}_2(\mathbb{R})$ as the *commutator subgroup* of $\bar\Gamma_{\mathrm{mod}} = \mathrm{PSL}_2(\mathbb{Z})$.

$\bar\Gamma_{\mathrm{com}}$ is contained in the kernel of each character of $\bar\Gamma_{\mathrm{mod}}$, in particular in $\ker(\chi_2)$; see 13.1.3. As χ_2 is trivial on $\tilde Z$, it induces a character of $\bar\Gamma_{\mathrm{mod}} = \tilde\Gamma_{\mathrm{mod}}/\tilde Z$. The image $\chi_2(\bar\Gamma_{\mathrm{mod}})$ has 6 elements, hence the index of $\bar\Gamma_{\mathrm{com}}$ in $\bar\Gamma_{\mathrm{mod}}$ is at least 6. In [32], Chapter XI, §3E, on p. 362, we see that the index is equal to 6, and that $\pm\begin{pmatrix}1&n\\0&1\end{pmatrix}$, n mod 6, are representatives of $\bar\Gamma_{\mathrm{mod}}/\bar\Gamma_{\mathrm{com}}$. This leads to the fundamental domain $F_{\mathrm{com}} = \bigcup_{-2\le n\le 3}\begin{pmatrix}1&n\\0&1\end{pmatrix} F_{\mathrm{mod}}$ given in Figure 15.1.

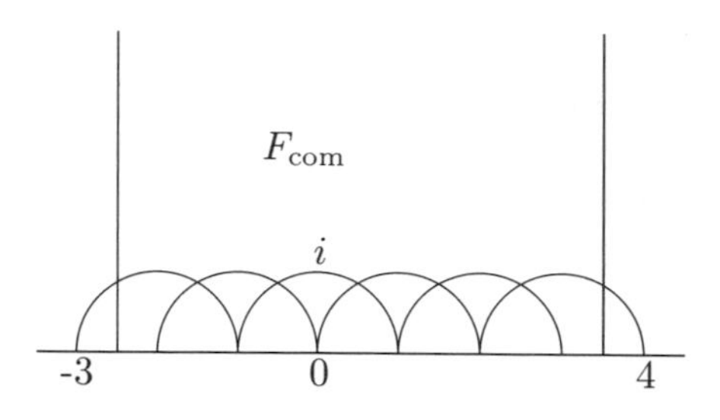

Figure 15.1 The fundamental domain F_{com} of $\bar{\Gamma}_{\mathrm{com}}$.

15.1.2 *Generators and relations.* In *loc. cit.*, on p. 363, we see also that $\bar{\Gamma}_{\mathrm{com}}$ is freely generated by $\pm\left(\begin{smallmatrix}2&1\\1&1\end{smallmatrix}\right)$ and $\pm\left(\begin{smallmatrix}1&1\\1&2\end{smallmatrix}\right)$. I prefer to use the free generators $D=\pm\left(\begin{smallmatrix}2&1\\1&1\end{smallmatrix}\right)$ and $C=\pm\begin{pmatrix}2&-1\\-1&1\end{pmatrix}$. In Figure 15.2 one finds a fundamental domain for which C, D, and $\pm\left(\begin{smallmatrix}1&6\\0&1\end{smallmatrix}\right)=DCD^{-1}C^{-1}$ give the boundary identifications.

15.1.3 *Commutator subgroup of the commutator subgroup.* $\bar{\Gamma}_{\mathrm{com}}$ itself has a commutator subgroup too. Let us denote it by $\bar{\Delta}$. As the abelianized group $(\bar{\Gamma}_{\mathrm{com}})^{\mathrm{ab}}$ is isomorphic to $\mathbb{Z}^2\times\{\pm\mathrm{Id}\}$, the group $\bar{\Delta}$ has infinite index in $\bar{\Gamma}_{\mathrm{com}}$. Hence $\bar{\Delta}\backslash\mathfrak{H}$ has infinite volume. It falls outside the class of cofinite discrete groups considered in this book. Note that $\pm\left(\begin{smallmatrix}1&6\\0&1\end{smallmatrix}\right)\in\bar{\Delta}$.

15.1.4 *Cusps.* Inspection of the fundamental domain F_{com} shows that $\bar{\Gamma}_{\mathrm{com}}$ has only one cuspidal orbit in $\mathbf{P}^1_{\mathbb{Q}}=\mathbb{Q}\cup\{i\infty\}$. We call it P in this chapter.

15.1.5 *Definition.* Γ_{com} is the full original of $\bar{\Gamma}_{\mathrm{com}}$ in $G=\mathrm{SL}_2(\mathbb{R})$. It is the direct product of $Z=\{\pm\mathrm{Id}\}$ and the group Γ^1_{com} freely generated by $\begin{pmatrix}2&-1\\-1&1\end{pmatrix}$ and $\left(\begin{smallmatrix}2&1\\1&1\end{smallmatrix}\right)$. The commutator subgroup Δ of Γ^1_{com} is isomorphic to its image $\bar{\Delta}$ in $\bar{\Gamma}_{\mathrm{com}}$.

15.1.6 *Definition.* $\tilde{\Gamma}_{\mathrm{com}}$ is the full original of $\bar{\Gamma}_{\mathrm{com}}$ in $\tilde{G}$. For Γ_{com} and $\tilde{\Gamma}_{\mathrm{com}}$ we shall use the name *commutator subgroup*, although this is not fully correct.

15.1.7 *Generators and relations.* We lift the generators C and D of $\bar{\Gamma}_{\mathrm{com}}$ to $\tilde{\Gamma}_{\mathrm{com}}$ by defining $\gamma=n(-1)wn(1)w^{-1}$ and $\eta=n(1)wn(-1)w^{-1}$. They generate a free subgroup $\tilde{\Gamma}^1_{\mathrm{com}}$ of $\tilde{\Gamma}_{\mathrm{com}}$, isomorphic to Γ^1_{com} and to $\bar{\Gamma}_{\mathrm{com}}$. As $\tilde{\Gamma}_{\mathrm{com}}$ is the direct product of $\tilde{\Gamma}^1_{\mathrm{com}}$ and $\tilde{Z}$, we obtain generators γ, η, and ζ of $\tilde{\Gamma}_{\mathrm{com}}$, with centrality of ζ as the only relation.

15.1.8 *Canonical generators.* $\tilde{\Gamma}_{\mathrm{com}}$ has one cuspidal orbit $P=\tilde{\Gamma}_{\mathrm{com}}\cdot i\infty$. The character χ_2 does not vanish on the elliptic classes in Γ_{mod}. Hence $\tilde{\Gamma}_{\mathrm{com}}$ has no elliptic orbits. It has genus $g=1$, number of cusps $p=1$ and number of elliptic orbits $q=0$. The fundamental domain F_{com} in Figure 15.1 is not canonical. Figure 15.2 shows a canonical fundamental domain. It is obtained in the following way:

$$l_1 \text{ the geodesic from } i\infty \text{ to } i-3,\ \pi_1=n(6),\ g_1=a(6)$$

$$a_1 \text{ the geodesic from } i \text{ to } \frac{3}{2}+\frac{i}{2},\ \gamma_1=\gamma=n(-1)wn(1)w^{-1}$$

$$b_1 \text{ the geodesic from } -\frac{3}{2}+\frac{i}{2} \text{ to } i,\ \eta_1=\eta=n(1)wn(-1)w^{-1}.$$

Note that $\gamma\eta\gamma^{-1}\eta^{-1}=n(-6)w^2=\pi_1^{-1}\zeta$.

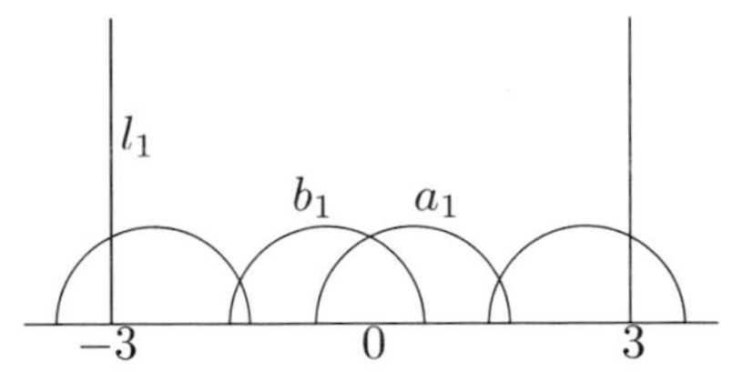

Figure 15.2 A canonical fundamental domain for Γ_{com}. The circles all have radius $\frac{1}{2}\sqrt{5}$. Their centers are at $-\frac{5}{2}$, $-\frac{1}{2}$, $\frac{1}{2}$, and $\frac{5}{2}$. They intersect in $\frac{-3+i}{2}$, i, and $\frac{3+i}{2}$.

15.1.9 *Homomorphisms.* To describe the linear space $\mathcal{V}$ of group homomorphisms $\tilde\Gamma_{\text{com}} \to \mathbb{C}$ we use the direct product decomposition $\tilde\Gamma_{\text{com}} = \tilde\Gamma^1_{\text{com}} \times \tilde Z$. The free generators γ and η of $\tilde\Gamma^1_{\text{com}}$ are commutators in $\tilde\Gamma_{\text{mod}}$. The homomorphism $\alpha : \tilde\Gamma_{\text{mod}} \to \mathbb{C}$ vanishes on $\tilde\Gamma^1_{\text{com}}$, see (13.1). As α has the non-zero value π on ζ, it provides us with a basis of $\hom(\tilde Z,\mathbb{C})$. We choose λ and $\bar\lambda$ as a basis for $\hom(\tilde\Gamma^1_{\text{com}},\mathbb{C})$, where $\lambda : \tilde\Gamma^1_{\text{com}} \to \mathbb{C}$ is the homomorphism defined by

$$\lambda(\gamma)=\rho\omega, \qquad \lambda(\eta)=\bar\rho\omega,$$
$$\omega=\frac{\sqrt{\pi}\,\Gamma(\frac{1}{6})}{6\sqrt{3}\,\Gamma(\frac{2}{3})}, \qquad \rho=e^{\pi i/3}.$$

Indeed, this defines two linearly independent elements of $\hom(\tilde\Gamma^1_{\text{com}},\mathbb{C})$. (In 15.3.3 we shall see that ω is a reasonable number in this context.)

So, $\alpha,\lambda,\bar\lambda$ form a basis of $\mathcal{V}$. Not all values of λ are real. We can use $\alpha, \operatorname{Re}\lambda, \operatorname{Im}\lambda$ to obtain a basis of $\mathcal{V}_r$.

15.1.10 *Characters.* The characters χ_r of $\tilde\Gamma_{\text{mod}}$, see 13.1.3, vanish on the commutators γ and η, hence on $\tilde\Gamma^1_{\text{com}}$. So the χ_r, with $r\in\mathbb{C} \bmod 2\mathbb{Z}$, describe the characters of the factor $\tilde Z$ of $\tilde\Gamma_{\text{com}}=\tilde\Gamma^1_{\text{com}}\times\tilde Z$. Of course χ_r belongs to the weights in $r+2\mathbb{Z}$. All characters of $\tilde\Gamma^1_{\text{com}}$ are of the form $\chi_{v,w} : \delta \mapsto e^{iv\lambda(\delta)+iw\overline{\lambda(\delta)}}$. They belong to weight 0.

In this way $(r,v,w)\mapsto \chi_r\chi_{v,w}$ gives an isomorphism

$$\mathbb{C}^3 \bmod \left(2\mathbb{Z}\times\hat\Lambda\right) \to \mathcal{X}(\tilde\Gamma_{\text{com}}),$$

where $\hat{\Lambda} = \frac{2\pi i}{\omega\sqrt{3}}(\mathbb{Z} + \mathbb{Z}\rho)$, and where $2\mathbb{Z} \times \hat{\Lambda}$ is embedded into $\mathbb{C}^3$ by $(m, v) \mapsto (m, v, \bar{v})$. (The lattice $\Lambda = \frac{\omega^2\sqrt{3}}{2\pi i}\hat{\Lambda}$ is the period lattice of the elliptic curve $X_{\mathrm{com}} = \tilde{\Gamma}_{\mathrm{com}}\backslash\mathfrak{H}^*$. It will be discussed in Section 15.3.)

15.1.11 *Decomposition.* The minimal choice of the set $\mathcal{P}$ of exceptional points (see Section 3.5) is $\mathcal{P} = \{P\}$, with $P = \tilde{\Gamma}_{\mathrm{com}} \cdot i\infty$. In the fundamental domain in Figure 15.2 we see that we can take the cut-off values $\tilde{A}_P = \frac{1}{6}$, $A_P = \frac{7}{6}$, if $\mathcal{P} = \{P\}$.

15.1.12 *The cells of continuation* are determined by r in the coordinates for which (v, w, r) corresponds to $v\lambda + w\bar{\lambda} + r\alpha \in \mathcal{V}$. The possibilities are $-2 < r < 2$ (null cell), and $-2 < r < 0$ and $0 < r < 2$ (minimal cells). The cells of continuation of $\tilde{\Gamma}_{\mathrm{com}}$ are not bounded.

15.2 Automorphic forms for $\tilde{\Gamma}_{\mathrm{com}}$

15.2.1 *Holomorphic cusp form of weight* 2. The space of holomorphic cusp forms on $\mathfrak{H}$ for Γ_{com} of weight 2 is spanned by $\eta(z)^4$. The function $\eta_4 : p(z)k(\theta) \mapsto y\eta(z)^4 e^{2i\theta}$ is an element of the space of cusp forms $S_2(\tilde{\Gamma}_{\mathrm{mod}}, \chi_2, \mathbf{c}_0, \frac{1}{2})$; see 13.1.7. (The growth condition $\mathbf{c}_0$ associates $\{0\}$ to $\tilde{\Gamma}_{\mathrm{mod}} \cdot i\infty$.) As χ_2 is trivial on $\tilde{\Gamma}_{\mathrm{com}}$, we have $\eta_4 \in S_2(\tilde{\Gamma}_{\mathrm{com}}, 1, \mathbf{c}_0, \frac{1}{2})$ (now $\mathbf{c}_0$ refers to $P \mapsto \{0\}$).

k	0	1	2	3	4	5	...
$c(k)$	1	-4	2	8	-5	-4	...

The product expression $\eta(z)^4 = e^{\pi iz/3}\prod_{m=1}^{\infty}(1 - e^{2\pi imz})^4$ implies that $\eta(z)^4$ has an expansion $\sum_{k=0}^{\infty} c(k)e^{2\pi ikz + \pi iz/3}$, with all $c(k) \in \mathbb{Z}$. Thus we obtain

$$\begin{aligned} g_P p(z)k(\theta) &\mapsto 6y\eta(6z)^4 e^{2i\theta} \\ &= \sum_{n\equiv 1 \bmod 6,\, n>0} \frac{3}{2\pi n} c\left(\frac{n-1}{6}\right) \omega_2(P, n, \tfrac{1}{2}; g_P p(z)k(\theta)). \end{aligned}$$

15.2.2 *Primitive of* η^4. Consider the holomorphic function $H : \mathfrak{H} \to \mathbb{C}$ given by

$$H(z) = -2\pi i \int_{i\infty}^{z} \eta(\tau)^4 \, d\tau;$$

the path of integration should approach $i\infty$ vertically. We shall see in 15.3.2 that it satisfies $H(\delta \cdot z) = H(z) + \lambda(\delta)$ for all $\delta \in \tilde{\Gamma}_{\mathrm{com}}$. (Actually, this is the reason that we used λ in 15.1.9.) The expansion of η^4 leads to

$$H(z) = \sum_{k=0}^{\infty} \frac{-6c(k)}{6k+1} e^{2\pi i(k+1/6)z}.$$

15.2.3 *Automorphic forms of weight* 0. For each $v \in \mathbb{C}$ the function $e^{ivH} : p(z)k(\theta) \mapsto e^{ivH(z)}$ is $\tilde{\Gamma}_{\text{com}}$-equivariant for the character $\chi_{v,0}$, and has weight 0. It satisfies $e^{ivH}(p(z)k(\theta)) = 1 + \mathcal{O}(e^{-\pi y/3})$ as $y \to \infty$. This implies that $e^{ivH} - 1$ is regular at P, and $e^{ivH} \in A_0(\chi_{v,0}, \mathbf{c}_0, \frac{1}{2})$. Moreover, $e^{ivH}(p(z)) - 1$ has an expansion in positive powers of $e^{\pi i z/3}$.

$$\begin{aligned}
\tilde{F}_{P,0} e^{ivH} &= \mu_0(P, 0, -\tfrac{1}{2}) \\
\tilde{F}_{P,1} e^{ivH} &= -6iv\omega_0(P, 1, \tfrac{1}{2}) \\
\tilde{F}_{P,n} e^{ivH} &= 0 \quad \text{for all } n < 0.
\end{aligned}$$

Similarly $e^{iw\bar{H}}$ is in $A_0(\chi_{0,w}, \mathbf{c}_0, \frac{1}{2})$, and

$$\begin{aligned}
\tilde{F}_{P,0} e^{iw\bar{H}} &= \mu_0(P, 0, -\tfrac{1}{2}) \\
\tilde{F}_{P,n} e^{iw\bar{H}} &= 0 \quad \text{for all } n > 0 \\
\tilde{F}_{P,-1} e^{iw\bar{H}} &= -6iw\omega_0(P, -1, \tfrac{1}{2}).
\end{aligned}$$

This means that $e^{ivH} - e^{iw\bar{H}}$ is a cusp form if we choose v and w such that $\chi_{v,0} = \chi_{0,w}$. This condition amounts to $v = -\bar{w} \in \hat{\Lambda} = \frac{2\pi i}{\omega\sqrt{3}}(\mathbb{Z} + \mathbb{Z}\rho)$. For $v = -\bar{w} = 0$ we get of course the zero automorphic form. But for $v \in \hat{\Lambda}$, $v \neq 0$, we obtain $e^{ivH} - e^{-i\overline{vH}} \in S_0(\chi_{v,0}, \mathbf{c}_0, \frac{1}{2})$ with

$$\begin{aligned}
\tilde{F}_{P,1}\left(e^{ivH} - e^{-i\overline{vH}}\right) &= -6iv\omega_0(P, 1, \tfrac{1}{2}) \\
\tilde{F}_{P,-1}\left(e^{ivH} - e^{-i\overline{vH}}\right) &= 6i\bar{v}\omega_0(P, -1, \tfrac{1}{2}).
\end{aligned}$$

This is a strange type of cusp form of weight 0. It belongs to a non-unitary character.

15.2.4 *Automorphic functions.* The quotient $X_{\text{com}} = \tilde{\Gamma}_{\text{com}} \backslash \mathfrak{H}^*$ is a Riemann surface of genus 1. Hence the field $A_0(\bar{\Gamma}_{\text{com}})$ of automorphic functions should be of the form $\mathbb{C}(\mathbf{x}, \mathbf{y})$, where $\mathbf{x}, \mathbf{y} \in A_0(\bar{\Gamma}_{\text{com}})$ satisfy a relation $\mathbf{y}^2 = 4\mathbf{x}^3 - \mathbf{g}_2\mathbf{x} - \mathbf{g}_3$, with $\mathbf{g}_2, \mathbf{g}_3 \in \mathbb{C}$. (Automorphic functions are holomorphic automorphic forms of weight zero, for the multiplier system 1, with possibly singularities in $\mathfrak{H}$, and satisfying some growth condition at the cusps; see [53], §2.1.)

Such $\mathbf{x}$ and $\mathbf{y}$ may be exhibited explicitly. The choice is not unique, and influences the $\mathbf{g}_j$. Take the modular forms g_2 and g_3 for Γ_{mod}:

$$\begin{aligned}
g_2(z) &= 60G_4(z) = (2\pi)^4 \left(\frac{1}{12} + 20 \sum_{n=1}^{\infty} \sigma_3(n) e^{2\pi i n z} \right) \\
g_3(z) &= 140G_6(z) = (2\pi)^6 \left(\frac{1}{216} - \frac{7}{3} \sum_{n=1}^{\infty} \sigma_5(n) e^{2\pi i n z} \right)
\end{aligned}$$

$$\sigma_a(n) = \sum_{d|n} d^a;$$

see, e.g., [53], Theorem 2.9. They satisfy $g_2^3 - 27g_3^2 = (2\pi)^{12}\eta^{24}$. Hence

$$\mathbf{x} = 12(2\pi)^{-4} g_2 \eta^{-8}, \qquad \mathbf{y} = 432(2\pi)^{-6} g_3 \eta^{-12},$$

defines modular functions $\mathbf{x}$ and $\mathbf{y}$ for $\bar{\Gamma}_{\mathrm{com}}$ satisfying $\mathbf{y}^2 = 4(\mathbf{x}^3 - 1728)$. With this choice, $\mathbf{g}_2 = 0$ and $\mathbf{g}_3 = 6912 = 4 \cdot 12^3$.

See also [28], Chapter 18, §5, Theorems 7 and 8 on p. 254.

15.3 The period map

X_{com} is a compact Riemann surface of genus 1. Hence there is a biholomorphic isomorphism $X_{\mathrm{com}} \cong \mathbb{C} \bmod \Lambda$, for some lattice Λ in $\mathbb{C}$. We use this isomorphism at several places in this chapter. So it is worthwhile to go into the determination of the lattice and the isomorphism.

15.3.1 *Choice of the isomorphism.* There is some freedom in the choice of the isomorphism. The lattice may be multiplied by an element of $\mathbb{C}^*$, and the isomorphism $X_{\mathrm{com}} \to \mathbb{C} \bmod \Lambda$ may be followed by a translation in $\mathbb{C} \bmod \Lambda$.

To get rid of this freedom we impose the condition that $\mathbf{x}$ and $\mathbf{y}$ correspond to the Weierstrass functions $\wp$ and $\wp'$ for Λ, see [53], §4.2. As η has no zeros in $\mathfrak{H}$, the automorphic functions $\mathbf{x}$ and $\mathbf{y}$ are infinite only at the cusps. This implies that $P = \tilde{\Gamma}_{\mathrm{com}} \cdot i\infty$ is mapped to $0 + \Lambda \in \mathbb{C} \bmod \Lambda$.

$\mathfrak{H}^* = \mathfrak{H} \cup \mathbb{Q} \cup \{i\infty\}$ is simply connected. Hence the isomorphism $X_{\mathrm{com}} \to \mathbb{C} \bmod \Lambda$ lifts to a holomorphic map $\mathfrak{H}^* \to \mathbb{C}$; we call it $\tilde{H}$ for the moment. We want it to satisfy

$$\mathbf{x}(z) = \wp(\tilde{H}(z), \Lambda), \qquad \mathbf{y}(z) = \wp'(\tilde{H}(z), \Lambda).$$

$\tilde{H}$ is determined up to an additive constant in Λ. We determine it completely by prescribing $\tilde{H}(i\infty) = 0$.

All holomorphic 1-forms on $\mathbb{C} \bmod \Lambda$ come from multiples of du, where u denotes the coordinate on $\mathbb{C}$. We obtain on $\mathfrak{H}$

$$du \circ \tilde{H} = d\tilde{H}(z) = \frac{d\wp(\tilde{H}(z), \Lambda)}{\wp'(\tilde{H}(z), \Lambda)} = \frac{d\mathbf{x}(z)}{\mathbf{y}(z)}.$$

By a computation of the expansions of η^4, g_2, g_3, $\mathbf{x}$, and $\mathbf{y}$ at the cusp we check that $\mathbf{y}^{-1} d\mathbf{x} = -2\pi i\, \eta(z)^4\, dz$. Hence $\tilde{H}$ extends the function H defined in 15.2.2. We drop the tilde from the notation, and use

$$H(\xi) = -2\pi i \int_{i\infty}^{\xi} \eta(z)^4 dz \qquad \text{for all } \xi \in \mathfrak{H}^*. \tag{15.1}$$

The transformation behavior of η^4 implies $H(\xi+1)=\rho H(\xi)$, with $\rho=e^{\pi i/3}$.

15.3.2 *Homomorphism.* Let $\gamma\in\tilde{\Gamma}_{\mathrm{com}}$. As H is a lift of the isomorphism $X_{\mathrm{com}}\rightarrow\mathbb{C}\bmod\Lambda$, the function $\mathfrak{H}^*\rightarrow\mathbb{C}:\xi\mapsto H(\gamma\cdot\xi)-H(\xi)$ is constant; its value is in Λ. We denote this value by $\lambda(\gamma)$.

$\lambda:\tilde{\Gamma}_{\mathrm{com}}\rightarrow\Lambda$ is a group homomorphism. Its kernel is the subgroup $\tilde{Z}\times\tilde{\Delta}$, where $\tilde{\Delta}$ denotes the commutator subgroup of $\tilde{\Gamma}^1_{\mathrm{com}}$. We use the same name for the induced homomorphisms $\Gamma_{\mathrm{com}}\rightarrow\Lambda$ and $\bar{\Gamma}_{\mathrm{com}}\rightarrow\Lambda$. In a moment we shall see that it is equal to the homomorphism defined in 15.1.9.

15.3.3 *Computation of* $H(0)$. To determine the lattice Λ, and the homomorphism λ, we compute some values of H.

As $0\in P$, the difference $H(0)-H(i\infty)=H(0)$ is an element of Λ. We can compute it by integration on the positive imaginary axis. In (15.1) we see that the value is negative. An analysis of the zeros and signs of g_2, g_3, η^4, $\mathbf{x}$ and $\mathbf{y}$ on this path shows that $\mathbf{x}(it)=(\frac{1}{4}\mathbf{y}(it)^2+1728)^{1/3}$, and that $\mathbf{y}(it)$ runs from ∞ to $-\infty$.

$$H(0)=\int_{i\infty}^{0}\frac{d\mathbf{x}(z)}{\mathbf{y}(z)}=\int_{y=\infty}^{-\infty}\frac{1}{y}d\sqrt[3]{\frac{1}{4}y^2+1728}.$$

Some computations lead to $H(0)=-\omega$, with

$$\omega=\frac{1}{6\sqrt{3}}\mathsf{B}\left(\frac{1}{6},\frac{1}{2}\right)=\frac{\sqrt{\pi}\Gamma(\frac{1}{6})}{6\sqrt{3}\Gamma(\frac{2}{3})}>0.$$

15.3.4 *Computation of* $H(i)$ *and* $H(\rho)$. From $\lambda(DC)=H(DC\cdot i)-H(i)=H(3+i)-H(i)=-2H(i)$ and $\lambda(DC)=H(2)-H(1)=(\rho^2-\rho)H(0)=\omega$ it follows that $H(i)=-\frac{1}{2}\omega$ and $\lambda(\eta\gamma)=\omega$.

Similarly $C\cdot\rho=\rho-2$ and $C\cdot 0=-1$ lead to $\lambda(C)=-\frac{1}{2}(3+i\sqrt{3})H(\rho)=\rho\omega$, $H(\rho)=-\frac{1}{3}(\rho+1)\omega$, and $\lambda(\gamma)=\rho\omega$.

We have obtained $\lambda(\gamma)=\rho\omega$, and $\lambda(\eta)=(1-\rho)\omega$. These are the values we used in 15.1.9 to define λ. Moreover, the lattice Λ is spanned by these two images of generators of $\tilde{\Gamma}_{\mathrm{com}}$. Hence $\Lambda=\omega(\mathbb{Z}\cdot\rho+\mathbb{Z})$.

15.3.5 *Image of the fundamental domain* F_{com}. The image $\lambda(\tilde{\Gamma}_{\mathrm{com}})$ is the lattice $\Lambda=\omega(\mathbb{Z}\cdot\rho+\mathbb{Z})$. The image $H(F_{\mathrm{com}})$ should be a fundamental domain for the action of Λ in $\mathbb{C}$ by translation. The relation $H(\xi+1)=\rho H(\xi)$ implies that this image is symmetric for the rotation $u\mapsto\rho u$.

Take $l\in\mathbb{Z}$ and $t>0$, and note that $\eta(it)^4>0$, $\eta(l+it)^4=\rho^l\eta(it)^4$ and $e^{-\pi i/6}\eta(\frac{1}{2}+it)^4>0$, $\eta(\frac{1}{2}+l+it)^4=\rho^l\eta(\frac{1}{2}+it)^4$. Hence the vertical lines from ∞ to $i+l$ are mapped to straight line segments from 0 to $\rho^l H(i)$ and the vertical lines from ∞ to $\rho+l$ to straight line segments from 0 to $\rho^l H(\rho)$.

All end points $H(i+l)$ and $H(\rho+l)$ are on the regular hexagon with center 0 and one corner at $-\frac{1}{3}(\rho-1)\omega$. To show that $H(F_{\mathrm{com}})$ is equal to this hexagon we show that the image of the arc between ρ and $\rho-1$ of the unit circle is mapped to a vertical line segment. As $\overline{\eta(iy-x)} = \eta(iy+x)$, we have for $\varphi \in \mathbb{R}$:

$$\begin{aligned} \overline{e^{i\varphi}\eta(e^{i\varphi})^4} &= e^{-i\varphi}\eta(e^{i(\pi-\varphi)})^4 = e^{-i\varphi}\eta(-1/e^{i\varphi})^4 \\ &= -e^{-i\varphi}e^{2i\varphi}\eta(e^{i\varphi})^4 = -e^{i\varphi}\eta(e^{i\varphi})^4. \end{aligned}$$

Hence

$$H(e^{i\varphi}) - H(i) = -2\pi i \int_{\pi/2}^{\varphi} \eta(e^{it})^4 ie^{it}\,dt = 2\pi \int_{\pi/2}^{\varphi} e^{it}\eta(e^{it})^4 dt$$

is purely imaginary.

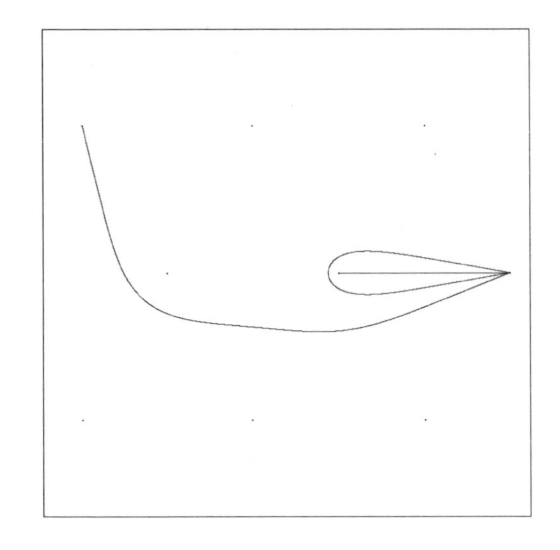

Figure 15.3 The image under H of three vertical lines in the upper half plane. All image curves start in 0. The image of $\operatorname{Re} z = 0$ runs to $-\omega$. The closed curve is the image of $\operatorname{Re} z = \frac{1}{6}$. The curve with end point $(-3+\rho)\omega$ is the image of $\operatorname{Re} z = \frac{4}{11}$. The dots are points of Λ.

15.3.6 *Images of vertical lines.* Let $\xi \in \mathbb{Q}$. In the same way as above, we see that for $\xi \in \frac{1}{2} + \mathbb{Z}$ the image under H of the vertical line in $\mathfrak{H}$ with real part ξ are straight line segments in $\mathbb{C}$ from 0 to $H(\xi)$. For other $\xi \in \mathbb{Q}$ the image $H(\xi + i(0,\infty))$ is a curve in $\mathbb{C} \smallsetminus \Lambda$ between the lattice points 0 and $H(\xi)$. In general, it is not straight. For instance, $H(1/6) = \lambda(C^{-1}D^{-1}CD) = 0$, hence $H(i(0,\infty) + 1/6)$ is a closed curve. See Figure 15.3.

15.3.7 *The symbol C.* For computations it is convenient to work with $C : \mathbf{P}^1_{\mathbb{Q}} \to \mathbb{Z} + \mathbb{Z}\rho$ defined by $C(\xi) = -\frac{1}{\omega}H(\xi)$. Note that $\mathbf{P}^1_{\mathbb{Q}} = \{\infty\} \cup \mathbb{Q} \subset \mathfrak{H}^*$. Quick recursive computation of C is possible with the relations

$$\begin{aligned}&C(\infty)=0, \quad C(0)=1,\\&C(\xi+1)=\rho C(\xi),\\&C(\xi)+C(-1/\xi)=1,\\&\overline{C(\xi)}=C(-\xi),\end{aligned}$$

which follow from the equalities $H(i\infty)=0$, $H(0)=-\omega$, and$H(\xi+1)=\rho H(\xi)$, and the transformation behavior of η^4, under $W=\pm\begin{pmatrix}0&1\\-1&0\end{pmatrix}$ and under conjugation.

We shall write $C(d,c)=C(d/c)$. To express λ in C we use for $\begin{pmatrix}a&b\\c&d\end{pmatrix}\in\Gamma_{\mathrm{com}}$:

$$\begin{aligned}\lambda\left(\pm\begin{pmatrix}a&b\\c&d\end{pmatrix}\right)&=-\lambda\begin{pmatrix}d&-b\\-c&a\end{pmatrix}\\&=-H\left(\begin{pmatrix}d&-b\\-c&a\end{pmatrix}\cdot i\infty\right)+H(i\infty)=\omega C(-d/c)=\omega\overline{C(d,c)}.\end{aligned}$$

I view C as analogous to the Dedekind sum S. In both cases there is a homomorphism ($\alpha:\tilde{\Gamma}_{\mathrm{mod}}\to\mathbb{C}$, respectively $\lambda:\tilde{\Gamma}_{\mathrm{com}}\to\mathbb{C}$) defined recursively by expressing group elements in generators. On the other hand, group elements can be given as the lift $\widetilde{\begin{pmatrix}a&b\\c&d\end{pmatrix}}$ of a matrix. S, respectively C, describes α, respectively λ, in terms of the matrix elements. An analogon of Dedekind's beautiful expression for $S(d,c)$ as a sum is lacking for C.

15.3.8 Proposition.

i) $C(d,c)=\mathcal{O}(\log c)$ *for* $d,c\in\mathbb{Z}$, $(d,c)=1$, $c\to\infty$.

ii) If $C(\xi)=0$ *for* $\xi\in\mathbf{P}^1_{\mathbb{Q}}$, *then* $\xi=\infty$, *or* $\xi=d/c$ *with* $(d,c)=1$ *and* $c\equiv 0 \bmod 6$.

Remark. There are many $\xi=d/6c$, $(d,6c)=1$, for which $C(d/6c)\neq 0$.
Proof. To obtain part i) we show by induction on n that $|C(d/c)|\leq n+1$ for all $c,d\in\mathbb{Z}$ with $(d,c)=1$ and $1\leq c\leq 2^n$.

For $n=0$ this is clear. If $1\leq c\leq 2^n$, then take $d_1\in\mathbb{Z}$ determined by $d_1\equiv\pm d \bmod c$ and $|d_1|\leq c/2$. Then $|C(d/c)|=|C(d_1/c)|=|1-C(-c/d_1)|\leq 1+n$ by the induction hypothesis.

To prove part ii) consider $d,c\in\mathbb{Z}$, $(d,c)=1$ with $C(d/c)=0$. There are $a,b\in\mathbb{Z}$ such that $\gamma=\begin{pmatrix}a&b\\c&d\end{pmatrix}\in\Gamma_{\mathrm{com}}$. From $\lambda(\gamma)=-H(-d/c)=\omega\overline{C(d/c)}=0$ it follows that γ is an element of $\Delta\times Z$, where Δ is the commutator subgroup of $\bar{\Gamma}_{\mathrm{com}}$. Let $\Gamma(6)$ denote the principal congruence subgroup of the $\begin{pmatrix}a&b\\c&d\end{pmatrix}\in\Gamma_{\mathrm{mod}}$ satisfying $\begin{pmatrix}a&b\\c&d\end{pmatrix}\equiv\begin{pmatrix}1&0\\0&1\end{pmatrix} \bmod 6$. As $\bar{\Gamma}_{\mathrm{com}}/\overline{\Gamma(6)}$ is abelian, we have $\bar{\Delta}\subset\overline{\Gamma(6)}$.

15.4 Poincaré series

The characters $\chi_{v,w}$ of $\tilde{\Gamma}_{\text{com}}$ belong to even weights. They describe one factor of the character group $\mathcal{X}(\Gamma_{\text{com}})$. The other factor is described by the χ_r. These we have seen already for the full modular group. In this and the next sections we consider mainly families of automorphic forms of weight 0, with $(v,w) \mapsto \chi_{v,w}$ as the family of characters. It is convenient to consider these automorphic forms as functions on $\mathfrak{H}$.

In this section we consider Poincaré series of weight 0.

15.4.1 *Notation.* For $v, s \in \mathbb{C}$, $\operatorname{Re} s > \frac{1}{2}$, and $n \in \mathbb{Z} = \mathcal{C}_P(\chi_{v,\bar{v}})$, we denote the Poincaré series by

$$P^n(v,s) = P_0(P, n, \chi_{v,\bar{v}}, s).$$

For $n = 0$ this is an Eisenstein series.

15.4.2 *Comparison with the modular case.* The subgroup Δ_P for $\tilde{\Gamma}_{\text{com}}$ is equal to $\{ n(6l)k(\pi m) : l, m \in \mathbb{Z} \}$, whereas the corresponding subgroup for the modular group is given by $\Delta_{\tilde{\Gamma}_{\text{mod}} \cdot i\infty} = \{ n(l)k(\pi m) : l, m \in \mathbb{Z} \} = \bigsqcup_{j \bmod 6} n(j)\Delta_P$. As $\tilde{\Gamma}_{\text{mod}} = \bigsqcup_{j \bmod 6} n(j)\tilde{\Gamma}_{\text{com}}$, any system of representatives for $\Delta_P \backslash \tilde{\Gamma}_{\text{com}}$ is also a system of representatives for $\Delta_{\tilde{\Gamma}_{\text{mod}} \cdot i\infty} \backslash \tilde{\Gamma}_{\text{mod}}$.

Proposition 5.1.6 defines

$$\begin{aligned} P_0(\Gamma_{\text{mod}} \cdot i\infty, n, 1, s; g) &= \sum_{\gamma \in \Delta_P \backslash \tilde{\Gamma}_{\text{com}}} \mu_0(\Gamma_{\text{mod}} \cdot i\infty, n, s; \gamma g) \\ P^{6n}(0, s; g) &= \sum_{\gamma \in \Delta_P \backslash \tilde{\Gamma}_{\text{com}}} \mu_0(P, 6n, s; \gamma g). \end{aligned}$$

As $g_P = a(6)$ and $g_{\Gamma_{\text{mod}} \cdot i\infty} = 1$, we have $\mu_0(P, 6n, s) = 6^{-s-1/2}\mu_0(\Gamma_{\text{mod}} \cdot i\infty, n, s)$, and

$$P^{6n}(0,s) = 6^{-s-1/2} P_0(\tilde{\Gamma}_{\text{mod}} \cdot i\infty, n, 1, s).$$

15.4.3 *Fourier expansion.* The characters $\chi_{v,w}$ are trivial on the subgroup $\tilde{\Gamma}^0_{\text{com}} = \tilde{Z} \times \tilde{\Delta}$ of $\tilde{\Gamma}_{\text{com}}$. Note that $\tilde{\Gamma}^0_{\text{com}}$ has infinite index in $\tilde{\Gamma}_{\text{com}}$, and that its image under $\gamma \mapsto \hat{\gamma}$ has infinite index in Γ_{com}.

As $\Delta_P \subset \tilde{\Gamma}^0_{\text{com}}$, we obtain

$$P^n(v,s;g) = \sum_{\gamma \in \tilde{\Gamma}^0_{\text{com}} \backslash \tilde{\Gamma}_{\text{com}}} \chi_{v,\bar{v}}(\gamma)^{-1} \sum_{\delta \in \Delta_P \backslash \tilde{\Gamma}^0_{\text{com}}} \mu_0(P, n, s; \delta\gamma g). \tag{15.2}$$

The inner sum is a Poincaré series for $\tilde{\Gamma}^0_{\text{com}}$. As $\tilde{\Gamma}^0_{\text{com}}$ has a fundamental domain with infinite volume, this group falls outside the class of discrete groups considered in this book.

$\chi_{v,\bar{v}}$ is a character of $\tilde{\Gamma}^0_{\text{com}} \backslash \tilde{\Gamma}_{\text{com}} \cong \mathbb{Z}^2$. We may view Equation (15.2) as the Fourier series expansion of the function $v \mapsto P^n(v,s;g)$ on $\mathbb{C} \bmod \hat{\Lambda}$.

15.4.4 *Kloosterman sums.* We turn to the generalized Kloosterman sums that occur in the Dirichlet series describing the Fourier coefficients of the Poincaré series $P^n(v,s)$.

$\tilde{\Gamma}_{\text{com}}$ has one cuspidal orbit. So we may find $\widetilde{\left(\begin{smallmatrix} a & b \\ c & d \end{smallmatrix}\right)} \in \tilde{\Gamma}_{\text{com}}$ for all $c, d \in \mathbb{Z}$, with $(c,d) = 1$. Of course, not every inverse a of $d \bmod c$ will give an element of $\tilde{\Gamma}_{\text{com}}$. It is easy to show that $a = 12cS(d,c) + 3c - d$ is a good choice.

For $\left(\begin{smallmatrix} a & b \\ c & d \end{smallmatrix}\right) \in \mathrm{SL}_2(\mathbb{R})$

$$g_P \widetilde{\begin{pmatrix} a & b \\ c & d \end{pmatrix}} g_P^{-1} \in \tilde{\Gamma}_{\text{com}} \iff \begin{pmatrix} 1 & 6b \\ c/6 & d \end{pmatrix} \in \Gamma_{\text{com}}.$$

The variable c in Proposition 5.2.9 runs through $6\mathbb{N}$. For a given $\frac{c}{6} \in \mathbb{N}$ all $d \in \mathbb{Z}$ with $(\frac{c}{6}, d) = 1$ occur, hence the d in the definition of generalized Kloosterman sums in Equation (5.1) on p. 80 run over $d \bmod c$ such that $(d, \frac{c}{6}) = 1$. In the sequel we let c run through $\mathbb{N}$, *i.e.*, we replace c by $6c$.

We define $S(v;m,n;c) = \mathrm{S}_\chi(P,m;P,n;6c)$, for $c \in \mathbb{N}$, $n, m \in \mathbb{Z}$, $v \in \mathbb{C}$, with $\chi = \chi_{v,\bar{v}}$; see Equation (5.1). We obtain

$$S(v;m,n;c) \quad = \sum_{d \bmod 6c,\ (d,c)=1} \chi_{v,\bar{v}}\left(\delta_{(c,d)}\right)^{-1} e^{2\pi i(na+md)/6c},$$

where the $\delta_{(c,d)} = \widetilde{\left(\begin{smallmatrix} a & b \\ c & d \end{smallmatrix}\right)}$ have been chosen in $\tilde{\Gamma}_{\text{com}}$. Use $\lambda\left(\delta_{(c,d)}\right) = \omega\overline{C(d,c)}$ (see 15.3.7), to obtain

$$S(v;m,n;c) \quad = \sum_{d \bmod 6c,\ (d,c)=1} e^{2i\omega \operatorname{Re}(\bar{v}C(d,c)) + 2\pi i(na+md)/6c}.$$

As far as we see, this does not reduce to a simpler expression for general v. For $v = 0$ one can obtain expressions in terms of Kloosterman sums for the full modular group with character χ_{2n}.

15.4.5 *Fourier coefficients of Eisenstein series.* Proposition 5.2.9 gives the Fourier coefficients of the Eisenstein series $P^0(v,s)$:

$$\begin{aligned} & c_0(P,m;P,0;\chi_{v,\bar{v}},s) \\ = \quad & \dot{\Psi}^m(v,2s+1)\frac{\sqrt{\pi}}{6^{1+2s}\Gamma(s+\frac{1}{2})} \cdot \begin{cases} \Gamma(s) & \text{if } m = 0 \\ \pi^s |m|^{s-1/2} & \text{if } m \neq 0, \end{cases} \end{aligned}$$

with the following series, converging for $\operatorname{Re} u > 2$,

$$\dot{\Psi}^m(v,u) = \sum_{c=1}^{\infty} c^{-u} \sum_{d \bmod 6c,\ (d,c)=1} e^{-2i\omega \operatorname{Re}(\bar{v}C(d,c)) + \pi i m d/3c}. \tag{15.3}$$

In general we have to be content with this formula. If

$$v \in \hat{\Lambda}_6 = \frac{2\pi i}{\omega\sqrt{3}}\left(\frac{1}{2}\rho\mathbb{Z} + \frac{1}{6}(2\rho - 1)\mathbb{Z}\right) \supset \hat{\Lambda} = \frac{2\pi i}{\omega\sqrt{3}}(\mathbb{Z} + \rho\mathbb{Z}),$$

one can obtain explicit formulas of the type $\dot{\Psi}^m(v,u) = \frac{\zeta(u-1)}{\zeta(u)}R_v(u)$, with R_v a rational function in 2^{-u} and 3^{-u}. ($R_0(u) = 6$ for the trivial character, and $R_v(2) = 0$ otherwise.) This is due to the fact that $\hat{\Lambda}(6)$ is the lattice in $\mathbb{C}$ dual to the principal congruence subgroup $\Gamma(6)$ under $(v,\gamma) \mapsto \chi_{v,\bar{v}}(\gamma)$. Actually, $\hat{\Lambda}_6$ is the largest lattice dual to a congruence subgroup contained in Γ_{com}.

15.5 Eisenstein family of weight 0

We consider the Eisenstein family in weight zero. The theory of Part I gives us information on its singularities, and those of the meromorphic continuation of the Dirichlet series $\dot{\Psi}^m(v,u)$ that occurs in its Fourier coefficients (see 15.4.5).

15.5.1 *Parameter space.* We work on the parameter space $\mathbb{C}^3 = V \times \mathbb{C}$, with $\chi(v,w,s) = \chi_{v,w}$, and $s(v,w,s) = s$. The weights are the constant functions with value in $2\mathbb{Z}$. We choose the weight $(v,w,s) \mapsto 0$. We take $\mathcal{P} = \{P\}$, and the growth condition $\mathbf{c}_0 : P \mapsto \{0\}$.

15.5.2 *How large is V_0?* The seminorm $\|.\|_c$ on V, used in the Lemmas 8.4.11 and 9.1.5, is zero. In view of Lemma 9.1.5, the dependence on v and w of the quantity $b_{1,b}\|\varphi\|_b + b_{1,c}\|\varphi\|_c + b_{2,c}\|\varphi\|_c^2$, with $\varphi = v\lambda + w\bar{\lambda}$, can be made arbitrarily small. This implies that the pseudo Casimir operator is well defined for all parameter values in V.

In the proof of Theorem 10.2.1, the V-part of the domain of the meromorphic continuation of Poincaré series is the union of V_0's centered at points of the relevant cell of continuation. Here this cell is the real space V_r. Thus, we obtain $\mathbb{C}^3$ as the domain of the Poincaré families in weight 0.

15.5.3 *Eisenstein family.* In this chapter we denote by E the Eisenstein family

$$E(v,w,s) = E_0(J(0), P; v\lambda + w\bar{\lambda}, s).$$

Consider a fixed $v \in \mathbb{C}$. The family $s \mapsto E(v,\bar{v},s)$ is given by the Eisenstein series $P^0(v,s) = P_0(P,0,\chi_{v,\bar{v}},s)$ on the region $\operatorname{Re} s > \frac{1}{2}$.

We use the notation $C^n(v,w,s) = C_0(P,n;J(0),P;v\lambda + w\bar{\lambda},s)$ for the n-th Fourier coefficient, with $n \in \mathbb{Z} = \mathcal{C}_P(\chi_{v,w})$.

15.5.4 *Functional equations.* The reader may like to derive the following equalities from 10.3.5 and the uniqueness of E:

$$\begin{aligned} E(v,w,-s) &= C^0(v,w,-s)E(v,w,s) \\ C^0(v,w,-s)C^0(v,w,s) &= 1 \end{aligned}$$

$$\begin{aligned} C^0(v,w,-s)C^n(v,w,s) &= C^n(v,w,-s) \quad \text{for } n \neq 0 \\ E(v,w,s;-\bar{z}) &= E(w,v,s;z) \\ C^n(v,w,s) &= C^{-n}(w,v,s) \\ E(v,w,s;z+1) &= E(\rho v,\rho^{-1}w,s;z) \quad \text{with } \rho = e^{\pi i/3} \\ \rho^n C^n(v,w,s) &= C^n(\rho v,\rho^{-1}w,s) \\ \overline{E(\bar{v},\bar{w},\bar{s})} &= E(-w,-v,s) \\ \overline{C^n(\bar{v},\bar{w},\bar{s})} &= C^{-n}(-w,-v,s). \end{aligned}$$

15.5.5 *Lower bound for the cuspidal spectrum.* The singularities of Eisenstein families with varying character may be related to the presence of cusp forms. There is a well known method of bounding the cuspidal spectrum, see, e.g., [21], Proposition 2.1 on p. 511. We use it to obtain:

15.5.6 Proposition. *If* $s \in \Sigma^0(\chi_{v,\bar{v}},0)$, *then* $\frac{1}{4} - s^2 \geq \pi^2/84$.

Remarks. This leaves open the existence of cuspidal exceptional eigenvalues of continuous series type in the interval $[\pi^2/84, 1/4)$. For $\chi_{v,\bar{v}} \neq 1$, non-cuspidal exceptional eigenvalues $\frac{1}{4} - s^2$ with $s \in \Sigma^e(\chi_{v,\bar{v}},0)$ may occur in $[0,\frac{1}{4})$, corresponding to $0 < s < \frac{1}{2}]$. One can get more information in the congruence case $v \in \hat{\Lambda}_6$: if one carries out the computations alluded to in 15.4.5, one obtains $\Sigma^e(1,0) = \{\frac{1}{2}\}$, and $\Sigma^e(\chi_{v,\bar{v}},0) = \emptyset$ for $v \in \hat{\Lambda}_6 \setminus \hat{\Lambda}$.

Proof. Take $U = \{g_P \cdot z : y \geq \frac{\sqrt{3}}{12}, -\frac{5}{12} \leq x \leq \frac{7}{12}\}$. This set U contains the fundamental domain F_{com} in Figure 15.1 on p. 276, and is contained in 7 translates of F_{com}. Let $f \in S_0(\chi_{v,\bar{v}},\mathbf{c}_0,s)$. We obtain with Lemma 6.3.7:

$$\begin{aligned} 7\left(\frac{1}{4} - s^2\right)\|f\|_0^2 &\geq \int_U \frac{1}{8}\sum_{\pm} \left|\mathbf{E}^{\pm} f(p(z))\right|^2 \, d\mu(z) \\ &= \sum_{n\neq 0} \int_{\sqrt{3}/12} \left|F_{P,n}\mathbf{E}^{\pm} f(y)\right|^2 \frac{dy}{y^2} \\ &\geq \sum_{n\neq 0} \left(2\pi|n|\frac{\sqrt{3}}{12}\right)^2 \int_{\sqrt{3}/12} |F_{P,n} f(y)|^2 \frac{dy}{y^2} \\ &\geq \frac{\pi^2}{12}\|f\|_0^2. \end{aligned}$$

15.5.7 *Singularities of E at real points.* Proposition 12.5.8 applies to the present situation. It lists the possibilities for singularities of Eisenstein families on a restricted parameter space.

a) $\operatorname{Re} s = 0$. A singularity at $(v,\bar{v},s)$ may occur if $s \in \Sigma^0(\chi_{v,\bar{v}})$. If it does, then E is of indeterminate type at $(v,\bar{v},s)$, and $\dot{E}(v,\bar{v})$ is holomorphic and non-zero at s.

b) $0 < s \leq \frac{1}{2}\sqrt{1-\pi^2/21} \approx 0.364$. If a singularity at $(v, \bar{v}, s)$ occurs, then $s \in \Sigma(\chi_{v,\bar{v}})$. If $s \in \Sigma^e(\chi_{v,\bar{v}})$, then $\dot{E}(v, \bar{v})$ is singular at s, hence E is singular as well. If $s \notin \Sigma^e(\chi_{v,\bar{v}})$, then the presence of a singularity of E at $(v, \bar{v}, s)$ implies that E is of indeterminate type at this point. But, even in the indeterminate case, part ii) of Proposition 12.5.8 implies that $(v, \bar{v}, s)$ may be approached by points $(v_1, \overline{v_1}, s_1)$ for which $s_1 \in \Sigma^e(\chi_{v_1,\overline{v_1}})$.

c) $\frac{1}{2}\sqrt{1-\pi^2/21} < s \leq \frac{1}{2}$. Proposition 15.5.6 implies that there are no cusp forms with the eigenvalue $\frac{1}{4} - s^2$. Hence the space $A_0(\chi_{v,\bar{v}}, \mathbf{c}_0, s)$ is spanned by the value or the residue of $\dot{E}(v, \overline{v})$ at s, see Proposition 11.2.6. Proposition 12.2.12 shows that E has determinate type at $(v, \bar{v}, s)$. Hence E is singular at $(v, \bar{v}, s)$ if and only if its restriction $\dot{E}(v, \bar{v})$ is singular at s, and that occurs if and only if $s \in \Sigma^e(\chi_{v,\bar{v}})$.

Suppose we have such a singularity at a point $(v_0, \overline{v_0}, s_0)$. As $\dot{E}(v_0, \overline{v_0})$ has a first order pole at s_0, and as E has determinate type at $(v_0, \overline{v_0}, s_0)$, there is a holomorphic function a on a neighborhood of $(v_0, \overline{v_0})$ in V such that $a(v_0, \overline{v_0}) = 0$, and $(v, w, s) \mapsto (s - s_0 - a(v, w))E(v, w, s)$ is primitive at $(v_0, \overline{v_0}, s_0)$; compare Lemma 12.3.4.

In all cases (except $s = 0$) the singularity of E is of the same order as that of the Fourier coefficient C^0. This follows from the fact that if $s \neq 0$ the sections $\mu_0(P, 0)$ and $\tilde{\mu}_0(P, 0)$ constitute a suitable basis for the space of Fourier terms of order zero.

15.5.8 *Restriction of E to $s = 0$.* We see in part iv) of Proposition 12.5.8 that $(v, w) \mapsto E(v, w, 0)$ exists as a meromorphic family of automorphic forms. If E would have a singularity at $(0, 0)$, then there would be non-zero cusp forms in $S_0(1, \mathbf{c}_0, \frac{1}{2})$. But this space is the direct sum $\bigoplus_{0 \leq r \leq 10,\, r \text{ even}} S_0(\Gamma_{\text{mod}}, \chi_r, \mathbf{c}_0, \frac{1}{2})$ of spaces of cusp forms for the full modular group. These spaces are trivial; see the discussion and references in 13.3.4. We conclude that E is holomorphic at $(0, 0, 0)$, and has determinate type at this point.

Near $s = 0$ a sensible description of the Fourier term of order zero is

$$F_{P,0}E(v, w, s) = 2s\lambda_0(P, 0, s) + (1 + C^0(v, w, s))\mu_0(P, 0, -s),$$

see Lemma 7.6.14. In 15.4.5 we see that $\dot{C}(0, 0; 0) = C^0(0, 0, 0) = -1$. The functional equation $C^0(v, w, s)C^0(v, w, -s) = 1$ implies that $C^0(v, w, 0) = -1$ on each neighborhood of $(0, 0)$ in $\mathbb{C}^2$ on which C^0 is holomorphic. Hence $C^0(v, w, 0) = -1$ as an identity between meromorphic functions. Proposition 10.2.14 shows that $E(v_0, \overline{v_0}, 0) = 0$ for all $(v_0, \overline{v_0})$ at which $(v, w) \mapsto E(v, w, 0)$ is holomorphic. Hence the family $(v, w) \mapsto E(v, w, 0)$ is the zero family.

15.5.9 *Behavior of E near $s = \frac{1}{2}$.* As $\dot{E}(0, 0)$ has a pole at $s = \frac{1}{2}$, there exists a holomorphic function $a_{1/2}$ on a neighborhood of $(0,0)$ in $\mathbb{C}^2$ such that

$$\tilde{E} : (v,w,s) \mapsto \left(s - \frac{1}{2} - a_{1/2}(v,w) \right) E(v,w,s)$$

is primitive at $(0,0,\frac{1}{2})$. Hence for $(v,\bar{v})$ near $(0,0)$ the one-dimensional Eisenstein family $\dot{E}(v,\bar{v})$ is singular at $s = \frac{1}{2} + a_{1/2}(v,\bar{v})$. In particular, $a_{1/2}(v,\bar{v}) < 0$ for all $v \neq 0$ in a neighborhood of 0. This is the first example we meet where the existence of $s \in \Sigma_c^e(\chi,l)$ is proved.

In Section 15.6 we shall discuss the restriction to $s = \frac{1}{2}$.

15.5.10 *Dirichlet series.* The $\dot{\Psi}^n$ defined in (15.3) have a meromorphic continuation Ψ^n to $\mathbb{C}^3$, as follows from Proposition 5.2.9, Section 10.3, and the computations in 15.4.4:

$$\begin{aligned} \Psi^0(v,w,u) &= \frac{6^u \Gamma(u/2)}{\pi^{1/2}\Gamma((u-1)/2)} C^0(v,w,(u-1)/2) \\ \Psi^n(v,w,u) &= 6^u \Gamma(u/2) \pi^{-u/2} |n|^{1-u/2} C^n(v,w,(u-1)/2) \quad \text{if } n \neq 0. \end{aligned}$$

We shall express the derivatives of Ψ^n in terms of the series

$$\Phi^n(a,b,u) = \sum_{c=1}^{\infty} c^{-u} \sum_{d \bmod 6c,\ (d,c)=1} C(d,c)^a \overline{C(d,c)^b} e^{\pi i n d/3c}.$$

From the logarithmic growth of the C-symbol, see Proposition 15.3.8, it follows that the series for Φ^n converge absolutely for $\operatorname{Re} u > 2$. The relation $C(d+c,c) = \rho C(d,c)$ implies that $\Phi^n(a,b)$ is zero if $a - b + n \not\equiv 0 \bmod 6$.

15.5.11 *Differentiation.* We differentiate the series for $\dot{\Psi}^n(x+iy,u)$ in (15.3) on p. 285 with respect to the real variables x and y. The logarithmic growth of the C-symbol implies that this can be done term by term. The effect is that the term of order (c,d) is multiplied by a factor:

differentiation	factor
∂_x	$-2i\omega \operatorname{Re} C(d,c)$
∂_y	$-2i\omega \operatorname{Im} C(d,c)$
$\frac{1}{2}(\partial_x - i\partial_y)$	$-i\omega \overline{C(d,c)}$
$\frac{1}{2}(\partial_x + i\partial_y)$	$-i\omega C(d,c)$

But the function Ψ^n is holomorphic at points $(x+iy, x-iy, u)$, hence

$$\begin{aligned} \frac{1}{2}(\partial_x - i\partial_y)\dot{\Psi}^n(x+iy,u) &= \partial_v \Psi^n(x+iy, x-iy, u) \\ \frac{1}{2}(\partial_x + i\partial_y)\dot{\Psi}^n(x+iy,u) &= \partial_w \Psi^n(x+iy, x-iy, u), \end{aligned}$$

where ∂_v and ∂_w denote differentiation with respect to the complex variables. This differentiation may be carried out repeatedly; $\partial_v^b \partial_w^a$ corresponds to multiplication by $(-i\omega)^{a+b} C(d,c)^a \overline{C(d,c)^b}$ in the term of order (d,c). If we apply this at $x+iy = 0$, then we obtain

$$\Phi^n(a,b,u) = (-i\omega)^{-a-b} \partial_v^b \partial_w^a \Psi^n(0,0,u).$$

This gives the meromorphic continuation to $\mathbb{C}$ of $u \mapsto \Phi^n(a,b,u)$ for all integral $a, b \geq 0$, and also functional equations, that get more and more complicated as $a+b$ increases.

The right-most pole of $\Phi^n(a,b)$ may occur at $u = 2$, or more to the left. See 15.6.23 for the behavior at $u = 2$.

15.5.12 *No distribution results.* In [8] I obtained distribution results for Dedekind sums from the Dirichlet series for the full modular group that are analogous to to the series $\Phi^n(a,b)$ here; compare Section 13.6. Here we work on a restricted parameter space. This implies that the position of the right-most pole of $\Phi^n(a,b)$ does not move to the right with increasing $a+b$. Moreover, the order of that pole gets higher than one. I cannot obtain distribution results for the C-symbol by the method shown in Section 13.6.

15.6 Harmonic automorphic forms

The Eisenstein family E in weight zero has a well defined restriction to $s = \frac{1}{2}$. This restriction can be viewed as a meromorphic family in (v,w) of functions on $\mathfrak{H}$ that transform according to the character $\chi_{v,w}$ of $\bar{\Gamma}_{\text{com}}$. As the eigenvalue is 0, the functions in this family are harmonic. We shall construct this family in terms of Jacobi theta functions. D. Zagier helped to obtain very explicit formulas.

The description in this section is in a sense orthogonal to what has been done in the previous sections. The family $(v,w) \mapsto E(v,w,\frac{1}{2})$ has been obtained rather indirectly, by meromorphic continuation of a family that has a series representation for other values of the parameters. Here it is described independently.

15.6.1 *Parameter space.* We use the parameter space $W = \mathbb{C}^2$ with $\chi : (v,w) \mapsto \chi_{v,w}$ and $s : (v,w) \mapsto \frac{1}{2}$. On W we have $\mathbf{wt} = 2\mathbb{Z} \subset \mathcal{O}(W)$; we shall work in weight 0. We use $\mathcal{P} = \{P\}$ with $P = \Gamma_{\text{com}} \cdot i\infty$. Hence $\mathcal{C}_P = \mathbb{Z} \subset \mathcal{O}(W)$. The growth condition is $\mathbf{c}_0$, given by $\mathbf{c}_0(P) = \{0\}$.

15.6.2 *Harmonic forms on $\mathfrak{H}$.* Any section $f \in \mathcal{A}_0(\mathbf{c}_0; \Omega)$, with $\Omega \subset W$, corresponds to a function $\tilde{f} \in C^\infty(\Omega \times \mathfrak{H})$ that satisfies

i) $\tilde{f}(v,w,\gamma \cdot z) = \chi_{v,w}(\gamma) \tilde{f}(v,w,z)$ for all $\gamma \in \bar{\Gamma}_{\text{com}}$.

ii) $(v,w) \mapsto \tilde{f}(v,w,z)$ is holomorphic on Ω for each $z \in \mathfrak{H}$

iii) $-y^2(\partial_x^2 + \partial_y^2)\tilde{f}(v,w,z) = 0$.

iv) The function $z \mapsto \tilde{f}(v,w,z) - \int_0^6 \tilde{f}(v,w,z+x')\,dx'$ is in $L^2(S_{P,2}, d\mu)$, and depends holomorphically on (v,w) in L^2-sense. ($S_{P,2} = \{ z \in \mathfrak{H} : 0 \leq x \leq 6, y \geq 2 \}$.)

This is a straightforward consequence of the definitions in 7.3.1 and 7.5.1.

15.6.3 *Reformulation in terms of a coordinate at the cusp.* Condition iii) implies that $\tilde{f}$ is *harmonic* as a function of z, i.e., $\partial_{\bar{z}}\partial_z \tilde{f} = 0$. As $\chi_{v,w}(\overline{\pi_P}) = 1$ for all (v,w), we may view $\tilde{f}$ on $S_{P,2}$ as a function of $t = e^{\pi i z/3}$ on $0 < |t| < e^{-2\pi/3}$; we use the same symbol $\tilde{f}$ for this function. In this coordinate $\partial_{\bar{t}}\partial_t \tilde{f} = 0$.

The scalar product in $L^2(S_{P,2})$ is given by

$$\begin{aligned} (g_1, g_2) &= \int_{S_{P,2}} g_1(z)\overline{g_2(z)}\frac{dz \wedge d\bar{z}}{-2iy^2} \\ &= \int_{0<|t|<e^{-2\pi/3}} g_1(t)\overline{g_2(t)}\,(|t|\log|t|)^{-2}\,\frac{dt \wedge d\bar{t}}{-2i}. \end{aligned}$$

Let g be harmonic on $0 < |t| < e^{-2\pi/3}$, and square integrable on $S_{P,2}$. Let $0 < |t_0| < \frac{1}{2}e^{-2\pi/3}$. By the mean value theorem for harmonic functions we have $g(t_0) = \frac{1}{2\pi}\int_0^{2\pi} g(t_0 + \delta e^{i\varphi})\,d\varphi$ for each $\delta < |t_0|$. Integration gives $\frac{1}{8}|t_0|^2 g(t_0) = \frac{1}{2\pi}\int_{|t-t_0|<|t_0|/2} g(t)\frac{dt\wedge d\bar{t}}{-2i}$. Hence $g(t_0) = \mathcal{O}\left(|t_0|^{-2}\|g\|_2\|\varphi\|_2\right)$, with φ the function equal to $(|t|\log|t|)^2$ if $|t - t_0| < \frac{1}{2}|t_0|$, and zero elsewhere.

$$\begin{aligned} \|\varphi\|_2^2 &\leq \int_{|t_0|/2<|t|<3|t_0|/2} (|t|\log|t|)^{4-2}\,\frac{dt \wedge d\bar{t}}{-2i} \\ &= 2\pi \int_{|t_0|/2}^{3|t_0|/2} x^3 (\log x)^2\,dx \\ &= \mathcal{O}\left(|t_0|^4 (\log|t_0|)^2\right). \end{aligned}$$

This implies that $|g(t_0)| = \mathcal{O}\left(\|g\|_2 \cdot \big|\log|t_0|\big|\right)$.

As $\partial_t g$ is holomorphic, it has a Laurent series in t on the annulus $0 < |t| < e^{-2\pi/3}$. Similarly, $\partial_{\bar{t}} g$ has a Laurent series in $\bar{t}$. Integration of these series gives a representation $g(t) = c_0 \log|t| + \sum_{n=-\infty}^{\infty} a_n t^n + \sum_{n=-\infty}^{\infty} b_n \bar{t}^n$, that converges absolutely on compact subsets of the annulus. (The terms with $\log t$ and $\log \bar{t}$ have the same factor, as g is continuous on the whole annulus.) The estimate $g(t_0) = \mathcal{O}(\log|t_0|)$ rules out the terms with negative powers of t or $\bar{t}$. The sums with t^n and $\bar{t}^n$ give square integrable functions. So the square integrability of g implies $c_0 = 0$.

In this way we conclude that a harmonic function g which is square integrable on $S_{P,2}$ has a harmonic extension to $|t| < e^{-2\pi/3}$. So, for harmonic functions, regularity at P implies the existence of a harmonic extension to P (with respect to the coordinate t). So condition iv) is equivalent to iv)′ The function $z \mapsto \tilde f(v,w,z) - \int_0^6 \tilde f(v,w,z+x')\,dx'$ extends to $\Omega \times \{ t \in \mathbb{C} : |t| < e^{-2\pi/3} \}$ as a C^∞-function in (v,w,t), that is harmonic in $t = e^{\pi i z/3}$ and holomorphic in (v,w).

The coefficient of order zero has the form

$$\begin{aligned} \int_0^6 \tilde f(v,w,z+x')\,dx' &= c_1\mu_0(P,0,\tfrac12;p(z/6)) + c_2\mu_0(P,0,-\tfrac12;p(z/6)) \\ &= c_1\,(y/6)^1 + c_2\,(y/6)^0 = -\frac{1}{2\pi}c_1 \log|t| + c_2, \end{aligned}$$

with c_1 and c_2 holomorphic on Ω.

15.6.4 *Harmonic forms on* $\mathbb{C} \smallsetminus \Lambda$. We use the holomorphic map $H : \mathfrak{H} \to \mathbb{C} \smallsetminus \Lambda$ defined in 15.2.2 by integration of the holomorphic cusp form η^4. As $H(\gamma \cdot z) = H(z) + \lambda(\gamma)$ and $\chi_{v,w}(\gamma) = e^{iv\lambda(\gamma) + iw\overline{\lambda(\gamma)}}$, each function f satisfying $f(\gamma \cdot z) = \chi_{v,w}(\gamma) f(z)$ for all $\gamma \in \bar\Gamma_{\mathrm{com}}$ is invariant under $\ker(\lambda)$, and corresponds uniquely to a function $h : \mathbb{C} \smallsetminus \Lambda \to \mathbb{C}$, by the equation $f(z) = h(H(z))$. In this way every section $f \in \mathcal{A}_0(\mathbf{c}_0;\Omega)$ may be written in the form $f(v,w,p(z)) = h(v,w;H(z))$, where $h \in C^\infty(\Omega \times (\mathbb{C} \smallsetminus \Lambda))$ satisfies

i)″ $h(v,w;u+\lambda) = e^{iv\lambda + iw\bar\lambda} h(v,w;u)$ for all $\lambda \in \Lambda$.

ii)″ $(v,w) \mapsto h(v,w;u)$ is holomorphic on Ω for each $u \in \mathbb{C} \smallsetminus \Lambda$.

iii)″ $\partial_{\bar u}\partial_u h(v,w;u) = 0$.

iv)″ There exists $a \in \mathcal{O}(W)$ such that $h(v,w;u) - a(v,w)\log|u|$ has a C^∞-extension to $\Omega \times \{ u \in \mathbb{C} : |u| < \omega \}$.

Condition iv)″ needs some explanation: ω has been introduced in Section 15.3, it is the positive number such that $\Lambda = \omega(\mathbb{Z} + e^{\pi i/3}\mathbb{Z})$. Invert $u = H(z) = \sum_{k=0}^\infty \frac{-6c(k)}{6k+1} t^{6k+1} = -6t + \ldots$, see 15.2.1, to obtain $t = -\frac16 u + \ldots$, and hence

$$-\frac{1}{2\pi}c_2 \log|t| = -\frac{1}{2\pi}c_2 \left(\log\left|\frac{u}{6}\right| + (\text{harmonic in } u) \right).$$

15.6.5 *Uniqueness.* Let us suppose that h is such a family on a non-empty open set $\Omega \subset W$, for which a as in condition iv)″ is identically zero. Then $\partial_u h$ and $\partial_{\bar u} h$ satisfy conditions i)″ and ii)″, and $u \mapsto \partial_u h(v,w;u)$ is holomorphic, whereas $u \mapsto \partial_{\bar u} h(v,w;u)$ is antiholomorphic. This implies that we have obtained a theta function and the complex conjugate of a theta function; see, e.g., [28], Ch. 18, §1. For each $(v,w) \in \Omega$ consider $f(u) = e^{i(\bar w - v)u}\partial_u h(v,w;u)$. Then $f(u+\lambda) = e^{i\bar w\lambda + iw\bar\lambda} f(u)$ for $\lambda \in \Lambda$. As $|e^{i\bar w\lambda + iw\bar\lambda}| = 1$, the holomorphic function f is bounded, hence constant. It can be non-zero only if $\lambda \mapsto e^{i\bar w\lambda + iw\bar\lambda}$ is trivial, or, equivalently, if $w \in \hat\Lambda$. This implies that $\partial_u h(v,w)$ is zero for almost all $(v,w) \in \Omega$, hence it vanishes completely. In the same way we obtain $\partial_{\bar{u}} h = 0$. Thus we are left with a constant function $u \mapsto h(v, w; u)$ for each $(v, w) \in \Omega$. Condition i)$''$ implies that $h = 0$.

This means that for any non-trivial family h satisfying the conditions i)$''$–iv)$''$, the function a in iv)$''$ does not vanish, and determines h. Any other family h_1 on the same Ω is related to h by $a_1 h = a h_1$. Note that this is a special case of the Maass-Selberg relation.

15.6.6 *Construction.* We start the construction of a family of this type. The method works for any cofinite group $\bar{\Gamma}$ with one cusp and genus 1. Here we can use the fact that the lattice is invariant under complex conjugation.

15.6.7 *Jacobi theta functions.* Consider for $u \in \mathbb{C}$, and $\tau \in \mathfrak{H}$, the *theta function*

$$\theta(u, \tau) = q^{1/8} \left(e^{u/2} - e^{-u/2} \right) \prod_{n \geq 1} \left((1 - q^n)(1 - q^n e^u)(1 - q^n e^{-u}) \right),$$

with $q = e^{2\pi i \tau}$. As a function of u it is an odd holomorphic function, with first order zeros at the points of $2\pi i(\mathbb{Z} + \mathbb{Z}\tau)$, and no other zeros. It satisfies $\theta(u + 2\pi i, \tau) = -\theta(u, \tau)$, and $\theta(u + 2\pi i\tau, \tau) = -e^{-\pi i \tau} e^{-u} \theta(u, \tau)$. The theta function θ in Section 14.1 is given by $\theta(\tau) = -i e^{\pi i \tau/4} \theta(\pi i(\tau + 1), \tau)$.

Here we are interested in the lattice $\mathbb{Z} + \mathbb{Z}\rho$, with $\rho = e^{\pi i/3} = \frac{1}{2} + \frac{i}{2}\sqrt{3}$. We put $\theta_\rho(u) = \theta(u, \rho)$, and use throughout this section $q = e^{2\pi i \rho} = -e^{-\pi\sqrt{3}}$, and $q^\alpha = e^{2\pi i \rho \alpha}$ for all $\alpha \in \mathbb{C}$.

15.6.8. In [61], §3, Zagier uses the theta function θ_ρ to build

$$F_\rho(u, v) = \frac{\theta_\rho'(0)\theta_\rho(u + v)}{\theta_\rho(u)\theta_\rho(v)}.$$

F_ρ is meromorphic in (u, v), and satisfies (among other relations)

$$\begin{aligned} F_\rho(u, v) &= -F_\rho(-u, -v) = F_\rho(v, u), \\ F_\rho(u + 2\pi i, v) &= F_\rho(u, v), \\ F_\rho(u + 2\pi i\rho, v) &= e^{-v} F_\rho(u, v). \end{aligned}$$

We define

$$h^+(v, w; u) = \frac{2\pi i}{\omega} e^{i(v+w)u} F_\rho(2\pi i u/\omega, -w\omega\sqrt{3}).$$

This function is meromorphic in (v, w, u), and satisfies

i) $h^+(v, w; u + \lambda) = e^{iv\lambda + iw\bar{\lambda}} h^+(v, w; u)$ for all $\lambda \in \Lambda$.

ii) $h^+(v + \mu, w + \bar{\mu}; u) = h^+(v, w; u)$ for all $\mu \in \frac{2\pi i}{\omega^2\sqrt{3}}\Lambda = \hat{\Lambda}$.

iii) It has singularities along the planes $u = \lambda$, with $\lambda \in \Lambda$, and the planes $w = \mu$, with $\mu \in \hat{\Lambda}$.

iv) $(v, w, u) \mapsto h^+(v, w; u) - \frac{1}{u}$ is holomorphic at $(v_0, w_0, 0)$ if $w_0 \notin \hat{\Lambda}$.

So the function $u \mapsto h^+(v,w;u)$ (with $w \notin \hat{\Lambda}$) satisfies some of the conditions i)″–iv)″. It turns out to be a useful ingredient in constructing families of functions completely satisfying those conditions.

15.6.9 *An equivariant 1-form.* $\overline{\theta_\rho(\bar{u})} = e^{-\pi i/4}\theta_\rho(u)$, and

$$h^-(v,w;u) = \overline{h^+(-\bar{w},-\bar{v};u)} = \frac{2\pi i}{\omega} e^{i(v+w)\bar{u}} F_\rho(2\pi i\bar{u}/\omega, -v\omega\sqrt{3})$$

is meromorphic in $(v,w,\bar{u})$. We define a closed 1-form $\kappa(v,w)$ on $\mathbb{C} \smallsetminus \Lambda$ by

$$\kappa(v,w;u) = h^+(v,w;u)\,du + h^-(v,w;u)\,d\bar{u}.$$

It satisfies $\kappa(v,w;u+\lambda) = e^{iv\lambda+iw\bar{\lambda}}\kappa(v,w;u)$ for all $\lambda \in \Lambda$, and the difference $\kappa(v,w;u) - u^{-1}du - \bar{u}^{-1}d\bar{u}$ has an extension to a neighborhood of $u=0$ as a closed 1-form for $v,w \notin \hat{\Lambda}$. As $d\,2\log|u| = u^{-1}du + \bar{u}^{-1}d\bar{u}$, the 1-form $\kappa(v,w)$ on $\mathbb{C} \smallsetminus \Lambda$ is exact. Each potential is annihilated by $\partial_u\partial_{\bar{u}}$ on $\mathbb{C} \smallsetminus \Lambda$. Hence, if we can pick the constant of integration in such a way that the potential transforms according to $\lambda \mapsto e^{iv\lambda+iw\bar{\lambda}}$ under $u \mapsto u+\lambda$, we have obtained a meromorphic family of functions satisfying the conditions in i)″–iv)″.

The following computations give an explicit expression for a potential. This approach was shown to me by D. Zagier. I am grateful for his permission to use it here. (I am responsible for the way the compuations are worked out here.)

15.6.10 *A primitive of h^+.* In the sequel we use $\xi = e^{2\pi iu/\omega}$, $\eta = e^{-w\omega\sqrt{3}}$. Non-integral powers of ξ are defined by $\xi^\mu = e^{2\pi i\mu u/\omega}$.

In [61], §3, at the start of the proof of the theorem, we find the expressions

$$\begin{aligned} F_\rho(2\pi iu/\omega, -w\omega\sqrt{3}) &= \frac{\xi\eta - 1}{(\xi-1)(\eta-1)} - \sum_{m,n\geq 1} \left(\xi^m\eta^n - \xi^{-m}\eta^{-n}\right) q^{mn} \\ &= \sum_{m=-\infty}^{\infty} \frac{\xi^m}{\eta q^m - 1}, \end{aligned}$$

valid for $|q| < |\xi| < 1$, $|q| < |\eta| < |q|^{-1}$, $\eta \neq 1$. The latter expression converges for $|q| < |\xi| < 1$, $\eta \notin q^{\mathbb{Z}}$. Hence it represents the function $(v,w,u) \mapsto F_\rho(2\pi iu/\omega, -w\omega\sqrt{3})$ on a larger region. For all u,v,w satisfying $0 < \operatorname{Im} u < \frac{1}{2}\sqrt{3}\omega$, $w \notin \hat{\Lambda}$:

$$h^+(v,w;u)\,du = \sum_{m=-\infty}^{\infty} \frac{\xi^{\mu+m-1}}{\eta q^m - 1} d\xi$$

with $\mu = (v+w)\omega/2\pi$.

For $\mu \notin \mathbb{Z}$ a potential of $h^+(v,w;u)\,du$ on the region $0 < \operatorname{Im} u < \frac{1}{2}\omega\sqrt{3}$ is given by

$$G_\mu(u,w) = \sum_{m=-\infty}^{\infty} \frac{1}{\mu+m}\frac{\xi^{\mu+m}}{\eta q^m - 1}.$$

It satisfies the relations

$$\begin{aligned} \overline{G_\mu(u,w)} &= G_{\bar\mu}(-\bar u,\bar w)\\ G_\mu(u+\omega,w) &= e^{2\pi i\mu}G_\mu(u,w)\\ G_\mu(u,w+2\pi i(n+m\rho)/\omega\sqrt{3}) &= G_{\mu+m}(u,w) \quad \text{for } n,m\in\mathbb{Z}. \end{aligned}$$

15.6.11 *A potential.* $h(v,w;u) = G_{(v+w)\omega/2\pi}(u,w) + G_{-(w+v)\omega/2\pi}(-\bar u,-v)$ gives a potential of $\kappa(v,w;u)$ on the region $0<\operatorname{Im}u<\frac12\omega\sqrt3$, for $v,w\notin\hat\Lambda$ and $(v+w)\omega/2\pi\notin\mathbb{Z}$. It satisfies $h(v,w;u+\omega) = e^{i(v+w)\omega}h(v,w;u)$. Any potential of $\kappa(v,w;u)$ extends as a function on $\mathbb{C}\smallsetminus\Lambda$. We need to check the property $h(v,w;u+\rho\omega) = e^{iv\omega\rho+iw\omega\bar\rho}h(v,w;u)$ on a non-empty open region containing both u and $u+\rho\omega$.

15.6.12 *Extension of* G_μ. We rewrite the series for G_μ:

$$\begin{aligned} G_\mu(u,w) &= A_\mu(\xi,\eta)+B_\mu(\xi,\eta)-C_\mu(\xi)\\ A_\mu(\xi,\eta) &= \sum_{m=-\infty}^{0}\frac{1}{m+\mu}\frac{\xi^{m+\mu}}{\eta q^m-1}\\ B_\mu(\xi,\eta) &= \sum_{m=1}^{\infty}\frac{1}{m+\mu}\frac{\xi^{m+\mu}\eta q^m}{\eta q^m-1}\\ C_\mu(\xi,\eta) &= \sum_{m=1}^{\infty}\frac{\xi^{m+\mu}}{m+\mu}\\ &= \sum_{m=1}^{\infty}\left(\frac{1}{m+\mu}-\frac1m+\frac{\xi^m}{m}+\int_0^{2\pi iu/\omega}e^{mx}(e^{\mu x}-1)\,dx\right)\\ &= -\gamma-\frac{\Gamma'(1+\mu)}{\Gamma(1+\mu)}-\log(1-\xi)+\int_0^{2\pi iu/\omega}\frac{1-e^{\mu x}}{1-e^{-x}}dx. \end{aligned}$$

To get the last step one uses $\sum_{m=1}^N\frac{1}{m+\mu} = \frac{\Gamma'(\mu+1+N)}{\Gamma(\mu+1+N)}-\frac{\Gamma'(\mu+1)}{\Gamma(\mu+1)} = \log(\mu+1+N)-\frac{\Gamma'(\mu+1)}{\Gamma(\mu+1)}+\mathcal{O}\left(\frac{1}{\mu+1+N}\right) = \log N-\frac{\Gamma'(\mu+1)}{\Gamma(\mu+1)}+\mathcal{O}\left(\frac1N\right)$, and the fact that Euler's constant is given by $\gamma = \lim_{N\to\infty}\left(\sum_{m=1}^N\frac1m-\log N\right)$.

This new expression has the advantage of representing G_μ on a larger region. In fact, $A_\mu(\xi,\eta)$ converges for $|\xi|>|q|$, $\eta\notin q^{\mathbb{Z}}$, $\mu\notin\mathbb{Z}$, and defines $A_\mu(\xi,\eta)$ as a holomorphic function in (v,w,u) on this region. For $B_\mu(\xi,\eta)$ the same holds, with the condition on ξ replaced by $|\xi|<|q|^{-1}$. In the last expression for C_μ we see that it is holomorphic on the region $\mu\notin-\mathbb{N}$, $\xi\in\mathbb{C}\smallsetminus[1,\infty)$, $|q|<|\xi|<|q|^{-1}$. In this way we have obtained the holomorphic extension of $u\mapsto G_\mu(u,w)$ to the region

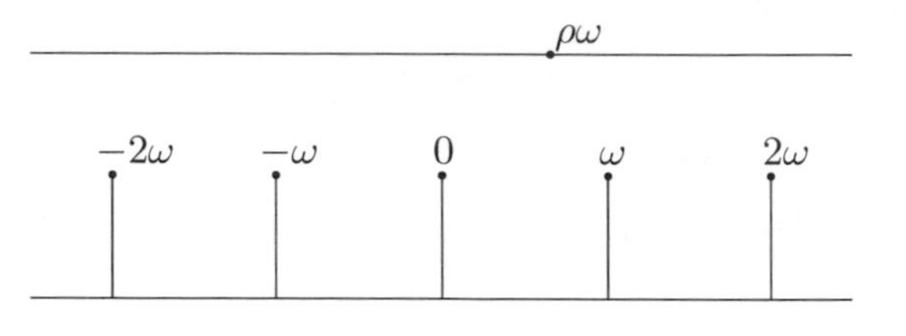

$$\left\{ u \in \mathbb{C} : |\operatorname{Im} u| < \frac{1}{2}\omega\sqrt{3}, \text{ and if } \operatorname{Im} u \le 0 \text{ then } \operatorname{Re} u \notin \omega\mathbb{Z} \right\}.$$

Note that $A_\mu(\xi,\eta)$ and $B_\mu(\xi,\eta)$ are single-valued in u, but not in ξ, due to the factor ξ^μ. The term $C_\mu(\xi)$ is even multivalued in u, as the path of integration influences the outcome. But on the simply connected region just described the three terms are single-valued in u. All terms except $\log(1-\xi) = \log(1-e^{2\pi iu/\omega})$ are holomorphic in u in a neighborhood of 0. We take the logarithm with $-\pi < \arg \le \pi$.

To obtain the behavior of $G_\mu(u,w)$ under $u \mapsto u + \rho\omega$ we take $0 < \pm \operatorname{Re} u < \frac{1}{2}\omega$, $-\frac{1}{2}\omega\sqrt{3} < \operatorname{Im} u < 0$.

$$\begin{aligned} G_\mu(u+\rho\omega, w) &= \sum_{m=-\infty}^{\infty} \frac{1}{m+\mu} \frac{(q\xi)^{m+\mu}}{\eta q^m - 1} \\ &= \eta^{-1} q^\mu \left(\sum_{m=-\infty}^{0} \frac{\xi^{m+\mu}}{m+\mu} + A_\mu(\xi,\eta) + B_\xi(\xi,\eta) \right) \\ &= \eta^{-1} q^\mu \left(A_\mu(\xi,\eta) + B_\mu(\xi,\eta) + \frac{\xi^\mu}{\mu} - C_{-\mu}(\xi^{-1}) \right). \end{aligned}$$

As $e^{iv\rho\omega + iw\bar\rho\omega} = \eta^{-1} q^\mu$, we obtain

$$G_\mu(u+\rho\omega, w) - e^{iv\rho\omega+iw\bar\rho\omega} G_\mu(u,w) = \eta^{-1} q^\mu \left(\frac{\xi^\mu}{\mu} - C_{-\mu}(\xi^{-1}) + C_\mu(\xi) \right).$$

We check that this is equal to $\pi e^{iv\rho\omega+iw\bar\rho\omega}\left(-\cot((v+w)\omega/2) \mp i\right)$ by considering the terms in $\frac{\xi^\mu}{\mu} - C_{-\mu}(\xi^{-1}) + C_\mu(\xi) = \frac{\Gamma'(1-\mu)}{\Gamma(1-\mu)} - \frac{\Gamma'(1+\mu)}{\Gamma(1+\mu)} + \log(1-\xi^{-1}) - \log(1-\xi) + \int_0^{2\pi iu/\omega} \frac{1-e^{\mu x}}{1-e^{-x}}\,dx - \int_0^{-2\pi iu/\omega} \frac{1-e^{-\mu x}}{1-e^{-x}}\,dx + \frac{\xi^\mu}{\mu}$ separately.

$\frac{\Gamma'(1-\mu)}{\Gamma(1-\mu)} - \frac{\Gamma'(1+\mu)}{\Gamma(1+\mu)} = -\frac{1}{\mu} + \pi\cot\pi\mu$ follows easily from $\Gamma(1+a)\Gamma(1-a) = \pi a/\sin\pi a$. On the regions to which we have restricted u the quantity $\log(1-\xi^{-1}) - \log(1-\xi)$ is holomorphic, and turns out to be equal to $\log(-\xi^{-1}) = \pm\pi i - 2\pi i\frac{u}{\omega}$. The two integrals give $\int_0^{2\pi iu/\omega}\left(\frac{1-e^{\mu x}}{1-e^{-x}} + \frac{1-e^{\mu x}}{1-e^{x}}\right)dx = \int_0^{2\pi iu/\omega}(1-e^{\mu x})\,dx = 2\pi i\frac{u}{\omega} - \frac{\xi^\mu}{\mu} + \frac{1}{\mu}$. Together these results yield the quantity we have indicated.

Now we turn to $G_\mu(-\bar{u},-v)$. We obtain the equality of $G_{-\mu}(-\overline{(u+\rho\omega)},-v)$ and $e^{2\pi i\mu}G_{-\mu}(-\bar{u}+\rho\omega,-v)$ from $-\bar{\rho}=\rho-1$. We apply the result obtained above to get

$$\begin{aligned} G_{-\mu}&\left(-\overline{(u+\rho\omega)},-v\right)-e^{iv\rho\omega+iw\bar{\rho}\omega}G_{-\mu}(-\bar{u},-v)\\ &= e^{i(v+w)\omega}\left(G_{-\mu}(-\bar{u}+\rho\omega,-v)-e^{-iw\rho\omega-iv\bar{\rho}\omega}G_{-\mu}(-\bar{u},-v)\right)\\ &= \pi e^{iv\rho\omega+iw\bar{\rho}\omega}\left(-\cot((v+w)\omega/2)\mp i\right).\end{aligned}$$

This implies that the extension of $u\mapsto h(v,w;u)$ to $\mathbb{C}\setminus\Lambda$ has the transformation behavior

$$h(v,w;u+\lambda)=e^{iv\lambda+iw\bar{\lambda}}h(v,w;u)\quad\text{for all }\lambda\in\Lambda.$$

For u satisfying $|\operatorname{Im}u|<\frac{1}{2}\omega\sqrt{3}$, and if $\operatorname{Im}u\le 0$ then $\operatorname{Re}u\notin\omega\mathbb{Z}$, we have obtained a rather explicit description of $h(v,w;u)$. This region contains a fundamental domain for the action of Λ in $\mathbb{C}$. Hence we have an explicit description everywhere.

The relation $G_\mu(u,w+2\pi i(n+m\rho)/\omega\sqrt{3})=G_{\mu+m}(u,w)$ implies that

$$h(v+\nu,w+\bar{\nu};u)=h(v,w;u)\quad\text{for all }\nu\in\hat{\Lambda}.$$

15.6.13 *Singularities.* This construction gives $h(v,w;u)$ for (v,w) satisfying $v,w\notin\hat{\Lambda}$ and $\mu=(v+w)\omega/2\pi\notin\mathbb{Z}$. The singularities along $v=v_0\in\hat{\Lambda}$ and along $w=w_0\in\hat{\Lambda}$ are already present in the 1-form κ. We shall show that h has no singularity along $v+w=\frac{2\pi l}{\omega}$, with $l\in\mathbb{Z}$. This cannot be trivially concluded from the behavior of κ. The choice of a potential implies choosing a constant of integration, which depends on (v,w), and might produce additional singularities.

15.6.14 *Singularities in (v,w).* Fix $u\in\mathbb{C}$ with $|\operatorname{Im}u|<\frac{1}{2}\omega\sqrt{3}$, $|\operatorname{Re}u|<\omega$, such that $\xi=e^{2\pi iu/\omega}\notin[1,\infty)$. As we already have obtained the transformation behavior of h under Λ, it suffices to study the singularities in (v,w) for such u. Consider $v,w\notin\hat{\Lambda}$, $\mu=(v+w)\omega/2\pi\notin\mathbb{Z}$; put $\eta_1=e^{-w\omega\sqrt{3}}$, $\eta_2=e^{v\omega\sqrt{3}}$. From the expression for G_μ in 15.6.12 we obtain

$$\begin{aligned} h(v,w;u) &= h_0(u)-h_1(v,w;u)+h_2(v,w;u)\\ h_0(u) &= \log\left(1-e^{2\pi iu/\omega}\right)+\log\left(1-e^{-2\pi i\bar{u}/\omega}\right)\\ h_1(v,w;u) &= \int_0^{2\pi iu/\omega}\frac{1-e^{\mu x}}{1-e^{-x}}\,dx+\int_0^{-2\pi i\bar{u}/\omega}\frac{1-e^{-\mu x}}{1-e^{-x}}\,dx\\ h_2(v,w;u) &= -\sum_{m=1}^{\infty}\left(\frac{1}{m+\mu}+\frac{1}{m-\mu}-\frac{2}{m}\right)\\ &\quad +A_\mu(\xi,\eta_1)+B_\mu(\xi,\eta_1)+A_{-\mu}(\bar{\xi},\eta_2)+B_{-\mu}(\bar{\xi},\eta_2).\end{aligned}$$

The term $h_0(u)$ is the only one that is not harmonic in u in a neighborhood of 0. The contribution of h_1 is C^∞ in (v,w,u), holomorphic in $(v,w) \in \mathbb{C}^2$, and harmonic in u for $|\operatorname{Re} u| < \omega$. For singularities in (v,w) we have to turn to $h_2(v,w;u)$.

The idea is simple: fix (v_0,w_0), and consider the series expansions of A and B, together with the additional sum in the expression for h_2. If we take away the terms with a singularity at (v_0,w_0), the remainder represents a harmless function. Here *harmless* means smooth in (v,w,u), holomorphic in (v,w) at (v_0,w_0), and harmonic in u on $|\operatorname{Im} u| < \frac{1}{2}\omega\sqrt{3}$, $|\operatorname{Re} u| < \omega$.

15.6.15 *No singularities at* $\mu \in \mathbb{Z}$. Let $v_0, w_0 \notin \hat{\Lambda}$, $v_0 + w_0 = 2\pi l/\omega$, $l \in \mathbb{Z}$. For $l = 0$ we obtain

$$\begin{aligned} h_2(v,w;u) &= \frac{1}{\mu}\left(\frac{e^{2\pi i u\mu/\omega}}{\eta_1 - 1} - \frac{e^{2\pi i \bar{u}\mu/\omega}}{\eta_2 - 1}\right) + (\text{harmless}) \\ &= \frac{1}{\mu}\left(\frac{1}{\eta_1 - 1} - \frac{1}{\eta_2 - 1}\right) + (\text{harmless}) \\ &= \frac{2\eta_1(e^{2\pi\mu\sqrt{3}} - 1)}{\mu(\eta_1 - 1)(\eta_2 - 1)} + (\text{harmless}) = (\text{harmless}). \end{aligned}$$

We have used that $\eta_1, \eta_2 \notin q^{\mathbb{Z}}$ for (v,w) near (v_0,w_0), and that $\eta_2\eta_1^{-1} = e^{2\pi\mu\sqrt{3}}$. The cases $l > 0$ and $l < 0$ are left to the reader. Thus $h(v,w;u)$ is holomorphic in (v,w) on $\mathbb{C}^2 \setminus \left((\hat{\Lambda} \times \mathbb{C}) \cup (\mathbb{C} \times \hat{\Lambda})\right)$.

15.6.16 *Singularities along lines given by* $\hat{\Lambda}$. Let $v_0 \notin \hat{\Lambda}$, $\lambda = \frac{2\pi i}{\omega\sqrt{3}}(n + l\rho)$. We show that $(v,w) \mapsto (w - \lambda)h(v,w;u)$ is holomorphic at (v_0, λ) with a non-zero value. At $(v,w) = (v_0,\lambda)$ we have $\mu \notin \mathbb{Z}$, $\eta_2 \notin q^{\mathbb{Z}}$, but $\eta_1 = q^{-l}$. Near (v_0,λ) we obtain

$$\begin{aligned} h_2(v,w;u) &= \frac{1}{\mu + l} \cdot \frac{e^{2\pi i u(\mu+l)/\omega}}{\eta_1 q^l - 1} \begin{Bmatrix} 1 & \text{if } l \le 0 \\ \eta_1 q^l & \text{if } l \ge 1 \end{Bmatrix} + (\text{harmless}) \\ &= \frac{2\pi}{e^{-(w-\lambda)\omega\sqrt{3}} - 1} \cdot \frac{e^{i((v+w)\omega + 2\pi l)u/\omega}}{(v+w)\omega + 2\pi l} \\ &\quad \cdot \begin{Bmatrix} 1 & \text{if } l \le 0 \\ e^{-(w-\lambda)\sqrt{3}} & \text{if } l \ge 1 \end{Bmatrix} + (\text{harmless}) \\ &= \frac{2\pi}{e^{-(w-\lambda)\omega\sqrt{3}} - 1} \cdot \frac{e^{i(v-\bar{\lambda})u}}{(v - \bar{\lambda})\omega} + (\text{harmless}) \\ &= -\frac{2\pi}{\omega^2\sqrt{3}} \cdot \frac{e^{i(v-\bar{\lambda})u}}{v - \bar{\lambda}} \cdot \frac{1}{w - \lambda} + (\text{harmless}). \end{aligned}$$

Hence $(v,w) \mapsto (w - \lambda)h(v,w;u)$ is holomorphic at (v_0,λ) with value

$$-\frac{2\pi}{\omega^2\sqrt{3}} \cdot \frac{e^{i(v_0-\bar{\lambda})u}}{v_0 - \bar{\lambda}}.$$

Similarly the function $(v, w) \mapsto (v - \lambda)h(v, w; u)$ is holomorphic at (λ, w_0), for $\lambda \in \hat{\Lambda}$ and $w_0 \notin \hat{\Lambda}$, with value

$$-\frac{2\pi}{\omega^2\sqrt{3}} \cdot \frac{e^{i(w_0-\bar{\lambda})\bar{u}}}{w_0 - \bar{\lambda}}.$$

15.6.17 *The singularities at intersection points.* The most interesting points are $(v_0, w_0) = (\lambda, \nu) \in \hat{\Lambda}^2$. Working as in the previous case leads to a complicated computation. The answer can be obtained more easily from the previous cases.

Let $(\lambda, \nu) \in \hat{\Lambda}$. We transfer the concepts of Section 12.2 to this situation. There is a function ψ, holomorphic near (λ, ν), such that $\psi \cdot h$ is a primitive germ at (λ, ν). This function ψ is well defined up to a unit in $\mathcal{O}_{(\lambda,\nu)}$. Remark 12.2.6 shows that the germ of $N(\psi)$ at (λ, ν) is the germ of the zero set of $(v, w) \mapsto (v-\lambda)(w-\nu)$. We conclude from Lemma 12.2.5 that we can take $\psi(v, w) = (v - \lambda)(w - \nu)$. Near (λ, ν) we write

$$\begin{aligned}(\psi \cdot h)(v, w, u)& \\ = \quad & A(u) + (v - \lambda)B(v, u) \\ & \quad + (w - \nu)C(w, u) + (v - \lambda)(w - \nu)D(v, w, u).\end{aligned}$$

Along the lines $w = \nu$, respectively $v = \lambda$, we obtain

$$\begin{aligned}A(u) + (v - \lambda)B(v, u) &= \frac{-2\pi}{\omega^2\sqrt{3}} \frac{v - \lambda}{v - \bar{\nu}} e^{i(v-\bar{\nu})u} \\ A(u) + (w - \nu)C(w, u) &= \frac{-2\pi}{\omega^2\sqrt{3}} \frac{w - \nu}{w - \bar{\lambda}} e^{i(w-\bar{\lambda})\bar{u}}.\end{aligned}$$

For $\nu = \bar{\lambda}$ we conclude that h has determinate type at (λ, ν), and $(\psi \cdot h)(\lambda, \nu, u) = \frac{-2\pi}{\omega^2\sqrt{3}}$. For $\nu \neq \bar{\lambda}$ we obtain $A(u) = 0$, $B(v, u) = \frac{-2\pi}{\omega^2\sqrt{3}} \frac{e^{i(v-\bar{\nu})u}}{v - \bar{\nu}}$, and $C(w, u) = \frac{-2\pi}{\omega^2\sqrt{3}} \frac{e^{i(w-\bar{\lambda})\bar{u}}}{w - \bar{\lambda}}$. Now h is of indeterminate type at (λ, ν).

15.6.18. We have shown that conditions i)″–iii)″ in 15.6.4 hold for h on $\mathbb{C}^2$ away from $(\hat{\Lambda} \times \mathbb{C}) \cup (\mathbb{C} \times \hat{\Lambda})$, and also for suitable multiples of h at the singularities in $(\hat{\Lambda} \times \mathbb{C}) \cup (\mathbb{C} \times \hat{\Lambda})$.

As $\log(1 - e^{2\pi i u/\omega}) + \log(1 - e^{-2\pi i \bar{u}/\omega}) - 2\log|u|$ is harmonic at $u = 0$, the computations we have performed show that h satisfies condition iv)″ as well. The same holds for the multiples of h near the points where h has a singularity in (v, w).

15.6.19 *Constant term.* Before formulating the results of the computations in terms of functions on the upper half plane, we compute the constant term in the u-expression of $h(v, w; u) - 2\log|u|$ at $u = 0$.

Put $\gamma_0(v,w) = \lim_{u\to 0} h_2(v,w;u)$. For $v,w \notin \hat\Lambda$, and $\mu = (v+w)\omega/2\pi$, $q = -e^{-\pi\sqrt{3}}$, $\eta_1 = e^{-w\omega\sqrt{3}}$, $\eta_2 = e^{v\omega\sqrt{3}}$:

$$\begin{aligned}
&\gamma_0(v,w) \qquad\qquad (15.4)\\
&= -\sum_{m=1}^{\infty}\left(\frac{1}{m+\mu}+\frac{1}{m-\mu}-\frac{2}{m}\right)+\frac{1}{\mu}\left(\frac{1}{\eta_1-1}-\frac{1}{\eta_2-1}\right)\\
&\quad+\sum_{m=1}^{\infty}\left(\frac{1}{m+\mu}\cdot\frac{\eta_1 q^m}{\eta_1 q^m-1}-\frac{1}{m+\mu}\cdot\frac{1}{\eta_2 q^{-m}-1}\right.\\
&\qquad\qquad\left.+\frac{1}{\mu-m}\cdot\frac{1}{\eta_1 q^{-m}-1}-\frac{1}{\mu-m}\cdot\frac{\eta_2 q^m}{\eta_2 q^m-1}\right)\\
&= \sum_{m=-\infty}^{\infty}\left(\frac{1}{\mu+m}\cdot\frac{1}{\eta_1 q^m-1}-\frac{1}{\mu+m}\cdot\frac{1}{\eta_2 q^{-m}-1}+\alpha(m)\right),
\end{aligned}$$

with $\alpha(0)=0$, $\alpha(m)=\frac{1}{|m|}$ if $m\in\mathbb{Z}\smallsetminus\{0\}$.

$$\begin{aligned}
&\lim_{u\to 0}\left(\log\frac{1-e^{2\pi i u/\omega}}{u}+\log\frac{1-e^{-2\pi i\bar u/\omega}}{\bar u}\right) = 2\log(2\pi)-2\log\omega,\\
&\lim_{u\to 0} h_1(v,w;u)=0.
\end{aligned}$$

The constant term at $u=0$ is equal to

$$2\log(2\pi)-2\log\omega+\gamma_0(v,w) \qquad \text{for } v,w\notin\hat\Lambda.$$

15.6.20 *Restriction of the Eisenstein family.* $(v,w,z)\mapsto -\frac{1}{4\pi}h(v,w;H(z))$ defines a meromorphic family in $(\mathcal{M}\otimes_{\mathcal{O}}\mathcal{A}_0(\mathbf{c}_0))(\mathbb{C}^2)$ that is holomorphic on $\mathbb{C}^2\smallsetminus\big((\hat\Lambda\times\mathbb{C})\cup(\mathbb{C}\times\hat\Lambda)\big)$. For each point in its polar set we have found a holomorphic ψ such that multiplication by ψ gives a primitive multiple $\tilde e$. In the table we denote by v_0 and w_0 elements of $\mathbb{C}\smallsetminus\hat\Lambda$, and by λ and ν elements of $\hat\Lambda$.

$(\tilde v,\tilde w)$		$\psi(v,w)$	$\tilde e(\tilde v,\tilde w;z)$
(λ,w_0)		$v-\lambda$	$\frac{1}{2\omega^2\sqrt3}\cdot e^{i(w_0-\overline{\lambda})\overline{H(z)}}(w_0-\overline{\lambda})^{-1}$
(v_0,ν)		$w-\nu$	$\frac{1}{2\omega^2\sqrt3}\cdot e^{i(v_0-\overline{\nu})H(z)}(v_0-\overline{\nu})^{-1}$
(λ,ν)	$\lambda\neq\overline{\nu}$	$(v-\lambda)(w-\nu)$	0
(λ,ν)	$\lambda=\overline{\nu}$	$(v-\lambda)(w-\nu)$	$\frac{1}{2\omega^2\sqrt3}$

To see that this family is the restriction $(v,w)\mapsto E(v,w,\frac12)$ of the Eisenstein family in weight zero, we have to show that its zero order Fourier term at P is given by $\mu_0(P,0,\frac{1}{2})+(\text{something})\cdot\mu_0(P,0,-\frac{1}{2})$. For $(v,w)\in(\mathbb{C}\smallsetminus\hat{\Lambda})^2$:

$$\begin{aligned}
&-\frac{1}{4\pi}h(v,w;H(6z))\\
&= -\frac{1}{4\pi}\left(2\log|H(6z)|+2\log(2\pi)-2\log\omega+\gamma_0(v,w)+\mathcal{H}(H(6z))\right)\\
&= -\frac{1}{2\pi}\left(\log 6-2\pi y+\mathcal{H}(e^{2\pi iz})\right)+\frac{1}{2\pi}\log\frac{\omega}{2\pi}-\frac{1}{4\pi}\gamma_0(v,w)+\mathcal{H}(e^{2\pi iz})\\
&= y+\frac{1}{2\pi}\log\frac{\omega}{12\pi}-\frac{1}{4\pi}\gamma_0(v,w)+\mathcal{H}(e^{2\pi iz}),
\end{aligned}$$

where $\mathcal{H}(t)$ denotes 'harmonic in t on a neighborhood of 0, with value 0 at $t=0$'.

This gives the desired identification, and also the conclusion that the restriction $(v,w)\mapsto C^0(v,w,\frac{1}{2})$ is given by $(v,w)\mapsto\frac{1}{2\pi}\log\frac{\omega}{12\pi}-\frac{1}{4\pi}\gamma_0(v,w)$.

15.6.21 *Primitive multiples.* If we consider the definitions in Section 12.2, we conclude that the table above gives primitive multiples of $(v,w)\mapsto E(v,w,\frac{1}{2})$ at various points. This family of automorphic forms is of indeterminate type at the points $(\lambda,\nu)\in\hat{\Lambda}^2$ with $\lambda\neq\overline{\nu}$, and has determinate type elsewhere.

Consider, in the indeterminate case, the local curves j_1 and j_2 given by $j_1: b\mapsto(\lambda+b,\nu)$ and $j_2: b\mapsto(\lambda,\nu+b)$. Denote the primitive multiple by $f(v,w)=(v-\lambda)(w-\nu)E(v,w,\frac{1}{2})$. We find

$$\begin{aligned}
f\circ j_1(b;z) &= \frac{1}{2\omega^2\sqrt{3}}b\frac{e^{i(b+\lambda-\overline{\nu})H(z)}}{b+\lambda-\overline{\nu}}\\
f[j_1] &= \frac{1}{2\omega^2\sqrt{3}}\frac{1}{\lambda-\overline{\nu}}e^{i(\lambda-\overline{\nu})H(z)},
\end{aligned}$$

and similarly

$$f[j_2]=\frac{1}{2\omega^2\sqrt{3}}\frac{1}{\nu-\overline{\lambda}}e^{\overline{i(\nu-\overline{\lambda})H(z)}}.$$

One may check that the value set $s\{(\lambda,\nu)\}$ consists of all non-zero linear combinations of $f[j_1]$ and $f[j_2]$. In particular, this value set contains the cusp form $(\lambda-\overline{\nu})f[j_1]-(\nu-\overline{\lambda})f[j_2]$.

15.6.22 *Singularity of E at* $(0,0,\frac{1}{2})$. In 15.5.9 we defined the function $a_{1/2}$ such that $\tilde{E}:(v,w,s)\mapsto(s-\frac{1}{2}-a_{1/2}(v,w))E(v,w,s)$ gives a primitive germ at $(0,0,\frac{1}{2})$. This function describes non-cuspidal exceptional eigenvalues of continuous series type.

From 11.2.7 it follows that $\tilde{E}(0,0,\frac{1}{2})=\lim_{s\to1/2}(s-\frac{1}{2})E(0,0,s)=\frac{1}{2\pi}$. The equality $\tilde{E}(0,0,\frac{1}{2})=\frac{1}{4\pi}\lim_{(v,w)\to(0,0)}a_{1/2}(v,w)h(v,w,H(z))$ implies $a_{1/2}(v,w)=-\frac{\omega^2\sqrt{3}}{\pi}vw\sum_{p,q\geq0}B_{p,q}v^pw^q$, with $B_{0,0}=1$. As $a_{1/2}(v,\bar{v})\leq0$ for v near 0, we have $B_{p,q}=\overline{B_{q,p}}$.

We shall show that $B_{1,0} = B_{0,1} = 0$. We have, modulo quadratic terms in v and w:

$$\begin{aligned}
&\tilde{E}(v, w, \frac{1}{2}; z) \\
&\equiv \frac{\omega^2\sqrt{3}}{\pi} (B_{0,0} + vB_{1,0} + wB_{0,1}) \frac{1}{2\omega^2\sqrt{3}} \left(1 + e^{ivH(z)} + e^{iw\overline{H(z)}}\right) \\
&\equiv \frac{1}{2\pi} \left(1 + v(B_{1,0} + i\,H(z)) + w(B_{0,1} + i\,\overline{H(z)})\right).
\end{aligned}$$

Let us denote by $\tilde{C}^n$ the Fourier coefficients of $\tilde{E}$.

$$\tilde{C}^n(v, w, \frac{1}{2}) \equiv \begin{cases} \frac{1}{2\pi}(1 + vB_{1,0} + wB_{0,1}) & \text{if } n = 0 \\ v\frac{-3ic(k)}{\pi n} & \text{if } n = 6k+1,\ k \geq 0 \\ w\frac{3ic(k)}{\pi n} & \text{if } n = -(6k+1),\ k \geq 0 \\ 0 & \text{in all other cases.} \end{cases}$$

15.6.23 *Dirichlet series with the C-symbol.* Use

$$\tilde{C}^n(v, w, \frac{1}{2}) \equiv \lim_{s \to 1/2} \tilde{C}^n(0, 0, s) + v \lim_{s \to 1/2} \partial_v \tilde{C}^n(0, 0, s) + w \lim_{s \to 1/2} \partial_w \tilde{C}^n(0, 0, s).$$

With 15.5.10 and 15.5.11, this gives

$$\tilde{C}^n(v, w, \frac{1}{2}) \equiv \frac{\pi}{72} \lim_{u \to 2} (u - 2) \left(\Phi^n(0, 0, u) - i\omega v \Phi^n(0, 1, u) - i\omega w \Phi^n(1, 0, u)\right).$$

As $\Phi^n(a, b) = 0$ unless $a - b + n \equiv 0 \bmod 6$, we obtain $B_{1,0} = B_{0,1} = 0$ from the case $n = 0$. For $n \neq 0$, the $\Phi^n(1, 0)$ and $\Phi^n(0, 1)$ are holomorphic at $u = 2$, except in the cases $n = \pm(6k+1)$ with $k \geq 0$; the $+$ is for $\Phi^n(1, 0)$, and the $-$ for $\Phi^n(0, 1)$. In these cases the order of the pole is one, and the residue equals $\pm 216c(k)/\pi^2\omega n$; the $c(k)$ are the Fourier coefficients of η^4, see 15.2.1.

It may be interesting to consider higher expansions for the case $n = 0$. Then we need to expand $\gamma_0(v, w)$. But there is no hope to obtain $a_{1/2}$ completely in this way.

15.7 Maass forms and singularities of the Eisenstein family

Finally we consider perturbation of cusp forms in weight zero. We discuss the relation with singularities of the Eisenstein family; for the modular case this has been discussed in Proposition 2.19 of [7].

Phillips and Sarnak, [45], have studied the persistence of cusp forms under perturbation of the group, and have given a criterion in terms of values of L-functions. We do not pursue this question for perturbation of the multiplier system.

15.7.1 *Maass forms of weight zero.* Here we mean by a *Maass form* a non-zero element of the space $S_0(\chi_{v,\bar{v}}, \mathbf{c}_0, s)$, for some $v, s \in \mathbb{C}$, with $\mathbf{c}_0(P) = \{0\}$, $\mathcal{P} = \{P\}$.

We have seen in Proposition 15.5.6 that this can occur only if $\frac{1}{4} - s^2 \geq \pi^2/84$. For the congruence case $v \in \hat{\Lambda}_6$, see 15.4.5, the trace formula of Selberg shows that there are lots of Maass forms. For $\chi_{v,\bar{v}} = 1$ this follows from the modular case, see [21], formula (2.11) on p. 511.

For general v there is no explicit knowledge of the scattering matrix that allows us to separate the cuspidal contribution to the spectral density from the contribution of the continuous spectrum; see [21], Proposition 13.7 on p. 207. As far as I know, nothing has been proved for general v.

15.7.2 *Families of cusp forms.* One may ask whether all, or some, Maass forms are obtained as linear combinations of values of families of cusp forms. If $f \in S_0(\chi_{v_0,\overline{v_0}}, \mathbf{c}_0, s_0)$ then f might be a linear combination of values $g(v_0, \overline{v_0}, 0)$ of $g \in \mathcal{S}_l(\mathbf{c}_0, \Omega)$, with Ω a neighborhood of $(v_0, \overline{v_0}, 0)$ in $\mathbb{C}^3$. We consider $\mathbb{C}^3$ as a parameter space with $\chi(v, w, r) = \chi_{v,w} \cdot \chi_r$ and s some holomorphic function on Ω with $s(v_0, \overline{v_0}, 0) = s_0$; we take $l(v, w, r) = r$ and $\mathbf{c}_0(P) = \{l\}$.

There is evidence that Maass forms are very special objects, bound to arithmetical situations; see, e.g., [46] and [13]. For the trivial character, this would imply that $\mathcal{S}_l(\mathbf{c}_0, \Omega)$ is the zero sheaf for all parameter spaces Ω through the point $(0, 0, 0, s_0) \in \mathcal{V} \times \mathbb{C}$ with dimension larger than zero. If this is true, Proposition 15.7.9 below would suggest that all Maass forms for the trivial character are due to singularities of the Eisenstein family.

15.7.3 *One-dimensional perturbation.* Consideration of higher dimensional parameter spaces seems much too difficult. We restrict ourselves in this section to a one-dimensional situation.

We choose $\varphi_0 \in \mathcal{V}_r$, and apply Chapters 8 and 9 with $\chi_0 = \chi_{v_0,\overline{v_0}}$ and $V_r = \mathbb{R} \cdot \varphi_0$, $V = \mathbb{C} \cdot \varphi_0$. We suppose that $S_0(\chi_0, \mathbf{c}_0, s_0) \neq \{0\}$. To avoid technical complications we assume that $s_0 \in i\mathbb{R}$, $s_0 \neq 0$.

15.7.4 *Remark.* For $\varphi_0 = \alpha$ and $v_0 = 0$, we know that there are cusp forms that occur in families. Indeed, in [7], Proposition 2.14, it is shown that each of the spaces $S_0(\tilde{\Gamma}_{\mathrm{mod}}, \chi_m, \mathbf{c}_0, s_0)$, with $m = 2, 4, 6, 8, 10$, has an orthonormal basis consisting of the values at 0 of families of cusp forms depending on the weight. Moreover, for each of these m there are infinitely many s_0 for which this space is non-zero (see Theorem 6.5.5). As these spaces $S_0(\tilde{\Gamma}_{\mathrm{mod}}, \chi_m, \mathbf{c}_0, s_0)$ are contained in $S_0(\tilde{\Gamma}_{\mathrm{com}}, 1, \mathbf{c}_0, s_0)$, this gives examples for $\tilde{\Gamma}_{\mathrm{com}}$.

15.7.5 *Pseudo Casimir operator.* In Lemma 9.1.6 and Remark 9.1.7 we have seen that there is an open neighborhood of 0 in $V = \mathbb{C} \cdot \varphi_0$, such that on V_0 the family $\mathfrak{s}$ of sesquilinear forms in ${}^aH$ satisfies the conditions for a holomorphic family of forms of type (a) in the sense of [25], Ch. VII, §4.2. As $\dim V = 1$, we may apply the theory in §4.2 of *loc. cit.* to see that $\varphi \mapsto {}^aA(\varphi)$ is a selfadjoint holomorphic family of operators in ${}^aH$, defined on V_0. It has compact resolvent, see Proposition 9.2.2.

Let J be an open interval containing 0 such that $r\varphi_0 \in V_r$ for all $r \in J$. Theorem 3.9 and Remark 4.22 in Ch. VII of [25] show that all eigenvalues and eigenvectors of ${}^aA(\varphi)$ with $\varphi \in J \cdot \varphi_0$ occur in real analytic families. This implies that there is a finite number of holomorphic functions $\lambda_1, \ldots, \lambda_m$ on a neighborhood Ω of J, with real values on J and with $\lambda_k(0) = \frac{1}{4} - s_0^2$, and for each k a finite number of L^2-holomorphic maps $p_{k,1}, \ldots, p_{k,n_m} : \Omega \to {}^aH$, such that ${}^aA(r\varphi_0)p_{k,l}(r) = \lambda_k(r)p_{k,l}(r)$ for all $r \in \Omega$, $k = 1, \ldots, n_m$, and such that $p_{1,1}(0), \ldots, p_{1,n_1}(0), p_{2,1}(0), \ldots, p_{m,n_m}(0)$ form an orthonormal basis of $\ker({}^aA(0) - \frac{1}{4} + s_0^2)$. Moreover, all eigenvalues $\frac{1}{4} - s^2$ of ${}^aA(r\varphi_0)$ with (r, s) near $(0, s_0)$ are obtained in this way.

The λ_k and $p_{k,l}$ need not be holomorphic on the whole of $\{ r \in \mathbb{C} : r\varphi_0 \in V_0 \}$. As we consider a neighborhood of the eigenvalue $\frac{1}{4} - s_0^2$ only, there is a common domain Ω. If we would take a more global point of view, the intersection of the domains on which the various eigenfamilies are holomorphic would not necessarily be open.

The λ_k may be supposed to be different holomorphic functions, with different values at all points except 0. The eigenvectors $p_{k,l}$ may be chosen to be orthonormal at each real point $r \in J$.

15.7.6 *Families of automorphic forms.* To obtain a relation with automorphic forms, we like to use $s = \sqrt{\frac{1}{4} - \lambda}$ as the spectral parameter. This is no problem: we have assumed $s_0 \neq 0$.

We choose roots $s_k = \sqrt{\frac{1}{4} - \lambda_k}$ for $k = 1, \ldots, m$. Take $\chi_k(r) = \chi_0 \cdot \mathbf{exp}\,(r\varphi_0)$ to obtain parameter spaces $\Omega_k = (\Omega, \chi_k, s_k)$ for $k = 1, \ldots, m$. We work with the weight $l : w \mapsto \frac{1}{\pi} r\varphi_0(\zeta)$ and with $\mathbf{c}_0(P) = \{l\}$ (note that $\mathcal{C}_P = \mathbf{wt}$ for $\tilde{\Gamma}_{\text{com}}$). Proposition 8.3.6 and the Lemmas 8.5.2, 8.5.4, and 9.2.3 show that the $p_{k,l}$ are also holomorphic as maps $\Omega \to {}^aD$, and that there are $f_{k,l} \in \mathcal{A}_l(\mathbf{c}_0, \Omega_k)$ such that $p_{k,l}(r) = {}^{(a)}(e^{-irt_{\varphi_0}} f_{k,l}(r))$.

15.7.7 *Arranging the Fourier coefficients.* The $F_{P,l} f_{k,l}$ are multiples of $\beta_{P,l}$. (Consult 7.6.2 for the definition of $\beta_{P,l}$. Lemma 8.5.2 explains why $\alpha_{P,l}$ does not occur.) If $F_{P,l} f_{k,l} \neq 0$ for some l, then we perform a linear transformation of the $f_{k,l}$ with k fixed and $l = 1, \ldots, n_k$, such that $F_{P,l} f_{k,l} = 0$ for $l = 2, \ldots, n_k$. The coefficients of this transformation may be chosen holomorphic in r on a neighborhood of 0, and for $r = 0$ the transformation may be assumed to be unitary. See, e.g., [5], Lemma 7.14, for a proof of this fact based on results in Chapter II of [25].

After that transformation, and after going over to a smaller Ω, we renumber the λ_k and $f_{k,l}$ to obtain:

- $h_k \in \mathcal{A}_l(\mathbf{c}_0, \Omega_k)$ for $k = 1, \ldots, \tilde{m} \leq m$ with $F_{P,l} h_k$ not the zero function,
- $g_{k,l} \in \mathcal{S}_l(\mathbf{c}_0, \Omega_k)$ for $k = 1, \ldots, m$, $l = 1, \ldots, \tilde{n}_k$,

such that the ${}^{(a)}h_k(0)$ and the $g_{k,l}(0)$ form an orthonormal basis of $\ker({}^aA(0) - \frac{1}{4} + s_0^2)$. (Truncation has no effect on the cusp forms $g_{k,l}(0)$.)

Here one sees an advantage of working with one cusp. With more than one cusp this step causes complications.

The h_k and their number $\tilde{m}$ depend on the truncation data. But the families of cusp forms $g_{k,l}$ are independent of the choice of $a(P)$ and of the transformation function t_{φ_0}. The $g_{k,l}$ might be absent, whereas the family ${}^aA$ is bound to have holomorphic families of eigenfunctions.

15.7.8 *Decomposition of the space of cusp forms.* We define $S_0^f(\chi_0, \mathbf{c}_0, s_0 \,|\, \varphi_0)$ as the linear subspace of $S_0(\chi_0, \mathbf{c}_0, s_0)$ spanned by the $g_{k,l}(0)$; its orthogonal complement we call $S_0^e(\chi_0, \mathbf{c}_0, s_0 \,|\, \varphi_0)$.

The space $S_0^f(\chi_0, \mathbf{c}_0, s_0 \,|\, \varphi_0)$ is the space of Maass forms that are generated by families of cusp forms along $\chi_0 \cdot \mathbf{exp}(\mathbb{C}\varphi_0)$. The orthogonal complement $S_0^e(\chi_0, \mathbf{c}_0, s_0 \,|\, \varphi_0)$ is a subspace of the kernel of ${}^aA(0) - \frac{1}{4} + s_0^2$; hence it is contained in the space spanned by the $h_k(0)$.

This decomposition may be different for various directions φ_0 of the parameter space.

15.7.9 Proposition. *Let $E : (r, s) \mapsto E_l(J(0), P; 2\,\mathrm{Re}(v_0\lambda) + r\varphi_0, s)$ be the restriction of the Eisenstein family to $\Omega \times \mathbb{C}$.*

The space $S_0^e(\chi_0, \mathbf{c}_0, s_0 \,|\, \varphi_0)$ is non-zero if and only if E is not holomorphic at $(0, s_0)$.

Remarks. This result generalizes Proposition 2.19 in [7]. It means that the occurrence of cusp forms that are not values of families of cusp forms along the line $2\,\mathrm{Re}(v_0\lambda) + \mathbb{C} \cdot \varphi_0$ is equivalent to the occurrence of singularities of the Eisenstein family along $(2\,\mathrm{Re}(v_0\lambda) + \mathbb{C} \cdot \varphi_0) \times \mathbb{C}$.

Proof. We start with the easy implication. Suppose that E is holomorphic at $(0, s_0)$. Then $F_{P,l}E$ is holomorphic at $(0, s_0)$ as well. The value $E(0, s_0)$ is equal to $\dot{E}_l(J(0), P; 2\,\mathrm{Re}(v_0\lambda); s_0)$. As $E(0, s_0) \neq 0$, it has a non-zero Fourier coefficient of order zero, and we may choose $a(P)$ such that $F_{P,l}E(0, s_0; a(P)) \neq 0$. Suppose that $\tilde{m} > 0$ for this choice of $a(P)$. Let $1 \leq q \leq \tilde{m}$. The Maass-Selberg relation implies that $h_q(0) = cE(0, s_0) + g$ with $c \in \mathbb{C}$ and $g \in S_0(\chi_0, \mathbf{c}_0, s_0)$. This gives $F_{P,l}h_q(0)(a(P)) \neq 0$, in contradiction to ${}^{(a)}h_q(0) \in {}^aD$. We conclude that $\ker({}^aA(0) - \frac{1}{4} + s_0^2)$ is spanned by the $g_{k,l}(0)$, hence it is equal to $S_0^f(\chi_0, \mathbf{c}_0, s_0 \,|\, \varphi_0)$.

Suppose that E has a singularity at $(0, s_0)$. As $\dot{E}(0)$ is holomorphic at s_0, the family E is of indeterminate type at $(0, s_0)$. Again we choose $a(P)$ such that $F_{P,l}\dot{E}(0; s_0; a(P)) \neq 0$. Take $\psi \in \mathcal{O}_{(0,s_0)}$ such that ψE is a primitive multiple of E. There are $A, B \in \mathcal{O}_{(0,s_0)}$ such that $F_{P,l}(\psi E) = A\alpha_{P,l} + B\beta_{P,l}$. As E is of indeterminate type, we have $A(0, s_0) = B(0, s_0) = 0$. From Proposition 10.2.14 we conclude that A and B have no common factor in $\mathcal{O}_{(0,s_0)}$. Consider an irreducible factor A_1 of A. Lemma 9.4.13 shows that $B(r, s) \neq 0$ for many $(r, s) \in N(A_1) \cap$

$(\mathbb{R} \times \mathbb{C})$, with $r \neq 0$ and (r,s) near $(0,s_0)$. For these (r,s) we have $\psi E(r,s) \neq 0$, and ${}^{(a)}\big(e^{-irt_{\varphi_0}}\psi E(r,s)\big) \in \ker({}^aA(r\varphi_0) - \frac{1}{4} + s^2)$. This shows that A_1 is of real type at $(0,s_0)$, see 12.1.9. Lemma 12.1.10 shows that there is a linear local curve $r \mapsto (r, \eta(r))$ through $N(A_1)$. Now $r \mapsto {}^{(a)}(e^{-irt_{\varphi_0}}\psi E)(r,\eta(r))$ is a family of eigenfunctions of ${}^aA$, hence up to a sign the function η is equal to one of the s_k, and the family $r \mapsto {}^{(a)}(\psi E)(r,\eta(r))$ may be expressed as a holomorphic linear combination of the corresponding ${}^{(a)}h_k$ and $g_{k,l}$. As it is not a cuspidal eigenfamily, we have $1 \leq k \leq \tilde{m}$, and the coefficient of h_k in the linear combination does not vanish identically. The Maass-Selberg relation implies that $h_k(0)$ is equal to a multiple of $\dot{E}(0;s_0)$ plus a cusp form. As ${}^{(a)}h_k(0) \in {}^aD$, the coefficient of $\dot{E}(0;s_0)$ is zero (here we use that $F_{P,l}\dot{E}(0;s_0;a(P)) \neq 0$). Hence $h_k(0)$ is a (non-zero) element of $S_0^e(\chi_0, \mathbf{c}_0, s_0 \,|\, \varphi_0)$. This completes the proof.

Bibliography

[1] S.S. Abhyankar: *Lectures on Expansion Techniques in Algebraic Geometry,* Lecture Notes **57**, Tata Institute, Bombay, 1977

[2] H. Baumgärtel: *Analytic Perturbation Theory for Matrices and Operators,* Birkhäuser-Verlag, 1985

[3] L. Bers, F. John, M. Schechter: *Partial Differential Equations,* Lectures in Appl. Math., Proc. Summer Seminar, Boulder, Cd., 1957; Interscience, 1964

[4] R.W. Bruggeman: *Fourier Coefficients of Automorphic Forms,* Lecture Notes in Math. **865**, Springer-Verlag, 1981

[5] R.W. Bruggeman: *Modular Forms of Varying Weight. I,* Math. Z. **190** (1985) 477–495

[6] R.W. Bruggeman: *Modular Forms of Varying Weight. II,* Math. Z. **192** (1986) 297–328

[7] R.W. Bruggeman: *Modular Forms of Varying Weight. III,* J. für die reine und angew. Math. **371** (1986) 144–190

[8] R.W. Bruggeman: *Eisenstein Series and the Distribution of Dedekind Sums,* Math. Z. **202** (1989) 181–198

[9] R.W. Bruggeman: *On the distribution of Dedekind sums,* 82–89 in Proc. Int. Conf. Automorphic Functions and their Applications, Khabarovsk, 1988, editors N. Kuznetsov and V. Bykovsky; Inst. Appl. Math. USSR, Ac. Sciences, Khabarovsk, 1990

[10] R.W. Bruggeman: *Dedekind sums for Hecke groups,* preprint

[11] Y. Colin de Verdière: *Pseudo-Laplaciens I,* Ann. Inst. Fourier, Grenoble **32**-3 (1982) 275–286

[12] Y. Colin de Verdière: *Pseudo-Laplaciens II,* Ann. Inst. Fourier, Grenoble **33**-2 (1983) 87–113

[13] J.-M. Deshouillers & H. Iwaniec: *The Non-vanishing of Rankin-Selberg zeta-functions at special points,* The Selberg Trace Formula and Related Topics, Proc. AMS-IMS-SIAM Joint Summer Research Conf., Brunswick, Maine, 1984; Contemp. Math. **53** (1986) 51–95

[14] M. Eichler: *Einführung in die Theorie der algebraischen Zahlen und Funktionen,* Birkhäuser-Verlag, 1963

[15] S.S. Gelbart: *Automorphic Forms on Adele Groups,* Ann. of Math. Studies **83**, Princeton Univ. Press & Univ. of Tokyo Press, 1975

[16] A. Good: *Local Analysis of Selberg's Trace Formula*; Lecture Notes in Math. **1040**, Springer-Verlag, 1983

[17] H. Grauert, R. Remmert: *Coherent Analytic Sheaves,* Grundl. math. Wiss. **265**; Springer-Verlag 1984

[18] Harish-Chandra: *Discrete Series for Semisimple Lie Groups. II,* Acta Math. **116** (1966) 1–111

[19] R. Hartshorne: *Algebraic geometry,* Graduate Texts in Math. **52**, Springer-Verlag, 1977

[20] E. Hecke: *Über die Bestimmung Dirichletscher Reihen durch ihre Funktionalgleichung,* Math. Ann. **112** (1936) 664–699 (p. 591–626 in E. Hecke: *Mathematische Werke,* Vandenhoeck & Ruprecht, 1970)

[21] D.A. Hejhal: *The Selberg trace formula for* $PSL(2,\mathbb{R})$, vol. 2, Lecture Notes in Math. **1001**, Springer-Verlag, 1983

[22] D.A. Hejhal: *A continuity Method for Spectral Theory on Fuchsian Groups,* 107–140 in Modular forms, Proc. Symp. Durham, 1983, ed. R.A. Rankin, Ellis Horwood, 1984

[23] L. Hörmander: *An Introduction to Complex Analysis in Several Variables,* D. van Nostrand Company, 1966

[24] N. Jacobson: *Basic Algebra I,* Freeman & Co., San Francisco, 1974

[25] T. Kato: *Perturbation Theory for Linear Operators,* Grundl. math. Wiss. **132**, Springer-Verlag, 1984

[26] S. Lang: *Real and Functional Analysis,* Grundl. math. Wiss. **142**, Springer-Verlag, 1993

[27] S. Lang: *Algebraic Number Theory,* Addison-Wesley, 1970

[28] S. Lang; *Elliptic Functions,* Addison-Wesley, 1973

[29] S. Lang: $SL_2(\mathbb{R})$; Addison-Wesley, 1975

[30] S. Lang: *Introduction to Modular Forms,* Grundl. math. Wiss. **222**, Springer-Verlag, 1976

[31] P.D. Lax & R.S. Phillips: *Scattering Theory for Automorphic Functions,* Ann. of Math. Studies **87**, Princeton University Press, 1976

[32] J. Lehner: *Discontinuous Groups and Automorphic Functions,* Math. Surveys VIII, A.M.S., Providence R.I. 1964

[33] Y.L. Luke: *Mathematical Functions and their Approximations,* Academic Press, 1975

[34] H. Maass: *Über eine neue Art von nichtanalytischen automorphen Funktionen und die Bestimmung Dirichletscher Reihen durch Funktionalgleichungen,* Math. Ann. **121** (1949) 141–183

[35] H. Maass: *Modular functions of one complex variable,* Tata Inst., Bombay; revised edition Springer-Verlag, 1983

[36] W. Magnus, F. Oberhettinger, R.P. Soni: *Formulas and Theorems for the Special Functions of Mathematical Physics,* 3. ed., Grundl. math. Wiss. **52**, Springer-Verlag, 1966

[37] R. Matthes: *Fourierkoeffizienten vektorwertiger reell-analytischer Spitzenformen zum Gewicht $\frac{1}{2}$ und verallgemeinerte Kloostermansche Summen,* thesis, Kassel, 1988

[38] T. Miyake: *Modular Forms,* Springer-Verlag, 1989

[39] R. Miatello & N.R. Wallach: *Automorphic Forms Constructed from Whittaker Vectors,* J. Funct. Anal. **86** (1989) 411–487

[40] H. Neunhöffer: *Über die analytische Fortsetzung von Poincaréreihen,* Sitzungsb. der Heidelberger Ak. der Wiss., Math.-naturw. Klasse, 2. Abh., 1973, 33–90

[41] D. Niebur: *A Class of Nonanalytic Automorphic Functions,* Nagoya Math. J. **52** (1973) 133–145

[42] H. Petersson: *Zur analytischen Theorie der Grenzkreisgruppen, I,* Math. Ann. **115** (1937) 23–67

[43] H. Petersson: *Zur analytischen Theorie der Grenzkreisgruppen, II,* Math. Ann. **115** (1937) 175–204

[44] H. Petersson: *Über den Bereich absoluter Konvergenz der Poincaréschen Reihen,* Acta Math. **80** (1948) 23–63

[45] R.S. Phillips & P. Sarnak: *On Cusp Forms for Cofinite Subgroups of* $PSL(2,\mathbb{R})$, Invent. math. **80** (1985) 339–364

[46] R.S. Phillips & P. Sarnak: *The Weyl Theorem and the Deformation of Discrete Groups,* Comm. Pure and Appl. Math. **38** (1985) 853–866

[47] H. Poincaré: *Mémoire sur les Fonctions Fuchsiennes,* Acta Math. **1** (1882) 193–294

[48] L. Pukánszky; *The Plancherel Formula for the Universal Covering Group of* $SL(R, 2)$, Math. Ann. **156** (1964) 96–143

[49] R.A. Rankin: *Modular Forms and Functions,* Cambridge University Press, 1977

[50] W. Roelcke: *Das Eigenwertproblem der automorphen Formen in der hyperbolischen Ebene, I,* Math. Ann. **167** (1966) 292–337

[51] W. Roelcke: *Das Eigenwertproblem der automorphen Formen in der hyperbolischen Ebene. II,* Math. Ann. **168** (1967) 261–324

[52] B. Schoeneberg: *Elliptic Modular Functions,* Grundl. math. Wiss. **203**; Springer-Verlag, 1974

[53] G. Shimura: *Introduction to the arithmetic theory of automorphic functions*; Iwanami Shoten and Princeton University Press, 1971

[54] C.L. Siegel: *Some Remarks on Discontinuous Groups,* Ann. Math. **46** (1945) 708–718

[55] C.L. Siegel: *Die Funktionalgleichungen einiger Dirichletscher Reihen,* Math. Z. **63** (1956) 363–373

[56] L.J. Slater: *Confluent Hypergeometric Functions,* Cambridge, at the University Press, 1960

[57] A. Terras: *Harmonic Analysis on Symmetric Spaces and Applications I,* Springer-Verlag, 1985

[58] E.C. Titchmarsh; *The Theory of the Riemann Zeta-Function,* Oxford, at the Clarendon Press, 1951

[59] A.B. Venkow: *Spectral theory of automorphic functions (Russian),* Proc. Steklow Inst. Math. 153, Leningrad, 1981; AMS transl. Proc. Steklov Inst. Math. 1982, 4

[60] A.B. Venkov: *Spectral Theory of Automorphic Functions,* Math. and its Appl., Soviet Series **51**, Kluwer, 1990

[61] D.B. Zagier: *Periods of modular forms and Jacobi theta functions,* Invent. math. **104** (1991) 449–465

Index

Monographs in Mathematics

The foundations of this outstanding book series were laid in 1944. Until the end of the 1970s, a total of 77 volumes appeared, including works of such distinguished mathematicians as Carathéodory, Nevanlinna and Shafarevich, to name a few. The series came to its name and present appearance in the 1980s. According to its well-established tradition, only monographs of excellent quality will be published in this collection. Comprehensive, in-depth treatments of areas of current interest are presented to a readership ranging from graduate students to professional mathematicians. Concrete examples and applications both within and beyond the immediate domain of mathematics illustrate the import and consequences of the theory under discussion.

Published in the series since 1983

Volume 78 **H. Triebel, Theory of Function Spaces I**
1983, 284 pages, hardcover, ISBN 3-7643-1381-1.

Volume 79 **G.M. Henkin/J. Leiterer, Theory of Functions on Complex Manifolds**
1984, 228 pages, hardcover, ISBN 3-7643-1477-X.

Volume 80 **E. Giusti, Minimal Surfaces and Functions of Bounded Variation**
1984, 240 pages, hardcover, ISBN 3-7643-3153-4.

Volume 81 **R.J. Zimmer, Ergodic Theory and Semisimple Groups**
1984, 210 pages, hardcover, ISBN 3-7643-3184-4.

Volume 82 **V.I. Arnold / S.M. Gusein-Zade / A.N. Varchenko, Singularities of Differentiable Maps – Vol. I**
1985, 392 pages, hardcover, ISBN 3-7643-3187-9.

Volume 83 **V.I. Arnold / S.M. Gusein-Zade / A.N. Varchenko, Singularities of Differentiable Maps – Vol. II**
1988, 500 pages, hardcover, ISBN 3-7643-3185-2.

Volume 84 **H. Triebel, Theory of Function Spaces II**
1992, 380 pages, hardcover, ISBN 3-7643-2639-5.

Volume 85 **K.R. Parthasarathy, An Introduction to Quantum Stochastic Calculus**
1992, 300 pages, hardcover, ISBN 3-7643-2697-2.

Volume 86 **M. Nagasawa, Schrödinger Equations and Diffusion Theory**
1993, 332 pages, hardcover, ISBN 3-7643-2875-4.

Volume 87 **J. Prüss, Evolutionary Integral Equations and Applications**
1993, 392 pages, hardcover, ISBN 3-7643-2876-2.

Volume 88 **R.W. Bruggeman, Families of Automorphic Forms**
1994, 328 pages, hardcover, ISBN 3-7643-5046-6.